# Essentials of
# SAFETY
# AND
# HEALTH
# MANAGEMENT

*Edited by*

## RICHARD W. LACK, P.E., CSP, CPP, CHCM

Safety Officer

San Francisco International Airport

San Francisco, California

**LEWIS PUBLISHERS**

Boca Raton  London  New York  Washington, D.C.

**Library of Congress Cataloging-in-Publication Data**

Lack, Richard W.
    Essentials of safety and health management / by Richard W. Lack.
        p.   cm.
    Includes bibliographical references and index.
    ISBN 1-56670-054-X (permanent paper)
    1. Industrial hygiene--Management.   2. Industrial safety--Management.
I. Title.
HD7261.L25 1996
658.4'08—dc20                                                                           95-52101
                                                                                                      CIP

© 1996 by CRC Press LLC
Lewis Publishers is an imprint of CRC Press LLC

No claim to original U.S. Government works
International Standard Book Number 1-56670-054-X
Library of Congress Card Number 95-52101
Printed in the United States of America        3  4  5  6  7  8  9  0
Printed on acid-free paper

# FOREWORD

There are many books that address the topic of safety and health management. Few of them accumulated the outstanding group of authors whose work and experience are contained in *Essentials of Safety and Health Management*. The painstaking effort made to develop an authoritative book will soon become apparent to you as you explore new dimensions of these old topics. The welcome addition of international developments is particularly noteworthy. You hold an education in your hands. The influence of this book on your work in safety and health will be considerable.

**Joseph LaDou, M.D.**
Division of Occupational and Environmental Medicine
University of California
San Francisco, California

# FOREWORD

The profession and practice of occupational safety is very different than it was just a few years ago. The workforce makeup and psyche is very different as well. The practice of safety, founded in sound technical principles, is today carried out in an environment of rapidly changing management approaches.

The safety practitioner finds their responsibilities broadening and the clean lines delineating the various disciplines in occupational safety and health blurring. The need to be competent in the broad practice of occupational safety spurs the professional to learn new technologies and products, and new applications of familiar technology.

The rapidly developing alliances and exchanges among world nations and marketplaces has expanded the safety practitioner's world so that as global cooperation develops, so does global competition. The safety professional is in need of at least a working knowledge of the technical standards and cultural approaches of other countries to occupational safety and health.

Fast-paced change and broadening responsibilities dictate that the professional remain competent through professional development. This volume is encyclopedic in breadth, providing an update of some fundamental topics, while introducing new and innovative subjects. This book prepares the professional to meet the challenges outlined above and can also serve the manager in need of an overview of occupational safety and health management.

This collection of topics and authors explores new approaches to the broadly accepted elements of accident prevention, while providing an overview of additional information necessary for the professional to maintain competency and confidence in a rapidly evolving profession.

*Essentials of Safety and Health Management* may be the *only* book the professional will need to add to their bookcase in order to remain current and meet the challenges of the changing workplace, workforce, and management environment. This remarkable collection of theory, application, and fact makes a significant contribution to the body of knowledge in the Science of Safety.

**Margaret Mock Carroll, CSP, P.E.**
Sandia National Laboratory
Albuquerque, New Mexico

# PREFACE

This book was born in the minds of two outstanding educators and professionals, Marion Gillen, RN, MPH, and Barbara Plog, MPH, CIH, CSP. Both are associated with the Center for Occupational and Environmental Health (COEH) of the University of California at Berkeley. At the time, Marion was serving as continuing education coordinator responsible for developing occupational health and safety and environmental related courses. Barbara is an instructor for the School of Public Health at the University of California at Berkeley, where she teaches courses in industrial hygiene and safety. More recently, Marion has commenced studies for her Ph.D., and Barbara has taken over duties as Director of Continuing Education at COEH.

During this time, while Marion and Barbara were discussing the book with Lewis Publishers, Marion was also working with me on a new course to be called, "The Essentials of Safety Management."

Some time later, Marion and Barbara decided that their work and personal study commitments were such that a book project was no longer feasible. Accordingly, Marion invited me to take over the project. It sounded such an exciting and challenging opportunity that I immediately decided to investigate the possibilities.

First, I approached the members of our faculty for the "Essentials of Safety Management" course. All were enthusiastically willing to join me on this project! The pioneering group of co-authors who first joined this project were: Jim Arnold, Gabe Gillotti, Tom Hanley, Kathleen Kahler, Bob Lapidus, Steve McConnell, and Herman Woessner.

Subsequently, I decided that the scope of our book needed expanding to include certain topics not covered in our existing two-day course. A number of other professionals were therefore invited to join the group and the complete panel may be found in the list of contributing authors.

Each of the co-authors has had wide experience in the various chapter subjects he or she is contributing. Several have already published books and/or papers and magazine articles. Many are also frequently speakers and instructors at professional seminars and courses.

I designed this book with the objective of serving a wide range of readers. First, the book is intended to serve as a desk reference source, with primary emphasis on the *management* aspects of each subject area. An overview of the subject elements is provided, together with examples of their practical application in the workplace. In addition, each subject chapter includes a Further

Reading section. Here are listed selected recommended books, articles, etc., to which readers may go for more in-depth information.

Second, the book is intended to serve both the needs of academia in preparing career professionals, and those already in practice. It will also serve as a useful reference for those managers who have responsibilities in the safety and health field.

Some readers may question why there are so many chapters devoted to management and communications-related subjects. It is the opinion of the faculty of the "Essentials of Safety Management" course that if safety and health practitioners are to perform their job functions effectively, it is essential they possess a comprehensive knowledge of the state-of-the-art in the fields of management and communications, and continuously develop their skills in these areas.

Finally, it is intended that the audience for this book should extend beyond the shores and borders of the United States. Obviously, we cannot address the many thousands of technical, legal, and regulatory aspects for all the countries of this globe. We have, however, included a chapter on International aspects. The author of this chapter, Kathy Seabrook, has provided many valuable insights, particularly with regard to European countries. Furthermore, the contents of the entire book have been designed so that their scope is virtually universal and the principles and procedures can be applied worldwide.

On behalf of the publisher and co-authors, we look forward to hearing from you. Your comments and suggestions for the book's improvement will be a most welcome aid in planning future editions.

**Richard W. Lack**
San Francisco, California

# ACKNOWLEDGMENTS

It would be impossible for me to list every person who has contributed to the production of this book. Nevertheless, I do want to mention a few who have made extraordinarily significant contributions.

First of all, my wife, Phillippa, who in spite of her work assignments, craft projects, and the numerous affairs of our household, took the time to proofread my chapters and also provided invaluable counsel on grammar, layout, etc.

My thanks also to my employer, the Airports Commission, San Francisco International Airport, and in particular, my supervisor, Assistant Deputy Director, Mel Leong, who provided much needed support and encouragement for me to take on this project. Two Airport staffers, Imelda Quesada and Leticia Aguilar, contributed outstanding support with all my correspondence to the publisher and co-authors, and typing of the book outline and table of contents. My thanks especially to Leticia Aguilar, who typed all my chapters and spent many hours working up the diagrams and example forms on her computer.

My contribution to this book would not have been possible had it not been for the guidance and teaching gained over many years from my foremost mentor, Homer K. Lambie. It is my principal goal to capture and preserve Homer's unique philosophies on safety management systems. They have inspired and guided several generations of safety professionals and countless managers during his 40 years of work, primarily with the Kaiser Aluminum and Chemical Corporation.

On a personal basis, my sincere thanks to another of my mentors, Dr. John Grimaldi, for his support and advice which has been invaluable particularly in developing my material for the book. My thanks also to Margaret Carroll for her support of this project and her thoughtful review and comments on my chapters. I have already mentioned Marion Gillen and Barbara Plog in the Preface. It goes without saying that this book would not have happened had it not been for their efforts, perception, and support.

My sincere thanks go to all the contributing authors. Their time and effort cannot be measured in value, the book is the true monument to their work.

I would like to specially acknowledge John Northey and others concerned on the editorial staff of Paramount Publishing Limited. Their support and acceptance of my earliest writings gave me a much needed opportunity to get started as a writer in our professional field. In this book, I have drawn from several of these papers, and this is so indicated in the relevant chapters.

My grateful acknowledgment to Lewis Publishers for their enthusiastic and unflagging support of this project. Their guidance and patience extended to a complete newcomer to the world of publishing was greatly appreciated.

Finally, my thanks to many fellow professionals in the American Society of Safety Engineers, National Safety Management Society, and the Institution of Occupational Safety and Health for their guidance, support, and encouragement.

# THE EDITOR

**Richard W. Lack, P.E., CSP, CPP,** has over 34 years of occupational safety and health experience, and for 18 of those years, he also held security and fire protection management responsibilities.

Currently, Mr. Lack is employed by the City and County of San Francisco, Airports Commission, as Safety Officer. In this position, he is responsible in providing consulting services for the continuous improvement of the airport's safety, health, and environmental programs. The airport has over 1,200 employees engaged in maintaining the facility and overseeing the operations of its numerous tenants.

Prior to this position, Mr. Lack was Risk Control Engineer with Castle & Cooke, Inc., a worldwide organization with individual companies in food production and distribution. Mr. Lack functioned as an advisor to the company's various business units both in the U.S. and overseas.

As Manager, Safety and Loss Prevention, for the $2 billion Great Plains coal gasification project at Beulah, North Dakota, Mr. Lack was fortunate to be able to participate in the design and implementation of safety, security, loss control, and fire protection programs from the "grass roots." Prior to this assignment, Mr. Lack pursued a 20-year career first with Kaiser Aluminum in Jamaica, West Indies, and then with Reynolds Aluminum in the U.S. This experience included mining, railroad, port operations, chemical plant, and construction. During this period, he held managerial positions in safety and loss prevention.

Mr. Lack was educated at private English schools. He is a Registered Professional Engineer in Safety Engineering, State of California, a Certified Safety Professional, Certified Hazard Control Manager, and Certified Protection Professional. Mr. Lack is a professional member of all the recognized safety and security organizations in the U.S. and U.K. His articles have appeared in professional journals in both countries, and he is a contributing author to the publication, *The Safety Management Handbook,* by the Bureau of Business Practice.

Mr. Lack has been active in professional societies for many years. He has been a member of the National Safety Management Society since 1977, and was President of the Golden Gate Chapter for the year 1988. Mr. Lack has been a professional member of the American Society of Safety Engineers since 1972 and served as Administrator of the Management Division, 1988–1989. He was elected to the Board of ASSE in the position of Vice President, Divisions

1990–1992. In the American Society for Industrial Security, Mr. Lack served as Founder Chairman of the Arkansas Chapter 1980–1981 and Chairman of the North Dakota Chapter in 1986. Mr. Lack has been a member of the Institute of Personnel and Development (U.K.) since 1961 and the Institution of Occupational Safety and Health (U.K.) since 1968.

# CONTRIBUTORS

## Project Director

**Richard W. Lack, P.E., CSP, CPP**
Safety Officer
Airports Commission
San Francisco International Airport
San Francisco, CA

## Contributing Authors

**John D. Adams, Ph.D.**
Education Program Manager
Sun University
Sun Microsystems, Inc.
Mountain View, CA

**Yvonne F. Alexander, M.A.**
Alexander Communications
San Francisco, CA

**James R. Arnold, Esq.**
Severson and Werson
San Francisco, CA

**Michael-Laurie Bishow, Ph.D.**
Contra Costa Training Institute
Martinez, CA

**Marc Bowman, B.S.**
Woodward-Clyde Consultants
Oakland, CA

**Ken Braly, M.S.**
San Jose, CA

**Roger L. Brauer, Ph.D., P.E., CSP**
Technical Director
Board of Certified Safety Professionals
Tolono, IL

**Margaret Mock Carroll, CSP, P.E.**
Engineering Manager
Sandia National Laboratory
Albuquerque, NM

**Gerard C. Coletta, M.S.**
Vice President/Manager
Risk Control Consulting
Sedgwick James of California
San Francisco, CA

**Kenneth M. Colonna, CSP**
Manager Safety/Environmental
Coca-Cola Consolidated
Charlotte, NC

**Barbara K. Cooper, MSPH, CIH, CSP**
Davis, CA

**Ellen E. Dehr, CIH**
Industrial Hygienist
Port of San Francisco
San Francisco, CA

**Michael L. Fischman, M.D., MPH**
Fischman Occupational and
  Environmental Medical Group
Walnut Creek, CA

**Robert S. Fish, Ph.D.**
San Jose, CA

**E. Scott Geller, Ph.D.**
Professor, Department of Psychology
Virginia Polytechnic Institute
 and State University
Blacksburg, VA

**Gabriel J. Gillotti**
Director
Voluntary Programs and Outreach
United States Department of
 Labor–OSHA
San Francisco, CA

**Thomas A. Hanley, M.A.**
Regional Manager
Division of Occupational Safety &
 Health
Anaheim, CA

**Mark D. Hansen, CSP, CPE, P.E.**
Safety Manager
Dixie Chemical Company, Inc.
Pasadena, TX

**Susan Bade Hull, Esq.**
Alvarado, Smith, Villa, & Sanchez
Los Angeles, CA

**Kathleen Kahler, MPH, CHES**
Inter-Regional Environmental Safety
 and Health Services
Kaiser Foundation Hospitals
Oakland, CA

**Joseph LaDou, M.D., M.S.**
Division of Occupational and
 Environmental Medicine
University of California
San Francisco, CA

**Robert A. Lapidus, CSP**
Safety Management Consultant
Lapidus Consulting
Pleasant Hill, CA

**Peter C. Lyon, Esq.**
Severson and Werson
San Francisco, CA

**Steven M. McConnell, CSP, REA**
Safety Engineer
Lawrence Livermore National
 Laboratory
Livermore, CA

**Donald L. Morelli, M.S., CPE**
San Carlos, CA

**Barbara Newman, M.A., M.F.C.C.**
Los Angeles, CA

**Richard C. Nugent, CSP**
Applied Extrusion Technologies,
 Inc.
Covington, VA

**Mary E. O'Connell, RN, COHN**
Health and Safety Specialist
 Occupational Safety/Health
Goldman Insurance
San Francisco, CA

**Neva Nishio Petersen, MPH, CIH**
Safety Manager
Integrated Device Technology, Inc.
Salinas, CA

**Lucy O Reinke, M.D., MPH**
Fischman Occupational and
 Environmental Medical Group
Walnut Creek, CA

**Peter B. Rice, CSP, CIH**
Vice President Industrial Hygiene
 and Safety
Harding Lawson Associates
Novato, CA

**Anne Durrum Robinson, M.A.**
Consultant Human Resource
  Development
Creativity, Communication,
  Common Sense
Austin, TX

**Roberta V. Romberg, Esq.**
Severson and Werson
San Francisco, CA

**Kathy A. Seabrook, CSP, RSP (UK)**
Global Business Solutions
Mendham, NJ

**Michal F. Settles, Ed.D.**
Department Manager, Human
  Resources
Bay Area Rapid Transit District
Oakland, CA

**Carolyn R. Shaffer, M.A.**
Growing Community Associates
Berkeley, CA

**Steven I. Simon, M.S. and Rosa
  Antonia Simon, Ph.D.**
Culture Change Consultants
Seal Beach, CA

**Robin Spencer, B.A., CHMM, REA**
Orinda, CA

**George E. Swartz, CSP**
Director, Safety and Occupational
  Health
Midas International Corporation
Chicago, IL

**Paula R. Taylor, M.A.**
Paula Taylor & Associates
Oakland, CA

**Sondra Thiederman, Ph.D.**
Cross-Cultural Communications
San Diego, CA

**Janice L. Thomas, D.P.A.**
Circle Safety and Health Consultants
Gum Spring, VA

**Michael D. Topf, B.S., M.A.**
President
Topf Organization
Rosemont, PA

**Joan F. Woerner, REA, REHS**
Safety and Risk Manager
Anheuser-Busch, Inc.
Fairfield Brewery
Fairfield, CA

**Herman Woessner, M.S., M.A., CSP**
President
Safe Risk Corporation
Chalmette, LA

**Holland A. Young**
Department of Aviation
City of Austin
Austin, TX

**Bonita B. Zahara**
Zahara and Associates
Spokane, WA

# CONTENTS

## SECTION III. HEALTH PROGRAM MANAGEMENT ASPECTS

## SECTION IV. SAFETY AND HEALTH MANAGEMENT — REGULATORY COMPLIANCE ASPECTS

## SECTION VIII. TRAINING ASPECTS

## SECTION IX. INTERNATIONAL DEVELOPMENTS

## SECTION X. STANDARDS OF COMPETENCE

## SECTION XI. AFTERWORD — THE FUTURE

# Section I
# Introduction

# 1           WHERE HAVE WE BEEN?

**Richard W. Lack**

From my perspective, this question goes back over a professional experience spanning more than 30 years both in the U.S. and overseas. Historically, we have seen a world recovering from a series of major conflicts, starting with World War II in the 1940s. This was followed by the Korean War in the 1º50s and the Vietnam War in the 1960s. These disastrous chapters in our history have since been followed by continued tensions and outbreaks of hostility which have erupted in many different parts of our globe. Such situations continue today, when we consider the troubles in the Balkans and the Middle East.

Politically speaking, in recent years we have seen some dramatic changes from the "cold war", "iron curtain", and "Berlin wall" days.

The earlier effect on the economies of the U.S. and other developed nations was an emphasis on heavy engineering and manufacturing. This was necessary to produce the arms and technology for a massive war effort.

Many of us began our careers in military service and followed this by entering heavy industry. It might have been a steel mill, a shipbuilding yard, a chemical refinery, or a major engineering works. In my case, it was mining bauxite, the raw material for aluminum.

The technologies developed originally for military purposes have in many cases been turned toward something much more beneficial for mankind. Witness the development of the war-time rocket, which was the forefather of our space program.

Developments in communications and computer technologies have helped to bring us rapidly into the "Information Age".

To adjust to all these changes, industries in the developed nations have been forced to make major shifts. Many have disappeared altogether; others have been radically restructured.

1-56670-054-X/96/$0.00+$.50

Another more recent trend has been the public's concern for preserving and protecting our environment.

Against this backdrop of political, historical, and economic influences, where do we stand in terms of losses, of both people and property? According to the National Safety Council, accidental deaths in 1991 were the lowest since 1924. The death total in 1993 was 90,000 (in 1990 the toll was 93,000).

The breakdown for all unintentional injuries in 1993 as recorded by the National Safety Council in their *Accident Facts,* 1994, is as follows:

| Injuries | Deaths | Deaths per 100,000 Persons | Disabling |
|---|---|---|---|
| All classes | 90,000 | 34.9 | 18,200,000 |
| Motor vehicle | 42,000 | 16.3 | 2,000,000 |
|   Public nonwork | 38,300 | | 1,800,000 |
|   Work | 3,500 | | 200,000 |
|   Home | 200 | | |
| Work | 9,100 | 3.5 | 3,200,000 |
|   Non-motor vehicle | 5,600 | | 3,000,000 |
|   Motor vehicle | 3,500 | | 200,000 |
| Home | 22,500 | 8.7 | 6,600,000 |
|   Non-motor vehicle | 22,300 | | 6,600,000 |
| | 200 | | |
| Public | 20,000 | 7.8 | 6,600,000 |

From National Safety Council, *Accident Facts,* 1994 edition, Itasca, IL: National Safety Council, 1994.

These accidents were estimated to cost $407.5 billion in medical expenses, insurance, lost wages, and property damage.

Dealing with work accidents in 1993, highlights from *Accident Facts* are

- Work accident costs were $111.9 billion, which breaks down to a cost per worker of $940 and cost per disabling injury of $27,000.
- Total time lost due to accidents in 1993 was 115 million days.
- Although the total case incidence rate has plateaued, the lost workdays incidence rate is climbing.*

Preliminary estimates published by the National Safety Council in their May 1993 issue of *Safety & Health* indicate the following facts:

- From 1982 to 1992, the worker accidental death rate declined 33% from 12 to 8 per 100,000 workers.
- The 1992 toll of off-the-job accidents was 31,000 deaths and about 2.4 million disabling injuries.
- As in earlier years, accidents were the fourth leading cause of death, exceeded only by heart disease, cancer, and stroke.
- The most common accident types causing death in 1992 were motor vehicle (48%), falls (15%), and solid and liquid poisonings (7%).

* Author's note: Although no cause is offered for this increase, the generally accepted reason is believed to be the explosion of stress-related illnesses in recent years.

A National Institute of Occupational Safety and Health (NIOSH) study reported in the April 28, 1993 issue of the *San Francisco Chronicle* indicates that the leading causes of work-related deaths are motor vehicle accidents (23%), machine-related incidents (14%), homicide (12%), falls (10%), electrocutions (7%), and being struck by falling objects (7%).

This same report was publicized by the National Safety Council in their August 1993 issue of *Safety & Health.* Other key points mentioned in this article were

- NIOSH says that 41% of women who die on the job are murdered. Most of those deaths take place in settings like convenience stores, where women work in a service setting.
- The construction industry has the highest percentage of deaths: 18% of the total.
- A National Highway Traffic Safety Administration study found that motor vehicle accidents accounted for 26.7% of on-the-job deaths in 1989. On-the-job motor vehicle accidents cost employers $40 billion per year at an average per accident of $22,500.

According to the *Injury Fact Book,* the percentages of injury deaths by manner of death in 1986 were

### Unintentional Injuries

31%  Motor vehicle crashes
32%  Other unintentional injuries

### Intentional Injuries

21%  Suicide
14%  Homicide
 2%  Unknown

In the book *Violence in America,* it is noted that in 1987 there were 20,000 homicides and 31,000 suicides in the U.S. Assaultive violence was reported to cause from 19,000 to 23,000 deaths a year.

In the *Report to Congress 1989: Cost of Injury,* among the findings of this study were the following highlights:

- In 1985, about 57 million persons were injured, which is 1 in 4 U.S. residents.
- The total lifetime cost of injury imposed an estimated burden of $158 billion on the U.S. economy.

These figures show that although injury rates are on a downward trend, they represent a staggering burden in terms of economic loss, inestimable human suffering, and misery. The loss far exceeds any losses sustained by war and other forms of hostilities.

It is an interesting fact of the development of our civilization when we consider that more Americans die in occupational and/or traffic accidents every year than died in the entire 8 years of the Vietnam War (approximately 55,000 killed). The wonder is that, with a slaughter of humanity of this magnitude, there has not been a greater public outrage.

Occasionally, some catastrophic event such as an explosion at a mine or industrial plant will attract attention. Such events have been a "driver" for increased legislation, especially during the past 20 years, and these developments are discussed in another chapter of this book. More stringent safety regulations have certainly had some impact, but opinions seem to vary widely as to the degree.

Greatly increased penalties and criminal sanctions have certainly gotten the attention of the business world. According to the U.S. Occupational Safety and Health Administration, some of the most expensive settlements to date were as follows:

| Date | Company | Proposed Penalties | Alleged Violations |
|------|---------|--------------------|--------------------|
| Oct. 31, 1991 | IMC Fertilizer | $10 million | Safety, emergency response |
| Aug. 29, 1991 | Citgo Petroleum Corp. | $5.8 million | Equipment, training |
| Aug. 22, 1990 | Phillips 66 | $4 million | Chemicals, safety |

As mentioned earlier, total injuries may be down, but time lost is up. Allied to this are skyrocketing medical costs. According to a 1990 study for the California Legislature, America's national health spending has increased from a total of $248 billion in 1980 to $647 billion in 1990.

Quoted in this report from a U.S. Department of Commerce source which is based on 1988 expenditures, health care is the largest industry in the U.S. economy at 10% of gross national product (GNP). This exceeds defense (9%), food and beverage (5%), and education (4%).

In closing this introduction, here is a quote from this report to ponder:

In the past, occupational health and safety focused on injuries and illnesses with a direct association to the occupational work environment. Nowadays, the distinction between general health problems which have their origins elsewhere are brought into the workplace. They reflect health problems endemic in modern society at large — abuse of drugs and alcohol, cigarette smoking, poor nutritional habits, lack of aerobic exercise, traumatic injury, obesity, psychological and unhealthy stress associated with problems of coping with life in general.

United States
Department
of Labor

Bureau of Labor Statistics        Washington, D.C. 20212

Technical information:                                    USDL - 95 - 288
G. Toscano  (202) 606-6175
Media contact:                                        FOR RELEASE:  10 a.m. EDT
K. Hoyle  (202) 606-5902                               Thursday, August 3,1995

## NATIONAL CENSUS OF FATAL OCCUPATIONAL INJURIES, 1994

A total of 6,588 fatal work injuries were reported in 1994, 4 percent more than the previous year's total, according to the Census of Fatal Occupational Injuries, Bureau of Labor Statistics, U.S. Department of Labor. The higher fatality count in 1994 largely reflects an increase over 1993 in the number of workers killed in transportation incidents, primarily highway and commercial airline crashes. Catastrophes that result in multiple worker deaths, such as fires, explosions, and aircraft crashes, can cause year-to-year fluctuations in fatality totals.

The BLS census uses multiple data sources to identify, verify, and profile fatal work injuries. Key information about each workplace fatality (occupation and other worker characteristics, equipment being used, and circumstances of the event) is obtained by cross-referencing source documents, such as death certificates, workers' compensation records, and reports to federal and state agencies. This method assures counts are as complete and accurate as possible.

Profiles from the 1994 fatality census

Highway traffic incidents and homicides led all other events that resulted in fatal work injuries in 1994. These two events totaled over a third of the work injury deaths that occurred during the year. (See table 1 and chart 1.)

Highway deaths accounted for 20 percent of the 6,588 fatal work injuries in 1994. Slightly over half of highway fatality victims were driving or riding in a truck, half of which collided with another vehicle and a quarter jackknifed or overturned. Transport-related incidents occurring on private property (such as tractors or forklifts overturning), aircraft crashes, and workers being struck by vehicles each accounted for about 6 percent of the worker fatalities. Rail and water transport together accounted for another 3 percent of the deaths.

Homicide was the second leading cause of job-related deaths, accounting for 16 percent of the total. Robbery was the primary motive for workplace homicide. About half the victims worked in retail establishments, such as grocery stores and eating and drinking establishments, where cash is readily available. Taxicab drivers, police, and security guards were other occupations with high numbers of worker homicides. Four-fifths of the homicide victims were shot; others were stabbed, beaten, or strangled. While highway traffic incidents were the leading manner of death for male workers, homicide was the leading cause of death for female workers, accounting for 35 percent of their fatal work injuries.

Falls accounted for 10 percent of the fatal work injuries. The construction industry, primarily special trade contractors such as roofing, painting, and structural steel erection, accounted for almost half of the falls. One-fifth of the falls were from or through roofs; falls from scaffolding and from ladders each accounted for about one-eighth.

Nine percent of the fatally injured workers were struck by various objects, a fourth of which were falling trees, tree limbs, and logs. Other objects that struck workers included machines and vehicles slipping into gear or falling onto workers, and various building materials, such as pipes, beams, metal plates, and lumber.

Electrocutions accounted for 5 percent of the worker deaths in 1994. About a third of these fatalities resulted from the worker or equipment being used coming in contact with overhead power lines.

**Occupation highlights (table 2 and chart 2):**

* Occupations with large numbers of worker fatalities included truck drivers, farm workers, sales supervisors and proprietors, and construction laborers.

* Specific events or exposures responsible for workers' deaths varied considerably among occupations. Highway crashes and jackknifings accounted for about two-thirds of the truck drivers' deaths, while almost three-fourths of the fatalities among sales supervisors and proprietors resulted from homicide. Half of the farm workers' deaths occurred in vehicle-related incidents. Falls accounted for one-fourth of the construction laborers' deaths.

**Industry highlights (table 3):**

* Major industry groups with the largest number of fatal work injuries were agricultural crop production, special trades construction contractors (for example, roofing and electrical work), and trucking and warehousing.

* Industry divisions with large numbers of fatalities relative to their employment include agriculture, forestry, and fishing; construction; transportation and public utilities; and mining.

**Worker characteristics highlights (table 4):**

* Men, the self-employed, and older workers suffer fatal injuries more often than their employment shares would suggest. Differences in the industries and occupations of these worker groups explain in part their high relative risk of fatal injury on the job.

* The types of events responsible for workers' deaths varied among worker groups. Highway crashes were most often cited for wage and salary workers, men, whites, and workers less than 18 years old and those between 45 and 64. Highway incidents and homicides each accounted for nearly a fifth of the deaths among workers 20-44. Homicide was the leading manner of death for self-employed workers, women, blacks, Asians and Pacific Islanders, Hispanics, and workers between 18 and 19 years old. Workers 65 and older were killed more frequently in nonhighway transportation incidents, such as tractor rollovers, than in any other event.

**Other highlights:**

* On average about 18 fatal work injuries occurred each day in 1994.

* Eighty-five percent of the fatally injured workers died the day they were injured; 97 percent died within 30 days.

Included in tables 3 and 4 are 1994 annual average employment data collected in the BLS Current Population Survey. By comparing the percent distributions of fatalities and employment, the user can evaluate the relative risk of a job-related fatality for a given occupation, industry, or worker characteristic. For example, the construction industry accounted for about 16 percent of the fatality total, which was about 3 times greater than its share of total employment of 6 percent. While employment can be used to evaluate the relative risk of a fatal work injury, other measures, such as employee exposure hours, also can be used.

Background of the program

The Census of Fatal Occupational Injuries, part of the BLS safety and health statistics program, provides the most complete count of fatal work injuries available because it uses multiple state and federal data sources. This is the third year that the fatality census has been conducted in all 50 states and the District of Columbia. The BLS fatality census is a federal/state cooperative venture in which costs are shared equally. State-specific data on workplace fatalities are available from the state agencies participating with BLS in the census program. A list of participating agencies and their telephone numbers is available from BLS by calling 202-606-6175.

The Survey of Occupational Injuries and Illnesses, conducted since 1972, profiles worker and case characteristics of serious, nonfatal, workplace injuries and illnesses resulting in lost worktime in addition to presenting frequency counts and incidence rates by industry. Copies of the 1993 news release are available from BLS by calling 202-606-6304. Incidence rates for 1994 by industry will be published in December 1995. Information on 1994 worker and case characteristics will be published in April 1996.

## Chart 1:  The manner in which workplace fatalities occurred, 1994

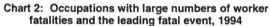

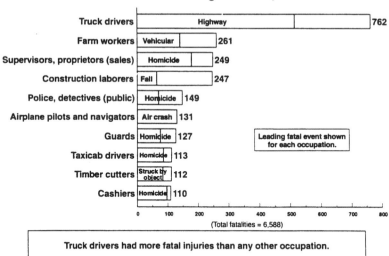

Event or exposure
(Total fatalities = 6,588)

Highway vehicle incidents and homicides led all other fatal events.

## Chart 2:  Occupations with large numbers of worker fatalities and the leading fatal event, 1994

| Occupation | Leading fatal event | Fatalities |
|---|---|---|
| Truck drivers | Highway | 762 |
| Farm workers | Vehicular | 261 |
| Supervisors, proprietors (sales) | Homicide | 249 |
| Construction laborers | Fall | 247 |
| Police, detectives (public) | Homicide | 149 |
| Airplane pilots and navigators | Air crash | 131 |
| Guards | Homicide | 127 |
| Taxicab drivers | Homicide | 113 |
| Timber cutters | Struck by object | 112 |
| Cashiers | Homicide | 110 |

Leading fatal event shown for each occupation.

0    100   200   300   400   500   600   700   800
(Total fatalities = 6,588)

Truck drivers had more fatal injuries than any other occupation.

SOURCE: Bureau of Labor Statistics, U.S. Department of Labor, Census of Fatal Occupational Injuries, 1994

**Table 1. Fatal occupational injuries by event or exposure, 1992-1994**

| Event or exposure | Fatalities | | | |
|---|---|---|---|---|
| | 1992 | 1993[2] | 1994 | |
| | Number | Number | Number | Percent |
| Total | 6,217 | 6,331 | 6,588 | 100 |
| **Transportation incidents** | 2,484 | 2,501 | 2,740 | 42 |
| Highway | 1,158 | 1,243 | 1,336 | 20 |
| Collision between vehicles, mobile equipment | 578 | 657 | 650 | 10 |
| Moving in same direction | 78 | 99 | 117 | 2 |
| Moving in opposite directions, oncoming | 201 | 244 | 229 | 3 |
| Moving in intersection | 107 | 123 | 143 | 2 |
| Vehicle struck stationary object or equipment | 192 | 190 | 255 | 4 |
| Noncollision | 301 | 336 | 370 | 6 |
| Jack-knifed or overturned--no collision | 213 | 237 | 272 | 4 |
| Nonhighway (farm, industrial premises) | 436 | 392 | 407 | 6 |
| Overturned | 208 | 212 | 225 | 3 |
| Aircraft | 353 | 282 | 424 | 6 |
| Worker struck by a vehicle | 346 | 365 | 383 | 6 |
| Water vehicle | 109 | 120 | 92 | 1 |
| Railway | 66 | 86 | 81 | 1 |
| **Assaults and violent acts** | 1,281 | 1,329 | 1,308 | 20 |
| Homicides | 1,044 | 1,074 | 1,071 | 16 |
| Shooting | 852 | 884 | 925 | 14 |
| Stabbing | 90 | 95 | 60 | 1 |
| Self-inflicted injury | 205 | 222 | 210 | 3 |
| **Contact with objects and equipment** | 1,004 | 1,045 | 1,015 | 15 |
| Struck by object | 557 | 566 | 589 | 9 |
| Struck by falling object | 361 | 346 | 371 | 6 |
| Struck by flying object | 77 | 82 | 67 | 1 |
| Caught in or compressed by equipment or objects | 316 | 309 | 280 | 4 |
| Caught in running equipment or machinery | 159 | 151 | 147 | 2 |
| Caught in or crushed in collapsing materials | 110 | 138 | 132 | 2 |
| **Falls** | 600 | 618 | 661 | 10 |
| Fall to lower level | 507 | 533 | 577 | 9 |
| Fall from ladder | 78 | 76 | 85 | 1 |
| Fall from roof | 108 | 120 | 129 | 2 |
| Fall from scaffold | 66 | 71 | 89 | 1 |
| Fall on same level | 62 | 49 | 62 | 1 |
| **Exposure to harmful substances or environments** | 605 | 592 | 638 | 10 |
| Contact with electric current | 334 | 325 | 346 | 5 |
| Contact with overhead powerlines | 140 | 115 | 132 | 2 |
| Contact with temperature extremes | 33 | 38 | 50 | 1 |
| Exposure to caustic, noxious, or allergenic substances | 127 | 115 | 131 | 2 |
| Inhalation of substances | 83 | 68 | 84 | 1 |
| Oxygen deficiency | 111 | 111 | 110 | 2 |
| Drowning, submersion | 78 | 89 | 90 | 1 |
| **Fires and explosions** | 167 | 204 | 202 | 3 |
| **Other events or exposures**[3] | 76 | 43 | 24 | - |

[1] Based on the 1992 BLS Occupational Injury and Illness Classification Structures.
[2] The BLS news release issued August 10,1994, reported a total of 6,271 fatal work injuries for calendar year 1993. Since then, an additional 60 job-related fatalities were identified, bringing the total job-related fatality count for 1993 to 6,331.
[3] Includes the category "Bodily reaction and exertion."
NOTE: Totals for major categories may include subcategories not shown separately. Percentages may not add to totals because of rounding. Dashes indicate less than 0.5 percent or data that are not available or that do not meet publication criteria.
SOURCE: Bureau of Labor Statistics, U.S. Department of Labor, in cooperation with state and federal agencies, Census of Fatal Occupational Injuries, 1994.

Table 2. Fatal occupational injuries by occupation and major event or exposure, 1994.

| Occupation[1] | Fatalities | | Major event or exposure[2] (percent) | | | |
|---|---|---|---|---|---|---|
| | Number | Percent | Highway[3] | Homicide | Struck by object | Fall to lower level |
| Total ........................................................ | 6,588 | 100 | 20 | 16 | 9 | 9 |
| **Managerial and professional specialty** .......................... | 768 | 12 | 21 | 19 | 2 | 6 |
| Executive, administrative, and managerial ...................... | 486 | 7 | 17 | 24 | 2 | 8 |
| Professional specialty ............................................. | 282 | 4 | 29 | 11 | 4 | 4 |
| **Technical, sales, and administrative support** ............... | 943 | 14 | 19 | 45 | 2 | 2 |
| Technicians and related support occupations ................. | 209 | 3 | 9 | 5 | - | 2 |
| Airplane pilots and navigators...................................... | 131 | 2 | - | - | - | - |
| Sales occupations.................................................... | 588 | 9 | 19 | 63 | 2 | 2 |
| Supervisors and proprietors, sales occupations.......... | 249 | 4 | 8 | 72 | 3 | 3 |
| Sales workers, retail and personal services................ | 252 | 4 | 16 | 72 | - | - |
| Cashiers.............................................................. | 110 | 2 | - | 96 | - | - |
| Administrative support occupations, including clerical .... | 146 | 2 | 33 | 28 | 5 | 3 |
| **Service occupations.** ............................................... | 601 | 9 | 16 | 41 | 2 | 6 |
| Protective service occupations..................................... | 332 | 5 | 20 | 45 | 1 | 2 |
| Firefighting and fire prevention occupations, including supervisors ............................................................. | 56 | 1 | 7 | 5 | - | - |
| Police and detectives including supervisors................ | 149 | 2 | 32 | 47 | - | - |
| Guards, including supervisors ..................................... | 127 | 2 | 10 | 60 | - | 3 |
| **Farming, forestry, and fishing** ............................... | 944 | 14 | 11 | 2 | 19 | 6 |
| Farming operators and managers ................................. | 382 | 6 | 11 | 1 | 13 | 7 |
| Other agricultural and related occupations .................... | 360 | 5 | 17 | 3 | 9 | 9 |
| Farm workers, including supervisors....................... | 261 | 4 | 18 | 2 | 8 | 4 |
| Forestry and logging occupations................................. | 137 | 2 | 4 | - | 74 | - |
| Timber cutting and logging occupations ..................... | 112 | 2 | 4 | - | 75 | - |
| Fishers, hunters, and trappers.................................... | 65 | 1 | - | - | - | - |
| Fishers ................................................................. | 55 | 1 | - | - | - | - |
| **Precision production, craft, and repair** ...................... | 1,090 | 17 | 11 | 4 | 12 | 24 |
| Mechanics and repairers ............................................ | 294 | 4 | 11 | 5 | 21 | 9 |
| Construction trades.................................................... | 614 | 9 | 10 | 2 | 6 | 36 |
| Carpenters and apprentices .................................... | 87 | 1 | 8 | 5 | 12 | 47 |
| Electricians and apprentices.................................... | 99 | 2 | 8 | - | - | 19 |
| Painters.............................................................. | 46 | 1 | - | - | - | 46 |
| Roofers .............................................................. | 53 | 1 | 8 | - | - | 72 |
| Structural metal workers.......................................... | 48 | 1 | - | - | 15 | 63 |
| **Operators, fabricators, and laborers** ............................ | 2,055 | 31 | 32 | 9 | 10 | 7 |
| Machine operators, assemblers, and inspectors............ | 256 | 4 | 4 | 4 | 17 | 9 |
| Transportation and material moving occupations........... | 1,169 | 18 | 50 | 10 | 7 | 2 |
| Motor vehicle operators ......................................... | 925 | 14 | 61 | 12 | 6 | 2 |
| Truck drivers.................................................... | 762 | 12 | 68 | 2 | 6 | 2 |
| Driver-sales workers.......................................... | 29 | - | 62 | 28 | - | - |
| Taxicab drivers and chauffeurs .......................... | 113 | 2 | 17 | 76 | - | - |
| Material moving equipment operators........................ | 172 | 3 | 9 | - | 16 | 5 |
| Handlers, equipment cleaners, helpers, and laborers ..... | 630 | 10 | 9 | 8 | 13 | 16 |
| Construction laborers............................................. | 247 | 4 | 6 | - | 11 | 24 |
| Laborers, except construction .................................. | 229 | 3 | 7 | 6 | 19 | 11 |
| **Military**.................................................................. | 109 | 2 | 10 | 4 | 6 | 3 |

[1] Based on the 1990 Occupational Classification System developed by the Bureau of the Census.

[2] The figure shown is the percent of the total fatalities for that occupational group.

[3] "Highway" includes deaths to vehicle occupants resulting from traffic incidents that occur on the public roadway, shoulder, or surrounding area. It excludes incidents occurring entirely off the roadway, such as in parking lots and on farms; incidents involving trains; and deaths to pedestrians or other nonpassengers.

NOTE: Totals for major categories may include subcategories not shown separately. Percentages may not add to totals because of rounding. There were 78 fatalities for which there was insufficient information to determine an occupation classification. Dashes indicate less than 0.5 percent or data that are not available or that do not meet publication criteria.

SOURCE: Bureau of Labor Statistics, U.S. Department of Labor, in cooperation with state and federal agencies, Census of Fatal Occupational Injuries, 1994.

Table 3. Fatal occupational injuries and employment by industry, 1994

| Industry | SIC code | Fatalities | | Employment[2] (in thousands) | |
|---|---|---|---|---|---|
| | | Number | Percent | Number | Percent |
| Total | | 6,588 | 100 | 124,469 | 100 |
| **Private industry** | | 5,923 | 90 | 104,754 | 84 |
| **Agriculture, forestry and fishing** | | 847 | 13 | 3,496 | 3 |
| Agricultural production - crops | 01 | 441 | 7 | 1,008 | 1 |
| Agricultural production - livestock | 02 | 172 | 3 | 1,316 | 1 |
| Agricultural services | 07 | 162 | 2 | 163 | - |
| **Mining** | | 180 | 3 | 668 | 1 |
| Coal mining | 12 | 41 | 1 | 115 | - |
| Oil and gas extraction | 13 | 99 | 2 | 387 | - |
| **Construction** | | 1,027 | 16 | 6,948 | 6 |
| General building contractors | 15 | 189 | 3 | - | - |
| Heavy construction, except building | 16 | 247 | 4 | - | - |
| Special trades contractors | 17 | 591 | 9 | - | - |
| **Manufacturing** | | 787 | 12 | 20 050 | 16 |
| Food and kindred products | 20 | 78 | 1 | 1,749 | 1 |
| Lumber and wood products | 24 | 199 | 3 | 731 | 1 |
| **Transportation and public utilities** | | 944 | 14 | 7,069 | 6 |
| Local and interurban passenger transportation | 41 | 114 | 2 | 520 | - |
| Trucking and warehousing | 42 | 502 | 8 | 2,326 | 2 |
| Transportation by air | 45 | 98 | 1 | 755 | 1 |
| Electric, gas, and sanitary services | 49 | 88 | 1 | 1,096 | 1 |
| **Wholesale trade** | | 269 | 4 | 4,702 | 4 |
| **Retail trade** | | 797 | 12 | 20,909 | 17 |
| Food stores | 54 | 235 | 4 | 3,474 | 3 |
| Automotive dealers and service stations | 55 | 120 | 2 | 2,019 | 2 |
| Eating and drinking places | 58 | 181 | 3 | 6,316 | 5 |
| **Finance, insurance, and real estate** | | 112 | 2 | 7,900 | 6 |
| **Services** | | 844 | 13 | 33,012 | 27 |
| Business services | 73 | 253 | 4 | 4,999 | 4 |
| Automotive repair, services, and parking | 75 | 89 | 1 | 1,537 | 1 |
| **Government[3]** | | 665 | 10 | 19,715 | 16 |
| Federal (including resident armed forces) | | 209 | 3 | 4,901 | 4 |
| State | | 112 | 2 | 5,163 | 4 |
| Local | | 333 | 5 | 9,650 | 8 |
| Police protection | 9221 | 118 | 2 | - | - |

[1] Standard Industrial Classification Manual, 1987 Edition.

[2] The employment is an annual average of employed civilians 16 years of age and older, plus resident armed forces, from the BLS Current Population Survey, 1994.

[3] Includes fatalities to workers employed by governmental organizations regardless of industry.

NOTE: Totals for major categories may include subcategories not shown separately. Percentages may not add to totals because of rounding. There were 124 fatalities for which there was insufficient information to determine a specific industry classification, though a distinction between private sector and government was made for each. Dashes indicate less than 0.5 percent or data that are not available or that do not meet publication criteria.

SOURCE: Bureau of Labor Statistics, U.S. Department of Labor, in cooperation with state and federal agencies, Census of Fatal Occupational Injuries, 1994.

Table 4. Fatal occupational injuries and employment by selected worker characteristics, 1994

| Characteristics | Fatalities | | Employment (in thousands)[1] | | Most frequent event (percent of total) |
|---|---|---|---|---|---|
| | Number | Percent | Number | Percent | |
| Total................................ | 6,588 | 100 | 124,469 | 100 | Highway[2] (20 percent) |
| **Employee status** | | | | | |
| Wage and salary workers................... | 5,336 | 81 | 113,641 | 91 | Highway (22) |
| Self-employed[3]................................. | 1,252 | 19 | 10,828 | 9 | Homicide (21) |
| **Sex and age** | | | | | |
| Men....................................... | 6,067 | 92 | 67,690 | 54 | Highway (20) |
| Women ............................... | 521 | 8 | 56,779 | 46 | Homicide (35) |
| **Both sexes:** | | | | | |
| Under 16 years............................. | 25 | - | - | - | Highway (20) |
| 16 to 17 years.............................. | 42 | 1 | 2,511 | 2 | "        (26) |
| 18 to 19 years.............................. | 112 | 2 | 3,749 | 3 | Homicide (24) |
| 20 to 24 years.............................. | 545 | 8 | 13,204 | 11 | "        (19) / Highway (18) |
| 25 to 34 years.............................. | 1,558 | 24 | 32,829 | 26 | Highway (19) / Homicide (18) |
| 35 to 44 years.............................. | 1,608 | 24 | 33,882 | 27 | "        (19) / Homicide (18) |
| 45 to 54 years.............................. | 1,304 | 20 | 23,383 | 19 | "        (23) |
| 55 to 64 years.............................. | 858 | 13 | 11,229 | 9 | "        (23) |
| 65 years and over............................. | 517 | 8 | 3,681 | 3 | Nonhighway (22) |
| **Race** | | | | | |
| White ....................................... | 5,420 | 82 | 106,285 | 85 | Highway (21) |
| Black....................................... | 702 | 11 | 13,102 | 11 | Homicide (30) |
| Asian or Pacific Islander..................... | 181 | 3 | - | - | Homicide (61) |
| American Indian, Aleut, Eskimo ......... | 40 | 1 | - | - | "        (18) |
| Other or unspecified............................. | 245 | 4 | - | - | "        (24) |
| **Hispanic origin** | | | | | |
| Hispanic[4]....................................... | 611 | 9 | 10,867 | 9 | Homicide (22) |

[1] The employment is an annual average of employed civilians 16 years of age and older, plus resident armed forces, from the BLS Current Population Survey, 1994.

[2] "Highway" includes deaths to vehicle occupants resulting from traffic incidents that occur on the public roadway, shoulder, or surrounding area. It excludes incidents occurring entirely off the roadway, such as in parking lots and on farms; incidents involving trains; and deaths to pedestrians or other nonpassengers.

[3] Includes paid and unpaid family workers, and may include owners of incorporated businesses, or members of partnerships.

[4] Persons identified as Hispanic may be of any race. Hispanic employment does not include resident armed forces.

NOTE: Totals may include subcategories not shown separately. Percentages may not add to totals because of rounding. Dashes indicate less than 0.5 percent or data that are not available or data that do not meet publication criteria.

SOURCE: Bureau of Labor Statistics, U.S. Department of Labor, in cooperation with state and federal agencies, Census of Fatal Occupational Injuries, 1994.

**TECHNICAL NOTES**

Definitions

For a fatality to be included in the census, the decedent must have been employed (that is working for pay, compensation, or profit) at the time of the event, engaged in a legal work activity, or present at the site of the incident as a requirement of his or her job. These criteria are generally broader than those used by federal and state agencies administering specific laws and regulations. (Fatalities that occur during a person's commute to or from work are excluded from the census counts.)

Data presented in this release include deaths occurring in 1994 that resulted from traumatic occupational injuries. An injury is defined as any intentional or unintentional wound or damage to the body resulting from acute exposure to energy, such as heat or electricity, or kinetic energy from a crash; or from the absence of such essentials as heat or oxygen caused by a specific event, incident, or series of events within a single workday or shift. Included are open wounds, intracranial and internal injuries, heatstroke, hypothermia, asphyxiations, acute poisonings resulting from a short-term exposure limited to the worker's shift, suicides and homicides, and work injuries listed as underlying or contributory causes of death.

Information on work-related fatal illnesses are not reported in the BLS census and are excluded from the attached tables because the latency period of many occupational illnesses and the difficulty of linking illnesses to work makes identification of a universe problematic. Partial information on fatal occupational illnesses, compiled separately, is available for 1991-1993 in BLS Report 891.

Measurement techniques and limitations

Data for the Census of Fatal Occupational Injuries are compiled from various federal, state, and local administrative sources--including death certificates, workers' compensation reports and claims, reports to various regulatory agencies, medical examiner reports, and police reports--as well as news reports. Multiple sources are used because studies have shown that no single source captures all job-related fatalities. Source documents are matched so that each fatality is counted only once. To ensure that a fatality occurred while the decedent was at work, information is verified from two or more independent source documents, or from a source document and a follow-up questionnaire. Approximately 30 data elements are collected, coded, and tabulated, including information about the worker, the fatal incident, and the machinery or equipment involved.

*Identification and verification of work-related fatalities.*

Because some state laws and regulations prohibit enumerators from contacting the next-of-kin, it was not possible to independently verify work relationship (whether a fatality is job related) for 258 fatal work injuries in 1994; however, the information on the initiating source document for these cases was sufficient to determine that the incident was likely to be job-related. Data for these fatalities, which primarily affected self-employed workers, are included in the Census of Fatal Occupational Injuries counts. An additional 56 fatalities submitted by states were not included because the initiating source document had insufficient information to determine work relationship, which could not be verified by either an independent source document or a follow-up questionnaire.

States may identify additional fatal work injuries after data collection close-out for a reference year. In addition, other fatalities excluded from the published count because of insufficient information to determine work relationship may be subsequently verified as work related. States have up to one year to update their initial published state counts. This procedure ensures that fatality data are disseminated as quickly as possible and that no legitimate case is excluded from the counts.

*Federal/state agency coverage*

The Census of Fatal Occupational Injuries includes data for all fatal work injuries, whether they are covered by the Occupational Safety and Health Administration (OSHA) or other federal or state agencies or are outside the scope of regulatory coverage. Thus, any comparison between the BLS census counts and those released by other agencies should take into account the different coverage requirements and definitions being used.

Several federal and state agencies have jurisdiction over workplace safety and health. OSHA and affiliated agencies in states with approved safety programs cover the largest portion of America's workers. However, injuries and illnesses occurring in several other industries, such as coal, metal, and nonmetal mining and water, rail, and air transportation, are excluded from OSHA coverage because they are covered by other federal agencies, such as the Mine Safety and Health Administration, the U.S. Coast Guard, the Federal Railroad Administration, and the Federal Aviation Administration. Fatalities occurring in activities regulated by federal agencies other than OSHA accounted for about 19 percent of the fatal work injuries for 1994.

Fatalities occurring among several other groups of workers are generally not covered by any federal or state agencies. These groups include self-employed and unpaid family workers, which accounted for about 19 percent of the fatalities; laborers on small farms, accounting for about 2 percent of the fatalities; and state and local government employees in states without OSHA-approved safety programs, which account for about 3 percent. (Approximately one-half of the states have approved OSHA safety programs, which cover state and local government employees.)

ACKNOWLEDGMENTS: BLS thanks the participating states for their efforts in collecting accurate, comprehensive, and useful data on fatal work injuries. BLS also appreciates the efforts of all federal, state, local, and private sector agencies that submitted source documents used to identify fatal work injuries. Among these agencies are the Occupational Safety and Health Administration; the National Transportation Safety Board; the Department of Justice (Bureau of Justice Assistance); the Mine Safety and Health Administration; the Department of Defense; the Employment Standards Administration (Federal Employees' Compensation and Longshore and Harbor Workers' divisions); the Department of Energy; the National Association of Chiefs of Police; state vital statistics registrars, coroners, and medical examiners; state departments of health, labor, and industries, and workers' compensation agencies; state and local police departments; and state farm bureaus.

## REFERENCES

### Books

Baker, S. P., O'Neil, B., Ginsburg, M. J. et al. *The Injury Fact Book.* New York: Oxford University Press, 1992.

Foege, W. H., Baker, S. P., Davis, J. H. et al. *Injury in America: A Continuing Public Health Problem.* Washington, D.C.: National Academy Press, 1985.

Mellius, J. M., Althafer, C. A., Perry, W. H. et al. *Proposed National Strategies for the Prevention of Leading Work-Related Diseases and Injuries, Part I.* Washington, D.C.: Association of Schools of Public Health, 1988.

National Safety Council. *Accident Facts,* 1994 edition. Itasca, IL: National Safety Council, 1994.

Polakoff, P. L., O'Rourke, P. F. et al. *Healthy Worker-Healthy Workplace: The Productivity Connection.* Sacramento, CA: California Legislature, 1990.

Rice, D. P., MacKenzie, E. J. et al. *Cost of Injury in the United States: A Report to Congress.* 1989. Atlanta: Centers for Disease Control, 1989.

Rosenberg, M. L., Fenley, M. A. et al. *Violence in America — A Public Health Approach.* New York: Oxford University Press, 1991.

Sauter, S. L., Althafer, C. A., Cahill, L. M. et al. *Proposed National Strategies for the Prevention of Leading Work-Related Diseases and Injuries, Part II.* Washington, D.C.: Association of Schools of Public Health, 1988.

### Other Publications

Castelli, J. Safety and health watch. *Saf. Health,* 148(2), 153-156, August 1993.

Hoskin, A. F. Council updates work-death statistics. *Saf. Health,* 147(5), 80-81, May 1993.

# 2

# WHAT WORKS TODAY? A SAFETY PROFESSIONAL'S VIEWPOINT

**Richard W. Lack**

Against the background previously discussed, let us now turn to our profession and the organizations we work in to discuss what issues we face and what works today. In attempting to address this question, I decided to involve the right brain in helping to focus on the issues. To do this, I used the "mind mapping" approach. See Figure 1 for what I came up with.

Most safety professionals, I believe, would agree that the past few years have been an exciting and challenging experience. We have had to master many new skills along the way. As an example, in my experience, after starting in a traditional safety field, subsequent responsibilities picked up have been security, fire protection, health, including industrial hygiene, and hazardous materials. Other added responsibility areas have included ergonomics, indoor air quality, asbestos, lead, and bloodborne pathogens. And the list goes on....

The "mind map" demonstrates the complexity of the issues facing our profession today. In spite of these complex, ever-changing issues, I believe "prevention" must still be our watchword. Prevention of injury, illness, and/or loss is accomplished by establishing a basic systematic approach that all levels in your organization will understand and effectively implement.

To achieve effective prevention, every aspect of your safety and health management systems must be designed to find and control hazards. This will in turn help to protect your organization's assets — both people and property — from the various risks to which they may be exposed.

From my viewpoint, in spite of all the tremendous changes in technology, the principal challenges facing members of the safety and health profession rarely change. In fact, they have been with us since the beginning of time. "They", of course, are *people!*

1-56670-054-X/96/$0.00+$.50
© 1996 by CRC Press, Inc.

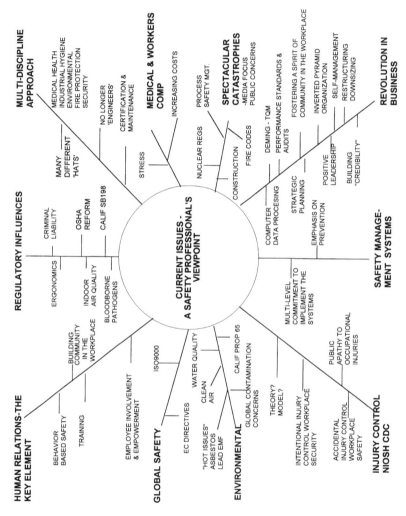

**Figure 2-1**  Current issues — a safety professional's viewpoint.

We should never lose sight of our need to continuously improve our people-relationship skills. This includes both communication and management skills. The overall mission, therefore, of the authors involved in the production of this book is dedicated to providing you, the reader, with the information and techniques you need to help you successfully meet these challenges.

## FURTHER READING

### Books

Grose, V. L. *Managing Risk — Systematic Loss Prevention for Executives.* Englewood Cliffs, NJ: Prentice Hall, 272–366, 1987.

Brauer, R. L. *Safety and Health For Engineers.* New York: Van Nostrand Reinhold, 79–91, 1990.

Manuele, F. A. *On the Practice of Safety.* New York: Van Nostrand Reinhold, 169–181, 1993.

### Other Publications

Lack, R. W. Systips: Getting back to the basics — or learning from history so we are not forced to repeat our mistakes. *National Safety Management Society Insights Into Management,* 2 (5/6), 6, November/December 1990.

# Section II
# Safety Program Management Aspects

# 3     SAFETY PROGRAM ORGANIZATION

**Richard W. Lack**

## TABLE OF CONTENTS

## LEGISLATIVE ASPECTS

The continuing loss experience of a wide spectrum of organizations and industries has resulted in an increasing stream of legislative initiatives requiring employers to establish a minimum safety and health organization.

Naturally, as readers are well aware, there are many employers who have outstanding professional staffs and who have established and maintain highly effective programs which produce superior results. Unfortunately, these employers are in the minority when one looks at the broad picture.

In the U.S., an early initiative was the Occupational Safety and Health Administration's (OSHA) Safety and Health Program Management Guidelines of 1989. These guidelines described the elements of an effective safety program, including management and employee responsibilities. In the original draft of these guidelines, there was a reference to a "competent person" for conducting work-site analysis. In the final version, this term was removed. However, in the commentary there is a discussion of the relative competence for the various approaches to work-site analysis.

The California State Law SB 198 of 1991 requires all employers to nominate an Injury and Illness Prevention Program Administrator who has the authority and responsibility for implementing and maintaining the employer's Injury and Illness Prevention Plan.

In Europe, the EC Directive *Management of Health and Safety at Work,* as adopted by the British Health and Safety Commission in 1992, contains some more specific references.

Regulation 6 in the HSE Regulations states in part:

(1) Every employer shall, subject to paragraphs (6) and (7) (self-employed or business partnerships), appoint one or more competent persons to assist him in undertaking the measures he needs to take to comply with the requirements and prohibitions imposed upon him by or under the relevant statutory provisions.

The approved Code of Practice relevant to this regulation discusses various options open to employers, such as appointing a health and safety department for large employers or hiring external services for smaller employers. There is also reference to the degree of competence needed for persons appointed to carry out the requirements of the regulations.

This trend can be expected to continue, especially in the area of competence, where there are several bills pending in various state legislatures requiring licensing and registration of safety and health professionals.

## ECONOMIC ASPECTS

In spite of these legislative "drivers", during recent years the business world has been suffering from the combined impacts of recession and increasing global competition. These pressures have resulted in a veritable tidal wave of reorganization in one form or another.

Familiar phrases summarizing the response of businesses both large and small are "downsizing", "restructuring", or the latest one, "reengineering". Added to this is the compelling need to control costs, especially costs not directly related to production, such as accidental losses or overheads.

Because of these economic "drivers", the trend has been for the larger organizations to reduce corporate or divisional safety and health staffs and/or combine responsibilities with other functions such as environmental, security, and fire protection.

On the positive side, there has been a trend among smaller organizations toward establishing a safety and health function. This trend is expected to continue, and there should also be an increasing trend of firms seeking assistance from outside consultants.

## MANAGEMENT RESPONSIBILITIES OF THE SAFETY AND HEALTH POSITIONS

The "technical" duties and responsibilities of this position will be many and varied and as such are clearly beyond the scope of this publication. What is a critical concern are the administrative or management responsibilities of this position.

In 1966, the American Society of Safety Engineers (ASSE) published a paper, *The Scope and Functions of the Safety Professional.* This summarizes the functional activities of safety and health professionals. The paper is periodically reviewed, and no doubt it may be expanded or amended in the future in the light of changing conditions.

During 1992 and 1993, the ASSE Government Affairs Committee prepared a working draft position paper, *Management Responsibilities of the Safety Professional.* In this paper, the principal management responsibilities for the safety and health professional were established as follows:

1. Assist the line leadership in assessing the effectiveness of the unit safety and health programs.
2. Provide guidance to line leadership and employees so that they understand the programs and how to implement them.
3. Assist line leadership in the identification and evaluation of high-risk hazards and develop measures for their control.
4. Provide staff engineering services on the safety and health aspects of engineering.
5. Maintain working relations with regulatory agencies.
6. Attain and maintain a high level of competence in all related aspects of the profession.
7. Represent the organization in the community on matters of safety and health.

Expanding this list could be an endless task; however, these seven broad responsibility areas are a good starting point.

The key to an effective management system is to ensure that the person in this position functions as a system or program administrator, not an implementor.

When problems exist, it is usually found that the organization's line management has failed to understand the philosophy and role of the safety and health function. Conversely, those involved in the safety and health functions must be sure they are not immersing themselves in technical details at the expense of administering and providing support for the management system.

A particular event has stuck in my memory: when I was a young safety professional, I attended the National Safety Congress in Chicago for the first time in 1964. Concurrent with this Congress, Homer Lambie had assembled a large group of Kaiser Aluminum safety professionals, and one of the special sessions which Mr. Lambie had organized for our group was a luncheon meeting with John Grimaldi, then the Safety Director of the General Electric Company. From my notes on his talk, I quote Dr. Grimaldi's remarks as follows: "Insofar as the safety professional at GE is concerned, an effective safety professional is one that persuades management action rather than promotes employee awareness. The safety professional should develop decision-making information for managers to apply rather than assume responsibility for corrective action." In other words, the safety professional teaches methods of solving problems rather than providing answers. This meeting was held on October 26, 1964, and the principles are as current today as they were then 30 years ago. Certainly, I have done my best to consistently live by them throughout my professional career.

To clarify this point, let us consider a typical day in the life of a full-time safety and health professional. A normal day when nothing unusual occurs should include some or all of the following activities:

1. Meeting with one manager or supervisor to discuss the status and progress of his/her safety and health programs
2. Conducting an audit or survey with or without members of management to assess the effectiveness of the unit's programs
3. Reviewing any loss event that may have occurred and assisting line management with the investigation of those considered significant
4. Working up programs such as training presentations, new or revised procedures, committee meetings, and assignments
5. Conducting training programs
6. Attending committee and other meetings
7. Conducting research and providing professional consultation
8. Developing and guiding staff members, if in a supervisory role.

This discussion now leads us to the final element of this chapter, which is to look at actually fitting the safety and health function into the management organization.

Obviously, the variables in any organization are immense: the unit's nature of business, overall mission, leadership style, and the employee relations "climate" will all contribute to differences in the management structure of the organization. In spite of these differences, it is still highly desirable from an effectiveness standpoint for the unit leadership to position their staff support functions in a direct reporting relationship with the chief executive or principal leader of the organization. An example "ideal" organization chart is shown in Figure 3-1.

In smaller units, the safety function may be grouped together with several others. The key is that safety and health is a management system, and to be effective it must be managed from the highest level in any unit organization.

The responsibility of the individual in this position is to bring to that organization a system that is designed to prevent unplanned (or, in the case of security, deliberate) accidents or incidents that can produce adversity of one kind or another for the organization. Systems necessary to achieve this objective are discussed in other chapters.

Many businesses are going though a process of "reengineering" in an effort to streamline their organization and related systems. The benefits of this process are cost reductions, quality improvements, time to market advantage,

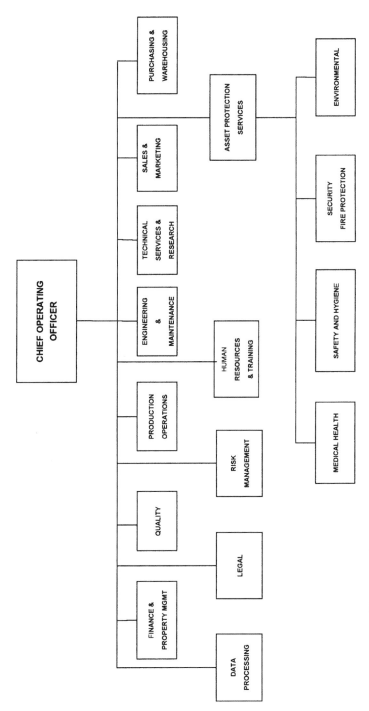

**Figure 3-1** This chart indicates the position of safety in an "ideal" corporate or major facility structure.

and greater employee job satisfaction and empowerment. These benefits lead in turn to a wide host of opportunities for improvement in the unit's day-to-day operations. Examples could be increased revenues, increased production, improved service, increased growth potential, enhanced communications and decision making, and — above all — reduced risks.

When all these factors are considered, it is clear that the safety and health professionals will need to demonstrate superior qualities in the fields of professional competence, systems design and administration, management, and communications skills. Professional competence and skills are discussed in more detail by Dr. Roger Brauer in Chapter 40. Management and communications skills are discussed in several other chapters.

## FURTHER READING

### Books

Bird, F. E., Jr. and Germain George, L. *Practical Loss Control Leadership.* Loganville, GA: Institute Publishing, Division of International Loss Control, 1986.

Hammer, M and Champy, J. *Re-Engineering the Corporation.* New York: Harper Business, 1993.

Petersen, D. *Techniques of Safety Management.* New York: Aloray, 1989.

Thomen, J. R. *Leadership in Safety Management.* New York: John Wiley & Sons, 1991.

### Other Publications

Health and Safety Commission. *Management of Health and Safety at Work Regulations 1992: Approved Code of Practice.* London: British Health and Safety Commission, 1992.

U.S. Department of Labor, Occupational Safety and Health Administration (OSHA). *Safety and Health Program Management Guidelines: Issuance of Voluntary Guidelines. Fed. Regist.,* 54(16), January 26, 1989.

# ENGINEERING DESIGN AND CONSTRUCTION — SAFETY MANAGEMENT ASPECTS

**4**

**Steven M. McConnell**

## TABLE OF CONTENTS

## INTRODUCTION

In his presentation to the attendees at the 1994 American Society of Safety Engineers conference held in Las Vegas, Mr. Joseph Dear, Assistant Secretary of Labor, U.S. Occupational Safety and Health Administration (OSHA) commented that every day over 60 million Americans go to work ... and 17 do not come home. Data show that the construction industry is second only to the mining industry in terms of fatalities per 100,000 workers. Frequency and severity rates in construction far exceed those of most other industries as well.

It is not surprising that many people have come to accept that injuries, even fatalities, are a part of doing construction work. That concept represents a paradigm ... and it is time for us to realize the need for a shift in this paradigm.

The term "construction" represents a broad scope of activity including new building, renovation, demolition, and tenant improvement work. Projects, like contractors, run the spectrum in size from very small to extremely large. The amount and effectiveness of training and education for individual construction workers varies greatly, sometimes more significantly in nonunion environments. Workers come and go with regularity on a typical construction project. Payment to the contractor for work performed on a project is often based on adherence to strict project schedules, sometimes tempting the contractor to shortcut safety measures in an attempt to gain time with respect to the schedule. Combine these issues with the many inherently dangerous tasks performed on a construction project and it is easy to conclude that managing safety in the construction industry is a difficult and dynamic process. However, the safety manager can be successful by employing sound management techniques.

Safety managers share a common frustration associated with reacting to events, often the result of a design error or oversight. This premise is encountered in construction as well as other industries and illustrates the importance of "safety" being associated with the design phase of a job.

Architects and engineers play an important part in a job's safety potential both during and after construction. Input from the safety department, in the form of comments from design reviews, etc., is also helpful. However, because of the volume of drawings and submittals associated with most projects, you, the safety manager, often miss the opportunity to provide valuable input.

When time (or ability) constraints prevent your full involvement in the review process, the conceptual/design stage of a job remains the best opportunity to identify and control many of the serious potential hazards that may exist. Each project is unique and may present a different set of concerns, but basic questions should be asked of the design/engineering firm and/or the contractor to ensure that those aspects of the job are assessed. Examples of the type of questions to be asked are illustrated in Table 4-1.

While it is recognized that the method for managing the safety program may vary depending on whether the client or the contractor is being represented, this chapter will attempt to speak generally with respect to construction safety management techniques. Some specific guidance will be given representing the viewpoints of both the client and the contractor.

## THE CLIENT'S PERSPECTIVE

If you manage the safety function for the client, you are interested in completing the job on time and within budget while minimizing job-related

**Table 4-1 Conceptual/Design Phase Questions**

- Is trenching or excavation work in excess of 5 feet required for the job?
- What type of soil is found on the site?
- Will lifting equipment such as boom trucks, cherry pickers, or cranes be operated in close proximity to overhead electrical lines?
- Has the need to use hazardous chemicals been evaluated and their use eliminated where possible or less hazardous materials substituted where possible?
- Is there a potential for employee exposure to ACM or lead and, if so, can the exposure be eliminated or reduced?
- Has adequate space been designated for a material storage area or "boneyard"?
- Are stairs, ramps, building parapets, etc., designed to code?

accidents and injuries. It is important to stress the safety-related aspects of the job with the contractor(s) from the beginning, the bid package stage. Safety-related requirements expected of the contractor(s) should be given to those bidding on the project at the pre-bid meeting. The project manager for your company must support you in this process by emphasizing project safety concerns when discussing project details. His or her efforts to impress how important safety is on the job is essential to the overall success of the program. This is the appropriate time to go beyond the boiler plate message and specify what is expected in terms of safety on the job and what safety-related documentation/data will be required from the contractor, including the information you want included in their bid for the project. Some typical items to request and discuss are illustrated in Table 4-2.

It is generally recommended to manage the safety program using only one set of site safety rules — yours! The contractor(s) must be made aware of this concept, and any requirements that exceed OSHA requirements should be highlighted in the bid process. This practice will eliminate surprises later and allow the contractor to accurately bid on the project. This may represent your best opportunity to convince the contractor that your company is serious about job site safety and health. Provide copies of specific requirements to the contractor; post them on site as well once the job begins.

## THE CONTRACTOR'S PERSPECTIVE

The safety representative for the contractor also wants the job to be completed on schedule, within budget, and without accidents and injuries. A good safety record makes it easier for the company to obtain work and obviously translates into higher profits. Safety performance has become a key job qualification for contractors, and it is usually the safety manager who is responsible for providing safety performance-related information requested by the client; he/she should be well versed with regard to the company's safety performance and programs. The safety manager can best serve his or her company by anticipating client needs and developing a custom program for the subject project. While the custom program may include many generic

**Table 4-2  Items Requested from Contractor**

- Your general expectations regarding project safety
- Contractor's EMR for the past 3 years
- Copy of contractor's OSHA citations for the past 3 years
- Information regarding fatalities or serious injuries for the past 3 years, including control measures implemented
- Contractor's written safety program if one exists (required by State law in California)
- Methods to be used to control anticipated hazards:
  - Job site inspection program
  - Accident investigation program
  - Fall protection
  - Lockout, tagout program
  - General electrical safety
  - Hazard communication program
  - Confined space program
  - Respiratory protection program
  - Scaffold plan
  - Trenching/excavation plan
  - Employee safety orientation program
  - Employee training/education program
  - Subcontractor orientation program
  - Names of key project personnel
  - Name of "competent" person(s)
  - Disciplinary action program
  - Client oversight/participation in safety plan

**Table 4-3  Items Presented by the Contractor**

- Company's safety policy statement
- Written safety program/custom program to follow
- Performance history, including accident statistics
- Company's OSHA record
- Job site hazard identification/control program
- Hazards anticipated on the job
- Methods used to control anticipated hazards
- Oversight of subcontractor safety
- Disciplinary action program
- Employee orientation/education program
- Methods to document safety program efforts

components, it is important for it to be directed toward, and address the needs of, the subject project. The safety manager should be prepared to provide and discuss the things requested by the client. Typical items to be presented by the contractor appear in Table 4-3.

The contractor's safety representative should be sincere in his/her attempt to explain the company's safety program and performance record, and cooperation with the client's representative should be stressed. This relationship will last throughout the project, and it helps to get started on the right foot.

## THE MULTIPLE EMPLOYER SITE

It is extremely common for a prime contractor to use several subcontractors to complete a project. Clients sometimes bid on projects in phases, which

## Monthly Safety Performance Review

Project Name: _____

Client Contact: _____

Project Manager: _____          Month: _____

ES & H Manager: _____          Year: _____

Jobsite Telephone: _____        T.C. Rate: _____

OSHA Telephone: _____          LWD Rate: _____

### Contractor Performance

| Contractor | Site Contact / Phone | T.C. Rate / LWD Rate |
|---|---|---|
| _____ | _____ | _____ |
| _____ | _____ | _____ |
| _____ | _____ | _____ |
| _____ | _____ | _____ |
| _____ | _____ | _____ |
| _____ | _____ | _____ |

TOTALS _____

### RATES

$$T.C. = \frac{\text{Total Recordable OSHA Cases} \times 200{,}000}{\text{Total Hours Worked}}$$

$$LWD = \frac{\text{\# LWD Cases} \times 200{,}000}{\text{Total Hours Worked}}$$

**Figure 4-1**  Monthly safety performance review.

ensures a multiple employer site. The need for "one set of rules" becomes even more important in these conditions.

Managing safety on these projects becomes even more difficult, at least administratively. Establishing an effective paper trail is necessary to ensure that appropriate documentation exists and that the appropriate people are informed. Guidelines must be developed to ensure that safety-related documentation is maintained accurately and is appropriately distributed. It is a good practice to maintain safety performance-related statistics (both for the project overall and by individual contractor), updated on a regular basis, perhaps monthly. Figure 4-1 illustrates a form that can be used to summarize safety performance for both the job and the individual contractor on a monthly basis.

A method must be developed to ensure the consistent enforcement of safety requirements and the fair, uniform use of disciplinary action. Intentions regarding enforcement (or oversight) efforts should be clearly communicated to the contractor(s) before actual work begins on the project. Again, there should be no surprises once the work begins.

It is important to recognize that responsibility, and subsequent liability, for job site safety can be shared by more than one contractor or company, at least in terms of an OSHA inspection. For example, on a particular job site one contractor is assigned the responsibility of installing and maintaining handrails (on stairs, overhead work platforms, etc.) regardless of who employs the individual(s) using the work area. For example, suppose OSHA visits the site, and the inspector observes a pipe fitter (who is not using fall protection) welding a pipe approximately 20 feet above ground while standing on a job-built wooden platform that is missing part of the required handrails. In this case there may be three citations issued. Those potentially cited would include

- The prime or general contractor because he or she is viewed to have "control" of the job
- The subcontractor assigned the responsibility for work platforms because the omission of a portion of the handrails created a potential serious hazard
- The subcontractor employing the pipe fitter because he or she allowed the individual to work in the area where the unsafe condition existed

In other scenarios the owner/client may share liability — particularly in third-party lawsuits. Indemnifications are not always viewed as viable in the court-room. It is important for the safety professional to discuss the liability issue with the company's senior management and legal counsel in order to under-stand exactly where their company fits into the situation and how oversight efforts will be handled.

## GENERAL SITE SAFETY

Whether you represent the client or the contractor, certain events must occur in order to implement an effective site safety program. The next logical step following contractor selection is the preconstruction meeting. This type of meeting, which has been common for years, is held primarily to discuss the scope of the work in detail with each contractor. In recent years safety has become more and more of an agenda item for these meetings. Potential contractors were informed in the pre-bid stage that certain safety-related information and programs would be required of them — this is the time for delivery. The scope of the work, as it relates to safety, is usually addressed in detail at the preconstruction meeting; specific attention is given to antici-pated hazards that may create potential for serious or fatal injury. These hazards include those items listed in Table 4-1 and other project-specific tasks that may be identified. The method(s) the contractor will use to control these hazards should be discussed in detail and agreed upon. It is necessary to reach agreement on compliance efforts at this stage. In many cases it is possible to achieve compliance in a number of ways; whether you represent the client or contractor you want to establish the method(s) for ensuring compliance on a particular job.

Companies use a variety of techniques to establish performance criteria on a job. An effective strategy, used by the Brown and Root Building Company (based in Houston, Texas), requires a contractor to provide a model for compliance. For example, the subcontractor with the most scaffold work on a job is required to build a scaffold model (full size, two tiers high by 16 feet long) which is erected adjacent to the construction office. The model serves as the standard for scaffolding used on that job. Subcontractors have no excuse for erecting scaffolding that does not meet the job standard. The same system could be used to establish job site standards in other areas, including job-built stairs, ladders, and trenching operations. Everyone benefits when such standards are established for the job **before** breaking ground on the project ... and worker safety is enhanced.

The preconstruction meeting is also an effective opportunity to address safety-related documentation requirements, client-contractor interface, and common safety concerns such as housekeeping and related requirements. This meeting provides the opportunity to review specific documentation requirements; Figure 4-2 illustrates the type of documentation that may be introduced at this stage.

An additional use of the preconstruction meeting is to establish the contractor's responsibility to adequately orient the subcontractors to site safety requirements before they begin work on the site. The objective is to minimize accidents and injuries on the project; subcontractors cannot be overlooked in the process. All of these efforts combine to establish the mechanics and criteria of the site safety program. When these criteria are addressed at or before the preconstruction meeting the job can begin without confusion about safety requirements for the job.

## THE DISCIPLINARY ACTION PROGRAM

At a time when behavior-based training and positive recognition programs are gaining popularity, the implementation of a fair, progressive disciplinary action program should not be overlooked. The typical construction worker "booms" from one job to the next with regularity; most never align themselves with one company long enough to develop loyalty, whether they want to or not. For them, violating a safety requirement may seem trivial, no big deal. In fact, some may do so purposely because they view it as the "macho" thing to do. "Walking the steel" rather than "cooning a beam" is a classic example. Such

TRENCH & EXCAVATION INSPECTION LOG

Date _____ Inspector _____Job _____

Trench / Excavation Type: _____

Soil Description: _____

Visual Soil Test Made: Yes☐ No ☐ Type _____

Protective System Used: Sloping☐  Box☐

    Wood Shoring  ☐ Hydraulic Shoring  ☐

Conditions:  Wet☐   Dry ☐  Submerged ☐

Hazardous Atmosphere?  Yes  ☐ No  ☐
(Confined Space Procedure Mandatory If Yes)

Trench / Excavation Dimensions:

Length _____ Width _____ Depth _____

GENERAL ITEMS - WRITE COMMENTS ON BACK

| Item | Yes | No |
| --- | --- | --- |
| Adequate access | ☐ | ☐ |
| Spoil pile away from edge | ☐ | ☐ |
| Utilities or structures secured | ☐ | ☐ |
| Materials stockpiled away from edge | ☐ | ☐ |
| DOT vests worn as needed | ☐ | ☐ |
| Barricades used where needed | ☐ | ☐ |

**Figure 4-2** Trench and excavation inspection log.

behavior, of course, can result in catastrophic consequences for the worker. Even short of a death or serious injury this type of behavior can be detrimental to the company. When an OSHA inspector observes a noncompliance he/she usually looks to cite the employer — maybe others, as previously discussed.

An effective, consistently implemented disciplinary program can be beneficial in two ways. First, the action (though viewed negatively) may reduce the potential for future noncompliances by the employee and co-workers. Second, documentation related to the program and its enforcement site-wide may persuade OSHA not to issue a particular citation. Also, if a citation is issued, this documentation sometimes may be used effectively during the appeal process. It may help get the violation removed or the monetary fine reduced.

The disciplinary program should be communicated to all contractors before they begin work on the project. The more successful programs include progressive disciplinary action. In such programs the employee receives more significant discipline with each successive violation. A typical progression of discipline is shown below:

- First offense ............................................. Verbal warning, documented
- Second offense ........................................ Written warning
- Third offense ........................................... Suspension of employee
- Fourth offense ......................................... Termination

In practice, disciplinary programs are implemented in a variety of ways. Some companies eliminate the verbal warning, choosing to issue a written warning on the first offense. Depending on the severity of the noncompliance, some companies suspend or terminate employees on the first offense. Regardless of how such a program is implemented, it is important that all parties understand it from the beginning of the job. The primary key to success is having the program written, communicated, and enforced uniformly on the job. Enforcement actions should be documented in writing.

## PROGRAM GOALS AND MEASUREMENTS

Similar to safety and health programs in other industries, successful construction safety programs have measurable goals and performance measures. In many cases program goals may come from the home office, but on-site management may also have input or even develop additional goals themselves. It is helpful to get your job superintendent or top official involved in the formation of a site-specific safety goal.

The importance of goals and measurements is well documented, but it has not always been common to establish them in the construction industry. With the realization that workers' compensation costs were reducing profits and that higher EMRs meant less work coming in, companies have escalated their loss control efforts. Attention to meaningful goals and performance measures is a part of that escalation.

At Bechtel, for example, a "zero accidents program" was implemented in January 1993. Not every company chooses to implement goals that are this ambitious, but positive results can be realized nonetheless. People pay attention to things that get measured.

Program measurements should be established early and communicated to all contractors at the start of the job. Contractors should be responsible for the safety performance of their short-term subcontractors because they oversee their work daily and are responsible for orienting them to the safety requirements of the project site. This relationship must be understood by all parties.

It is common to see job site statistics, in the form of frequency and severity rates, reported both as total job performance and by individual contractor (refer to Figure 4-1). In many cases other meaningful data are measured as well. These may include performance measurements like total injuries, a company's or supervisor's compliance to holding safety meetings, training/educating employees, safety sampling, conducting required site safety inspections, and hazard abatement. A number of measurements can be used to assess safety performance — for management and nonmanagement employees.

Whatever the measurement(s), specific criteria should be established so that accountability for safety performance can be appropriately assigned. Timely feedback should be given to employees and management detailing experience on the job, including a comparison to project goals. Evaluation of the safety and health program is recognized as the final process of hazard control and is important to those interested in the program's effectiveness.

## HAZARD IDENTIFICATION AND CONTROL

A basic component of the program, hazard identification and control, is at the heart of the loss control effort. Activities on construction projects occur rapidly and can result in potential serious hazards at any time. It is common to have several operations underway at one time on a construction site. Identifying hazards and keeping track of abatement efforts can be a difficult task. Systems may be established, either manual or computerized, to record and track site hazards (see Figure 4-3). Desired data can be obtained (and tracked) for the site as a whole, by building, by contractor, by supervisor, etc. The information to be gathered is determined by you, the safety manager, but the reader is cautioned to determine the uses ahead of time to be sure the needed data are obtained. Figure 4-3 illustrates a hazard tracking log. Although the system can be set up either manually or on computer, a tracking system set up on your computer offers obvious advantages to you.

The tracking system provides a "closed loop" system for managing job site hazards. If inspected, this information can be used to illustrate your hazard control efforts — an inspector will be impressed by the fact that potential hazards and corrective measures are routinely identified and monitored until acceptable corrective action is achieved.

The system can also help identify problem areas or contractors on the job. With this knowledge specific action can be taken to address such problem areas on a proactive basis.

The purpose of tracking hazard identification and abatement efforts is threefold:

Hazard Identification and Tracking Log

| Date Found | Reference | Contractor | General Description | Action(s) Taken | Date Closed |
|---|---|---|---|---|---|
|  |  |  |  |  |  |
|  |  |  |  |  |  |
|  |  |  |  |  |  |
|  |  |  |  |  |  |
|  |  |  |  |  |  |
|  |  |  |  |  |  |
|  |  |  |  |  |  |
|  |  |  |  |  |  |
|  |  |  |  |  |  |
|  |  |  |  |  |  |
|  |  |  |  |  |  |
|  |  |  |  |  |  |

**Figure 4-3** Hazard identification and tracking log.

- Potential problem areas can be identified.
- The data lend themselves to "trending" site performance.
- Documentation serves as a "closed loop" approach to hazard control.

Inherently, someone is identified as being responsible for locating job site hazards. It is often you, the safety representative. It may be appropriate to provide yourself with assistance. Teams can be established (labor/management safety committees, etc.) to inspect sites; supervisors and even the job superintendent can participate in site safety walk-throughs. This activity is encouraged, but when it occurs, make sure the purpose of the walk-through is understood to be inspection of the site for safety. Too often these efforts fail when the key individual gets hung up on or sidetracked with a schedule- or material-related problem because the focus is removed from site safety.

Inspection results usually should be documented; this will provide you with a history of the job and documentation that may later prove useful if OSHA visits the site.

Many safety professionals question whether a checklist should be used during the inspection process. Whereas this author generally discourages the use of checklists, it is acknowledged that they may be of use in some circumstances, particularly when those performing the inspection are not trained safety professionals or are new to the field of hazard control. In either case, the checklist should be viewed as a tool to help complete the inspection process, but not as the final inspection report. Checklists offer some value to remind inspectors of the various topics they need to consider on the job site during the inspection, but they can be difficult to complete, especially when set up in a yes/no format. For example, a typical checklist might include the following question: "Are **all** electrical cords maintained in good condition?" If the inspector observed 50 or 60 cords on the site and found only 1 in disrepair, he would still need to answer the question "no". It may be more useful if the checklist provides space for written comments. However, many such examples could be used to illustrate difficulties associated with the use of checklists which support the view that their value is highest when they are used as a tool to help the inspector remember items to look for and as a tool to use when writing the inspection report. An example of a construction-oriented checklist is located at the end of this chapter (see Appendix A).

## KEY PROGRAM COMPONENTS

To this point our focus has been on the management aspects of construction safety. What about specific components of the overall safety program? Safety managers are familiar with OSHA requirements, whether affected by federal or state regulations, and generally understand the need for a set of rules (called a code of safe practices in some states) that dictate compliance with applicable standards for the job. These requirements should be addressed in your site safety and health manual. But avoid becoming complacent; stay current with OSHA trends and revisions. Keep your management team informed with regard to regulatory activity. Keep your program up to date as well.

One way to be prepared for your job is to research the standards and develop checklists addressing key areas. These checklists can become part of your files and may be taken from job to job. They can be used as necessary to address key stages or components of a job prior to that need actually existing. In other words, they will help you be proactive on your job. When something comes up on your site you will be prepared. Table 4-4 illustrates some topics to consider for the development of a checklist.

Model programs reach far beyond a basic site safety plan and are often customized for a particular job. For example, if you are involved in a project

**Table 4-4  Checklists for Construction Safety**

- Job site medical provisions
- Emergency plan
- Contractor's accident reporting requirements
- Fire protection
- Temporary power for the site
- Fuel storage
- Chemicals storage
- Traffic routes
- Sites for contractors' field offices
- Sites for materials storage areas
- Job site security needs
- Capping/covering of impalement hazards
- Assured grounding program
- Requirements for reporting to client
- Fall protection program

requiring a lot of deep trenches, then you may launch a "safe trench" program focusing on proper shoring methods and sloping requirements. You may implement a requirement for a permit to be obtained prior to personnel entering and working below grade — that way you can ensure that a trench is inspected prior to personnel being allowed to enter it. You may also arrange to have "competent person" training/education available on site. The program can be made fun for employees by including calculation of things like the number of yards of soil moved without injury, the total distance of trenching completed without incident, or the number of manhours spent working in trenches without mishap. Be sure to use this information when promoting your program.

The most successful programs usually place emphasis on those areas where serious accident potential exists. Although construction sites are viewed as dangerous, in general there are many operations that present even greater potential for serious injury. These operations should be identified and given your attention. Table 4-5 illustrates many of the operations or processes that present serious accident potential; be proactive — anticipate these operations and plan for them. Your time and energy are well spent focusing on those areas that apply on your job.

**Table 4-5  Areas for Special Emphasis**

- Cranes and hoisting equipment
- Rigging materials and practices
- Electrical safety [assured grounding, lock out–tag out (LOTO)]
- Confined space work
- Trenching/excavations
- Exposure to asbestos or lead
- Heavy equipment operation
- Blasting
- Steel erection
- Roof work
- Scaffolding (particularly two-point suspended)
- Weather extremes

| JOB HAZARD ANALYSIS | | |
| --- | --- | --- |
| Contractor: _____        Date: _____ | | |
| Job Number: _____        Location: _____ | | |
| Phase Number: _____        Page ___ of ___ | | |
| Activity or Operation | Unsafe Condition or Process | Preventive Measure or Corrective Action |
| | | |
| | | |
| | | |
| | | |
| | | |

**Figure 4-4**  Job hazard analysis.

These areas and other identified high-risk processes can be assessed using the job hazard analysis or JHA technique. This well-known process is used to break down the components or steps of a task in order to identify where potential for injury exists and to establish acceptable control measures to reduce the injury potential. Figure 4-4 illustrates an example of a typical JHA form. This accident prevention method provides an opportunity to get workers involved in your accident prevention efforts and has been used successfully for years. Completed JHAs may be used to train and educate workers; they also make good safety meeting material.

## CONCLUSION

### Ten Points For Success

- Know your project — anticipate hazards.
- Understand applicable regulations and interpretations.
- Strive to manage your program proactively.
- Make yourself and your program visible to the workforce.
- Communicate project safety requirements early — during pre-bid and preconstruction phases.
- Establish time-bound, measurable goals for safety performance.
- Measure safety performance — provide timely feedback.
- Network with peers; keep yourself current in the field.
- Enforce safety program requirements in a fair, uniform manner.
- Involve others in the program — JHAs, etc.; establish yourself as the expert on safety and health.

By its nature managing safety in construction is a dynamic process. People and activities change on a regular basis; it is common to have multiple high-risk operations occurring at the same time. In many cases safety people manage safety efforts on multiple projects. The high accident rates in the construction industry have resulted in OSHA having a strong presence. Thousands of citations are issued annually. There are health hazards associated with construction work that previously were ignored or not even identified. The safety manager is often the industrial hygienist as well and is expected to be fundamentally sound in these principles.

In spite of these challenges the safety manager can perservere and be successful in the construction arena. A daily presence on the job site is fundamental to success. Early communication of project safety requirements is essential to set the job up properly. A strong presence in the pre-bid and preconstruction stages will yield positive results. Time-bound, measurable goals combined with periodic feedback to employees, contractors, and management are necessary components of the program.

All of these efforts can be referred to as "safety management". The secret to a successful construction safety program is to employ good management techniques. The safety manager must plan, organize, lead, and control. Indeed, managing safety involves the same principles as managing other functions in an organization. Good management techniques will integrate safety into the company culture and set the tone for safety performance on the project. Not unlike safety managers in other industries, the construction safety manager must strive to stay current with the regulatory community, safe practices in the construction field, and management techniques. The safety manager must establish himself/herself as the project authority, or expert, regarding safety and health. It is your ambition to have craftsmen and others on the management team recognize your expertise and seek your advice regarding safety and health questions. Professionalism must be practiced daily.

The Board of Certified Safety Professionals, BCSP, now offers a CSP specialty in construction safety. The achievement of this designation can solidify the expertise of the holder and will likely grow to become the norm for practitioners in this field. Of course, the designation itself does not make the individual a success; it merely establishes his or her knowledge base. It is the implementation of sound management techniques that leads to success in the field.

Many components of a safety program, and some discussed within this chapter, can be viewed as controversial; therefore, plans should be reviewed by the company's legal counsel prior to implementation. The approach to safety enforcement may vary depending on whether you represent the client or contractor, particularly in the areas of drug testing and giving direction to workers.

Whereas enforcement of safety programs in construction continues to be done with methods emphasizing the negative, the proactive use of positive recognition should not be overlooked. The negative techniques seem driven primarily by a fear of OSHA, or perhaps out of respect for the OSHA inspection, and the dynamic nature of a construction project. Still, the use of positive approaches can be woven into the program to maximize your overall effectiveness. Many companies manage a recognition program and reward safe performance. Criteria must be established and communicated to program participants. Some in the profession condemn this type of program because employees or contractors may be influenced to under report accident/injury experiences in order to meet objectives. Others suggest that such programs increase worker interest in safety and help to improve overall safety performance. Still others believe this type of program may have a positive psychological effect on employees which results in improved safety. If you choose to employ this tactic, be sure the criteria are fair and clearly communicated to all those involved. Monitor the results and provide feedback in terms of performance to all interested parties. Do not consider this type of program to be your savior; attach the program to the overall project or company goal. Make it meaningful and make it interesting ... but never encourage under reporting of injuries or illnesses. Many have had success with incentive programs when they have been used as short-term stimulants for the overall safety program. Most agree that the value of the award(s) offered does not determine the overall success of the incentive program — people simply like to achieve the award.

# Comprehensive Construction Project Survey

| | |
|---|---|
| Project: | _____ |
| Location: | _____ |
| Prime Contractor: | _____ |
| Inspector: | _____ |
| Date of Inspection: | _____ |

## 1. Records, Administration, General

|  | Yes | No | N/A |
|---|---|---|---|
| 1.1 Copy of OSHA regulations on site | | | |
| 1.2 Required signage, postings in place | | | |
| 1.3 Company safety plan or prejob safety plan on site | | | |
| 1.4 Contractor identifying and correcting hazards | | | |
| 1.5 Tool box safety meetings held weekly | | | |
| 1.6 Required permits obtained and posted | | | |
| 1.7 Confined space entry procedure followed | | | |

## 2. Housekeeping Subpart C 1926.25

| | Yes | No | N/A |
|---|---|---|---|
| 2.1 Stairs, work areas, walkways free of hazards | | | |
| 2.2 Proper containers provided for trash, rags, etc. | | | |
| 2.3 Scrap and debris removed daily | | | |
| 2.4 Nails bent over or removed from scrap lumber | | | |

## 3. First Aid, Sanitation & Illumination
## Subpart D 1926.50-51 & 56

| | Yes | No | N/A |
|---|---|---|---|
| 3.1 First aid kit available and complete | | | |
| 3.2 Drinking water, cups and receptacle provided | | | |
| 3.3 Adequate number portable toilets provided | | | |
| 3.4 Washing facilities provided | | | |
| 3.5 Work area lighting adequate | | | |

## 4. Occupational Health & Environmental
## Controls Subpart D 1926.52-59

| | Yes | No | N/A |
|---|---|---|---|
| 4.1 Hazard assessments completed where needed | | | |
| 4.2 Hazard Communication program satisfactory | | | |
| 4.3 MSDSs available for employee use | | | |
| 4.4 Hearing protection provided and worn | | | |
| 4.5 Exposures to gases, vapors, fumes, dusts, controlled | | | |
| 4.6 Asbestos work conforms to regulations | | | |
| 4.7 Hazardous materials properly labeled | | | |
| 4.8 Employees trained to work with hazardous materials | | | |
| 4.9 Work with lead conforms to regulations | | | |

# Construction Survey

## 5. Personal Protective Equipment Subpart E
   ## 1926.100 - 107

5.1 Hazard assessments completed where necessary
5.2 PPE provided and used where necessary
5.3 Hardhats worn by all personnel in construction zone
5.4 Safety harnesses, belts, lanyards inspected and used
5.5 Safety nets provided where needed
5.6 Employees trained in use of PPE

## 6. Fire Protection and Prevention
   ## Subpart F 1926.150-151

6.1 Flammable & combustible liquids in safety cans
6.2 Extinguishers properly placed and inspected
6.3 Hydrants clear and access open
6.4 Proper signage posted where required

## 7. Handling, Storage of Materials
   ## Subpart H 1926.251

7.1 Materials properly stored to prevent toppling
7.2 Maximun safe loads for floors posted and observed
7.3 Aisles kept clear and adequately sized
7.4 Mechanical aids used when possible

## 8. Rigging Equipment for Mateial Handling
   ## Subpart H 1926.251

8.1 Safe working load limits not exceeded
8.2 Wire rope not secured by knots
8.3 Wire rope clips not used to form eyes of slings
8.4 Custom lifting devices labeled for maximum load
8.5 Rigging equipment properly stored
8.6 Rigging equipment in good condition

## 9. Tools, Hand and Power Subpart I
   ## 1926.300-303

9.1 Tools maintained in good conditon
9.2 Damaged tools repaired or replaced, not used
9.3 All mechanical safeguards in use

# Construction Survey

## 9. Tools, Hand and Power (continued)

9.4 Power tools grounded or double insulated
9.5 Pneumatic tools equipped with attachment restraints and restraints on air line connections
9.6 Compressed air used at 30 psi or less
9.7 Powder actuated tool operators properly trained
9.8 Grinders provided with safety guards
9.9 Work rests and wheel guards on bench mounted grinders adjusted properly

## 10. Woodworking Tools Subpart I 1926.304 & Subpart O 1910.213

10.1 Portable circular saws equipped with upper and lower blade guards, lower guard returns automatically
10.2 Radial saws - upper hood enclosed, lower portion of blade guarded to full diameter of blade, non-kickback fingers or dogs, adjustable stops or other effective device to eliminate table over-run and automatic return provided
10.3 Belts, pulleys, gears, shafts & moving parts guarded
10.4 Hand fed circular rip and crosscut table saws - saw above table guarded by hood, provided with non-kickback fingers or dogs and spreader

## 11. Welding & Cutting Subpart J 1926.350-354

11.1 Gas cylinders secured in upright position
11.2 Valve caps in place when cylinders not in use
11.3 Special wrench available when required by cylinder
11.4 Cylinders transported properly
11.5 Fire wall or 20' clearance for oxygen & acetylene
11.6 Fire extinguisher immediately available
11.7 Valves closed, equipment purged when not in use
11.8 Welding curtains used where needed
11.9 Welding cables in good condition & properly insulated
11.10 Proper ventilation and/or respirators used

## 12. Electrical Subpart K 1926.400-417

12.1 All exposed live parts guarded
12.2 All conductors and equipment "approved"
12.3 Tools and cords in good condition
12.4 Proper working clearances provided
12.5 Ground fault circuit interrupters provided
12.6 Portable and/or cord plug connected equipment grounded

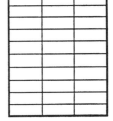

# Construction Survey

## 12. Electrical (continued)

12.7  Temporary lights provided guards
12.8  Cables & cords protected from damage
12.9  Outlet boxes covered
12.10 Flexible cords used in continuous lengths without
       splicing
12.11 Portable hand lights provided with guards
12.12 Lock-out, tag-out procedures followed

## 13. Scaffolds Subpart L 1926.451

13.1  Guardrails and toeboards on all open sides and ends of
       scaffold platforms 10 feet or more above the ground
13.2  Guardrails installed on all open sides and ends of
       scaffolding from 4 feet to 10 feet high with minimum
       horizontal dimension of 45 inches
13.3  Tube and coupler scaffold posts accurately spaced,
       erected on suitable base and maintained plumb, brace
       connections secure
13.4  Planks secured or not less than 6 inches nor more
       than 12 inches over the end support
13.5  Ladder or equivalent safe access provided
13.6  Scaffold tied into structure when required
13.7  Lifeline and harness provided and used by each
       worker on swinging and single-point adjustable
       suspension scaffolds

## 14. Floor, Wall Openings Subpart M 1926.500

14.1  Floor, roof & wall openings guarded by a standard
       railing & toeboards or a secured cover
14.2  Open sided floors & platforms 6 feet or more above
       ground guarded with standard railing & toeboard
       where necessary
14.3  Runways 4 feet or more above ground properly
       guarded
14.4  Standard railings consist of top rail, toeboard &
       posts, & have a vertical height of approximately
       42 inches
14.5  Anchor points & framing members for railings of
       all types capable of withstanding 200 pound load
       in any direction

# Construction Survey

### 15. Roofing Subpart M 1926.500

15.1 During built-up roof work employees protected against falling from unprotected sides by motion stopping safety system, or warning line 6 feet from edge and safety monitor

15.2 Guardrails 4 feet on either side of material access point at roof

15.3 Long sleeved clothing and adequate foot protection

15.4 Class BC fire extinguisher near kettle

15.5 Materials stored a minimum of 6 feet from edge of roof

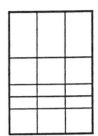

### 16. Cranes & Derricks Subpart N 1926.550

16.1 Crane certified annually

16.2 Daily visual inspection, mtce logs available

16.3 Signals and load capacities posted

16.4 Fire extingusiher provided and maintained

16.5 Swing radius of rear of rotating super-structure of crane barricaded

16.6 Power lines de-energized or safe distance maintained (groundman provided)

16.7 Safety latch provided on hook

16.8 Tag lines used to control loads

16.9 Personnel prohibited from riding on hook or loads

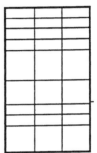

### 17. Aerial Lifts, Elevating & Rotating Work Platforms Subpart N 1926.556

17.1 Safety belt worn by each occupant of an aerial lift with lanyards attached to the boom or basket

17.2 Safety rails on all open sides of elevating work platforms

### 18. Earthmoving Equipment Subpart O 1926.602

18.1 Scissor points guarded on front-end loaders which constitute hazard to operator

18.2 Braking system, horn, seat belts (unless designed for stand-up operation) back-up alarm, overhead protection in good condition

# Construction Survey

## 19. Excavating, Trenching & Shoring Subpart P 1926.650-652

19.1    Name of competent person provided
19.2    Employees in excavations protected from cave-ins
19.3    Excavations 5 feet or greater in depth guarded by sloping, benching, shoring, shielding, or other equivalent means
19.4    Slopes excavated to angle of repose
19.5    Banks benched to proper heights
19.6    Spoil material pulled back at least 2 feet
19.7    Adjacent structure adequately shored / supported
19.8    Roads and sidewalks shored / supported
19.9    Ladder or ramp provided in trenches 4 feet or more in depth with no more than 25 feet lateral travel required for employees
19.10   Excavation adequately barricaded & lighted
19.11   Equipment a safe distance from excavation
19.12   Bridges over excavations equipped with safety rail
19.13   Daily excavation records available
19.14   Location of underground utilities complete

## 20. Concrete Work, Forms & Shoring Subpart O 1926.700-701

20.1    Forms propely installed & braced
20.2    Adequate shoring, plumbed & cross braced
20.3    Shoring in place until strength attained
20.4    Protruding re-bar guarded, capped or covered
20.5    Pipelines, hoses and tremies used as a part of concrete pumping systems secured by wire or equivalent means in addition to the regular couplings or connections
20.6    Form lumber organized once stripped

## 21. Steel Erection Subpart R 1926.750-752

21.1    Safety railing installed around floor periphery, 1/2 inch wire rope or equal
21.2    Ladder or other acceptable means of access provided
21.3    Safety harnesses (belts for positioning) used
21.4    "Christmas treeing" of steel prohibited
21.5    Safety nets provided where required
21.6    Means provided to keep bolts and drift pins from falling when being knocked out (during bolting)

# Construction Survey

## 22. Demolition Subpart T 1926.850-860

22.1 Engineering survey of structure to be demolished made and discussed prior to operations and approved by competent person
22.2 Public and employee protection provided
22.3 Floor openings covered
22.4 Material chutes enclosed or guarded
22.5 Adequate access ladders or stairs provided
22.6 Dust controlled adequately
22.7 Utilities cut off or controlled
22.8 Clear operating space for trucks and equipment

## 23. Power Transmission and Distribution Subpart V 1926.950-960

23.1 Existing conditions including energized lines & their voltages determined prior to starting work
23.2 Safe working & clearance distances maintained
23.3 Procedures followed to de-energize lines
23.4 Employees trained in first aid & CPR
23.5 PPE inspected and tested prior to use
23.6 Live-line tools certified, inspected before use
23.7 Portable metal or conductive ladders not used near energized lines or equipment
23.8 Aerial lift trucks near energized lines or equipment are grounded or barricaded, or insulated
23.9 Conductors and equipment treated as energized until tested or determined to be de-energized or grounded
23.10 Grounding procedures followed
23.11 Open manholes or vaults posted by warning signs & protected by barricades, temporary covers, or other guards
23.12 Manholes or vaults provided with forced ventilation or found to be safe by testing for oxygen & flammable gases or fumes prior to entry
23.13 Authorization obtained prior to starting work in energized substations
23.14 Temporary fence provided when existing substation fence is removed or expanded for construction

## 24. Stairways Subpart X 1926.1050-1060

24.1 Stairway or ladder provided at all personnel points of access where there is a break in elevation of 19 inches or more & no other means of access provided

# Construction Survey

## 24. Stairways (continued)

| | | |
|---|---|---|

24.2  Riser height and tread depth are uniform

24.3  Stairways free of slippery conditons or hazardous
projections

24.4  Stairs with 4 or more risers or rising more than
30 inches equipped with at least one handrail &
one stair rail system along each unprotected side
or edge

24.5  Stair rails not less than 36 inches from upper
surface of the stair rail system to the surface of
the tread

24.6  Midrails, screens, mesh or equivalent between top
of stair rail system & top of stairs

24.7  Handrails and the top of stair rail systems capable
of withstanding a force of at least 200 pounds

24.8  Unprotected sides and edges of stairway landings
provided with guardrails

24.9  Electrical cords & cables collected on one side of
stairs to reduce trip / fall hazard

## 25. Ladders Subpart X 1926.1050-1060

| | | |
|---|---|---|

25.1  Extension ladders extend 3 feet above landing (or grab
rail provided) & are secured against displacement

25.2  Non self supporting ladders used at an angle of
1:4

25.3  Personnel face ladder and use side rail(s)

25.4  Personnel avoid carrying objects that could come
loose and fall

25.5  Metal spreader or locking device on step ladders
to hold ladder open when in use

25.6  Ladders with broken or missing rungs, cleats,
steps, broken or split rails, corroded components
or other faulty or defective components tagged &
removed from service

25.7  Safety feet provided on entension ladders

NOTE: This checklist does not address all OSHA requirements for construction activities. For
additional requirements, and conditions not included such as blasting and explosives,
underground construction, ionizing radiation, roll-over protection, use of helicopters,
etc. refer directly to the applicable OSHA / ANSI standard.

## FURTHER READING

Many articles and books have been written to address safety in the construction industry. The following is a recommended list of further reading material.

### Books

Petersen, D. *Safety Management — A Human Approach.* Goshen, New York, Aloray, 1988.

Levitt, R. E. and Samelson, N. M. *Construction Safety Management.* New York, McGraw-Hill, 1994.

National Safety Council. *Accident Prevention Manual For Business and Industry, Engineering and Technology.* Itasca, IL, National Safety Council, 1992.

MacCollum, D. V. *Construction Safety Planning.* New York, Van Nostrand Reinhold, 1995.

### Other Publications

Minter, S. Building safety into construction. *Occup. Hazards*, September 1993.

Roughton, J. Managing a safety program through job hazard analysis. *Prof. Saf.* 37(1): 28–31, 1992.

Roychowdhury, M. A confined space entry program for any industry. *Prof. Saf.* 37(2): 16–21, 1992.

Gregory, E. Motivational management techniques for safety and health. *Prof. Saf.* 36(1): 29–33, 1991.

# 5

# OVERVIEW OF THE KEY ELEMENTS OF A SYSTEMATIC SAFETY MANAGEMENT PROGRAM

**Richard W. Lack**

## TABLE OF CONTENTS

## INTRODUCTION

A number of learned fellow members of the safety profession have recently been debating the so-called "paradigm shift" from the unsafe act/unsafe condition world of the mid-20th century, to quote my colleague Dan Petersen. In fact, an intriguing article on this issue was published in the American Society of Safety Engineers journal, *Professional Safety,* in December 1992. The writer, Gary Winn, provided a fascinating review of trends, including research background on the meaning of the word paradigm. He reluctantly concluded that although the safety profession appears to be getting close to a shift, there is no certainty as to when this will actually occur; "we're waiting impatiently", as he says!

From my perspective, safety practitioners owe a great deal to Dan Petersen for his outstanding research and pioneering work in the human behavior aspects of safety management.

There are many others who have made significant contributions in this field, and my list of all those whom I have studied is too long to mention here. I will, therefore, limit my recognition to those teachers who have had a major influence on my development and understanding of the humanistic element in safety. These teachers are Frank Bird, John Grimaldi, Homer Lambie, Dan Petersen, and Bill Pope.

The Dupont Company and Technica DNV (formerly the International Loss Control Institute) have developed safety management systems that I have observed to be very effective. However, I should point out that, like any management system, they require careful "tailoring" to your organization's traditions, climate, and nature of operations. Strong management support and comprehensive training of key personnel are essential if they are to be implemented effectively.

The behavior-based safety process is not new; in fact, Dan Petersen's studies trace it back to the 1930s. It has, of course, gone through many stages of change and development.

Tom Krause and others have developed systems that are reported to be producing significant results.

For safety practitioners considering programs of this type, I recommend a careful review of Scott Geller's chapter, "Managing the Human Element of Occupational Health and Safety" (Chapter 30). This will provide the reader with valuable insights into the psychological and motivational aspects of human work performance, and it also emphasizes how complex and confusing this field can be to the unwary. The Simons and Michael Topf also discuss "cultural" aspects of the accident prevention process in their chapters (32 and 33, respectively).

In summary, these systems deserve very thorough investigation and observation, preferably including trial on a pilot basis, before a decision is made on unit-wide adoption.

In the meantime, while this experimentation and debate continues, this chapter will, for practical reasons, focus on the more "traditional" elements of a sound safety management program. These elements that have stood the test of time have proven themselves to be effective in the continuous reduction of unintentional losses. Furthermore, these programs are very simple and flexible and can be implemented in a variety of ways. Thus they can fit equally well into an organization such as a municipality, which may be structured with multiple layers of hierarchy, versus, say, a hi-tech firm organized on a team basis with maximum employee decision-making empowerment.

In selecting the key elements of a safety management system, one could come up with a very long list if one wished to include every conceivable aspect. Frank Bird and the International Loss Control Institute have identified the following 20 program elements which are considered essential for a successful loss control effort. These are

1. Leadership and administration
2. Management training
3. Planned inspections
4. Task analysis and procedures
5. Accident/incident investigation
6. Task observation
7. Emergency preparedness
8. Organization rules
9. Accident/incident analysis
10. Employee training
11. Personal protective equipment
12. Health control and services
13. Program evaluation systems
14. Engineering controls
15. Personal communications
16. Group meetings
17. General promotion
18. Hiring and placement
19. Purchasing controls
20. Off-the-job safety

In the interest of simplicity, I have distilled these 20 programs down to an 8-point program as follows:

---

**Eight-Point Safety and Health Management System**

**Administrative Elements**
1. Manual (procedures and guidelines)
2. Committees and coordinators
3. Training, interest, and motivation

**Action Elements**
1. Inspections
2. Hazard control
3. Job hazard analysis
4. Safety meetings
5. Accident investigation

---

These key elements will be discussed in detail in the balance of this chapter. Many other related programs such as industrial hygiene, medical health, engineering controls, construction, and risk management are discussed in other chapters of this book.

## EVALUATING THE EFFECTIVENESS OF THE SAFETY MANAGEMENT SYSTEM

Many organizations still evaluate the effectiveness of their safety and health systems by their injury experience. This is one form of measurement, but it is not a reliable approach because it does not measure system failures, only the chance product of these failures.

A more reliable system, which is discussed in this and other chapters of this book, is to measure the quantity and quality of activity in each element of the safety system. This approach provides management with information as to potential weak links in the system, substandard performance, and training needs for those implementing the system.

## Primary Elements of an Effective Safety Management System

### 1. Safety and Health Manual (Procedures and Guidelines)

A manual serves as the foundation of an effective safety and health management system. The manual is to the system what a blueprint is to construction. Without some basic procedures and guidelines, the loss control effort will be uncoordinated and haphazard in its emphasis. Issues will be addressed as they come up, rather than on a systematic prevention-oriented basis.

Obviously, there are many different ways to organize a manual, and each unit's needs will be different. Important criteria for a manual are

- The manual must be "user friendly". That is, there should be a logical layout for the contents so people can easily find procedures.
- The index and numbering system must make the job of filing new or revised procedures an easy process.
- The index and numbering system must be expandable into a total classification system so future procedures can be fitted into the system.
- The index system must be such that it can be translated to a reference filing system where additional materials can be kept and located easily.

Taking these criteria into consideration, a system adapted from one originally developed by Homer Lambie, former Safety Director at Kaiser Aluminum and Chemical Corporation, has been found to work very satisfactorily. The core of this system is a classification index (see Figure 5-1). The three general sections for safety procedures require some explanation.

**Administration (100 Series Procedures)** — This section includes procedures primarily related to the planning, coordination, and evaluation of the unit's safety activities.

**Safety Engineering Standards (200 Series Procedures)** — This section includes procedures primarily related to the adoption of standards for a safe work environment.

**Service Programs (300 Series Procedures)** — This section includes procedures primarily related to the human element and the adoption of procedures aimed at the behavioral aspects of safety.

For the sake of consistency and control, there should be a standard layout method for each manual procedure. Again, every organization has different needs. A typical procedure layout is shown in Figure 5-2.

The Safety Engineering Standards section is subdivided into nine major sections which relate to the source areas of most typical workplace hazards. They are self-explanatory; however, a few guidelines are called for to help with selection of the appropriate classification.

First, it is important when using this system to keep in mind the question, "What is the center of attraction in this procedure?" For example, a mobile crane could be classified as either equipment or vehicle. Where you classify it will depend on the "center of attraction". Is your procedure concerned with the hoisting aspects or the mobile equipment aspects?

The source category descriptions, as previously noted, are self-explanatory. However, experience has shown that there is a need to emphasize the distinction between the categories Machines and Equipment and between Materials — Health and Materials — Traumatic.

**Machines** — This category includes all types of machines that produce or finish materials or objects. Examples are lathes, drill presses, milling machines, grinders, and extrusion presses.

**Equipment** — This category includes all types of systems, facilities, or equipment that convey or process materials or objects. Examples are cranes,

**SAFETY AND HEALTH MANUAL**

**CLASSIFICATION INDEX**

Issue Date:
Revised Date:

| TAB NO. | SECTION | PROCEDURE CLASSIF. RANGE | ASSIGNED NUMBER | SUBJECT | ISSUE DATE | REVISED DATE | AUTO REVIEW DATE |
|---|---|---|---|---|---|---|---|
| 1 | Administration | | 101 | OSHA Inspections | | | |
| | | 110 | | **Union Safety** | | | |
| | | 120 | | **Accident Investigation** | | | |
| | | | 121 | Occupational Injury & Illness Investigation | | | |
| | | | 122 | Vehicle Accident Investigation | | | |
| | | | 123 | Significant Incident (No Injury) Investigation | | | |
| | | 130 | | **Safety Reports** | | | |
| | | | 131 | Safety & Health Program Responsibilities | | | |
| | | | 132 | Statistical Injury Experience Reports | | | |
| | | | 133 | Safety Activity Reports | | | |
| | | 140 | | **Safety Manuals** | | | |
| | | 150 | | **Purchasing Safety** | | | |
| | | | 151 | Control of Purchasing Hazardous Materials | | | |
| | | 160 | | **Contractor Safety & Health** | | | |
| | | | 161 | Contractor Selection | | | |
| | | | 162 | Contractor Safety Program | | | |
| | | 170 | | **Professional Associations** | | | |
| 2 | Safety Engineering Standards | | 201 | Regulatory Safety Standards | | | |
| | | | 202 | National Safety Standards | | | |
| | | | 203 | Safety & Health Design Review & Engineering Projects | | | |
| | | 210 | | WORK AREA | | | |
| | | 200 | | Floors | | | |
| | | | 201 | Excavation Procedure | | | |
| | | | 202 | Standard for Scaffolds | | | |
| | | | 203 | Standard for Portable Ladders | | | |
| | | | 204 | Standard for Traffic Signs & Controls for Work in Roadways | | | |
| | | 210 | | Temperature | | | |
| | | 215 | | Illumination | | | |

**Figure 5-1** A classification index layout for a safety and health program manual.

hoists, escalators, elevators, chemical process plant, conveyors, ovens, and robotic equipment.

**Materials — Health** — This category includes all types of materials that may cause "health injuries or illnesses" resulting from an acute or chronic exposure effect. Examples include asbestos, cyanide, benzene, lead, carbon monoxide, and hydrogen sulfide.

| TAB NO. | SECTION | PROCEDURE CLASSIF. RANGE | ASSIGNED NUMBER | SUBJECT | ISSUE DATE | REVISED DATE | AUTO REVIEW DATE |
|---|---|---|---|---|---|---|---|
| | | 220 | | MACHINES | | | |
| | | 221 | | Mechanical Power & Transmission | | | |
| | | 222 | | Point of Operation | | | |
| | | 223 | | Nip Points | | | |
| | | 224 | | Production Machines | | | |
| | | 225 | | Metal Working Machines - Woodworking Machines | | | |
| | | 230 | | PORTABLE TOOLS | | | |
| | | 231 | | Hand Tools Manual | | | |
| | | 235 | | Hand Tools Power (Except Electrical) | | | |
| | | 240 | | EQUIPMENT | | | |
| | | | 241 | Inspection and Maintenance of Cranes & Hoisting Equipment | | | |
| | | | 242 | Inspection and Maintenance of Elevators | | | |
| | | | 243 | Inspection and Maintenance of Boilers and Pressure Vessels | | | |
| | | 244 | | Fans | | | |
| | | 245 | | Pumps, Pipes, Valves, Tanks | | | |
| | | 246 | | Conveyors | | | |
| | | 247 | | Gas Welding & Burning | | | |
| | | 250 | | MATERIALS - HEALTH | | | |
| | | 251 | | Toxic Materials | | | |
| | | | 255 | Standards for Chlorine Handling and Transportation | | | |
| | | 260 | | MATERIALS - TRAUMATIC | | | |
| | | 261 | | Compressed Gases & Liquids | | | |
| | | | 261.1 | Compressed Gas Storage & Transportation | | | |
| | | 262 | | Acids and Caustics | | | |
| | | | 262.1 | Standard for Nitric Acid Handling and Transportation | | | |
| | | | 262.2 | Emergency Eyewash/Showers | | | |
| | | | 263 | Flammables & Explosives | | | |
| | | | 264 | Molten Metals | | | |
| | | | 265 | Hot Materials | | | |
| | | | 266 | Flying Particles | | | |
| | | | 267 | Sharp Edges | | | |
| | | | 268 | Weight & Shape, Stacking | | | |

**Figure 5-1  continued**

**Materials — Traumatic** — This category includes all types of materials that may cause "traumatic injury" from the effect of material or impact contact. Examples include sodium hydroxide, sulfuric acid, flammables, welding sparks, blasting materials, oxygen, refractory materials and other abrasives, and materials with sharp edges.

This manual classification system can be easily adapted as a checklist for system audits. The Safety Engineering Standards section can also serve as a checklist for safety inspections and for hazard recognition.

| TAB NO. | SECTION | PROCEDURE CLASSIF. RANGE | ASSIGNED NUMBER | SUBJECT | ISSUE DATE | REVISED DATE | AUTO REVIEW DATE |
|---------|---------|------|------|---------|------|------|------|
| | | 270 | | ELECTRICAL | | | |
| | | | 271 | Motors & Generators | | | |
| | | | 272 | Transformers & Rectifiers | | | |
| | | | 272.1 | High Voltage Switching | | | |
| | | | 273 | Wiring Conductors | | | |
| | | | 273.1 | Electrical Grounding | | | |
| | | | 273.2 | Ground Fault Circuit Interrupters | | | |
| | | | 274 | Electrical Hand Tools | | | |
| | | | 275 | Electrical Welding & Burning | | | |
| | | 280 | | VEHICLE | | | |
| | | | 281 | Transporting Passengers Safety Regulations | | | |
| | | | 282 | Seat Belts | | | |
| | | | 283 | Power Driven Trucks | | | |
| | | | 284 | Forklifts | | | |
| | | | 285 | Railroads | | | |
| | | | 286 | Marine Vessels | | | |
| | | | 287 | Aircraft | | | |
| | | 290 | | MULTI-CATEGORY | | | |
| | | | 291 | Confined Space Entry | | | |
| | | | 292 | Lock Out - Tag Out | | | |
| | | | 293 | Hazardous Area Entry | | | |
| | | | 294 | Safe Work Permits | | | |
| 3 | Safety Service Programs | 300 | | INSPECTIONS | | | |
| | | | 301 | Safety Inspection Program | | | |
| | | 310 | | HAZARD CONTROL | | | |
| | | | 311 | Hazard Control Program | | | |
| | | | 311.1 | Safety Recommendations | | | |
| | | 320 | | JOB HAZARD ANALYSIS | | | |
| | | | 321 | Job Hazard Analysis Program | | | |
| | | 330 | | SAFETY MEETINGS | | | |
| | | | 331 | Safety Meetings Program | | | |
| | | 340 | | ACCIDENT EXPERIENCE REVIEWS | | | |
| | | | 341 | Significant Incident - Damage Control Program | | | |

**Figure 5-1  continued**

In summary, some form of safety and health manual is already a mandatory document in several countries. In the U.S. some states require it, and most likely it will be required under the forthcoming federal OSHA Reform Act.

The manual should be widely distributed in every organization. All members of the management staff, including first-line supervisors, should be included in the distribution. In organizations operating with work teams where the teams are responsible for conducting their own safety programs, each team should have a copy.

Certain procedures, such as hazard control or accident investigation, may require development of supplementary training guidelines for supervision. This type of document can be inserted in the manual. Alternatively, many

| TAB NO. | SECTION | PROCEDURE CLASSIF. RANGE | ASSIGNED NUMBER | SUBJECT | ISSUE DATE | REVISED DATE | AUTO REVIEW DATE |
|---|---|---|---|---|---|---|---|
| | | 350 | | SAFETY COMMITTEES & COORDINATORS | | | |
| | | | 351 | Executive Safety Committee | | | |
| | | | 352 | Labor-Management Safety Committee | | | |
| | | | 353 | Technical Safety Committees (Traffic, Electrical, Chemical, etc.) | | | |
| | | | 354 | Interest Motivation Safety Committee | | | |
| | | | 355 | Departmental Safety Committees | | | |
| | | 360 | | SAFETY TRAINING | | | |
| | | | 361 | New Employee Safety Orientation | | | |
| | | | 362 | New Supervisor Safety Orientation | | | |
| | | | 363 | Monthly Safety Training Meetings for Supervisors | | | |
| | | | 364 | Supervisory Training | | | |
| | | | 365 | Lineworker/Team Training | | | |
| | | | 366 | Safety Visuals Library | | | |
| | | 370 | | PERSONAL PROTECTIVE EQUIPMENT | | | |
| | | | 371 | Personal Protective Equipment Policy | | | |
| | | | 372 | Safety Shoe Policy | | | |
| | | | 373 | Prescription Eyewear Policy | | | |
| | | 380 | | AWARD AND RECOGNITION PROGRAMS | | | |
| | | 390 | | INTEREST AND INFORMATION PROGRAMS | | | |
| | | | 391 | Accident Experience Information | | | |
| | | | 392 | Hazard Control Ideas | | | |
| | | | 393 | Safety Idea Exchange | | | |
| | | | 394 | Safety Bulletins | | | |
| | | | 395 | Posters & Signs | | | |
| | | | 397 | Off-The-Job Safety Program | | | |
| 4 | Medical Health | | 401 | Medical Surveillance Program | | | |
| | | | 402 | Bloodborne Pathogen Exposure Control Program | | | |
| | | | 403 | Cumulative Trauma Disorders Control Program | | | |

**Figure 5-1  continued**

organizations find it more effective to develop a separate supervisor training manual in which supervisors can keep all such material as it may be developed from time to time.

Development of the contents of the safety manual should be the responsibility of top management, with professional advice and input provided by in-house safety personnel or outsourced to consultants.

Many of the procedures in the engineering standards section will be selected from those prepared by professional organizations such as the National Safety Council and the American National Standards Institute, trade associations,

| TAB NO. | SECTION | PROCEDURE CLASSIF. RANGE | ASSIGNED NUMBER | SUBJECT | ISSUE DATE | REVISED DATE | AUTO REVIEW DATE |
|---|---|---|---|---|---|---|---|
| 5 | Industrial Hygiene | | 501 | Respiratory Protection Program | | | |
| | | | 511 | Hearing Conservation Program | | | |
| | | | 521 | Employee Exposure Monitoring Program | | | |
| | | 530 | | Ventilation | | | |
| | | 540 | | Radiation | | | |
| | | | 551 | Asbestos Control Program | | | |
| | | | 561 | Lead Exposure Control Program | | | |
| 6 | Hazardous Materials | | 601 | Hazard Communication Program | | | |
| | | | 611 | Chemical Work Practices Guidelines | | | |
| | | | 621 | Process Safety Management Program | | | |
| | | | 631 | Laboratory Hazard Control Program | | | |
| 7 | Emergency Procedures | | 701 | Disaster Plan | | | |
| | | | 702 | Earthquake Emergency Plan | | | |
| | | | 703 | Fire | | | |
| | | | 704 | Tornado | | | |
| | | | 705 | Flood | | | |
| 8 | Fire Protection | 900 | | Fire Protection Standards | | | |
| | | | 911 | Fire Suppression Equipment Inspection & Maintenance | | | |
| | | | 912 | Fire Extinguisher Inspection & Maintenance | | | |
| | | | 921 | Fire Insurance Inspections | | | |
| | | | 931 | Fire Prevention Plan | | | |
| 9 | Security | 900 | | Security Standards | | | |
| | | | 911 | Security Surveys | | | |
| | | | 921 | Access Control Procedure | | | |
| | | | 931 | Threats and Violence Control Plan | | | |
| 10 | Workers Compensation | | 1001 | Insurance | | | |
| | | | 1002 | Claims Management | | | |
| | | | 1003 | Alternative Duty Policy | | | |

**Figure 5-1   continued**

state and federal agencies, product manufacturers, etc. They can be added to or modified as necessary to meet the unit's needs.

All procedures should be regularly reviewed and audited to ensure that they both are current and are being effectively implemented by the organization.

Once a manual is set up along the lines discussed and is efficiently maintained, it will provide the organization with a positive focus and a solid building block to help ensure continuously improving safety results.

## Safety Committees and Coordinators

### Safety Committees

The size of an organization, its management structure, and its overall employee relations "climate" will all be factors in determining the number and type of safety committees that best fit its needs. "Top-down" autocratic-style organizations will likely have very few committees of any kind. Others which

| ORGANIZATION NAME | DATE: |
|---|---|
| SAFETY AND HEALTH MANUAL | NUMBER: |
| POLICY PROCEDURES | MANUAL SECTION: |
| SUBJECT: | REFERENCE: |
| | Approval |

**I.   PURPOSE**

Description of _what_ the unit wants to be done.

**II.  OBJECTIVE**

Description of the reasons _why_ the procedure is necessary.

**III. DEFINITION**

Description of the principal terms referred to in the procedure. May also contain Policy Statements.

**IV. PROCEDURE**

Description _how_ the procedure will be implemented.

**Figure 5-2** Layout of a typical safety and health manual procedure.

are run in a more participatory or consensus style may have many different committees with a wide variety of responsibilities.

    Committees have been the subject of jokes for many years. My favorite is "a camel is a horse designed by a committee" (see below)!

Most of these jokes are probably a little exaggerated, but they do reflect some of the potential problems that can plague committees.

    First, what are the advantages of a safety committee? The overall purpose of a systematic safety and health program obviously is to prevent accidents. To achieve this purpose, the system must be focused on the target of finding and controlling hazards. The key advantages of the committee meeting approach in hazard discovery and control are

1. The pooling of a group's experience and expertise. This, together with the opportunity to brainstorm problems, results in the development of innovative and practical solutions.
2. The opportunity for groups of people to work together in committee meetings results in improved communications.
3. A committee's recommendations, especially on controversial issues, are usually accepted more positively by others in the organization.

Some of the pitfalls which must be avoided if your committees are going to be effective are

- Committees may try to take over the job of line operations.
- Committee members are unclear as to their duties and responsibilities.
- The purpose and objectives of the committee have not been clearly defined.
- The flow of problems to committees and their recommended solutions are undefined.

Recommended guidelines for committee operations are as follows.

*Committee Organization*

1. Membership should be between 3 and 12. Anything over this can result in the meeting discussions being dominated by the more aggressive members. The bigger the group, the more difficult it is to "draw out" quieter members who may have important ideas to contribute.
2. The committee should report to one individual.
3. Committee members are either appointed or selected by the group.
4. Subcommittees are responsible to the chairperson of the committee or some member designated by the chairperson.
5. The chairperson of the committee either is appointed or is nominated and selected by majority vote of the committee.
6. Committees should have the responsibility to recommend only. Their recommendations must be implemented by those persons having the necessary authority.
7. Committees should operate in a business-like fashion. Minutes should be recorded and distributed promptly after meetings. Agendas should be established, and the committee should meet at a regularly scheduled date and time.

See Appendix B for a sample Committee Method of Operation.

*Types of Committees*

There obviously can be a wide variety of safety committees. Broad categories for safety committees are as follows:

- Executive Safety Committee — This should be the principal unit policy advisory committee to the chief executive of the unit or organization. See Appendix A.

- Labor-Management safety committee — This committee, when appointed, should report to the unit's top manager. It will provide a forum for joint discussion on safety problems and recommendations for improvements.
- Safety Program Committees — These committees will report to the executive safety committee and will submit recommendations for the improvement of specific program elements such as
  —training
  —recognition and awards
  —hazard control
- Departmental Safety Committees — For large departments, it may be desirable for the department manager to appoint a committee which will provide recommendations for the improvement of departmental programs.
- Technical Safety Committees — Technical committees should be appointed as needs arise for the consideration of special technical problems. Examples are
  —Electrical hazard control
  —Chemical hazard control
  —Traffic safety
  —Crane safety committee
  —Ergonomics committee

Safety committees, if provided with the proper direction and structure, can be a very effective element in your overall system. Their effectiveness, as with any committee, is measured by the action taken as a result of their meetings to solve specific safety problems.

### Safety Coordinators

The safety coordinator program is designed to provide support and assistance to department management. The position will normally be a part-time assignment. Duties of a safety coordinator will include assisting the department manager with the administrative aspects of the various programs, as well as defining program weak points and making recommendations for their improvement.

The unit safety professional should meet regularly with all coordinators and provide them with specialized training and guidance.

A guideline procedure should be developed so that all concerned will understand the purpose, objectives, and procedures of this program. This guideline should state that safety coordinators will not relieve line supervisors of their basic responsibility for the safety of their assigned employees. Their primary function is to assist the unit manager with the administration of the unit's safety and health programs.

The duties and responsibilities of safety coordinators will vary with the unit's needs. Some examples are

1. Audit unit safety manuals to ensure they are kept up to date.
2. Ensure that supervisors complete required documentation for new employee safety orientation, hazard communication, and other mandatory safety training.
3. Assist with investigation of serious accidents and incidents.

4. Review unit hazard control logs.
5. Track control action on hazards pending correction.
6. Assist supervisors with administration of the job hazard analysis program.
7. Audit employee compliance with safety procedures.

Safety coordinators can provide vital support to their unit management. In addition, those assigned to these positions have an invaluable opportunity to increase their knowledge of the unit's operations and also of safety program management techniques. Development of their leadership and communications skills as safety coordinators will help to prepare them for future promotions.

## APPENDIX A: SAMPLE POLICY PROCEDURE FOR EXECUTIVE SAFETY COMMITTEE

### Purpose

The Executive Safety Committee will provide the Director with recommendations as to unit-wide programs and major problems.

### Objectives

1. To ensure continuous improvement of unit safety and health systems and programs.
2. To provide a means for the efficient and effective resolution of major potential problems that could adversely affect the success of the enterprise.

### Committee Duties and Responsibilities

1. A written method of operation (MO) should be developed which will outline how the committee will receive problems and other items for discussion.
2. Problems up for discussion should be ranked by priority, assigned a title, document control number, and date.
3. Wherever possible, committee members should have received the meeting agenda so they can come prepared to discuss the issues and contribute to developing their solution.

## APPENDIX B: SAMPLE SAFETY COMMITTEE METHOD OF OPERATION

### Procedure

### A. Committee organization

1. The Director is the control chairperson of the executive committee.
2. The Operations Manager is chairperson of the meeting.
3. The committee meetings will be held each month on the third Wednesday at 10:00 a.m.

## B. Committee Duties and Responsibilities

1. The committee will establish basic unit-wide programs.
2. Any deviations from standard policy will be reviewed by the committee.
3. The committee will normally deal only with multidepartment problems.
4. Problems for discussion will be circulated as agenda items so that members are prepared to discuss them and reach a conclusion.
5. Individual committee members or small subcommittees will be appointed by the chairperson to conduct periodic audits and surveys to assess the effectiveness of the unit safety and health programs.

## C. Committee Meeting Agenda

Meeting agendas will include the following:

| Agenda Item | Person Responsible |
|---|---|
| 1. Review of overall unit safety program activity and results | Safety manager |
| 2. Discussion on significant accident experience | Department managers |
| 3. Discussion on identified problems | Committee members |
| 4. Reports from subcommittees | Subcommittee chairpersons |
| 5. Review new or revised policy-procedures submitted for approval | Committee members |

## FURTHER READING

Boylston, R. P. *Managing Safety and Health Programs.* New York: Van Nostrand Reinhold, 1990. (This entire book has extensive references to committee activities and functions.)

Colvin, R. J. *The Guidebook to Successful Safety Programming.* Boca Raton, FL: Lewis Publishers, 51–61, 1992.

Thomen, J. R. *Leadership in Safety Management.* New York: John Wiley & Sons, 61–88, 1991.

Ferry, T. *Safety and Health Management Planning.* New York: Van Nostrand Reinhold, 82–86, 1990.

Ridley, J. et al. *Safety at Work.* London: Butterworths, 137–138, 294–295, 1983.

Grimaldi, J. and Simonds, R. *Safety Management.* Boston, MA: Irwin, 118–125, 1989.

## SAFETY AND HEALTH TRAINING

Training in this element of the system is usually provided by specialist personnel or employees who have received special instructor training. Much of this training is legally mandated by Occupational Safety and Health Administration (OSHA) and other regulations. The following is a list of the more common training requirements:

## HEALTH-RELATED TRAINING

- Hazard communication
- Hazardous waste operator
- Asbestos awareness
- Hearing conservation
- Infection exposure control
- Respiratory protection
- Ergonomic (proposed regulations)

The specific content and requirements of these training programs will be discussed in more detail in the chapter on health.

**Safety-Related Training**

| Program | Scope | Requirements |
|---|---|---|
| • Excavations | Employees who enter excavations | 1. Initial training for "competent persons" and awareness training for all employees who enter and work in excavations |
| • Confined space | Entry supervisors<br>Confined space attendants and authorized entrants | 1. Training, initial and annual<br>2. Basic first aid and CPR certificate for confined space attendants |
| • Lockout/tagout | Employees who work with/ around energized equipment | 1. Training (initial and annual) |
| • Motor vehicle accident prevention | Employees who drive for employer on business | 1. Training: behind the wheel every 2 years for frequent drivers<br>2. Preventability workshops for employees who have accidents |
| • Emergency response | All employees | 1. Annual training and retraining |

This is by no means a complete list because, depending on the industry, there are numerous other specialized training requirements. Some examples of this type of legally mandated safety training are

- Window cleaning operations
- Forklift truck operation
- Powered industrial trucks
- Mechanical power presses
- Welding and cutting
- Crane operation
- Scaffold erection and dismantling
- Powder actuated tools
- Fire extinguisher operation

The Federal Mine Safety and Health Act of 1977 went even further than OSHA and actually contains specific mandatory safety training requirements in the Act itself. Examples (quoted in part) are below.

**Sec. 115(a)**

Each operator of a coal or other mine shall have a Health and Safety Training program which shall be approved by the Secretary.

Each training program approved by the Secretary shall provide as a minimum that

(1) New miners having no underground experience shall receive no less than 40 hours of training if they are to work underground. Such training shall include instruction in the statutory rights of miners and their representatives under this Act, use of the self-rescue device and use of respiratory devices, hazard recognition, escape-ways, walk-around training, emergency procedures, basic ventilation, basic roof control, electrical hazards, first aid, and the health and safety aspects of the task to which he will be assigned.

(3) All miners shall receive no less than 8 hours of refresher training no less frequently than once each 12 months.

(4) Any miner reassigned to a new task in which he has had no previous experience shall receive training in accordance with a training plan approved by the secretary.

Keeping track of this training has become something of a paperwork nightmare. Fortunately, computer systems are now available which not only satisfy regulatory documentation requirements, but also track when employees are due for training or retraining. (Refer to Figure 5-3 for illustration of a sample format for tracking safety training completed.) Interactive-type programs are now available, and these are a big help with scheduling, especially in multishift operations. For more in-depth information on training systems and techniques, refer to the chapter on training.

Much of this training, especially new employee and supervisor orientation, requires close coordination between safety professionals and the human resource function. A sign-off checklist is commonly used to ensure that all elements are covered. See Figures 5-4 and 5-5 for sample formats.

Successfully completing this array of training always presents scheduling problems and other production and timing conflicts. To help minimize these problems and spread the training throughout the year, one system that works well is to establish a schedule of monthly meetings. All the required topics can thus be covered, plus other subjects as needs are identified. The monthly training session can be repeated a sufficient number of times to allow for shift schedules and coverage.

Besides organizing and/or conducting these training programs, the safety professional, or designated person having this responsibility, will also need to establish a reference/resource library to support these activities. The training reference/resource library should include

- An audio-visual library so that supervisors and team leaders can loan videos, tapes, slides, etc., for their training needs

- A reference library of books, pamphlets, and bulletins
- A library of professional journals, magazines, and periodicals
- A reference file of articles, accident experience information, hazard control ideas, and other safety information

The person maintaining this file should circulate new items to the proper persons concerned in the organization for their information and action as appropriate.

Preferably, the reference and audio-visual libraries should be established in a learning center, training room, conference room, or similar facility. In this complex all the visual aid projection equipment and other training materials will be stored as needed.

To summarize, the training phase must as a minimum include

- New employee orientation
- Supervisory/team leader training
- Backup management training
- Line employee training
- Specialized training and retraining according to task and operations needs
- Learning center and training equipment
- Training information and reference library

Training, both the specialized type, and the specific, are absolutely vital factors for the success of the safety and health effort. Unit management will need to commit increasing time and effort to emphasize training and ensure that it is built into the unit's goals and strategic planning process.

## Interest and Motivation

As mentioned earlier, after training has been provided, successful loss prevention programs must include ways and means to promote employee interest and motivate them to put their knowledge into action.

### *Interest Programs*

For the innovative professional, there are almost limitless ways of creating and maintaining employee interest in safety and health issues and the unit's programs. Some common approaches are

- Bulletin boards
- Posters and signs
- Distribution of magazines, bulletins, etc. — example: National Safety Council "Family Safety" and numerous other subscription bulletins
- Special in-house flyers or handouts
- Speeches and presentations

# SAFETY TRAINING RECORD

| DIVISION | SECTION | SHIFT | TIME STARTED | TIME ENDED | | | |
|---|---|---|---|---|---|---|---|
| | | | | | DATE STARTED | | |
| | | | | | DATE COMPLETED | | |

| | TOTAL TIME | | DATE | | |
|---|---|---|---|---|---|
| | HOURS | MINUTES | MO | DAY | YR |

## CHECK TRAINING COURSE COMPLETED

| NEW EMPLOYEE TRAINING | | ANNUAL REFRESHER TRAINING | |
|---|---|---|---|
| COURSE NO. | TITLE | COURSE NO. | TITLE |
| | MATERIALS HAZARD CONTROL | | HAZMAT PART I |
| | HAZARD COMMUNICATIONS | | HAZMAT PART II |
| | MSDS | | ASBESTOS |
| | HAZMAT EMERGENCY RESPONSE | | RESPIRATORY |
| | HAZARDOUS WASTE | | HEARING CONSERVATION |
| | ASBESTOS AWARENESS | | FIRST AID CPR |
| | RESPIRATORY PROTECTION | | DEFENSIVE DRIVER |
| | HEARING CONSERVATION | | BLOODBORNE PATHOGENS |
| | FIRST AID-CPR | | FIRE EXTINGUISHER |
| | DEFENSIVE DRIVER | | EARTHQUAKE PLAN |
| | BLOODBORNE PATHOGENS | | EMERGENCY/EVACUATION |
| | FIRE EXTINGUISHER | | CONFINED SPACE ENTRY |
| | EARTHQUAKE PLAN | | LOCK-OUT TAG-OUT |
| | EMERGENCY/EVACUATION | | |

| | | | | | | |
|---|---|---|---|---|---|---|
| | | CONFINED SPACE ENTRY | | | | |
| | | LOCK-OUT TAG-OUT | | | | NEW JHA |
| | | HAZ. CONTROL PROCED. GEN. SAFE. PRACT. | | JHA TRAINING (CHECK ONE) | | REVIEW OF OLD JHA |
| | | | | | | ON SPOT JHA |
| | | | | | | |
| | JHA NO. | JHA TITLE | | JHA NO. | JHA TITLE | |
| | | | | | | |
| | | | | | | |
| | | | | | | |
| | | | | | | |
| | | | | | | |

**ATTENDANCE LIST**

I acknowledge by my signature below that I have received the training stated on this report.

| NAME | SIGNATURE | NAME | SIGNATURE |
|---|---|---|---|
| | | | |
| | | | |
| | | | |
| | | | |
| | | | |
| | | | |
| | | | |

INSTRUCTOR NAME: ______ POSITION TITLE: ______

SIGNATURE: ______ SECTION: ______ DATE ______

**Figure 5-3** Sample format for safety training completed.

| DATE EMPLOYED | | |
|---|---|---|
| MO | DAY | YEAR |
| | | |

**NEW EMPLOYEE ORIENTATION CHECK LIST**

| NAME | JOB CLASSIFICATION | DEPT. |
|---|---|---|

| ITEM COVERED | INITIAL | ITEM COVERED | INITIAL |
|---|---|---|---|
| **HUMAN RESOURCES** | | **SECURITY** | |
| Introduction to Facility/Plant/Organization | | Procedures for Entering and Leaving Plant | |
| Date Schedule | | Parking Regulations | |
| Food Service | | ID Badge Regulations | |
| Probation Period | | Loss Reporting | |
| Role of Supervisor | | | |
| Performance Reviews | | **SAFETY** | |
| Pay Rates and Increases | | The Unit Safety and Health Program | |
| Overtime Pay | | Purpose and function of Safety & Health Dept. | |
| Payday | | In-Plant Medical Clinic | |
| Promotions | | Accident Reporting and Investigation | |
| Pay Deductions (Union NIS) | | Workers Compensation Procedures | |
| | | Special Plant Hazards | |
| Emergency Messages | | Special Programs (Asbestos, Infection Control) | |
| Reporting Absences | | Emergency Procedures | |
| Rules of Conduct | | Fire Prevention and Protection System | |
| Holidays | | Fire Reporting and Firefighting Techniques | |
| Vacation | | General Safety Rules | |
| Sick Leave | | Respiratory Protection | |
| Credit Union | | Hearing Conservation | |
| Change House Facilities & Regulations | | Safe Work Clothing and Practices | |
| Change House Locker Assigned | | "Employees Guide to Safety" issued | |
| Identification Badge Issued | | | |
| Issued all Personal Protective Equipment | | Shown 5-Point Emergency Care Film | |

The items appearing above were discussed with the employee on _____

EMPLOYEE SIGNATURE _____      HUMAN RESOURCES SUPERVISOR _____

HUMAN RESOURCES SUPERINTENDENT _____      SAFETY SUPERVISOR _____

**Figure 5-4**   Sample new employee orientation checklist.

It also promotes employee acceptance when you encourage employee involvement. Go out and find those with graphic or artistic skills to help illustrate your handouts. Seek out those with writing skills who would be happy to contribute short articles. Publicize accident and incident experiences, especially information on "near misses". Some organizations have, for example, a "brown bag lunch" program in which they arrange for in-house and outside invited speakers to talk on a variety of topics. Finally to encourage employees, top management must support these programs and use every opportunity to demonstrate their commitment.

NEW EMPLOYEES ORIENTATION FLOW SHEET

**Instructions**:
After you have completed your phase, sign in the space provided. After completing Section 4, return form to Personnel Office.

_____
Employee's Name

_____
Classification

1.  EMPLOYMENT

General information about Company
Character of employment
Schedule (working)
Rates of pay, hours, and overtime
Equal Employment Opportunities
Vacations, holidays
Importance of good attendance record
Union
Probationary period
Job bidding
Group insurance-life, hospital-surgical
Bond purchase
Proper clothing, safety shoes
Pension plan, S.U.B.
Employee's Association
Issue Security badge

_____
(Personnel Representative)

2.  PLANT PROTECTION

Locker room facilities
Master bulletin board
Telephone calls
Parking facilities
Security procedures

_____
(Safety/Security Representative)

3.  SAFETY AND HEALTH

a.  MSHA Safety Training (Total 3 Hours)
    Statutory rights
    Employee's handbook, safety manual
    Introduction to work environment Part I
    Hazard recognition
    Health/Safety of tasks assigned Part I
b.  Other Items:
    Cafeteria
    First-aid facilities
    Safety equipment
    Safety shoes, how to order
    Visit to storeroom
    Visit to Medical Department
    Medical Surveillance Program
    Point out high rise potential hazards
    and preventive measures.

_____
(Safety Representative)

4.  DEPARTMENT

a.  MSHA Safety Training (Total 5 Hours)
    Introduction to Work Environment Part II
        General Nature of work in department
        Relation of dept. to other depts.
        Location of lockers and rest rooms
        Brief tour of department
    Health/Safety of Tasks Assigned Part II
        Departmental safety practices
        Proper clothing for work
        Importance of good housekeeping
        Report injuries, treatment, & record
b.  Other Items:
        Nature of work assigned
        Work schedule, hours, lunch arrangements
        Overtime work and procedures
        Importance of good attendance record
        Call in when absent
        Tool checks
        Requisition of storeroom items
        Public address system
        Introduce to other employees

_____
Supervisor

**Figure 5-5**    Sample new employee orientation checklist to comply with MSHA requirements.

## *Motivation Programs*

Much has been written on employee motivation, and this aspect of human behavior is explored in depth in later chapters.

A highly recommended reading on this subject is Homer Lambie's article "Accident Control Through Motivation". It is as current today as when it was written 30 years ago. It really crystallizes the spirit of the human soul, its desires, and needs as they relate to accident prevention.

Safety motivation programs, for the purpose of this chapter, will include

- Award and recognition for outstanding safety results
- Award and recognition for outstanding safety program performance
- Incentive programs
- Award and recognition for outstanding safety service and/or heroism

Recognition of outstanding safety results is the most common of all motivation programs. These programs provide recognition and awards for injury-free records. The unit award can be on an individual basis, by section or department, or for the total organization.

A word of caution: experience has shown that when the award is based on all injuries there is a tendency for injuries to go "underground", i.e., to go into the nonoccupational care system rather than risk spoiling the group's chance for an award. To avoid this problem, many organizations base their records on lost workday cases. In this way, the measurement is on cases that are inherently more severe in terms of injury and thus less likely to be concealed.

In the writer's experience, a more effective recognition system is one that is based not only on the program results, but also on the program performance. This type of award is more positive in that it emphasizes the concept that a performance- and results-based award recognizes the group's accident prevention efforts as well as their superior injury-free results. Results alone may be just the product of luck.

Incentive programs come in such a variety of forms that it is clearly beyond the scope of this chapter to discuss them in detail. A well-thought-out incentive program that is properly promoted can be a useful aid to an effective safety program. For this reason, presented here is a consolidated list of hints and pitfalls to avoid that will be helpful in designing either a contest or a recognition program.

- Organization contests are often geared toward the *best* individual or group. This can result in others not being positively reinforced. One or a few winners and many losers results in a very negative impact.
- Contests or incentive programs deteriorate over time, and if the goal is too long range, the groups lose interest.
- Contests should be based on rewarding positive effort by *all* the contestants. Too many contests are based on "the luck of the draw".
- Contests should be short term, say 3 to 6 months in duration. The results are also more positive when all participants are rewarded with some tangible recognition.
- The reward should be a token, not a prize. People like something that they can associate with their organization, the contest, and their group. Poll your group to find their preferences.

- People love to see themselves in a picture. A plant that I was working at some years ago established a world safety record for the industry. The company photographer went around the plant and took dozens of pictures of people at work. This was made up into a huge two-page-spread collage and published in the local newspaper. A special reprint was also provided for employees. In addition, all employees were given an inscribed wall clock with a picture of the plant as a background.
- If you are going to select a results-only recognition program, make sure the accomplishment is recognizing a *superior* record — for example, the first 100 injury-free days for a new plant, or a million hours without a lost workday injury; for individuals, say 1, 3, and 5 years without a lost workday injury.

To summarize, the value of interest and motivation programs is measured by the extent to which they assist every employee in the organization to develop a strong sense of responsibility and the desire to continuously improve their skills in finding and controlling hazards.

## FURTHER READING

### Books

Bird, F. E. and Germain, G. L. *Practical Loss Control Leadership.* Loganville, GA: International Loss Control Institute, 239–262, 263–284 (Motivation), 1990.

Broadwell, M. M. *The Supervisor On-The-Job Training.* Reading, MA: Addison-Wesley, 1989.

Colvin, R. J. *The Guidebook to Successful Safety Programming.* Boca Raton, FL: Lewis Publishers, 113–131, 1992.

Gellerman, S. W. *Motivation in the Real World.* New York: Penguin Books, 1993.

Grimaldi, J. V. and Simonds, R. H. *Safety Management.* Boston, MA: Irwin, 185–191 (Motivation), 478–493 (Training), 1989.

Grund, E. V. *Lockout/Tagout,* Itasca, IL: National Safety Council, 1995.

Heath, E. D. and Ferry, T. S. *Training in the Work Place.* Goshen, NY: Aloray, 1990. (This entire book contains extensive information on safety training.)

Krause, T. R., Hidley, J. H., and Hodson, S. J. *The Behavior Based Safety Process.* New York: Von Nostrand Reinhold, 1990. (This book contains extensive research and information on motivation as it affects employee behavior.)

LaDou, J. (Ed.) et al. *Occupational Safety and Health.* Itasca, IL: National Safety Council, 1995.

Lambie, H. K. *Accident Control through Motivation. Selected Readings in Safety.* Macon, GA: Academy Press, 1973.

MacCollum, D. V. Construction Safety Planning. New York, NY: Van Nostrand Reinhold, 1995.

McSween, T. E. *The Values-Based Safety Process.* New York, Van Nostrand Reinhold, 1995.

Petersen, D. *Safe Behavior Reinforcement.* Goshen, NY: Aloray, 1989. (Extensive research and information on motivation as it affects employee behavior.)

Petersen, D. *Safety Management: A Human Approach.* Goshen, NY: Aloray, 59–63 (Training), 139–150 (Motivation), 205–215 (Training), 217–271 (Motivation), 1988.

Petersen, D. *Techniques of Safety Management: A Systems Approach.* Goshen, NY: Aloray, 141–158, 1989.

Pike, R. W. *Creative Training Techniques Handbook.* Minneapolis, MN: Lakewood Publications, 1989.

Robinson, A. D. Incentives and rewards. In *Human Resources Management and Development Handbook.* Tracey, W. R., Ed. New York: Amacon (American Management Association), Chapter 42, 1994.

## Other Publications

Jones, S. E. The key issues of safety and health. Occupational hazards. *NSMS Focus* 87–90, May 1991.

Lack, R. W. Industrial safety training — coping with the unsafe act problem in today's world. *Prof. Saf.* 24(2), 33–37, February 1979.

Lack, R. W. Is your safety program a paper tiger? *Am. Soc. Saf. Eng. Manage. Div. Newsl.* No. 9, 6–9, March 1985.

Lapidus, R. The psychology of safety: how to get people to perform in a safe manner. Issues in perspective. *Nat. Saf. Manage. Soc.* 1(4), 1–6, December 1986.

Swartz, G. Safety in supervisors salary review — a formal approach. *Prof. Saf.* 36(5), 21–24, May 1991.

Winn, G. L. In the crucible: testing for a real paradigm shift. *Prof. Saf.* 37(12), 30–33, December 1992.

Winn, G. L. Total quality? The "new" paradigm seems out of reach for safety managers. *Occup. Health Saf.* 63(10), 53–54, October 1994.

## SAFETY INSPECTIONS

### Introduction

On January 26, 1989, the U.S. Department of Labor OSHA issued their document "Safety and Health Program Management Guidelines; Issuance of Voluntary Guidelines". Among the recommended actions, there is a section dealing with inspections. This is quoted (in part) as follows:

(C)(2)  Worksite analysis (i) so that all hazards are identified:

   (A) Conduct comprehensive baseline worksite surveys for safety and health and periodic comprehensive update surveys;

   (B) Analyze planned and new facilities, processes, materials, and equipment; and

   (C) Perform routine job hazard analyses

(ii)    Provide for regular site safety and health inspections, so that new or previously missed hazards and failures in hazard controls are identified....

Comment: Identification at a work site of those safety and health hazards which are recognized in its industry is a critical foundation for safety and health protection. It is the general duty of the employer under the Occupational Safety and Health Act of 1970....

Personnel performing regular inspections should, however, possess a degree of experience and competence adequate to recognize hazards in the areas they review and to identify reasonable means for their correction or control. Such competence should normally be expected of ordinary employees who are capable of safely supervising or performing the operations of the specific workplace.

The frequency and scope of these "routine" inspections depends on the nature and severity of the hazards which could be present and the relative stability and complexity of work site operations.

The State of California Title 8, General Industry Safety Orders, Article 1, Section 3203, "Injury and Illness Prevention Program", went into effect in July 1991 and requires employers to establish, implement, and maintain an effective injury and illness prevention program.

This standard is quoted (in part).

(a)(4)   Include procedures for identifying and evaluating work place hazards including scheduled periodic inspections to identify unsafe conditions and work practices.
(b)    Records of the steps taken to implement and maintain the Program shall include:
(1)   Records of scheduled and periodic inspections required by Subsection (a)(4) to identify unsafe conditions and work practices, including person(s) conducting the inspection, the unsafe conditions and work practices that have been identified and action taken to correct the identified unsafe conditions and work practices....

The material that follows is based on a paper on a related subject published in 1980 by *Protection,* the official journal of the Institution of Industrial Safety Officers, U.K. (now the Institution for Occupational Safety and Health), and is printed with permission from Paramount Publishing Ltd.

The traditional approach to industrial safety and housekeeping, probably copied from the military, was the routine inspection. This approach required that senior members of management, usually second level supervision or higher, made regular inspections and pointed out to first-line supervisors what must be done. This method had some benefits in that it was a form of training for the supervisors in what sort of problems they should be looking for, but it is inefficient insofar as the second-line supervisor's time is concerned and very often ineffective because little effort is made to find out why the problems existed in the first place.

This section will, therefore, describe a more systematic approach to safety inspections which is in one form or another being practiced by many organizations around the world.

## The Key Parts of an Effective Safety Inspection System

An effective safety inspection system is built around three key elements: assignment of responsibility, the inspector's self-survey, and follow-up controls.

### Assignment of Responsibility

The first step in establishing a sound safety inspection system is based on the principle of total single responsibility. This requires that the entire work site, including fence, gates, yards, and grounds, be divided up and assigned to the lowest level of management.

To establish this part of the program, it is recommended that first a work site map be drawn up showing major department areas of responsibility, preferably subdivided down to subdepartment areas usually supervised by second-line supervision. This map should have the approval of department heads and top management and be kept current with unit expansions, management changes, etc.

Based on the plant safety inspection area responsibility map, second-line supervisors should then develop maps showing how their respective areas are subdivided among their first-line supervisors or work team leaders.

Confusion often arises here as to the difference between supervisors' program responsibility versus their job responsibility for safety and housekeeping. Let us take a simple example to see how this approach can be misunderstood. In this sketch, we show a typical production area subdivided among four supervisors:

| SMITH | JONES |
|-------|-------|
| BROWN | GREEN |

This department operates on a 40-hour, 7-days a week basis, and only one of the supervisors is on duty at any particular time. Supervisor Smith may, therefore, be tempted while on duty to polish his/her area and ignore problems in the other supervisors' areas. To overcome this problem, supervisors must be instructed that, on a daily basis, they have a *job* responsibility to maintain good safety and housekeeping in the total operating area under their supervision. Conversely, their *program* responsibility calls for them to periodically survey their assigned area to detect and solve safety and housekeeping problems. If such problems are not systematically eliminated, the result will be steady deterioration of the level of safety and housekeeping in the total department.

On a daily basis, supervisors must ensure *job* safety and housekeeping. Here are a few examples:

- Compliance with safety, health, and environmental procedures and job hazard analyses
- Cleanup of production spills
- Collection and proper storage of tools and equipment
- Proper disposal of waste and scrap

On a periodic basis (recommended minimum of once a month), supervisors should survey their assigned areas of responsibility for safety and housekeeping problems. Checklists can be used to assist inspectors in their survey for safety problems.

An example of this type is provided by Cal/OSHA Consultation and is illustrated in part in the following pages (Appendix C).

# Appendix C: Self-inspection Checklists

These checklists are by no means inclusive. You should add to them or delete items that do not apply to your operations. However, carefully consider each item as you come to it before making your decision.

## Employer Posting

☐ Is the Cal/OSHA poster *Safety and Health Protection on the Job* displayed in a prominent location where all employees are likely to see it?

☐ Are emergency telephone numbers posted where they can be readily found in case of emergency?

☐ Where employees may be exposed to any toxic substances or harmful physical agents, has appropriate information concerning employee access to medical and exposure records and Material Safety Data Sheets been posted or otherwise made readily available to affected employees?

☐ Are signs concerning exiting from buildings, room capacities, floor loading, exposures to x-ray, microwave, or other harmful radiation or substances posted where appropriate?

Are other California posters properly displayed, such as:

☐ Industrial Welfare Commission orders regulating wages, hours, and working conditions?

☐ Discrimination in employment prohibited by law?

☐ Notice to employees of unemployment and disability insurance?

☐ Payday notice?

☐ Summary of occupational injuries and illnesses posted in the month of February?

☐ Notice of compensation carrier?

## Permit Requirements

☐ Is a permit obtained for excavations which are 5 feet or deeper and into which a person is required to descend?

☐ Is a permit obtained for construction of any building, structure, scaffolding or falsework more than 3 stories high or the equivalent height?

☐ Is a permit obtained for demolition of any building, structure, or the dismantling of scaffolding or falsework more than 3 stories high or the equivalent height?

## Record Keeping

☐ Are all occupational injuries or illnesses, except minor injuries requiring only first aid, being recorded as required on the Cal/OSHA Form 200?

☐ Are employee medical records and records of employee exposure to hazardous substances or harmful physical agents current?

☐ Have arrangements been made to maintain required records for the legal period of time for each specific type of record? (Some records must be maintained for at least 40 years.)

☐ Are operating permits and records current for such items as elevators, air pressure tanks, liquefied petroleum gas tanks?

☐ Are carcinogen use reports filed with Cal/OSHA as required? (Contact the nearest Cal/OSHA office for the list of regulated carcinogens.)

☐ Are employee safety and health training records maintained?

☐ Is documentation of safety inspections and corrections maintained?

☐ Are safety committee meeting records maintained?

## Injury & Illness Prevention Program

☐ Do you have a written, effective injury and illness prevention program?

☐ Do you have a person who is responsible and has authority for overall activities of the injury and illness prevention program?

☐ Do you have a system for identifying and evaluating your workplace hazards?

☐ Do you systematically correct these hazards in a timely manner?

☐ Do you provide training in both general and specific safe work practices?

☐ Do you encourage employee participation in health and safety matters?

☐ Do you maintain an ongoing safety training program?

☐ Do you have a system in place that ensures employees will be recognized for safe and healthful work practices?

☐ Will employees be disciplined for unsafe safety or health acts?

☐ Is there a labor-management safety committee?

☐ If there is no safety committee, is there in place a system for communicating safety and health concerns to employees?

☐ On construction sites, is a Code of Safe Practices posted?

☐ Are "toolbox" meetings conducted every 10 days, or sooner if appropriate?

## Medical Services & First Aid

☐ Do you require each employee to have a pre-employment physical examination?

☐ Is there a hospital, clinic, or infirmary for medical care in proximity of your workplace?

☐ If medical and first aid facilities are not in proximity of your workplace, is at least one employee on each shift currently qualified to render first aid?

☐ Are medical personnel readily available for advice and consultation on matters of employee health?

☐ Are emergency phone numbers posted?

☐ Are first aid kits easily accessible to each work area, with necessary supplies available, periodically inspected and replenished as needed?

☐ Have first aid kit supplies been approved by a physician, indicating they are adequate for a particular area or operation?

☐ Are means provided for quick drenching or flushing of the eyes and body in areas where corrosive liquids or materials are handled?

## Fire Protection

☐ Do you have a fire prevention plan?

☐ Does your plan describe the type of fire protection equipment and/or systems?

☐ Have you established practices and procedures to control potential fire hazards and ignition sources?

☐ Are employees aware of the fire hazards of the materials and processes to which they are exposed?

☐ Is your local fire department well acquainted with your facilities, location and specific hazards?

☐ If you have a fire alarm system, is it certified as required?

☐ If you have a fire alarm system, is it tested at least annually?

☐ If you have interior stand pipes and valves, are they inspected regularly?

☐ If you have outside private fire hydrants, are they flushed at least once a year on a routine preventive maintenance schedule?

☐ Are fire doors and shutters in good operating condition?

☐ Are fire doors and shutters unobstructed and protected against obstructions, including their counterweights?

☐ Are fire door and shutter fusable links in place?

☐ Are automatic sprinkler system water control valves, air and water pressures checked weekly/periodically as required?

☐ Is maintenance of automatic sprinkler systems assigned to responsible persons or to a sprinkler contractor?

☐ Are sprinkler heads protected by metal guards, when exposed to physical damage?

☐ Is proper clearance maintained below sprinkler heads?

☐ Are portable fire extinguishers provided in adequate number and type?

☐ Are fire extinguishers mounted in readily accessible locations?

☐ Are fire extinguishers recharged regularly and noted on the inspection tag?

☐ Are employees periodically instructed in the use of extinguishers and fire protection procedures?

## Personal Protective Equipment & Clothing

☐ Are protective goggles or face shields provided and worn where there is any danger of flying particles or corrosive materials?

☐ Are approved safety glasses required to be worn at all times in areas where there is risk of eye injuries such as punctures, abrasions, contusions or burns?

☐ Are employees who need corrective lenses (glasses, contact lenses) in working environments with harmful exposures, required to wear only approved safety glasses, protective goggles, or to use other medically approved precautionary procedures?

☐ Are protective gloves, aprons, shields, or other means provided against cuts, corrosive liquids and chemicals?

☐ Are hard hats provided and worn where danger of falling objects exists?

☐ Are hard hats inspected periodically for damage to the shell and suspension system?

☐ Is appropriate foot protection required where there is risk of foot injuries from hot, corrosive, poisonous substances, falling objects, crushing or penetrating actions?

☐ Are approved respirators provided for regular or emergency use where needed?

☐ Is all protective equipment maintained in a sanitary condition and ready for use?

☐ Do you have eye wash facilities and a quick drench shower within a work area where employees are exposed to injurious corrosive materials?

☐ Where special equipment is needed for electrical workers, is it available?

☐ When lunches are eaten on the premises, are they eaten in areas where there is no exposure to toxic materials or other health hazards?

☐ Is protection against the effects of occupational noise exposure provided when sound levels exceed those of the Cal/OSHA noise standard?

☐ Are adequate work procedures, protective clothing and equipment provided and used when cleaning up spilled toxic or otherwise hazardous materials or liquids?

## General Work Environment

☐ Are all worksites clean and orderly?

☐ Are work surfaces kept dry or appropriate means taken to assure the surfaces are slip-resistant?

☐ Are all spilled materials or liquids cleaned up immediately?

☐ Is combustible scrap, debris and waste stored safely and removed from the worksite promptly?

☐ Is accumulated combustible dust routinely removed from elevated surfaces, including the overhead structure of buildings?

☐ Is combustible dust cleaned up with a vacuum system to prevent the dust going into suspension?

☐ Is metallic or conductive dust prevented from entering or accumulating on or around electrical enclosures or equipment?

☐ Are covered metal waste cans used for oily and paint-soaked waste?

☐ Are all oil and gas fired devices equipped with flame failure controls that will prevent flow of fuel if pilots or main burners are not working?

☐ Are paint spray booths, dip tanks and the like, cleaned regularly?

☐ Are the minimum number of toilets and washing facilities provided?

☐ Are all toilets and washing facilities clean and sanitary?

☐ Are all work areas adequately illuminated?

☐ Are pits and floor openings covered or otherwise guarded?

## Walkways

☐ Are aisles and passageways kept clear?

☐ Are aisles and walkways marked as appropriate?

☐ Are wet surfaces covered with non-slip materials?

☐ Are holes in the floor, sidewalk or other walking surface repaired properly, covered or otherwise made safe?

☐ Is there safe clearance for walking in aisles where motorized mechanical handling equipment is operating?

☐ Are spilled materials cleaned up immediately?

☐ Are materials or equipment stored in such a way that sharp projectives will not interfere with the walkway?

☐ Are changes of direction or elevations readily identifiable?

☐ Are aisles or walkways that pass near moving or operating machinery, welding operations or similar operations arranged so employees will not be subjected to potential hazards?

☐ Is adequate headroom provided for the entire length of any aisle or walkway?

☐ Are standard guardrails provided wherever aisle or walkway surfaces are elevated more than 30 inches above any adjacent floor or the ground?

☐ Are bridges provided over conveyors and similar hazards?

## Floor & Wall Openings

☐ Are floor openings guarded by a cover, guardrail, or equivalent on all sides (except at entrance to stairways or ladders)?

☐ Are toeboards installed around the edges of a permanent floor opening (where persons may pass below the opening)?

☐ Are skylight screens of such construction and mounting that they will withstand a load of at least 200 pounds?

☐ Is the glass in windows, doors, glass walls, which are subject to human impact, of sufficient thickness and type for the condition of use?

☐ Are grates or similar covers over floor openings, such as floor drains, of such design that foot traffic or rolling equipment will not be affected by the grate spacing?

☐ Are unused portions of service pits and pits not actually in use either covered or protected by guardrails or equivalent?

☐ Are manhole covers, trench covers and similar covers, plus their supports, designed to carry a truck rear axle load of at least 20,000 pounds when located in roadways and subject to vehicle traffic?

☐ Are floor or wall openings in fire resistive construction provided with doors or covers compatible with the fire rating of the structure, and provided with self-closing features when appropriate?

## Stairs & Stairways

☐ Are standard stair rails or handrails on all stairways having four or more risers?

☐ Are all stairways at least 22 inches wide?

☐ Do stairs have at least a 6'6" overhead clearance?

☐ Do stairs angle no more than 50 and no less than 30 degrees?

☐ Are stairs of hollow-pan type treads and landings filled to noising level with solid material?

☐ Are step risers on stairs uniform from top to bottom, with no riser spacing greater than 7-1/2 inches?

☐ Are steps on stairs and stairways designed or provided with a surface that renders them slip resistant?

☐ Are stairway handrails located between 30 and 34 inches above the leading edge of stair treads?

☐ Do stairway handrails have at least 1-1/2 inches clearance between the handrails and the wall or surface they are mounted on?

☐ Are stairway handrails capable of withstanding a load of 200 pounds, applied in any direction?

☐ Where stairs or stairways exit directly into any area where vehicles may be operated, are adequate barriers and warnings provided to prevent employees stepping into the path of traffic?

☐ Do stairway landings have a dimension measured in the direction of travel, at least equal to the width of the stairway?

☐ Is the vertical distance between stairway landings limited to 12 feet or less?

☐ Is a stairway provided to the roof of each building four or more stories in height, provided the roof slope is 4 in 12 or less?

## Elevated Surfaces

☐ Are signs posted, when appropriate, showing the elevated surface load capacity?

☐ Are surfaces elevated more than 30 inches above the floor or ground provided with standard guardrails?

☐ Are all elevated surfaces (beneath which people or machinery could be exposed to falling objects) provided with standard 4-inch toeboards?

☐ Is a permanent means of access and egress provided to elevated storage and work surfaces?

☐ Is required headroom provided where necessary?

☐ Is material on elevated surfaces piled, stacked or racked in a manner to prevent it from tipping, falling, collapsing, rolling or spreading?

☐ Are dock boards or bridge plates used when transferring materials between docks and trucks or rail cars?

## Exiting or Egress

☐ Are all exits marked with an exit sign and illuminated by a reliable light source?

☐ Are the directions to exits, when not immediately apparent, marked with visible signs?

☐ Are doors, passageways or stairways, that are neither exits nor access to exits and which could be mistaken for exits, appropriately marked "NOT AN EXIT", "TO BASEMENT", "STOREROOM", and the like?

☐ Are exit signs provided with the word "EXIT" in lettering at least 5 inches high and the stroke of the lettering at least 1/2 inch wide?

☐ Are exit doors side-hinged?

☐ Are all exits kept free of obstructions?

☐ Are at least two means of egress provided from elevated platforms, pits or rooms where the absence of a second exit would increase the risk of injury from hot, poisonous, corrosive, suffocating, flammable, or explosive substances?

☐ Are there sufficient exits to permit prompt escape in case of emergency?

☐ Are special precautions taken to protect employees during construction and repair operations?

☐ Is the number of exits from each floor of a building, and the number of exits from the building itself, appropriate for the building occupancy load?

☐ Are exit stairways which are required to be separated from other parts of a building, enclosed by at least two-hour fire-resistive construction in buildings more than four stories in height, and not less than one-hour fire resistive construction elsewhere?

☐ When ramps are used as part of required exiting from a building, is the ramp slope limited to 1 foot vertical and 12 feet horizontal?

☐ Where exiting will be through frameless glass doors, glass exit doors, storm doors and such, are the doors fully tempered and meeting safety requirements for human impact?

## Exit Doors

☐ Are doors which are required to serve as exits designed and constructed so that the way of exit travel is obvious and direct?

☐ Are windows which could be mistaken for exit doors, made inaccessible by means of barriers or railings?

☐ Are exit doors openable from the direction of exit travel, without the use of a key or any special knowledge or effort, when the building is occupied?

☐ Is a revolving, sliding or overhead door prohibited from serving as a required exit door?

☐ Where panic hardware is installed on a required exit door, will it allow the door to open by applying a force of 15 pounds or less in the direction of the exit traffic?

☐ Are doors on cold storage rooms provided with an inside release mechanism which will release the latch and open the door even if it's padlocked or otherwise locked on the outside?

☐ Where exit doors open directly onto any street, alley or other area where vehicles may be operated, are adequate barriers and warnings provided to prevent employees stepping into the path of traffic?

☐ Are doors that swing in both directions and are located between rooms where there is frequent traffic, provided with viewing panels in each door?

## Portable Ladders

☐ Are all ladders maintained in good condition, joints between steps and side rails tight, all hardware and fittings securely attached, and moveable parts operating freely without binding or undue play?

☐ Are non-slip safety feet provided on each ladder?

☐ Are non-slip safety feet provided on each metal or rung ladder?

☐ Are ladder rungs and steps free of grease and oil?

☐ Is it prohibited to place a ladder in front of doors opening toward the ladder except when the door is blocked open, locked or guarded?

☐ Is it prohibited to place ladders on boxes, barrels, or other unstable bases to obtain additional height?

☐ Are employees instructed to face the ladder when ascending or descending?

☐ Are employees prohibited from using ladders that are broken, missing steps, rungs, or cleats, broken side rails or other faulty equipment?

☐ Are employees instructed not to use the top step of ordinary stepladders as a step?

☐ When portable rung ladders are used to gain access to elevated platforms, roofs and the like, does the ladder always extend at least 3 feet above the elevated surface?

☐ Is it required that when portable rung or cleat type ladders are used, the base is so placed that slipping will not occur, or it is lashed or otherwise held in place?

☐ Are portable metal ladders legibly marked with signs reading "CAUTION" "Do Not Use Around Electrical Equipment" or equivalent wording?

☐ Are employees prohibited from using ladders as guys, braces, skids, gin poles, or for other than their intended purposes?

☐ Are employees instructed to only adjust extension ladders while standing at a base (not while standing on the ladder or from a position above the ladder)?

☐ Are metal ladders inspected for damage?

☐ Are the rungs of ladders uniformly spaced at 12 inches, center to center?

## Hand Tools & Equipment

☐ Are all tools and equipment (both company and employee-owned) used by employees at their workplace in good condition?

☐ Are hand tools such as chisels or punches, which develop mushroomed heads during use, reconditioned or replaced as necessary?

☐ Are broken or fractured handles on hammers, axes and similar equipment replaced promptly?

☐ Are worn or bent wrenches replaced regularly?

☐ Are appropriate handles used on files and similar tools?

☐ Are employees made aware of the hazards caused by faulty or improperly used hand tools?

Are appropriate safety glasses, face shields and similar equipment used while using hand tools or equipment which might produce flying materials or be subject to breakage?

Are jacks checked periodically to assure they are in good operating condition?

Are tool handles wedged tightly in the head of all tools?

Are tool cutting edges kept sharp so the tool will move smoothly without binding or skipping?

Are tools stored in a dry, secure location where they won't be tampered with?

Is eye and face protection used when driving hardened or tempered spuds or nails?

## Portable (Power Operated) Tools & Equipment

Are grinders, saws and similar equipment provided with appropriate safety guards?

Are power tools used with the correct shield, guard or attachment recommended by the manufacturer?

Are portable circular saws equipped with guards above and below the base shoe?

Are circular saw guards checked to assure they are not wedged up, thus leaving the lower portion of the blade unguarded?

Are rotating or moving parts of equipment guarded to prevent physical contact?

Are all cord-connected, electrically-operated tools and equipment effectively grounded or of the approved double insulated type?

Are effective guards in place over belts, pulleys, chains, sprockets, on equipment such as concrete mixers, air compressors and the like?

Are portable fans provided with full guards or screens having openings of 1/2 inch or less?

Is hoisting equipment available and used for lifting heavy objects, and are hoist ratings and characteristics appropriate for the task?

Are ground-fault circuit interrupters provided on all temporary electrical 15 and 20 ampere circuits, used during periods of construction?

Are pneumatic and hydraulic hoses on power-operated tools checked regularly for deterioration or damage?

## Abrasive Wheel Equipment Grinders

Is the work rest used and kept adjusted to within 1/8 inch of the wheel?

Is the adjustable tongue on the top side of the grinder used and kept adjusted to within 1/4 inch of the wheel?

Do side guards cover the spindle, nut, and flange and 75 percent of the wheel diameter?

Are bench and pedestal grinders permanently mounted?

Are goggles or face shields always worn when grinding?

Is the maximum RPM rating of each abrasive wheel compatible with the RPM rating of the grinder motor?

Are fixed or permanently mounted grinders connected to their electrical supply system with metallic conduit or by other permanent wiring method?

Does each grinder have an individual on and off control switch?

Is each electrically operated grinder effectively grounded?

Before new abrasive wheels are mounted, are they visually inspected and ring tested?

Are dust collectors and powered exhausts provided on grinders used in operations that produce large amounts of dust?

☐ Are splash guards mounted on grinders that use coolant, to prevent the coolant reaching employees?

☐ Is cleanliness maintained around grinders?

## Powder Actuated Tools

☐ Are employees who operate powder-actuated tools trained in their use, and carry valid operator cards?

☐ Do the powder-actuated tools being used have written approval of the Division of Occupational Safety and Health?

☐ Is each powder-actuated tool stored in its own locked container when not being used?

☐ Is a sign at least 7" by 10" with bold type reading "POWDER-ACTUATED TOOL IN USE" conspicuously posted when the tool is being used?

☐ Are powder-actuated tools left unloaded until they are actually ready to be used?

☐ Are powder-actuated tools inspected for obstructions or defects each day before use?

☐ Do powder-actuated tool operators have and use appropriate personal protective equipment such as hard hats, safety goggles, safety shoes and ear protectors?

## Machine Guarding

☐ Is there a training program to instruct employees on safe methods of machine operation?

☐ Is there adequate supervision to ensure that employees are following safe machine operating procedures?

☐ Is there a regular program of safety inspection of machinery and equipment?

☐ Is all machinery and equipment kept clean and properly maintained?

☐ Is sufficient clearance provided around and between machines to allow for safe operations, set up and servicing, material handling and waste removal?

☐ Is equipment and machinery securely placed and anchored when necessary, to prevent tipping or other movement that could result in personal injury?

☐ Is there a power shut-off switch within reach of the operator's position at each machine?

☐ Can electric power to each machine be locked out for maintenance, repair or security?

☐ Are the noncurrent-carrying metal parts of electrically operated machines bonded and grounded?

☐ Are foot-operated switches guarded or arranged to prevent accidental actuation by personnel or falling objects?

☐ Are manually operated valves and switches controlling the operation of equipment and machines clearly identified and readily accessible?

☐ Are all emergency stop buttons colored red?

☐ Are all pulleys and belts that are within 7 feet of the floor or working level properly guarded?

☐ Are all moving chains and gears properly guarded?

☐ Are splash guards mounted on machines that use coolant, to prevent the coolant from reaching employees?

☐ Are methods provided to protect the operator and other employees in the machine area from hazards created at the point of operation, ingoing nip points, rotating parts, flying chips, and sparks?

☐ Are machinery guards secure and arranged so they do not offer a hazard in their use?

☐ If special hand tools are used for placing and removing material, do they protect the operator's hands?

☐ Are revolving drums, barrels and containers required to be guarded by an enclosure that is interlocked with the drive mechanism, so that revolution cannot occur unless the guard enclosure is in place, so guarded?

☐ Do arbors and mandrels have firm and secure bearings and are they free from play?

☐ Are provisions made to prevent machines from automatically starting when power is restored after a power failure or shut-down?

☐ Are machines constructed so as to be free from excessive vibration when the largest size tool is mounted and run at full speed?

☐ If machinery is cleaned with compressed air, is air pressure controlled and personal protective equipment or other safeguards used to protect operators and other work-ers from eye and body injury?

☐ Are fan blades protected with a guard having openings no larger than 1/2 inch, when operating within 7 feet of the floor?

☐ Are saws used for ripping equipped with anti-kick back devices and spreaders?

☐ Are radial arm saws so arranged that the cutting head will gently return to the back of the table when released?

## Lockout Blockout Procedures

☐ Is all machinery or equipment capable of movement required to be de-energized or disengaged and blocked or locked out dur-ing cleaning, servicing, adjusting or set-ting up operations, whenever required?

☐ Is the locking-out of control circuits in lieu of locking-out main power disconnects prohibited?

☐ Are all equipment control valve handles provided with a means for locking out?

☐ Does the lock-out procedure require that stored energy (i.e. mechanical, hydraulic, air) be released or blocked before equip-ment is locked out for repairs?

☐ Are appropriate employees provided with individually keyed personal safety locks?

☐ Are employees required to keep personal control of their key(s) while they have safety locks in use?

☐ Is it required that employees check the safety of the lockout by attempting a start up after making sure no one is exposed?

Where the power disconnecting means for equipment does not also disconnect the electrical control circuit:

☐ Are the appropriate electrical enclosures identified?

☐ Is means provided to assure the control circuit can also be disconnected and locked out?

## Welding, Cutting & Brazing

☐ Are only authorized and trained person-nel permitted to use welding, cutting or brazing equipment?

☐ Do all operators have a copy of the appro-priate operating instructions and are they directed to follow them?

☐ Are compressed gas cylinders regularly examined for obvious signs of defects, deep rusting or leakage?

☐ Is care used in handling and storage of cylinders, safety valves, relief valves and the like, to prevent damage?

☐ Are precautions taken to prevent mixture of air or oxygen with flammable gases, ex-cept at a burner or in a standard torch?

☐ Are only approved apparatus (torches, regulators, pressure-reducing valves, acetylene generators, manifolds) used?

☐ Are cylinders kept away from sources of heat?

☐ Is it prohibited to use cylinders as rollers or supports?

☐ Are empty cylinders appropriately marked, their valves closed and valve-protection caps on?

☐ Are signs reading: **DANGER—NO SMOKING, MATCHES, OR OPEN LIGHTS,** or the equivalent, posted?

☐ Are cylinders, cylinder valves, couplings, regulators, hoses and apparatus kept free of oily or greasy substances?

☐ Is care taken not to drop or strike cylinders?

☐ Unless secured on special trucks, are regulators removed and valve-protection caps put in place before moving cylinders?

☐ Do cylinders without fixed hand wheels have keys, handles, or non-adjustable wrenches on stem valves when in service?

☐ Are liquefied gases stored and shipped valve-end up with valve covers in place?

☐ Are employees instructed to never crack a fuel-gas cylinder valve near sources of ignition?

☐ Before a regulator is removed, is the valve closed and gas released from the regulator?

☐ Is red used to identify the acetylene (and other fuel-gas) hose, green for oxygen hose, and black for inert gas and air hose?

☐ Are pressure-reducing regulators used only for the gas and pressures for which they are intended?

☐ Is open circuit (No Load) voltage of arc welding and cutting machines as low as possible and not in excess of the recommended limits?

☐ Under wet conditions, are automatic controls for reducing no-load voltage used?

☐ Is grounding of the machine frame and safety ground connections of portable machines checked periodically?

☐ Are electrodes removed from the holders when not in use?

☐ Is it required that electric power to the welder be shut off when no one is in attendance?

☐ Is suitable fire extinguishing equipment available for immediate use?

☐ Is the welder forbidden to coil or loop welding electrode cable around his/her body?

☐ Are wet welding machines thoroughly dried and tested before being used?

☐ Are work and electrode lead cables frequently inspected for wear and damage, and replaced when needed?

☐ Do means for connecting cables' lengths have adequate insulation?

☐ When the object to be welded cannot be moved and fire hazards cannot be removed, are shields used to confine heat, sparks and slag?

☐ Are fire watchers assigned when welding or cutting is performed, in locations where a serious fire might develop?

☐ Are combustible floors kept wet, covered by damp sand, or protected by fire-resistant shields?

☐ When floors are wet down, are personnel protected from possible electrical shock?

☐ When welding is done on metal walls, are precautions taken to protect combustibles on the other side?

☐ Before hot work is begun, are used drums, barrels, tanks and other containers so thoroughly cleaned that no substances remain that could explode, ignite or produce toxic vapors?

☐ Is it required that eye protection helmets, hand shields and goggles meet appropriate standards?

☐ Are employees exposed to the hazards created by welding, cutting or brazing operations protected with personal protective equipment and clothing?

☐ Is a check made for adequate ventilation in and where welding or cutting is performed?

☐ When working in confined spaces are environmental monitoring tests taken and means provided for quick removal of welders in case of an emergency?

## Compressors & Compressed Air

☐ Are compressors equipped with pressure relief valves and pressure gauges?

☐ Are compressor air intakes installed and equipped to ensure that only clean, uncontaminated air enters the compressor?

☐ Are air filters installed on the compressor intake?

☐ Are compressors operated and lubricated in accordance with the manufacturer's recommendations?

☐ Are safety devices on compressed air systems checked frequently?

☐ Before any repair work is done on the pressure system of a compressor, is the pressure bled off and the system locked out?

☐ Are signs posted to warn of the automatic starting feature of the compressors?

☐ Is the belt drive system totally enclosed to provide protection for the front, back, top and sides?

☐ Is it strictly prohibited to direct compressed air towards a person?

☐ Are employees prohibited from using highly compressed air for cleaning purposes?

☐ If compressed air is used for cleaning off clothing, is the pressure reduced to less than 10 psi?

☐ When using compressed air for cleaning, do employees use personal protective equipment?

☐ Are safety chains or other suitable locking devices used at couplings of high pressure hose lines where a connection failure would create a hazard?

☐ Before compressed air is used to empty containers of liquid, is the safe working pressure of the container checked?

☐ When compressed air is used with abrasive blast cleaning equipment, is the operating valve a type that must be held open manually?

☐ When compressed air is used to inflate auto tires, is a clip-on chuck and an inline regulator preset to 40 psi required?

☐ Is it prohibited to use compressed air to clean up or move combustible dust if such action could cause the dust to be suspended in the air and cause a fire or explosion hazard?

## Compressed Air Receivers

☐ Is every receiver equipped with a pressure gauge and with one or more automatic, spring-loaded safety valves?

☐ Is the total relieving capacity of the safety valve capable of preventing pressure in the receiver from exceeding the maximum allowable working pressure of the receiver by more than 10 percent?

☐ Is every air receiver provided with a drain pipe and valve at the lowest point for the removal of accumulated oil and water?

☐ Are compressed air receivers periodically drained of moisture and oil?

☐ Are all safety valves tested frequently and at regular intervals to determine whether they are in good operating condition?

☐ Is there a current operating permit issued by the Division of Occupational Safety and Health?

☐ Is the inlet of air receivers and piping systems kept free of accumulated oil and carbonaceous materials?

## Compressed Gas & Cylinders

☐ Are cylinders with a water weight capacity over 30 pounds equipped with means for connecting a valve protector device, or with a collar or recess to protect the valve?

☐ Are cylinders legibly marked to clearly identify the gas contained?

☐ Are compressed gas cylinders stored in areas which are protected from external heat sources such as flame impingement, intense radiant heat, electric arcs or high temperature lines?

☐ Are cylinders located or stored in areas where they will not be damaged by passing or falling objects, or subject to tampering by unauthorized persons?

☐ Are cylinders stored or transported in a manner to prevent them creating a hazard by tipping, falling or rolling?

☐ Are cylinders containing liquefied fuel gas stored or transported in a position so that the safety relief device is always in direct contact with the vapor space in the cylinder?

☐ Are valve protectors always placed on cylinders when the cylinders are not in use or connected for use?

☐ Are all valves closed off before a cylinder is moved, when the cylinder is empty, and at the completion of each job?

☐ Are low pressure fuel-gas cylinders checked periodically for corrosion, general distortion, cracks, or any other defect that might indicate a weakness or render them unfit for service?

☐ Does the periodic check of low pressure fuel-gas cylinders include a close inspection of the cylinder's bottom?

## Hoist & Auxiliary Equipment

☐ Is each overhead electric hoist equipped with a limit device to stop the hook travel at its highest and lowest points of safe travel?

☐ Will each hoist automatically stop and hold any load up to 125 percent of its rated load, if its actuating force is removed?

☐ Is the rated load of each hoist legibly marked and visible to the operator?

☐ Are stops provided at the safe limits of travel for trolley hoists?

☐ Are the controls of hoists plainly marked to indicate direction of travel or motion?

☐ Is each cage-controlled hoist equipped with an effective warning device?

☐ Are close-fitting guards or other suitable devices installed on hoists to assure hoist ropes will be maintained in the sheave groves?

☐ Are all hoist chains or ropes of sufficient length to handle the full range of movement for the application, while maintaining two full wraps on the drum at all times?

☐ Are nip points or contact points between hoist ropes and sheaves which are permanently located within 7 feet of the floor, ground or working platform, guarded?

☐ Is it prohibited to use chains or rope slings that are kinked or twisted?

☐ Is it prohibited to use the hoist rope or chain wrapped around the load as a substitute for a sling?

☐ Is the operator instructed to avoid carrying loads over people?

☐ Are only employees who have been trained in the proper use of hoists allowed to operate them?

## Industrial Trucks–Forklifts

☐ Are only trained personnel allowed to operate industrial trucks?

☐ Is substantial overhead protective equipment provided on high lift rider equipment?

☐ Are the required lift truck operating rules posted and enforced?

☐ Is directional lighting provided on each industrial truck that operates in an area with less than 2 foot candles per square foot of general lighting?

☐ Does each industrial truck have a warning horn, whistle, gong or other device which can be clearly heard above the normal noise in the area where operated?

☐ Are the brakes on each industrial truck capable of bringing the vehicle to a complete and safe stop when fully loaded?

☐ Will the industrial truck's parking brake effectively prevent the vehicle from moving when unattended?

☐ Are industrial trucks operating in areas where flammable gases or vapors, combustible dust or ignitable fibers may be present in the atmosphere, approved for such locations?

☐ Are motorized hand and hand/rider trucks so designed that the brakes are applied and power to the drive motor shuts off when the operator releases his/her grip on the device that controls the travel?

☐ Are industrial trucks with internal combustion engines, operated in buildings or enclosed areas, carefully checked to ensure such operations do not cause harmful concentration of dangerous gases or fumes?

## Spraying Operations

☐ Is adequate ventilation assured before spray operations are started?

☐ Is mechanical ventilation provided when spraying operation is done in enclosed areas?

☐ When mechanical ventilation is provided during spraying operations, is it arranged so that it will not circulate the contaminated air?

☐ Is the spray area free of hot surfaces?

☐ Is the spray area at least 20 feet from flames, sparks, operating electrical motors and other ignition sources?

☐ Are portable lamps used to illuminate spray areas suitable for use in a hazardous location?

☐ Is approved respiratory equipment provided and used when appropriate during spraying operations?

☐ Do solvents used for cleaning have a flash point of 100° F or more?

☐ Are fire control sprinkler heads kept clean?

☐ Are "NO SMOKING" signs posted in spray areas, paint rooms, paint booths and paint storage areas?

☐ Is the spray area kept clean of combustible residue?

☐ Are spray booths constructed of metal, masonry or other substantial noncombustible material?

☐ Are spray booth floors and baffles noncombustible and easily cleaned?

☐ Is infrared drying apparatus kept out of the spray area during spraying operations?

☐ Is the spray booth completely ventilated before using the drying apparatus?

☐ Is the electric drying apparatus properly grounded?

☐ Are lighting fixtures for spray booths located outside of the booth, and the interior lighted through sealed clear panels?

☐ Are the electric motors for exhaust fans placed outside booths or ducts?

☐ Are belts and pulleys inside the booth fully enclosed?

☐ Do ducts have access doors to allow cleaning?

☐ Do all drying spaces have adequate ventilation?

## Entering Confined Spaces

☐ Are confined spaces thoroughly emptied of any corrosive or hazardous substances, such as acids or caustics, before entry?

☐ Before entry, are all lines to a confined space, containing inert, toxic, flammable, or corrosive materials, valved off and blanked or disconnected and separated?

☐ Is it required that all impellers, agitators, or other moving equipment inside confined spaces be locked-out if they present a hazard?

☐ Is either natural or mechanical ventilation provided prior to confined space entry?

☐ Before entry, are appropriate atmospheric tests performed to check for oxygen deficiency, toxic substances and explosive concentrations in the confined space?

☐ Is adequate illumination provided for the work to be performed in the confined space?

☐ Is the atmosphere inside the confined space frequently tested or continuously monitored during conduct of work?

☐ Is there an assigned safety standby employee outside of the confined space, whose sole responsibility is to watch the work in progress, sound an alarm if necessary, and help render assistance?

☐ Is the standby employee, or other employees, prohibited from entering the confined space without lifelines and respiratory equipment, if there is any question as to the cause of any emergency?

☐ In addition to the standby employee, is there at least one other trained rescuer in the vicinity?

☐ Are all rescuers appropriately trained and using approved, recently inspected equipment?

☐ Does all rescue equipment allow for lifting employees vertically from a top opening?

☐ Are there trained personnel in First Aid and CPR immediately available?

☐ Is there an effective communication system in place whenever respiratory equipment is used and the employee in the confined space is out of sight of the standby person?

☐ Is approved respiratory equipment required if the atmosphere inside the confined space cannot be made acceptable?

☐ Is all portable electrical equipment used inside confined spaces either grounded and insulated, or equipped with ground fault protection?

☐ Before gas welding or burning is started in a confined space, are hoses checked for leaks, compressed gas bottles forbidden inside the confined space, torches lighted only outside the confined area, and the confined area tested for an explosive atmosphere each time before a lighted torch is to be taken into the confined space?

☐ If employees will be using oxygen-consuming equipment—such as salamanders, torches, furnaces—in a confined space, is sufficient air provided to assure combustion without reducing the oxygen concentration of the atmosphere below 19.5 percent by volume?

☐ Whenever combustion-type equipment is used in a confined space, are provisions made to ensure that the exhaust gases are vented outside of the enclosure?

☐ Is each confined space checked for decaying vegetation or animal matter which may produce methane?

☐ Is the confined space checked for possible industrial waste which could contain toxic properties?

☐ If the confined space is below the ground and near areas where motor vehicles are operating, is it possible for vehicle exhaust or carbon monoxide to enter the space?

## Environmental Controls

☐ Are all work areas properly illuminated?

☐ Are employees instructed in proper first aid and other emergency procedures?

☐ Are hazardous substances identified which may cause harm by inhalation, ingestion, skin absorption or contact?

☐ Are employees aware of the hazards involved with the various chemicals they may be exposed to in their work environment, such as ammonia, chlorine, epoxies, caustics?

☐ Is employee exposure to chemicals in the workplace kept within acceptable levels?

☐ Can a less harmful method or product be used?

☐ Is the work area's ventilation system appropriate for work being performed?

☐ Are spray painting operations done in spray rooms or booths equipped with an appropriate exhaust system?

☐ Is employee exposure to welding fumes controlled by ventilation, use of respirators, exposure time, or other means?

☐ Are welders and other workers nearby provided with flash shields during welding operations?

☐ If forklifts and other vehicles are used in buildings or other enclosed areas, are carbon monoxide levels kept below maximum acceptable concentration?

☐ Has there been a determination that noise levels in the facilities are within acceptable levels?

☐ Are steps being taken to use engineering controls to reduce excessive noise levels?

☐ Are proper precautions being taken when handling asbestos and other fibrous materials?

☐ Are caution labels and signs used to warn of asbestos?

☐ Are wet methods used, when practicable, to prevent emission of airborne asbestos fibers, silica dust and similar hazardous materials?

☐ Is vacuuming with appropriate equipment used whenever possible, rather than blowing or sweeping dust?

☐ Are grinders, saws and other machines that produce respirable dusts vented to an industrial collector or central exhaust system?

☐ Are all local exhaust ventilation systems designed and operating properly at the airflow and volume necessary for the application? Are the ducts free of obstructions or the belts slipping?

☐ Is personal protective equipment provided, used and maintained wherever required?

☐ Are there written standard operating procedures for the selection and use of respirators where needed?

☐ Are restrooms and washrooms kept clean and sanitary?

☐ Is all water provided for drinking, washing and cooking potable?

☐ Are all outlets for water not suitable for drinking clearly identified?

☐ Are employees' physical capacities assessed before being assigned to jobs requiring heavy work?

☐ Are employees instructed in the proper manner of lifting heavy objects?

☐ Where heat is a problem, have all fixed work areas been provided with spot cooling or air conditioning?

☐ Are employees screened before assignment to areas of high heat to determine if their health condition might make them more susceptible to having an adverse reaction?

☐ Are employees working on streets and roadways, where they are exposed to the hazards of traffic, required to wear a bright colored (traffic orange) warning vest?

☐ Are exhaust stacks and air intakes located so that contaminated air will not be recirculated within a building or other enclosed area?

☐ Is equipment producing ultra-violet radiation properly shielded?

# Flammable & Combustible Materials

☐ Are combustible scrap, debris and waste materials (i.e. oily rags) stored in covered metal receptacles and removed from the worksite promptly?

☐ Is proper storage practiced to minimize risks of fire and spontaneous combustion?

☐ Are approved containers and tanks used for the storage and handling of flammable and combustible liquids?

☐ Are all connections on drums and combustible liquid piping, vapor and liquid tight?

☐ Are all flammable liquids kept in closed containers when not in use (e.g. parts cleaning tanks, pans)?

☐ Are bulk drums of flammable liquids grounded and bonded to containers during dispensing?

☐ Do storage rooms for flammable and combustible liquids have explosion-proof lights?

☐ Do storage rooms for flammable and combustible liquids have mechanical or gravity ventilation?

☐ Is liquefied petroleum gas stored, handled and used in accordance with safe practices and standards?

☐ Are liquefied petroleum storage tanks guarded to prevent damage from vehicles?

☐ Are all solvent wastes and flammable liquids kept in fire-resistant, covered containers until they are removed from the worksite?

☐ Is vacuuming used whenever possible, rather than blowing or sweeping combustible dust?

☐ Are fire separators placed between containers of combustibles or flammables, when stacked one upon another, to assure their support and stability?

☐ Are fuel gas cylinders and oxygen cylinders separated by distance, fire resistant barriers or other means while in storage?

☐ Are fire extinguishers selected and provided for the types of materials, in areas where they are to be used?
**Class A:** Ordinary combustible material fires.
**Class B:** Flammable liquid, gas or grease fires.
**Class C:** Energized-electrical equipment fires.

☐ If a Halon 1301 fire extinguisher is used, can employees evacuate within the specified time for that extinguisher?

☐ Are appropriate fire extinguishers mounted within 75 feet of outside areas containing flammable liquids, and within 10 feet of any inside storage area for such materials?

☐ Is the transfer/withdrawal of flammable or combustible liquids performed by trained personnel?

☐ Are fire extinguishers mounted so that employees do not have to travel more than 75 feet for a class "A" fire or 50 feet for a class "B" fire?

☐ Are employees trained in the use of fire extinguishers?

☐ Are extinguishers free from obstructions or blockage?

☐ Are all extinguishers serviced, maintained and tagged at intervals not to exceed one year?

☐ Are all extinguishers fully charged and in their designated places?

☐ Is a record maintained of required monthly checks of extinguishers?

☐ Where sprinkler systems are permanently installed, are the nozzle heads directed or arranged so that water will not be sprayed into operating electrical switch boards and equipment?

☐ Are "NO SMOKING" signs posted where appropriate in areas where flammable or combustible materials are used or stored?

☐ Are "NO SMOKING" signs posted on liquefied petroleum gas tanks?

☐ Are "NO SMOKING" rules enforced in areas involving storage and use of flammable materials?

☐ Are safety cans used for dispensing flammable or combustible liquids at a point of use?

☐ Are all spills of flammable or combustible liquids cleaned up promptly?

☐ Are storage tanks adequately vented to prevent development of excessive vacuum or pressure as a result of filling, emptying, or atmosphere temperature changes?

☐ Are storage tanks equipped with emergency venting that will relieve excessive internal pressure caused by fire exposure?

☐ Are spare portable or butane tanks which are used by industrial trucks stored in accord with regulations?

# Hazardous
# Chemical Exposures

☐ Are employees trained in the safe handling practices of hazardous chemicals such as acids, caustics, and the like?

☐ Are employees aware of the potential hazards involving various chemicals stored or used in the workplace—such as acids, bases, caustics, epoxies, phenols?

☐ Is employee exposure to chemicals kept within acceptable levels?

☐ Are eye wash fountains and safety showers provided in areas where corrosive chemicals are handled?

☐ Are all containers such as vats and storage tanks labeled as to their contents—e.g. "CAUSTICS"?

☐ Are all employees required to use personal protective clothing and equipment when handling chemicals (i.e. gloves, eye protection, respirators)?

☐ Are flammable or toxic chemicals kept in closed containers when not in use?

☐ Are chemical piping systems clearly marked as to their content?

☐ Where corrosive liquids are frequently handled in open containers or drawn from storage vessels or pipe lines, is adequate means readily available for neutralizing or disposing of spills or overflows properly and safely?

☐ Have standard operating procedures been established and are they being followed when cleaning up chemical spills?

☐ Where needed for emergency use, are respirators stored in a convenient, clean and sanitary location?

☐ Are respirators intended for emergency use adequate for the various uses for which they may be needed?

☐ Are employees prohibited from eating in areas where hazardous chemicals are present?

☐ Is personal protective equipment provided, used and maintained whenever necessary?

☐ Are there written standard operating procedures for the selection and use of respirators where needed?

☐ If you have a respirator protection program, are your employees instructed on the correct usage and limitations of the respirators?

☐ Are the respirators NIOSH approved for this particular application?

☐ Are they regularly inspected and cleaned, sanitized and maintained?

☐ If hazardous substances are used in your processes, do you have a medical or biological monitoring system in operation?

☐ Are you familiar with the Threshold Limit Values or Permissible Exposure Limits of airborne contaminants and physical agents used in your workplace?

☐ Have control procedures been instituted for hazardous materials, where appropriate, such as respirators, ventilation systems, handling practices, and the like?

☐ Whenever possible, are hazardous substances handled in properly designed and exhausted booths or similar locations?

☐ Do you use general dilution or local exhaust ventilation systems to control dusts, vapors, gases, fumes, smoke, solvents or mists which may be generated in your workplace?

☐ Is ventilation equipment provided for removal of contaminants from such operations as production grinding, buffing, spray painting, and/or vapor degreasing, and is it operating properly?

☐ Do employees complain about dizziness, headaches, nausea, irritation or other factors of discomfort when they use solvents or other chemicals?

☐ Is there a dermatitis problem—do employees complain about skin dryness, irritation, or sensitization?

☐ Have you considered the use of an industrial hygienist or environmental health specialist to evaluate your operation?

☐ If internal combustion engines are used, is carbon monoxide kept within acceptable levels?

☐ Is vacuuming used, rather than blowing or sweeping dusts, whenever possible for clean-up?

☐ Are materials which give off toxic asphyxiant, suffocating or anethetic fumes, stored in remote or isolated locations when not in use?

## Hazardous Substances Communication

☐ Is there a list of hazardous substances used in your workplace?

☐ Is there a written hazard communication program dealing with Material Safety Data Sheets (MSDS), labeling and employee training?

☐ Who is responsible for MSDSs, container labeling, employee training?

☐ Is each container for a hazardous substance (i.e. vats, bottles, storage tanks) labeled with product identity and a hazard warning (communication of the specific health hazards and physical hazards)?

☐ Is there a Material Safety Data Sheet readily available for each hazardous substance used?

☐ How will you inform other employers whose employees share the same work area where the hazardous substances are used?

☐ Is there an employee training program for hazardous substances?

Does this program include:

☐ An explanation of what an MSDS is and how to use and obtain one?

☐ MSDS contents for each hazardous substance or class of substances?

☐ Explanation of "Right to Know"?

☐ Identification of where employees can see the employer's written hazard communication program and where hazardous substances are present in their work area?

☐ The physical and health hazards of substances in the work area, how to detect their presence, and specific protective measures to be used?

☐ Details of the hazard communications program, including how to use the labeling system and MSDSs?

☐ How employees will be informed of hazards of non-routine tasks, and hazards of unlabeled pipes?

## Electrical

☐ Are your workplace electricians familiar with the Cal/OSHA Electrical Safety Orders?

☐ Do you specify compliance with Cal/OSHA for all contract electrical work?

☐ Are all employees required to report as soon as practicable any obvious hazard to life or property observed in connection with electrical equipment or lines?

☐ Are employees instructed to make preliminary inspections and/or appropriate tests to determine what conditions exist before starting work on electrical equipment or lines?

☐ When electrical equipment or lines are to be serviced, maintained or adjusted, are necessary switches opened, locked out and tagged whenever possible?

☐ Are portable electrical tools and equipment grounded or of the double insulated type?

☐ Are electrical appliances such as vacuum cleaners, polishers, vending machines grounded?

☐ Do extension cords being used have a grounding conductor?

☐ Are multiple plug adaptors prohibited?

☐ Are ground-fault circuit interrupters installed on each temporary 15 or 20 ampere, 120 volt AC circuit at locations where construction, demolition, modifications, alterations or excavations are being performed?

☐ Are all temporary circuits protected by suitable disconnecting switches or plug connectors at the junction with permanent wiring?

☐ Is exposed wiring and cords with frayed or deteriorated insulation repaired or replaced promptly?

☐ Are flexible cords and cables free of splices or taps?

☐ Are clamps or other securing means provided on flexible cords or cables at plugs, receptacles, tools, equipment, and is the cord jacket securely held in place?

☐ Are all cord, cable and raceway connections intact and secure?

☐ In wet or damp locations, are electrical tools and equipment appropriate for the use, or location, or otherwise protected?

☐ Is the location of electrical power lines and cables (overhead, underground, underfloor, other side of walls) determined before digging, drilling or similar work is begun?

☐ Are metal measuring tapes, ropes, handlines or similar devices with metallic thread woven into the fabric prohibited where they could come in contact with energized parts of equipment or circuit conductors?

☐ Is the use of metal ladders prohibited in areas where the ladder or the person using the ladder could come in contact with energized parts of equipment, fixtures or circuit conductors?

☐ Are all disconnecting switches and circuit breakers labeled to indicate their use or equipment served?

☐ Are disconnecting means always opened before fuses are replaced?

☐ Do all interior wiring systems include provisions for grounding metal parts or electrical raceways, equipment and enclosures?

☐ Are all electrical raceways and enclosures securely fastened in place?

☐ Are all energized parts of electrical circuits and equipment guarded against accidental contact by approved cabinets or enclosures?

☐ Is sufficient access and working space provided and maintained about all electrical equipment to permit ready and safe operations and maintenance?

☐ Are all unused openings (including conduit knockouts) in electrical enclosures and fittings closed with appropriate covers, plugs or plates?

☐ Are electrical enclosures such as switches, receptacles, junction boxes provided with tight-fitting covers or plates?

☐ Are disconnecting switches for electrical motors in excess of two horsepower capable of opening the circuit when the motor is in a stalled condition without exploding? (Switches must be horsepower rated equal to or in excess of the motor hp rating.)

☐ Is low voltage protection provided in the control device of motors driving machines or equipment which could cause probable injury from inadvertent starting?

☐ Is each motor disconnecting switch or circuit breaker located within sight of the motor control device?

☐ Is each motor located within sight of its controller or the controller disconnecting means capable of being locked in the open position, or is separate disconnecting means installed in the circuit within sight of the motor?

☐ Is the controller for each motor in excess of two horsepower rated in horsepower equal to or in excess of the rating of the motor it serves?

☐ Are employees who regularly work on or around energized electrical equipment or lines instructed in cardio-pulmonary resuscitation (CPR) methods?

☐ Are employees prohibited from working alone on energized lines or equipment over 600 volts?

## Noise

☐ Are there areas in the workplace where continuous noise levels exceed 85 dBA? (To determine maximum allowable levels for intermittent or impact noise, see California Code of Regulations Title 8, Section 5097.)

☐ Are noise levels being measured using a sound level meter or an octave band analyzer, and records being kept?

☐ Have you tried isolating noisy machinery from the rest of your operation?

☐ Have engineering controls been used to reduce excessive noise levels?

☐ Where engineering controls are determined not feasible, are administrative controls (i.e. worker rotation) being used to minimize individual employee exposure to noise?

☐ Is there an ongoing preventive health program to educate employees in safe levels of noise and exposure, effects of noise on their health, and use of personal protection?

☐ Is the training repeated annually for employees exposed to continuous noise above 85 dBA?

☐ Have work areas where noise levels make voice communication between employees difficult been identified and posted?

☐ Is approved hearing protective equipment (noise attenuating devices) available to every employee working in areas where continuous noise levels exceed 85 dBA?

☐ If you use ear protectors, are employees properly fitted and instructed in their use and care?

☐ Are employees exposed to continuous noise above 85 dBA given periodic audiometric testing to ensure that you have an effective hearing protection system?

## Fueling

☐ Is it prohibited to fuel an internal combustion engine with a flammable liquid while the engine is running?

☐ Are fueling operations done in such a manner that likelihood of spillage will be minimal?

☐ When spillage occurs during fueling operations, is the spilled fuel cleaned up completely, evaporated, or other measures taken to control vapors before restarting the engine?

☐ Are fuel tank caps replaced and secured before starting the engine?

☐ In fueling operations is there always metal contact between the container and the fuel tank?

☐ Are fueling hoses of a type designed to handle the specific type of fuel?

☐ Is it prohibited to handle or transfer gasoline in open containers?

☐ Are open lights, open flames, or sparking or arcing equipment prohibited near fueling or transfer of fuel operations?

☐ Is smoking prohibited in the vicinity of fueling operations?

☐ Are fueling operations prohibited in building or other enclosed areas that are not specifically ventilated for this purpose?

☐ Where fueling or transfer of fuel is done through a gravity flow system, are the nozzles of the self-closing type?

## Identification of Piping Systems

☐ When non-potable water is piped through a facility, are outlets or taps posted to alert employees it is unsafe and not to be used for drinking, washing or personal use?

☐ When hazardous substances are transported through above-ground piping, is each pipeline identified at points where confusion could introduce hazards to employees?

☐ When pipelines are identified by color painting, are all visible parts of the line so identified?

☐ When pipelines are identified by color painted bands or tapes, are the bands or tapes located at reasonable intervals, and at each outlet, valve or connection?

☐ When pipelines are identified by color, is the color code posted at all locations where confusion could introduce hazards to employees?

☐ When the contents of pipelines are identified by name or name abbreviation, is the information readily visible on the pipe near each valve or outlet?

☐ When pipelines carrying hazardous substances are identified by tags, are the tags constructed of durable materials, the message carried clearly and permanently distinguishable, and tags installed at each valve or outlet?

☐ When pipelines are heated by electricity, steam or other external source, are suitable warning signs or tags placed at unions, valves, or other serviceable parts of the system?

## Material Handling

☐ Is there safe clearance for equipment through aisles and doorways?

☐ Are aisleways designated, permanently marked, and kept clear to allow unhindered passage?

☐ Are motorized vehicles and mechanized equipment inspected daily or prior to use?

☐ Are vehicles shut off and brakes set prior to loading or unloading?

☐ Are containers of combustibles or flammables, when stacked while being moved, always separated by dunnage sufficient to provide stability?

☐ Are dock boards (bridge plates) used when loading or unloading operations are taking place between vehicles and docks?

☐ Are trucks and trailers secured from movement during loading and unloading operations?

☐ Are dock plates and loading ramps constructed and maintained with sufficient strength to support imposed loading?

☐ Are hand trucks maintained in safe operating condition?

☐ Are chutes equipped with sideboards of sufficient height to prevent the materials being handled from falling off?

☐ Are chutes and gravity roller sections firmly placed or secured to prevent displacement?

☐ At the delivery end of rollers or chutes, are provisions made to brake the movement of the handled materials?

☐ Are pallets usually inspected before being loaded or moved?

☐ Are hooks with safety latches or other arrangements used when hoisting materials, so that slings or load attachments won't accidentally slip off the hoist hooks?

☐ Are securing chains, ropes, chockers or slings adequate for the job to be performed?

☐ When hoisting material or equipment, are provisions made to assure no one will be passing under the suspended loads?

☐ Are Material Safety Data Sheets available to employees handling hazardous substances?

## Transporting Employees & Materials

☐ Do employees who operate vehicles on public thoroughfares have operator's licenses?

☐ When seven or more employees are regularly transported in a van, bus or truck, is the operator's license appropriate for the class of vehicle being driven?

☐ Is each van, bus or truck used regularly to transport employees equipped with an adequate number of seats?

☐ When employees are transported by truck, are provisions provided to prevent their falling from the vehicle?

☐ Are vehicles used to transport employees equipped with lamps, brakes, horns, mirrors, windshields and turn signals in good repair?

☐ Are transport vehicles provided with handrails, steps, stirrups or similar devices, so placed and arranged that employees can safely mount or dismount?

☐ Are employee transport vehicles equipped at all times with at least two reflective-type flares?

☐ Is a fully-charged fire extinguisher, in good condition, with at least 4 B:C rating maintained in each employee transport vehicle?

☐ When cutting tools with sharp edges are carried in passenger compartments of employee transport vehicles, are they placed in closed boxes or containers which are secured in place?

☐ Are employees prohibited from riding on top of any load which can shift, topple, or otherwise become unstable?

## Control of Harmful Substances by Ventilation

☐ Is the volume and velocity of air in each exhaust system sufficient to gather the dusts, fumes, mists, vapors or gases to be controlled, and to convey them to a suitable point of disposal?

☐ Are exhaust inlets, ducts and plenums designed, constructed, and supported to prevent collapse or failure of any part of the system?

☐ Are clean-out ports or doors provided at intervals not to exceed 12 feet in all horizontal runs of exhaust ducts?

☐ Where two or more different types of operations are being controlled through the same exhaust system, will the combination of substances being controlled constitute a fire, explosion or chemical reaction hazard in the duct?

☐ Is adequate makeup air provided to areas where exhaust systems are operating?

☐ Is the intake for makeup air located so that only clean, fresh air, which is free of contaminates, will enter the work environment?

☐ Where two or more ventilation systems are serving a work area, is their operation such that one will not offset the functions of the other?

## Sanitizing Equipment & Clothing

☐ Is personal protective clothing or equipment, that employees are required to wear or use, of a type capable of being easily cleaned and disinfected?

☐ Are employees prohibited from interchanging personal protective clothing or equipment, unless it has been properly cleaned?

☐ Are machines and equipment, which process, handle or apply materials that could be injurious to employees, cleaned and/or decontaminated before being overhauled or placed in storage?

☐ Are employees prohibited from smoking or eating in any area where contaminates are present that could be injurious if ingested?

☐ When employees are required to change from street clothing into protective clothing, is a clean change room with separate storage facility for street and protective clothing provided?

☐ Are employees required to shower and wash their hair as soon as possible after known contact has occurred with a carcinogen?

☐ When equipment, materials or other items are taken into or removed from a carcinogen-regulated area, is it done in a manner that will not contaminate non-regulated areas or the external environment?

## Tire Inflation

☐ Where tires are mounted and/or inflated on drop center wheels, is a safe practice procedure posted and enforced?

☐ Where tires are mounted and/or inflated on wheels with split rims and/or retainer rings, is a safe practice procedure posted and enforced?

☐ Does each tire inflation hose have a clip-on chuck with at least 24 inches of hose between the chuck and an in-line hand valve and gauge?

☐ Does the tire inflation control valve automatically shut off the air flow when the valve is released?

☐ Is a tire restraining device such as a cage, rack or other effective means used while inflating tires mounted on split rims, or rims using retainer rings?

☐ Are employees strictly forbidden from taking a position directly over or in front of a tire while it's being inflated?

## Emergency Action Plan

☐ Are you required to have an emergency action plan?

☐ Does the emergency action plan comply with requirements of T8 CCR 3220 (a)?

☐ Have emergency escape procedures and routes been developed and communicated to all employers?

☐ Do employees, who remain to operate critical plant operations before they evacuate, know the proper procedures?

☐ Is the employee alarm system that provides a warning for emergency action recognizable and perceptible above ambient conditions?

☐ Are alarm systems properly maintained and tested regularly?

☐ Is the emergency action plan reviewed and revised periodically?

Do employees know their responsibilities:

☐ For reporting emergencies?

☐ During an emergency?

☐ For conducting rescue and medical duties?

## Infection Control

☐ Are employees potentially exposed to infectious agents in body fluids?

☐ Have occasions of potential occupational exposure been identified and documented?

☐ Has a training and information program been provided for employees exposed to or potentially exposed to blood and/or body fluids?

☐ Have infection control procedures been instituted where appropriate, such as ventilation, universal precautions, workplace practices, personal protective equipment?

☐ Are employees aware of specific workplace practices to follow when appropriate? (Hand washing, handling sharp instruments, handling of laundry, disposal of contaminated materials, reusable equipment.)

☐ Is personal protective equipment provided to employees, and in all appropriate locations?

☐ Is the necessary equipment (i.e. mouthpieces, resuscitation bags, other ventilation devices) provided for administering mouth-to-mouth resuscitation on potentially infected patients?

☐ Are facilities/equipment to comply with workplace practices available, such as hand-washing sinks, biohazard tags and labels, needle containers, detergents/disinfectants to clean up spills?

☐ Are all equipment and environmental and working surfaces cleaned and disinfected after contact with blood or potentially infectious materials?

☐ Is infectious waste placed in closable, leak proof containers, bags or puncture-resistant holders with proper labels?

☐ Has medical surveillance including HBV evaluation, antibody testing and vaccination been made available to potentially exposed employees?

How often is training done and does it cover:

☐ Universal precautions?

☐ Personal protective equipment?

☐ Workplace practices which should include blood drawing, room cleaning, laundry handling, clean-up of blood spills?

☐ Needlestick exposure/management?

☐ Hepatitis B vaccination?

## Ergonomics

☐ Can the work be performed without eye strain or glare to the employees?

☐ Does the task require prolonged raising of the arms?

☐ Do the neck and shoulders have to be stooped to view the task?

☐ Are there pressure points on any parts of the body (wrists, forearms, back of thighs)?

☐ Can the work be done using the larger muscles of the body?

☐ Can the work be done without twisting or overly bending the lower back?

☐ Are there sufficient rest breaks, in addition to the regular rest breaks, to relieve stress from repetitive-motion tasks?

☐ Are tools, instruments and machinery shaped, positioned and handled so that tasks can be performed comfortably?

☐ Are all pieces of furniture adjusted, positioned and arranged to minimize strain on all parts of the body?

## Ventilation for Indoor Air Quality

☐ Does your HVAC system provide at least the quantity of outdoor air required by the State Building Standards Code, Title 24, Part 2 at the time the building was constructed?

☐ Is the HVAC system inspected at least annually, and problems corrected?

☐ Are inspection records retained for at least 5 years?

## Crane Checklist

☐ Are the cranes visually inspected for defective components prior to the beginning of any work shift?

☐ Are all electrically operated cranes effectively grounded?

☐ Is a crane preventive maintenance program established?

☐ Is the load chart clearly visible to the operator?

☐ Are operating controls clearly identified?

☐ Is a fire extinguisher provided at the operator's station?

☐ Is the rated capacity visibly marked on each crane?

☐ Is an audible warning device mounted on each crane?

☐ Is sufficient illumination provided for the operator to perform the work safely?

☐ Are cranes of such design, that the boom could fall over backward, equipped with boomstops?

☐ Does each crane have a certificate indicating that required testing and examinations have been performed?

☐ Are crane inspection and maintenance records maintained and available for inspection?

# Appendix D: Training Requirements in California Code of Regulations Title 8, January 1991

Specific requirements for employee instruction or training are contained in Title 8 of the *California Code of Regulations*, and are listed sequentially here by their subject titles.

## CCR T8 Construction Safety Orders
5101 (a) Safety Instructions for Employees
1510 (c) Hazardous Materials Instruction
1512 (b) Appropriately Trained Person for First Aid
1512 (1) Written Plan (Emergency Medical Service, Appropriately Trained Persons)
1531 (c) Respiratory Protective Equipment
1532 Confined Space (see CCR T8, 5157 General Industry Safety Orders)
1637 (k)(1) Scaffold Erection and Dismantling
1662 (a) Boatswains Chairs
1585 (a)(1) Powder Actuated Tools, Operator and Instructor
1585 (b)(1) Qualifications
1599 (f) Vehicle Traffic Control, Flaggers, Barricades and Warning Signs
1739 (k)(l) Use of Fuel Gas (Liquid Propane)
1801 (a) Ionizing Radiation, High Voltage Electrical Safety Orders
2940 Work Procedures (Inspection of Safety Devices)

## CCR T8 General Industry Safety Orders
3203 (a)(1) Accident Prevention Program
3220 (g) Emergency Action Plan
3221 (d) Fire Prevention Plan
3282 (f) Window Cleaning Operations
3286 (f)(2) Boatswains Chairs
3314 (a) Cleaning, Repairing Servicing and Adjusting Prime Movers, Machinery and Equipment
3326 (c) Servicing Single, Split and Multi Piece Rims or Wheels
3333 (d) Blue Stop Signs (Railcars)
3400 (b) Medical Service and First Aid
3411 (c) Private Fire Brigades

3421 (c)(f) Tree-work Maintenance and Removal
3439 (b) Agricultural Operations, First Aid
3441 (a) Operation of Agricultural Equipment, Operating Instructions Marine Terminal Operations
3463 (b)(5)(A) Respiratory Protective Equipment (reference to 5144)
3464 (a)(1) Accident Prevention and First Aid
3472 (b)(1) Qualification of Machinery Operation
3638 (d) Elevating Work Platform
3648 (1)(7) Aerial Devices (Towering)
3657 (h) Elevating Employees with Lift Trucks
3664 (a)(1) Operating Rules (Industrial Trucks)
4203 (b) Power Press Operation
4243 (a)(6) Forging Machinery and Equipment
4355 (a)(2) Operating Rules for Compaction Equipment
4402 Pulp and Paper Mills
4445 (3) Hand-fed Engraving Press
4494 (a) Operating Rules, Laundry and Dry Cleaning
4799 (a) Training of Operators, Gas Systems and for Welding and Cutting
4848 (a) 21 Fire Prevention and Suppression Procedure
5006 (a) Crane, Hoists, Derrick Operators Qualifications
5099 (a) Control of Noise Exposure
5144 (c) Respiratory Protective Equipment
5154 (j)(1) Open Surface Tank Operations
5157 (b) Confined Spaces
5166 (a) Cleaning, Repairing or Altering Containers
5185 (a) Changing and Charging Storage Batteries
5190 (i) Cotton Dust
5194 (b)(1) Hazard Communication, Employee Information and Training
5208 (h)(1) Asbestos
5209 (d)(5)(j) Carcinogens
5210 (j) Vinyl Chloride

5211 (t)(1) Coke Oven Emissions
5212 (r)(1) 1, 2 Dibromo–3 Chloropropane (DBCP)
5213 (o)(1) Acrylonitrile (AN)
5214 (m)(1) Inorganic Arsenic
5215 (j)(1) 4, 4'–Methylenebis (2-Chloroaniline MBOCA)
5216 (l)(1) Lead
5217 (n) Formaldehyde
5218 (j)(3) Benzene
5219 (j) Ethylene Dibromide
5220 (i)(1) Ethylene oxide (EtO)
5221 (c) Fumigation: General
5229 Protection (Labels)
5239 Handle or Transport Explosives
5322 Manufacture of Explosives and Fireworks
5571 (g) Service Stations (Portable Fire Extinguishers)
6052 (d)(1) Diving Operations (Dive Team Training)

## Fire Protection
6151 (g)(1) Portable Fire Extinguishers
6165 (f)(2)(f) Standpipe and Hose Systems
6175 (a)(10) Fixed Extinguishers Systems

## Logging and Sawmills
6251 (d) First Aid Training

## Petroleum Safety Orders: Drilling and Production
6507 (a) Safety Training and Instruction

## Petroleum Safety Orders: Refining, Transportation, Handling
6760 (A) Safety Training and Instruction

## Mine Safety Orders and Instruction
6963 Safety Training and Instruction
6967 Certification of Safety Representative at Underground Mines
6968 (a)(b)(c) First Aid Training

7074 (d) Emergency Plan (Underground)
7083 (c) Mine Rescue Stations
7085 (a) Mine Rescue Training and Procedure
7150 (d) Qualified Hoistmen (Persons)
7201 Explosives

## Ship Building, Ship Repairing and Ship Breaking Safety Orders
8355 (b) Hazardous Work
8397.1(b) Radioactive Material

## Tunnel Safety Orders
8407 Safety Instructions for New Employees
8421 (a) First Aid Training
8430 (a) Rescue Apparatus (Trained Rescue Crew)
8455 (a) Mechanical Tunneling Methods and Equipment
8499 (b) Hoisting Engineers
8506 Explosives

## Labor Code Part 9: Tunnel and Mine Safety
7952 Trained DOSH Safety Engineers
7958 Trained Rescue Crews
7990 Explosive Blasters Licensed
7999 Gas Testers and Safety Representatives

## Telecommunication Orders
8603 Training

## Elevator Safety Orders
3003 (c) Certified Elevator Inspectors

## Aerial Passenger Tramway Safety Orders
3171.1(c)(g)(1) Operator Experience and Training

# Appendix E: Non-mandatory Checklist Evaluation Injury & Illness Prevention Programs

☐ Does the written injury and illness prevention program contain the elements required by Section 3203(a)?

☐ Is the person or persons with authority and responsibility for implementing the program identified?

☐ Is there a system for ensuring that employees comply with safe and healthy work practices (i.e. employee incentives, training and retraining programs, and/or disciplinary measures)?

☐ Is there a system that provides communication with affected employees on occupational safety and health matters (i.e. meetings, training programs, posting, written communications, a system of anonymous notification concerning hazards and/or health and safety committees)?

☐ Does the communication system include provisions designed to encourage employees to inform the employer of hazards at the worksite without fear of reprisal?

☐ Is there a system for identifying and evaluating workplace hazards whenever new substances, processes, procedures, or equipment are introduced to the workplace and whenever the employer receives notification of a new or previously unrecognized hazard?

☐ Were workplace hazards identified when the program was first established?

☐ Are periodic inspections for safety and health hazards scheduled?

☐ Are records kept of inspections made to identify unsafe conditions and work practices, if required?

☐ Is there an accident and near-miss investigation procedure?

☐ Are unsafe or unhealthy conditions and work practices corrected expeditiously, with the most hazardous exposures given correction priority?

☐ Are employees protected from serious or imminent hazards until they are corrected?

☐ Have employees received training in general safe and healthy work practices?

☐ Do employees know the safety and health hazards specific to their job assignments?

☐ Is training provided for all employees when the training program is first established?

☐ Is training provided to all new employees, and those given new job assignments?

☐ Are training needs of employees evaluated whenever new substances, processes, procedures or equipment are introduced to the workplace and whenever the employer receives notification of a new or previously unrecognized hazard?

☐ Are supervisors knowledgeable of the safety and health hazards to which employees under their immediate direction and control may be exposed?

☐ Are records kept documenting safety and health training for each employee by name or other identifier, training dates, type(s) of training and training providers?

☐ Does the employer have a labor-management safety and health committee?

☐ Does the committee meet at least quarterly?

☐ Is a written record of safety committee meetings distributed to affected employees and maintained for Division review?

☐ Does the committee review results of the periodic, scheduled worksite inspections?

☐ Does the committee review accident and near-miss investigations and, where necessary, submit suggestions for prevention of future incidents?

☐ When determined necessary by the committee does it conduct its own inspections and investigations, to assist in remedial solutions?

☐ Does the committee verify abatement action taken by the employer as specified in Division citations upon request of the Division?

Naturally, housekeeping problems will be many and varied, but for simplicity they may be divided into four major areas:

*Cleanliness* — Includes floors and waste collection in all areas.
*Orderliness* — Includes material and equipment storage, aisles and exits, fire protection, and emergency equipment.
*Appearance* — Includes windows, walls, doors, furniture, signs, and bulletin boards.
**Lighting** — Includes area lighting, emergency lighting, and warning lights.

Two other principles which require emphasis are

1. Area responsibility must be assigned to one person only on a 24-hour basis. This is most important, for the success of the system depends on each area being the responsibility of one person. Thus, any given area will reflect the ability and interest of one person to define and solve the safety and housekeeping problems that may exist in that area. Likewise, if problems do exist, it is clear who is responsible to act on the problems. Supervisors often object to this concept, pointing out that they have no authority over other people messing up their area when they are not at work. Nevertheless, experience has shown that this approach actually helps supervisors to develop as managers because they have to practice leadership abilities in working with others to get their cooperation.
2. Production problems such as spills, piles of scrap, and stacks of parts or equipment are not necessarily housekeeping problems, especially when they are placed in a predictable area for a predictable length of time. Obviously, these require attention, but they are not the kind of problems which supervisors should be concerned with when searching for substandard housekeeping conditions.

### Area Self-Survey

The self-survey approach is designed to teach the inspector to do what managers do — that is, go around and find their own problems. The system helps train inspectors to discover their own problems instead of having them pointed out by their supervisors.

Naturally, supervisors and team leaders will survey their areas frequently, but at some regular interval, such as once a month, they should be required to evidence their surveys in writing. This then is the self-survey, and in order to be an effective training tool it should incorporate the features illustrated in Figure 5-6.

### Source-Cause

As mentioned earlier, one of the most common problems with safety programs is that people tend to treat the symptom rather than get at the root of the problem. The result of this approach is that the problem tends to keep recurring. Here is a typical example of what we mean.

**SAFETY INSPECTION**

| DEPARTMENT | INSPECTION RESPONSIBILITY AREA/EQUIPMENT ASSIGNED | INSPECTED BY | DATE |
|---|---|---|---|

| PROBLEM | | CONTROL ACTION | | FOLLOW UP |
|---|---|---|---|---|
| Describe WHAT is wrong and where it is located | Give reasons WHY it exists (source/cause) | Priority Code | Describe actions taken or planned to control the problem and its source/cause | Code | Name/Date |

| | | | |
|---|---|---|---|

| PROBLEM | Is the problem clearly described and located? Are the reasons why the problem exists identified? | 10 30 |
|---|---|---|
| PRIORITY | Do the priorities reflect an understanding of the potential effects of the problems listed? | 10 |
| CONTROL | Did the action control the problems and the reasons for their existence? | 40 |
| FOLLOW-UP | Does the follow-up specify who will do what, when, to ensure action completed? | 10 |
| TOTAL | | 100 |

REVIEWER'S COMMENTS

Signature: Date:

**PROBLEM PRIORITY CODE**
E Emergency
A 1 week
B 1 Month
C 3 Months
D 6 Months

**FOLLOW-UP CODE**
C Corrected (Date)
S Scheduled (Date)
P Passed to (Name)

DISTRIBUTION:   SAFETY   ORIGINATOR   DEPARTMENT

Figure 5-6   Sample format for safety inspection report.

*Problem* — Tools are scattered on the tool storage floor (tripping hazard).
*Action* — Had employees tidy them up. Next week, the supervisor checks again — same problem! He berates the employees and once again they tidy them up, and for a few weeks the tool store stays reasonably tidy, but before long the problem is back again!

Applying the source-cause approach, the supervisor must investigate "why" the problem exists. He considers what thing might produce the problem (source) or what is the reason why the people do it the wrong way (cause).

Let us now take the above example:

*Problem* — Tools are scattered on the tool store floor.
*Source* — Insufficient storage shelving and racks. No labeling for each tool item.
*Cause* — Employees were not trained to store like items together.
*Action* — Source: Install shelves and racks and label storage areas.
Cause: Train employees how and where to store tools.

The important thing is for the supervisors to apply the source-cause approach to all problems. By considering the "whys" as well as the "whats", their control actions will be more permanent and/or long range.

| Problem | Control through |
|---|---|
| Source (origin of the problem) | Engineering |
| Cause (incorrect action of the people) | Training |

### Priority

Part of the training process for first-line supervisors is the evaluation of identified problems in order to determine their priority for correction. The system indicated on the sample format above is similar to that used in many typical hazard control programs. By using the same approach in both hazard control and housekeeping, supervisors do not have to learn and apply two different systems. Furthermore, many housekeeping problems can also be a hazard.

A simple evaluation system which can be used for both programs is as follows:

| | | Priority | | |
|---|---|---|---|---|
| | Criteria | High | Moderate | Low |
| *Frequency* — | Number of persons exposed to the problem | 3 | 1 | 0 |
| *Severity* — | Potential result of the problem in terms of injury, damage, adverse employee or public relations | 6 | 3 | 1 |
| *Probability* — | Is the problem *likely* to produce injury or adverse reaction? | 3 | 1 | 0 |

**Priority Rating**

| | | |
|---|---|---|
| E | Today | 10–12 points |
| A | 7 Days | 8–10 |
| B | 1 Month | 6–8 |
| C | 3 Months | 4–6 |
| D | 6 Months | 0–4 |

It should be pointed out that in considering control of any problem, and certainly one involving a hazard, the supervisor must first consider what immediate temporary control action can be taken pending a permanent solution.

*Example:*

> *Problem* — Hose was left lying on the floor.
> *Source* — No rack provided.
> *Cause* — Employees fail to coil hose up after use.
> *Immediate Temporary Control* — Coil up hose.
> *Priority Evaluation* — 7 Points — 1 month.
> *Permanent Control* — Write maintenance work ticket to install hose rack.
> *Long-Range Control* — Train employees to coil up hose on rack.

### Control

As pointed out above, for control of all problems, effective action must be taken to get at the source and/or the cause of the problem. For this reason, our self-survey report is laid out so that the reporter must indicate what will be done to control the "why"s.

### Follow-Up

Here the inspector should indicate what action has been taken, at the time the report is submitted, to process each of the stated control actions. Supervisors and team leaders should be trained that there are only three ways to process control of an identified problem. These are

> *Corrected* — Action has already been completed.
> *Scheduled* — Action such as job ticket or engineering request submitted but pending scheduling.
> *Passed* — Action is beyond the authority or ability of the inspector to resolve and has been passed to a higher level or to the proper person concerned for their action.

The letters C, S, or P should be indicated in the survey report, along with the date and name in the case of the *Passed* category.

This completes the description of the first-line supervisor's part in the self-survey system. However, to assist supervisors in improving their effectiveness, it is recommended that second-line supervisors be required to evaluate these reports and give their supervisors necessary "feedback".

TIME DISCOVERED: _____ DATE: _____     TIMEDISCOVERED:_____DATE: _____

                                              <u>TASK OBSERVATION</u>

                                              Job being performed:

☐ Unsafe Work Practice    ☐ Unsafe Condition     ☐ Safe Work Practice    ☐ Unsafe WorkPractice

DESCRIBE:                                     DESCRIBE:

                    Priority ☐ Rating                          Priority ☐ Rating

REPORTED BY: _____        REPORTED BY: _____

HAZARD SOURCE AND CAUSE CONTROL               HAZARD SOURCE AND CAUSE CONTROL

DESCRIBE:  (Immediate Temporary Control) Action   DESCRIBE: (Immediate Temporary Control)Action

DESCRIBE (Permanent Control) Action           DESCRIBE (Permanent Control) Action

TIME COMPLETED: _____             TIME COMPLETED: _____

COMPLETION Corrected within 24 hours ☐ or     COMPLETION Corrected within 24 hours ☐ or
Entered in Hazard Control Log ☐ Date: _____   Entered in Hazard Control Log ☐ Date: _____

**Figure 5-7**  Sample of a hazard checklist simplified by grouping hazard categories into major source areas.

To help second-line supervisors in this evaluation process, some organizations include a points rating system as well as written comments. See Figure 5-7 for a sample format for this evaluation process.

## *Follow-Up Control Systems*

The third key element of an effective industrial safety inspection system consists of follow-up controls. These are primarily management controls established at levels above the first-line supervisor or team leader. Management controls can be divided into those actions that are departmental and those that are plant-wide or multidepartment.

### Departmental Controls

Second-line supervisors, in addition to evaluating the survey reports of their supervisors, obviously need to ensure that their program is producing the desired results. Here are some of the actions that should be taken by the department levels above the first line.

1. Keep a department hazard log or follow-up file. Either all items listed in survey reports as pending should be transferred to a hazard log or the report should be kept in a pending file. This file should be regularly reviewed by the second-line supervisor to ensure that action has been taken as scheduled.
2. Perform department safety and housekeeping audits. Second-line supervisors should periodically inspect their department areas and compare their findings with what is reported in their supervisors' self-surveys. Exceptions should be discussed with the supervisors concerned to help them upgrade their abilities in the program.

### Unit Controls

Just as in many other aspects of the total enterprise, management must establish controls to measure the results of the plan and ensure that the program is effective. The following list is by no means comprehensive, but will outline those elements considered to be the minimum needed for a successful program.

**Goals and Objectives** — Each department and subdepartment should establish goals and objectives for the continuous improvement of their safety inspection programs. Progress on these goals should be reviewed by upper levels of management.

**Audits** — Periodic audits should be conducted to measure the results of the program. This can be done in any number of ways. In some organizations it is done by the safety department, in others by special committees. The audit group randomly checks self-surveys for action on listed undesirable conditions and notifies the management if and where problems exist. They will also make random inspections of areas to measure program effectiveness and determine training needs. Primarily, the audit group should be a program assessment committee rather than a problem discovery committee.

**Appearance Standards** — Here is a list of some recommended items:

1. Standard paint scheme for buildings and offices
2. Routine maintenance painting schedule for all buildings and facilities
3. Schedule and performance standards for custodial service
4. Standard layout for all signs
5. Beautification program for grounds and surrounding areas

6. Easy maintenance materials selected for floors (nonskid) and walls in heavy traffic areas
7. Scheduled roof maintenance plan
8. Scheduled engineering inspection of all buildings and facilities
9. Recycling and waste collection program and schedule with disposal containers well identified and distributed around the facility

Adoption of these kinds of programs automatically results in improved appearance and takes care of many problems that usually are beyond the authority of first-line supervisors to resolve.

## Interest and Motivation

With the best of procedures there is still a need to provide ways to promote employee interest and motivate them to use the available program. It is beyond the scope of this chapter to describe all of the numerous programs that have been tried in the past. Suffice it to say that experience has shown that the most outstanding programs often fail without motivation. It is a human need to seek recognition, so it is recommended that after installing a sound system some simple motivation program should be implemented. This might include quarterly plant inspections and awards for good safety and housekeeping, with appropriate publicity in the organization's newsletter.

## Summary and Conclusions

The purpose and objective of this section was to outline the major elements of an effective safety inspection system. The concept behind the system described is based on delegating the responsibility for safety and housekeeping down to the lowest level of management and then helping them do it. Supervisors systematically search for and control hazards and substandard housekeeping conditions in their areas. At the same time, supervisors are taught to consider the source and cause of the problems they discover so that the problems are less likely to recur.

## FURTHER READING

### Books

Boylston, R. P. *Managing Safety and Health Programs.* New York: Van Nostrand Reinhold, 61–71, 142–143, 167–188, 208–219, 1990.
Bird, F. E. and Germain, G. L. *Practical Loss Control Leadership.* Loganville, GA: International Loss Control Institute, 121–144, 1990.
Petersen, D. *Techniques of Safety Management.* Goshen, NY: Aloray, 159–169, 1989.
Colvin, R. J. *The Guidebook to Successful Safety Programming.* Boca Raton, FL: Lewis Publishers, 97–110, 1992.

Holt, A. St. J. and Andrews, H. *Principles of Health and Safety at Work.* Leicester, U.K.: IOSH Publishing, 105–111, 1993.

Grimaldi, J. V. and Simonds, R. H. *Safety Management.* Boston, MA: Irwin, 144–146, 1989.

Bureau of Business Practice. *The BBP Safety Management Handbook.* Waterford, CT: Prentice-Hall, 1986 (or latest edition).

## Other Publications

Cal/OSHA Consultation Service. *Guide to Developing Your Workplace Injury and Illness Prevention Program.* CS-1, Cal/OSHA Consultation Service, San Francisco, CA, 1993.

Lack, R. W. The frustrating problem of industrial housekeeping. *Protection* 17(11), 12–16, November 1980.

U.S. Department of Labor, Occupational Safety and Health Administration. OSHA Safety and Health Program Management Guidelines: Issuance of Voluntary Guidelines. *Fed. Regist.* 54(16), January 26, 1989.

## HAZARD CONTROL

### Introduction

The federal OSHA Safety and Health Program Management Guidelines state (in part):

(c)(3)(i) Hazard Prevention and Control. So that all current and potential hazards, however detected, are corrected or controlled in a timely manner, establish procedures for that purpose, using the following measures:

(A)  Engineering techniques where feasible and appropriate;
(B)  Procedures for safe work which are understood and followed by all affected parties, as a result of training, positive reinforcement, correction of unsafe performance, and if necessary, enforcement through a clearly communicated disciplinary system;
(C)  Provision of personal protective equipment; and
(D)  Administrative controls such as reducing the duration of exposure.

The State of California Cal/OSHA Regulations for Injury and Illness Prevention Program states (in part):

(2)  Include a system for ensuring that employees comply with safe and healthy work practices.
(3)  Include a system for communicating with employees in a form readily understandable by all affected employees on matters relating to occupational safety and health, including provisions designed to encourage employees to inform the employer of hazards at the worksite without fear of reprisal.

(6) Include methods and/or procedures for correcting unsafe or unhealthy conditions, work practices and work procedures in a timely manner based on the severity of the hazard:

(A) When observed or discovered; and

(B) When an imminent hazard exists which cannot be immediately abated without endangering employee(s) and/or property, remove all exposed personnel from the area except those necessary to correct the condition.

The material that follows is based on a paper that was published in 1976 by *Protection,* the official journal of the Institution of Industrial Safety Officers, U.K. (now the Institution for Occupational Safety and Health), and is printed with permission from Paramount Publishing Ltd.

The need for some type of a hazard control system is well understood by anyone concerned with the problem of occupational accidents. It has long been recognized that accidents can be prevented provided that the hazards which cause them can be identified and controlled.

From the author's experience, two major problems have to be overcome before any hazard control system can become effective. The first problem is that the people most concerned with the accident problem, i.e., the first-line supervisor and those at the work level, do not systematically look for hazards. The average supervisors, faced with the everyday problems of getting the job done, do not on their own initiative take time to search for unsafe acts or unsafe conditions on a daily basis. Hazard searching at most becomes a periodic ritual inspection or a "look now and then". The second problem is that when hazards are discovered they are seldom processed in a systematic and logical manner. Management, possibly reacting to union pressure, may spend vast sums of money to correct a relatively low-risk hazard while other, high-risk hazard problems await action buried in some department's maintenance backlog.

## The Hazard Control System

To help overcome the major problem areas mentioned above, a three-part hazard control system is recommended. Part one is the Supervisor's Hazard Control Notebook, part two is the Department Hazard Control Log, and part three consists of Safety Recommendations. The following sections provide brief descriptions of these three parts.

### Hazard Control Notebook

Each first-line supervisor or team leader is issued a Hazard Control (HC) Notebook. They are expected to carry this notebook on their person while at work and to evidence activity in the program by recording a minimum of one hazard control action for each shift worked. This approach has a double advantage. First, it motivates the supervisor to look for hazards more systematically. Second, it gets them more into the habit of writing down the hazards they discover so they are not forgotten in the rush. Employees are also encouraged to look for hazards and report them to their supervisor, who will then record the hazards in their notebook.

The intent should be for supervisors to evidence systematic hazard control activity. This can be either an observation or report of a physical unsafe condition or contact with a team member to discuss the safety aspects of a specific task. These observations should note not only unsafe work practices, but also positive recognition for the employee when he/she is seen following the proper procedure.

To help supervisors systematize their daily search for hazards, they can be trained to focus their surveys by hazard source/cause category. In Figure 5-7, a simple nine-point source/cause category guide is illustrated. In this way, a supervisor can pick one category per day for their search. This search can also be alternated with audits of job hazard analysis compliance.

After a hazard has been recognized and reported, the procedure requires that first-line supervisors take the following action steps: (1) write the hazard down in their notebook; (2) risk-rate the hazard and discuss with the person reporting it; (3) take immediate temporary control action (ITC); (4) rerate after ITC has been taken and communicate to the person reporting it; and (5) if permanent control has not been completed within 24 hours, the hazard is transferred to the department hazard control log.

A typical format for a supervisor's hazard control notebook is shown in Figure 5-8.

After the hazard has been recorded, the supervisor has to evaluate the potential effect of the hazard in order to decide on the priority for its control. Measuring the risk value of a hazard is a difficult thing for anyone because feelings and emotions usually become involved. The following system has proved to be a helpful guide.

### *Risk Rating System*

Consider each of the following criteria and assess points as appropriate:

a. Severity potential for injury or damage — high (6 points); moderate (2 points); low (1 point)
b. Frequency of employee exposure — high (3 points); moderate (2 points); low (1 point)
c. Probability of accident occurrence — high (3 points); moderate (2 points); low (1 point)
d. Violation of safety codes, standards, rules, etc. — yes (1 point); no (0 points)

Having established the risk factor on a points basis, the priority for control may be obtained from the following table:

**HAZARD SURVEY CHECKLIST**
Hazard = Unsafe Act or Unsafe Condition

HAZARD SOURCE/CAUSE CATEGORY

ITEMS IN AREAS
SURVEYED

1.  WORK AREAS - Floors, aisles, yards and grounds, roadways,
    ladders, platforms, scaffolds, temperature, illumination.

2.  MACHINES - All *production* or *finishing* machines: lathes,
    drill-presses, milling machines, grinders.
    *Point of Operation* and *Nip* points.

3.  EQUIPMENT - All equipment that *conveys* or *processes*:
    Fans, boilers and pressure vessels, ovens and heaters,
    compressors, and air-receivers, pumps, pipes, valves,
    tanks, hoses and connections, cranes, hoists, slings,
    escalators, elevators, conveyors, robots, gas welding
    and burning. *Rotating* equipment.

4.  PORTABLE TOOLS - All portable manual hand tools, all
    portable compressed air operated hand tools.

5.  MATERIALS - HEALTH - All types of materials that may
    produce *health* injury resulting from an acute or
    chronic toxic effect: Asbestos, lead, carbon monoxide,
    hydrogen sulfide, benzene, cyanide, arsenic.

6.  MATERIALS - TRAUMATIC - All types of materials that may
    produce a traumatic injury resulting from corrosive,
    explosive or impact contact: Acids and caustics, flammables and
    explosives, compressed gases and liquids, molten metals, hot
    materials, flying sparks, sharp edges, abrasive materials, weight-
    shape, stacked materials, radiation, noise. Also includes bodily
    movements in handling materials.

7.  ELECTRICAL - Motors and generators, transformers and
    rectifiers, wiring and conductors, electrical hand tools,
    electrical welding and burning. *Electrical guard*s - insulation,
    isolation, over-current protection and grounding. Ground-
    fault protection.

8.  VEHICLES - Manual hand trucks and dollies, automobiles,
    power trucks, forklifts, railroad equipment and rolling
    stock, marine vessels, aircraft.

9.  MULTI-CATEGORY - Critical procedures relating to more than
    one category:
    Excavations
    Confined Space Entry
    Lockout-Tagout
    Personal Protective Equipment

**Figure 5-8** A typical format for a supervisor's hazard control notebook.

| Priority Rating | Risk Rating |
|---|---|
| E — Emergency | 10-13 points |
| A — 1 week | 8-10 points |
| B — 1 month | 6-8 points |
| C — 3 months | 4-6 points |
| D — 6 months | 4 points |

Besides providing a system to assist supervisors to increase the frequency of their hazard surveys, the HC notebook is also a useful training tool. Experience with the program has shown that new supervisors usually find and record a majority of unsafe conditions. However, as they continue to develop their knowledge and skills in the principles and procedures of hazard control, the proportion of unsafe work practices increases.

A notebook can be assessed for the quantity and quality of the entries. A method used by some organizations is as follows:

Score

$$0 - 100 \quad \frac{\text{total no. hazards recorded}}{\text{total no. shifts worked}} \times \text{quality}$$

Example:

$$\frac{16 \text{ hazards recorded}}{20 \text{ shifts worked}} \times 80 \text{ quality} = 64 \text{ score}$$

Quality of the entries is obtained by randomly selecting five entries and assessing the average quality according to a criteria guideline. A simplified version of this system is as follows:

| | |
|---|---|
| Hazard description | 0–40 points |
| Risk rating | 0–10 points |
| Hazard control | 0–50 points |

Refer to Figure 5-9 for illustration of another quality evaluation technique.

Using this approach enables department managers to identify nonperforming supervisors and assist them to improve their skills in recognizing, understanding, and controlling hazards.

### Department Hazard Control Logbook

When permanent control action is not completed within 24 hours, the supervisor must enter the hazard in the Department Hazard Control Logbook (a typical format is illustrated in Figure 5-10).

It is important to note that the hazard must be entered in the log where the hazard exists. This may result in the supervisor recording the hazard in two logs. For example, a supervisor in Department A finds a hazard in Department B. In this case, the supervisor must enter the hazard in the Department B logbook and also in the Department A logbook.

The double entry of the hazard as outlined above is necessary in order to inform the proprietary department of the hazard so that control action is taken and also to enable follow-up by the supervisor who discovered the hazard.

After entering the hazard in the log, the supervisor is expected to follow up periodically during the span of time that the job should be done (priority

## EVALUATION OF HAZARD CONTROL ACTIVITY

| DESCRIPTION OF HAZARD | MAX. | CONTROL OF HAZARD | MAX. |
|---|---|---|---|
| Poor identification (location of unsafe condition or employee name not listed) Example: "Fork truck speeding" (No truck number, location, or identification) | -5 | Future control listed, but no scheduled date or follow-up | -5 |
| Desription vague, i.e., "Slipping hazard," instead of "Hydraulic oil on steps." | -10 | Questionable or vague control | -10 |
| Describes result instead of hazard, i.e., "Man receiving hand cuts from cable," instead of "working without gloves" | -15 | Improper or inadequate control (instruction only when unsafe condition could have been eliminated). | -15 |
| | | No control listed | -20 |

The supervisor's HC score is derived by multiplying the quantity score by the quality score. Example:

| | |
|---|---|
| Number of HC Entries | = 16 |
| Number of HC shifts worked | = 20 |
| Quantity score 16/20 | = 80% |
| Average quality score of 5 HC | = 80 |

Supervisor's HC Score　=　.80 x .80 = 64

**Figure 5-9** Example of a hazard notebook entry quality evaluation technique.

rating). When the hazard is not corrected within the priority rating period, the supervisor takes the problem to the next level for assistance and action.

The supervisors' risk ratings on a hazard should stand unless they participate in a rerating. If a hazard goes beyond the original scheduled date of correction, the hazard is rerated in the department with the consent of all concerned. If the originator disagrees with the rerating, then the hazard must be referred to a higher level for a final decision.

Experience with the logbook phase of this program has shown that follow-up must be maintained by management to ensure effectiveness. To provide department managers with a status of hazard control action, supervisors should provide a monthly report of outstanding hazards on an exception basis. In this report, only those hazards which have overrun their priority rating by 30 or more days are listed.

**AIRPORTS COMMISSION**
SAN FRANCISCO INTERNATIONAL AIRPORT
CITY AND COUNTY OF SAN FRANCISCO

# HAZARD CONTROL LOG

SECTION _____    LOCATION _____

| DATE ENTERED | HAZARD | CONTROL ACTION | | HAZARD DETECTED/ ENTERED BY | PRIOR-ITY * | WORK ORDER SERVICE REQUEST | DATE COMPLETED and SIGNATURE |
|---|---|---|---|---|---|---|---|
| | | ITC TAKEN | RECOMMENDED PERMANENT CONTROL | | | | |
| | | | | | | | |
| | | | | | | | |
| | | | | | | | |
| | | | | | | | |
| | | | | | | | |
| | | | | | | | |
| | | | | | | | |
| | | | | | | | |
| | | | | | | | |

REVIEWED BY _____ DATE _____

REVIEWED BY _____ DATE _____

PRIORITY
E-EMERGENCY          C-THREE MONTHS
A- ONE WEEK          D-SIX MONTHS
B-ONE MONTH          Subject to further review

*Priority rating must be
after ITC action has been taken.

**Figure 5-10** Sample format for hazard log.

Copies of all reports should also be sent to the safety department. This ensures that plant-wide or major problems can be monitored by safety and top management.

A hazard control program, like any other management system, needs follow-up. Some supervisors still do not conscientiously look every day for hazards, and then at the end of the month they merely fill in their notebook to meet the "paper" requirement. Second-line supervisors should be responsible for measuring the effectiveness of the system through random audits.

Hazard control is one of the five systematic action programs which are implemented by first-line supervisors. Of these five programs, hazard control is considered the most important and productive. In fact, it is fair to say that hazard control activity is the "barometer" of an organization's safety results. When this program is effectively implemented, the frequency of accidents will be steadily reduced.

### Safety Recommendations

Employees should be informed that safety and health hazards must be reported to their supervisor for investigation and processing through the Hazard Control System. In spite of this requirement, the reality is that for various reasons some employees may be reluctant to take certain safety problems to their supervisor. Recognizing this issue, OSHA requires employers to establish methods for employees to report hazards without fear of reprisal, even anonymously.

The Hazard Control System must, therefore, include a system of methods by which employees can report hazards. This serves as a relief valve and helps to point out problems which otherwise might go uncorrected. This need can be met by various means. Examples are

- A safety committee which has representatives from all departments
- An in-house safety department which is available to receive reports of problems
- A "hotline" on which employees can call in safety and health problems
- A formal safety recommendations (SR) program

One or more of these systems should be in place and known to all employees.

## FURTHER READING

### Books

Ridley, J. et al. *Safety at Work.* London: Butterworths, 142–153, 1977.
Manuele, F. A. *On the Practice of Safety.* New York: Van Nostrand Reinhold, 169–210, 1993.
Colvin, R. J. *The Guidebook to Successful Safety Programming.* Boca Raton, FL: Lewis Publishers, 77–93, 1992.

Gordon, H. L. et al. *A Management Approach to Hazard Control.* Bethesda, MD: Board of Certified Hazard Control Management, 1994.

Holt, A. St. J. and Andrews, H. *Principles of Health and Safety at Work.* Leicester, U.K.: IOSH Publishing, 76–87, 1993.

Grimaldi, J. V. and Simonds, R. H. *Safety Management.* Boston, MA: Irwin, 1989.

Asfahl, C. R. *Industrial Safety and Health Management.* Englewood Cliffs, NJ: Prentice-Hall, 1990.

Brauer, R. L. *Safety and Health for Engineers.* New York: Van Nostrand Reinhold, 1990.

## Other Publications

Institution of Occupational Safety and Health. Risk Assessment — A Practical Guide. *Saf. Health Pract. Suppl.,* May 1993.

Lack, R. W. A hazard control system for effective results. *Protection* 13(5), 2–4, June 1976.

U.S. Department of Labor Occupational Safety and Health Administration. OSHA Safety and Health Program Management Guidelines: Issuance of Voluntary Guidelines. *Fed. Regist.* 54 (16), January 26, 1989.

## JOB HAZARD ANALYSIS

### Introduction

The federal OSHA Safety and Health Program Management Guidelines state (in part):

(c)(4)(i) Ensure that all employees understand the hazards to which they may exposed and how to prevent harm to themselves and others from exposure to these hazards, so that employees accept and follow established safety and health protections.

The State of California Cal/OSHA Regulations for Injury and Illness Prevention Programs state (in part):

(7) Provide training and instructions;
    (B) To all new employees;
    (C) To all employees given new job assignments for which training has not previously been received;
    (D) Whenever new substances, processes, procedures or equipment are introduced to the workplace and represent a new hazard;
    (E) Whenever the employer is made aware of a new or previously unrecognized hazard; and
    (F) For supervisors to familiarize themselves with the safety and health hazards to which employees under their immediate direction and control may be exposed.

The material that follows is based on a paper published in 1980 by *Protection,* the official journal of the Institution of Industrial Safety Officers, U.K. (now the Institution for Occupational Safety and Health), and is printed with permission from Paramount Publishing Ltd.

Job hazard analyses (JHA) or job safety analyses (JSA) have been generally utilized by industry for many years as a key part of their accident prevention programs. In spite of this, industry in general is faced with a continued accident problem, particularly in the human element area. It is toward this area of safety that the JHA program is aimed. Why do JHAs sometimes fail to get the message across? What should be done differently to achieve better results? How can the "paper" aspect be handled? These and many other questions are being encountered today by safety professionals and others concerned with improving the effectiveness of industrial safety programs.

Much has already been written on the basics of JHA or JSA systems, and most readers will be very familiar with this type of program. Therefore, this section will be primarily concerned with identifying some of the problems with the program and outlining various methods for their solution.

## What Are the Problems?

The study of accidents where the key cause was found to be an unsafe work practice usually points up many, if not all, of these typical problems:

1. No written JHA existed for the job being performed.
2. The hazard involved was not included in the JHA.
3. The JHA was written as a generality (hazards not clearly described and/or remedies not specific).
4. The practice was not communicated to the employee.
5. The employee did not follow the specific methods outlined in the JHA.

Conversely, it has also been found that accidents rarely occur when there exists a well-written JHA containing a clear description of the hazards involved that has been systematically reviewed with and followed by the employee concerned.

The solution to these problems appears at first glance to be deceptively simple. However, the author has found in 30 years work with these programs that the problems are far from easy to solve out in the "real world" at the industrial work level.

## How to Improve the Effectiveness of Your Job Hazard Analysis Program

Suggested steps which have proven effective in overcoming the problems mentioned are described in the following sections.

## Assessment of the Program

A common objection among many first-line supervisors to the JHA program is, "why all the paper?" The feeling is that, apart from a few important safety rules contained in the employee handbook, their employees are very experienced and, therefore, verbal safe practice instructions are sufficient. Unfortunately, those of us involved in accident prevention are only too well aware that when a JHA is verbal, one may as well call it useless. In fact, if a JHA is not written, the best position to adopt is to assume it does not exist because the probability of it being communicated and implemented correctly is very low.

Another aspect of this resistance to the "paper" is that some supervisors feel the program will expose their weaknesses. These supervisors, who are afraid to expose themselves to professional critics, will need help to understand the principles and procedures of the JHA program and how the system will benefit them personally.

A different problem from the above is the organization of the paper. A plant at which the author worked for a number of years had mostly young, well-educated supervisors. These people had no objections to writing JHAs. In fact, they turned them out by the hundreds! The problem was how to find them when they were needed for later reference. Needless to state, if they could not find their JHAs, they obviously were not being utilized in any systematic manner.

Many companies have established classification systems to organize their JHAs. A basic JHA classification system should consist of the following elements:

1. *Types of JHAs* — To avoid excessive repetition, it is best to divide JHAs into three types:
   a. *General or common JHAs* will cover hazards associated with activities common to many jobs. Examples include manual lifting, using ladders, walking on slippery surfaces, etc.
   b. *Specific JHAs* will cover hazards related to specific tasks.
   c. *On-the-spot JHAs* are utilized whenever a job is being done which involves any high-risk hazard and for which there is no general or specific JHA. As the name implies, this JHA is usually a verbal discussion at the job location, and the supervisor should be encouraged to convert this JHA to writing for future reference.
2. *JHA Classification and Numbering system* — Here is a plant-wide system that also adapts well to computerization.

| | |
|---|---|
| Plant | one digit assigned |
| Major department | one digit assigned to each |
| Subdepartment | one digit assigned to each |
| Area or section | one digit assigned to each |
| Classification | one digit per classification table (see below) |
| Sequential number | two digits |

The classification element is broken down into the nine (9) major hazard source areas:

| | | |
|---|---|---|
| 1. | Work area | Includes ladders, platforms and scaffolds |
| 2. | Machines | Anything that "produces"; e.g., lathes, presses, saws, and grinders |
| 3. | Equipment | Anything that "conveys"; e.g., pumps, pipes, valves, tanks, cranes, and conveyors |
| 4. | Portable tools | All types of portable hand tools, excluding electrical |
| 5. | Materials — traumatic | Includes acids, caustics, flammables, explosives |
| 6. | Materials — toxic | Includes toxic and systemic poisons |
| 7. | Electrical | Includes all types of electrical conductors and equipment |
| 8. | Vehicle | Includes all types of power mobile equipment, railroads, and marine |
| 9. | Multicategory | |

Using this numbering system will provide a maximum of seven numbers. Here are a few examples:

1000601 would be a plant-wide general JHA related to materials-toxic.
0200601 would be a department-wide general JHA related to materials-toxic.
0220601 would be a subdepartment specific JHA related to materials-toxic.

Each first-line supervisor or section office should be provided with a JHA manual containing all the general and specific JHAs relating to their jobs. The manual should contain an index having a general layout as follows:

### JHA INDEX

DEPARTMENT _____
SUBDEPARTMENT _____          CLASSIFICATION
AREA OR SECTION _____
                                          _____

| JHA NO. | JSP TITLE | JHA TYPE | DATE ORIGIN | DATE REVISED | DATE REVIEWED |
|---|---|---|---|---|---|

3. Determining the Need for New JHAs — To help combat the problem of JHA coverage, first- and second-line supervisors should periodically review their JHAs to determine whether new JHAs may be needed. Sources of information for this review are

a. List all jobs supervised and note which ones are not covered. Write JHAs for any which involve high-risk hazards.
b. Accident experiences. Every accident should cause an automatic JHA review and revision as needed.
c. Task observations by the supervisor. This should be a systematic effort by the supervisor to look for and anticipate acts or conditions which might result in injury.
d. Reports and recommendations from the supervisor's crew.
e. New or modified operations.
f. Experiences of other plants or organizations.

### Reviewing the Quality of Written Job Hazard Analyses

After looking at your JHA coverage, the next step in upgrading the quality of your program should be to look at what your JHAs are saying. Start with the center column (hazard). The major weak point of the JHA program begins with the description of the hazard. A format for a JHA, including guidelines for the evaluation of its quality, is shown in Figure 5-11.

What do your JHAs tell your people? Do they provide a list of clearly described hazards and specific control actions, or do they lapse into meaningless generalities?

Many standard procedures on developing JHAs include such instructions for identifying hazards as the following examples:

1. Is there a danger of striking against, being struck by, or otherwise making injurious contact with an object?
2. Can a person be caught in or between objects?

It is suggested that a better approach toward identifying the hazard is to have the supervisor and his/her crew define what the person is likely to do wrong that could produce an accident. In other words, the hazard should clearly describe the actual or potential unsafe work practice.

The reason behind this approach is that studies of many accidents where an unsafe work practice was involved have indicated that supervisors often do not know what their people are doing wrong and what they should do differently to prevent accidents. By leading their people to describe the hazard, they are at the same time helping their people to recognize the risk of their actual or potential unsafe actions.

To illustrate this approach, here are some examples:

**Standard Approach**

| Step | Hazard |
| --- | --- |
| 1. Climbing ladder | 1a. Fall from ladder |
| | 1b. Ladder slipping |
| | 1c. Slip off ladder rung |

| JOB HAZARD ANALYSIS | | JHA NO. | | PAGE _____ of _____ PAGES |
|---|---|---|---|---|
| JOB CLASSIFICATION OF INVOLVED EMPLOYEES | | TYPE: | ☐ General    ☐ Specific | ☐ On-Spot |
| TASK/JOB TITLE | | | | |
| DEPARTMENT | PERSONAL PROTECTIVE EQUIPMENT REQUIRED | THE FOLLOWING GENERAL JHA's MUST BE READ AND UNDERSTOOD BEFORE STARTING JOB | | |
| AREA | | | | |

| BASIC JOB STEPS | POTENTIAL HAZARDS What the person is likely to do wrong that could cause an accident | CONTROL ACTION What the person must do to control the hazard. |
|---|---|---|
| | | |
| | | |
| | | |
| | | |
| | | |

| BASIC JOB STEPS | Does it describe what is done (not how it is done) using brief action statements in sequence? | 10 | REVIEWER'S COMMENTS | PREPARED BY: |
|---|---|---|---|---|
| POTENTIAL HAZARDS | Does it describe what may be done that could cause an accident (not the injury or the accident). | 40 | | |
| CONTROL ACTION | Does it describe specifically what the person must do to control the hazard? | 50 | | CHECK ONE ☐ New JHA written ☐ Old JHA revised ☐ Old JHA reviewed |
| TOTAL | | 100 | Signature:          Date: | |

DISTRIBUTION:   SAFETY      ORIGINATOR      DEPARTMENT

**Figure 5-11**   Example format for job hazard analysis which includes quality evaluation.

**Suggested Approach**

| Step | Hazard (Unsafe Work Practice) |
|------|-------------------------------|
| 1. Climbing ladder | 1a. Climbing with tools or equipment in one hand |
| | 1b. Failure to secure ladder |
| | 1c. Placing toe of shoe on ladder rung |

One of the advantages of this approach is that, once the supervisor has identified and clearly described what the person may do wrong, it is more likely that an equally specific hazard control action will also be stated. To illustrate this point, here is another example:

**Standard Approach — Survey Assistant**

| Step | Hazard | Control |
|------|--------|---------|
| Climbing steep rocky areas | Slip or fall | Use proper climbing technique |

**Suggested Approach**

| Step | Hazard (Unsafe Act) | Control |
|------|---------------------|---------|
| Climbing steep rocky areas | Placing full weight on a hand hold before testing it | Always test the hand hold by pulling with your full strength before you test it with your full weight |

In the author's experience, most supervisors using this method do have difficulty in describing the specific unsafe work practice. However, with constant practice plus regular training and guidance, their understanding of the principles and procedures improves and, as a result, the quality of their written efforts is steadily upgraded.

In fact, to be effective, a JHA program must be a constant ongoing system. All JHAs should be reviewed by the supervisor and crew at least once every year, and this is an opportunity to improve the JHA and make it more specific. In this process, it is most important for the supervisor to let the employees become involved in making suggestions for changes, even to the point of challenging them to find something that needs to be changed.

What type of a hazard should go into a JHA? A suggested guideline would be as follows:

1. The hazard should normally be an existing or potential high-risk unsafe work practice, since unsafe conditions will be dealt with through plant hazard control procedures.
2. It is important for supervisors to study the "potential". In other words, they should consider the job when conditions are not normal or under an emergency situation, as well as the routine. They should practice using their intuition and encourage their people to do the same so that they can anticipate mistakes the people might make which could produce severe injury.

3. Many JHAs are multi-page and list literally dozens of hazards, many of which are relatively minor in terms of their ultimate potential to produce injury. This tends to dilute the effectiveness of the total JHA. It is recommended that a JHA list only those hazards considered to be of a high-risk nature. Some suggested criteria for a high-risk hazard would be

   a. Any hazard that can produce permanent-partial disability or more
   b. Any hazard that can produce an injury resulting in days away from work
   c. Any hazard that can potentially produce severe injury or significant damage

### Obtaining Better Utilization of Job Hazard Analyses

A JHA is merely a piece of paper until somebody converts it into action. How can supervisors get their people to abide by their JHAs rather than take short-cuts? As we said earlier, the first step is to look at the hazard description and control action. "What are we telling our people?" The second step is that supervisors must adopt a more systematic approach. This would include regular review of all their JHAs and daily task observations to ensure compliance with the safe practices outlined in each JHA.

Supervisors often fail to do an effective job of communicating JHAs. It would be beneficial for them to constantly remember the old adage, "If the student hasn't learned, the teacher hasn't taught!" They should frequently test to make sure their people understand their JHAs. One way is to show an employee a JHA and ask, "What does this paragraph say?" The employee's answer will indicate

1. Whether the employee understands it
2. Whether the JHA needs revising

Supervisors should frequently test for JHA understanding because only when their people can repeat their JHAs can they be sure the message got across. They should ask questions such as, "What are the key parts of your JHA?" or, "Tell me the hazards of your JHA. I expect you to know all five of them."

By adopting the systematic approach, supervisors will avoid the trap of tolerating JHA violations and thus permitting their people to slip into old habits. The systematic approach should also be targeted according to the risk potential of the hazard. The higher the risk, the more repetition and emphasis supervisors must provide to keep their people from making judgments. They should use this repetitive process in order to make the impression deep enough that it becomes a part of the subconscious.

### Managing the Job Hazard Analysis Program

| Management Level | Responsibilities |
| --- | --- |
| First-line supervisor | 1. Minimum standard — write, revise, or review a minimum of one JHA per month.<br>2. Conduct on-spot JHAs as needed. |

|  |  |
|---|---|
|  | 3. Audit for crew compliance with their JHAs. |
| Second-level supervision | 1. Set up a development schedule for jobs involving high-risk hazards |
|  | 2. Set up department manual and index. |
|  | 3. Review and approve JHAs. |
|  | 4. Provide training and guidance. |
|  | 5. Audit effectiveness. |
| Department manager | 1. Provide a department procedure defining who does what to ensure that the plant JHA policy is implemented in the department. |
|  | 2. Audit the effectiveness of the program. |
|  | 3. Identify department-wide problems and set objectives for continuous improvement of the program |
|  | 4. Establish a department general JHA index and development schedule. |
|  | 5. Provide advanced training and guidance. |

## Summary and Conclusions

In the writer's opinion, based on years of work with JHA programs, such a program is still the most effective and practical approach for occupational safety and health education and training relative to job-specific hazards and their control. Most of the problems experienced with this program are concerned with the implementation of the process rather than the process itself.

The key to solving most of the problems with this program is to establish a constantly improving system which is aimed at total involvement of all levels concerned. To achieve this objective, the program must require systematic activity by supervisors and work teams. In addition, the backup management line and safety professionals must provide the necessary training, guidance, and management leadership to ensure continuous improvement.

JHAs are vital because some of the worst catastrophes are caused by small mistakes that could easily have been predicted by the supervisor and people involved.

## FURTHER READING

### Books

Boley, J. W. *A Guide to Effective Industrial Safety.* Houston, TX: Gulf Publishing, 83–91, 1977.

Colvin, R. J. *The Guidebook to Successful Safety Programming.* Boca Raton, FL: Lewis Publishers, 81–84, 1992.

Petersen, D. *Techniques of Safety Management.* Goshen, NY: Aloray, 56–60, 1989.

U.S. Department of Labor, Mine Safety and Health Administration. *Job Safety Analysis.* National Mine Health and Safety Academy Safety Manual No. 5, 1989.

U.S. Department of Labor, Occupational Safety and Health Administration. *Job Hazard Analysis.* OSHA 3071. 1987.

Bureau of Business Practice. *The BBP Safety Management Handbook.* Waterford, CT: Prentice-Hall, 1986 (or latest edition).

## Other Publications

Lack, R.W. Job Safe Practices. *Protection* 17(2), 9–11, February 1980.

U.S. Department of Labor, Occupational Safety and Health Administration. OSHA Safety and Health Program Management Guidelines: Issuance of Voluntary Guidelines. *Fed. Regist.* 54 (16), January 26, 1989.

## SAFETY MEETINGS

### Introduction

The federal OSHA Safety and Health Program Management Guidelines state (in part):

> (C)(1)(IV) Provide for and encourage employee involvement in the structure and operation of the program and in decisions that affect their safety and health, so that they will commit their insight and energy to achieving the safety and health program's goal and objectives.
>
> Comment ... Forms of participation which engage employees more fully in systematic prevention include (1) inspecting for hazards and recommending corrections and controls; (2) analyzing jobs to locate potential hazards and develop safe work procedures; (3) developing or revising general rules for safe work; (4) training newly hired employees in safe work procedures and rules, and/or training their co-workers in newly revised safety work procedures; (5) providing programs and presentations for safety meetings; and (6) assisting in accident investigations.

Note that these forms of employee participation embrace each one of the five safety action programs discussed in this section and underline the critical need for effective communications and employee involvement.

The State of California Cal/OSHA Regulations for Injury and Illness Prevention Programs state (in part):

> The program shall at a minimum: (3) include a system for communicating with employees in a form readily understandable by all affected employees on matters relating to occupational safety and health ... substantial compliance with this provision includes meetings, training programs, posting written communications, a system of anonymous notification by employees about hazards, labor/management safety and health committees, or any other means that ensures communication with employees.

These requirements reinforce government concerns that safety and health programs cannot be truly effective unless and until those at the work level are involved in the accident prevention process.

The Cal/OSHA Construction Safety Orders take this one step further by requiring employers to ensure that employees performing construction-related work attend a "tailgate" safety meeting once every 10 workdays.

Several later chapters in this book will discuss communication and training techniques, so this chapter will be confined to discussing safety meetings as an accident prevention system implemented at the work level by first-line supervisors or team leaders.

## Some Problems with Safety Meetings

In conventional safety meetings, you may find a group of up to 20 people sitting in a meeting which is being conducted by a supervisor who does not know what to say or how to say it. In this case, it is not surprising that the meeting turns into a "gripe" session and ends up nowhere.

Alternatively, the supervisor may have some relevant information, but lacks the communication skills to get the message across.

Over 30 years ago, Kaiser Aluminum and Chemical Corporation did a study of 10,000 safety meeting reports written by the company's supervisors. This study was lead by Homer K. Lambie, Corporate Safety Manager.

Each of these reports was evaluated against the question, "Is there any statement indicating the group will do something differently after the meeting to solve a single safety problem?" The result was shocking to the management at that time, for the study found that in practically every case nothing positive happened. The group just discussed generalities with no specific action conclusions.

Following this study, Mr. Lambie and others at Kaiser devised a simple five-step problem-solving approach which would help supervisors to plan and conduct more effective meetings. This system was named the K-5 by Kaiser and has since been adopted worldwide.

The reason that this technique has proven so effective is that, even without professional communications training or speaking skills, a supervisor has a simple tool that helps in selecting a topic and processing its solution with the group. The key is that the focus is on specific behavior rather than attitudes or generalities. A sample planning sheet report form and evaluation format are illustrated in Figures 5-12 and 5-13.

This method is a powerful tool to help meeting leaders once they understand the philosophy and the techniques. Training in how to use this system is most important because the concept is quite profound. The key point which must be stressed in this training is selection of a meeting topic (the problem). When leaders select a good specific problem, they will be able to process it to a positive action conclusion relatively easily. Conversely, if they select a problem which is vague or general, they will end up with an equally vague or general solution.

Some guidelines for preparation and conducting this type of meeting are as follows:

**SUPERVISORS PROBLEM SOLVING SAFETY MEETING**

| DEPARTMENT: | HINTS ON HOLDING THIS TYPE OF MEETING: | |
|---|---|---|
| SECTION: | A.   Before scheduled meeting, identify clearly and the behavioral safety problem which you and your group can solve. | |
| SUPERVISOR: | B.   Jot down a list of expected discussion points. | |
| DATE OF MEETING: | C.   Let your group know 2 or 3 days in advance the subject matter and time of meeting. | |
| EMPLOYEES ATTENDING: | 1.   PROBLEM (THE UNSAFE WORK PRACTICE) WE ARE DOING OR ARE LIKELY TO DO THE FOLLOWING: | |
| 1. | | |
| 2. | | |
| 3. | | |
| 4. | | |
| 5. | | |
| 6. | 2.   DISCUSSION - REASONS WHY THIS PROBLEM EXISTS. | THINGS WE CAN DO TO OVERCOME THESE OBSTACLES. |
| 7. | I. | I. |
| 8. | | |
| 9. | II. | II. |
| 10. | | |
| 11. | III. | III. |
| 12. | | |
| 13. | IV. | IV. |
| 14. | | |
| 15. | V. | V. |
| 16. | | |
| 17. | 3.   CONCLUSION: WHAT WE AS A GROUP WILL DO TO SOLVE OUR PROBLEM: | |
| 18. | | |
| 19. | | |
| 20. | | |
| NUMBER OF EMPLOYEEES IN CREW: | 4.   FOLLOW-UP: WHAT WE MUST DO TO ENSURE THAT OUR CONCLUSION IS PUT INTO ACTION. | |
| REVIEWED BY:<br>NAME:<br>DATE: | 5.   REMARKS: WHAT ELSE WAS SAID WHICH MUST BE ACTED UPON AFTER THIS MEETING? | |

"THE VALUE OF ANY MEETING IS MEASURED BY WHAT THE GROUP WILL DO DIFFERENTLY AFTER THE MEETING TO SOLVE A SPECIFIC SAFETY PROBLEM"

Distribution: Safety, Originator, Department

**Figure 5-12**   Example layout for a five-step problem-solving safety meeting report.

1. Problem (The Unsafe Work Practice)
   a. When the problem is of a general type, e.g., "using ladders improperly", such a problem must be dissected into the specific practices that make up this general problem. (How many and which unsafe work practices were involved in "improper use"?) Then decide on which unsafe work practice is the most critical in terms of injury potential.
   b. When the problem is one of "feelings" (attitude or an undesirable personal characteristic, e.g., "people are not safety-minded"), such a problem must be converted to the action that led to this judgment. (What did the members of the team do that led to the belief that they are not "safety-minded"?). Address their actions, not their feelings!

| EVALUATION OF PROBLEM-SOLVING SAFETY MEETING | | DEPARTMENT | | DATE | |
|---|---|---|---|---|---|
| | | SUPERVISOR | | | |

| | | POSSIBLE RATING | ACTUAL RATING 1 | 2 | 3 |
|---|---|---|---|---|---|
| **1** **THE PROBLEM** | A. The problem states clearly the action that must be discontinued or avoided. (NOT: an unsafe condition, an attitude or feeling, or a 'general' act or action.) .... | 0-20 | ....... | ...... | ....... |
| | B. Are most of the people doing or likely to do this? .. | 0-10 | ....... | ...... | ....... |
| | C. Can a change in *their* action control the problem?........... | 0-10 | | | |
| **2** **DISCUSSION** | REASONS: A. Are the reasons listed those which contribute to the problem? ............ | 0-5 | ....... | ...... | ....... |
| | B. Do the items address the question "What causes us to make these mistakes?" ................................... | 0-5 | ....... | ...... | ....... |
| | CONTROLS: A. Is each reason adequately controlled or eliminated? ............................ | 0-5 | ....... | ......... | ....... |
| | B. Is the necessary follow-up on the controls noted below? | 0-5 | | | |
| **3** **CONCLUSION** | A. Does it require a change in their behavior? ........... | 0-15 | ....... | ......... | ....... |
| | B. Did each employee agree to the action? ............... | 0-5 | | | |
| **4** **FOLLOW-UP** | Does this section specify WHO will do WHAT, WHEN, to see that the conclusion sticks? ................................... | 0-10 | | | |
| **5** **REMARKS** | Employees' significant comments about this or other hazards listed. ......................................................... | 0-5 | ....... | ......... | ....... |
| | FORM PREPARATION--GENERAL: The form is completely filled in, including date, department, leader's name, and names of all employees attending the meeting.................................. | 0-5 | | | |
| **TOTAL** | | | | | |

| REVIEWER'S COMMENTS: | ACTUAL RATING BY | |
|---|---|---|
| | 1 | SUPERVISOR |
| | | Date |
| | 2 | REVIEWER |
| | | Date |
| | 3 | NAME |
| | | Date |

RL4

**Figure 5-13**  Sample layout for quality evaluation of a five-step problem-solving meeting.

   c.  When the problem is an unsafe condition, convert it to the action which produced it and conduct a meeting with those people whose action is involved. Unsafe conditions are not usually the subject of safety meetings with a team and would most logically be handled relative to their sources, as follows:

| Unsafe Condition Source | Meeting with |
|---|---|
| Employee unsafe work practice | Team whose action produced it |
| Wear and tear | Production/engineering and/or maintenance |
| Design deficiencies | Engineering |

   d.  The rule of thumb is to meet with those whose action can solve the problem.

    e. The problem must be a description of what the members of the group are doing or likely to do wrong, and it must be one that will be controlled by a change in the actions of the members of the group.

2. Discussion
    a. *"Their reasons for the above action"* — The leader must list beforehand all the significant reasons the members of the group may have to justify the undesirable action. During the meeting, the leader should list any additional reasons that may be offered by the group.
    b. *"How do we overcome these reasons?"* — Listed here, in advance, are the facts to be considered and the steps necessary to overcome the reasons for the action. During the meeting, other controls may be added by the employees.

3. Conclusion — The conclusion should outline clearly the action of the members of the group necessary after they leave the meeting to control the problem. Individual commitment tends to ensure the desired action.

4. Follow-up — "other action needed to ensure that the conclusion is carried out". Here outline and schedule actions of identified individuals which facilitate or ensure the carrying out of the conclusion:
    a. On commitments made by the leader during the meeting
    b. On all other incomplete control measures, including checks to see that the unsafe work practice does not recur

5. Remarks — "related or unrelated comments or observations for action or review after this meeting". It is important to list in this section related or unrelated items, information, or observations for these reasons:
    a. To give the leader a means to avoid deviation from the "problem" being solved by postponing nonrelevant discussion
    b. To ensure review of those items not dealt with in the meeting

An objection to the five-step process frequently raised is, "We have so many problems — if we only take one per meeting, will we ever solve all of them?" To counter this argument, it must be pointed out that learning and using the five-step process will result in a transfer of problem-solving skills. With practice, leaders will find that they and their crews will automatically start defining and solving more and more problems of all types, not only safety issues.

In the chapter on training techniques, there will be much more information on how to prepare and conduct meetings. A brief checklist for the guidance of safety meeting leaders follows.

## Checklist for Presenting Successful Safety Meetings

1. Planning meetings
    • Regularly schedule meetings throughout the year.
    • Remind attendees of date, time, and place.
    • Preferably schedule meetings in the morning; if one must be held in the afternoon, it should not be immediately after lunch, and never after 3 p.m.

- Occasionally invite a guest such as a person from safety, industrial hygiene, security, or fire protection, a department manager, etc.
- Select a location where the group will be comfortable with minimal distractions.

2. Preparing for meetings
   - Identify the meeting topic (problem).
   - Research meeting subject and materials, including reasons why the meeting problem exists and methods to overcome the objections.
   - Check out visual aids such as a flip chart, VCR and monitor, or overhead projector (including a spare bulb).
   - Come early and set up the meeting room.
   - Be there to greet your attendees.
   - Where appropriate, provide refreshments.
   - Have a sign-in sheet.

3. Presenting
   - Start meetings on time!
   - Circulate a sign-in sheet.
   - Review the meeting agenda.
   - Set clear time limits. Set an ending time.
   - Review old business action items carried over from the previous meeting.
   - Introduce the subject.
   - Lead a discussion.
   - Reach an action conclusion.
   - Establish follow-up.
   - Present new business items and suggestions.
   - Close the meeting on a positive note.
   - Set the next meeting date and time.
   - Clean up the room and leave it in the condition in which you found it.

4. Other presentation hints
   - Avoid interrupting a speaker. Wait until he/she is finished before you comment.
   - Avoid judgmental comments: turn the question back to the group. ("What do the rest of you think?")
   - Avoid negative comments. Keep the meeting on a positive note.

5. Documenting meetings
   - Record all action taken or to be taken — who will do what, when.
   - Record all other suggestions or issues raised and note follow-up action.
   - Write up and distribute meeting reports promptly.

## Hints for Working with Difficult People in Meetings

| Participant's Meeting Behavior | Techniques for Handling |
| --- | --- |
| Overly talkative | • Pass the person's comments out to the group. Let the group take care of the problem as much as possible. |
| Highly argumentative | • Look for the positive points in their comments. Thank them and then move on. |

| | |
|---|---|
| | • Ask the group for their view. Let the group reject the idea.<br>• As a last resort, talk to the person after the meeting in private. |
| Side conversations | • Avoid embarrassing the side talkers.<br>• It is best to ignore side conversations unless they continue to the point where they have become an obvious distraction for the entire group; then pause in your presentation. As the meeting becomes quiet, usually side talkers will stop their discussion. |
| Rambling or off-the-subject issues | • Thank the person for their point and then bring the meeting back to the subject. |
| Gripes and "pet peeves" | • Ask the group for their comments.<br>• Point out that the meeting cannot change policy.<br>• Ask the person bringing up the issues how he/she would solve them.<br>• Record the points for later investigation and/or referral to the appropriate person.<br>• Discuss the issues with the individual in private after the meeting. |

## Summary and Conclusions

The benefits of group discussion and brainstorming have already been demonstrated by the use of "quality circles", and they can be equally effective in solving safety problems. What is often not appreciated is that the average first-line supervisor or work team leader needs help in improving his/her skills in meeting leadership. Providing systematic training and retraining will help these leaders become more effective as managers and specifically as meeting facilitators.

The safety meeting system works best when it is used to provide special emphasis on the control of unsafe work practices which involve high risk potential for serious injury and/or damage. As such, they are a critical element in the total accident prevention system.

## FURTHER READING

### Books

Bird, F. E. and German, G. L. *Practical Loss Control Leadership.* Loganville, GA: International Loss Control Institute, 187–217, 1990.

Doyle, M. and Strauss, D. *How to Make Meetings Work.* New York: Berkley Publishing Group, Jove Edition, 1982.

Fettig, A. *World's Greatest Safety Meeting Idea Book.* Battle Creek, MI: Growth Unlimited, 1990.

Corfield, A. *Safety Management.* London: International Institute of Risk and Safety Management, 185–200, 1988.

Bureau of Business Practice. *The BBP Safety Management Handbook.* Waterford, CT: Prentice-Hall, 1986 (or latest edition).

## Other Publications

U.S. Department of Labor, Occupational Safety and Health Administration. OSHA Safety and Health Program Management Guidelines: Issuance of Voluntary Guidelines. *Fed. Regist.* 54(16), January 26, 1989.

## ACCIDENT INVESTIGATION

### Introduction

The federal OSHA Safety and Health Program Management Guidelines state (in part):

(c)(2)(IV) Provide for investigation of accidents and "near miss" incidents, so that their causes and means for preventing repetitions are identified.

Comment: Accidents and incidents in which employees narrowly escape injury, clearly expose hazards. Analysis to identify their causes permits development of measures to prevent future injury or illness. Although a first look may suggest that "employee error" is a major factor, it is rarely sufficient to stop there. Even when an employee has disobeyed a required work practice, it is critical to ask, "Why?" A thorough analysis will generally reveal a number of deeper factors, which permitted or even encouraged an employee's action. Such factors may include a supervisor's allowing or pressuring the employee to take short cuts in the interest of production, inadequate equipment, or a work practice which is difficult for the employee to carry out safely. An effective analysis will identify actions to address each of the causal factors for an accident or "near miss" incident.

These words speak for themselves and really stand alone as a fitting introduction to this subject.

The State of California Cal/OSHA Regulations for Injury and Illness Prevention Programs require all employers to "include a procedure to investigate occupational injury or occupational illness".

The draft Cal/OSHA Model Injury and Illness Prevention Program for High Hazard Employers includes the following guidelines:

ACCIDENT/EXPOSURE INVESTIGATIONS Procedures for investigating workplace accidents and hazardous substance exposures include:

1. Visiting the accident scene as soon as possible;
2. Interviewing injured workers and witnesses;

3. Examining the workplace for factors associated with the accident/exposure;
4. Determining the cause of the accident/exposure;
5. Taking corrective action to prevent the accident/exposure from recurring; and
6. Recording the findings and corrective action taken.

## Definitions

Before we begin a discussion on the techniques used for accident investigation, it will be important to establish a few basic terms related to this process. These are defined in the following sections.

### Accident

In *Practical Loss Control Leadership* by Bird and Germain, an accident is defined as "an undesired event that results in harm to people, damage to property, or loss to process." A simplified version that I prefer because it includes "near miss"-type events is "any unplanned event which may result in injury and/or damage".

The British Health and Safety Executive in their publication *Successful Health and Safety Management* define the term accident as follows:

> Accident includes any undesired circumstances which give rise to ill health or injury; damage to property, plant, products, or the environment; production losses, or increased liabilities.

### Accident Investigation

The American Society of Engineers *Glossary of Terms* defines an accident investigation as follows:

> A determination by one or more qualified persons of the significant facts and background information relating to an accident, based on statements taken from involved persons, and inspection of the site, vehicles, machinery, or equipment involved.

I am indebted to Homer Lambie for his outstanding training on this subject and for the next two definitions.

### Accident Source

The accident source is the specific unsafe condition that was most closely related to or directly produced the accident.

### Accident Cause

The accident cause is the specific error/unsafe work practice of a person that either contributed to or directly caused the accident.

### Hazard

Fred Manuele, in his book *On the Practice of Safety,* defines the term hazard as the potential for harm or damage to people, property, or the environment. Hazards include the characteristics of things and the actions or inactions of people.

The British Management of Health and Safety at Work Regulations 1992 define it as follows:

> A hazard is something with the potential to cause harm (this can include substances or machines, methods of work, and other aspects of work organization).

My simplified version of these definitions is: A hazard is either an unsafe condition or an unsafe work practice (act).

One final definition that must be considered is significant incident (near miss). This is any accident involving a high potential for serious injury and/or damage. The ratio of accidents and incidents has been the subject of numerous studies. Three well-known ones are illustrated in Figure 5-14. The importance of investigating damage-only and near-miss events is underlined by these statistics. A later chapter will discuss a program for this type of investigation.

Obviously, it is vital for management to encourage employees to report all incidents as these are "warning flags" of weak links in the unit's accident prevention system and, if not addressed, eventually a more serious event will point out the need.

## Accident Investigation and Reporting

This section will provide an overview of a very broad and complex subject. Those readers desiring more in-depth information should consult the list of references at the end of the section.

Besides being required by law, the establishment of a written accident investigation and reporting policy and procedure is of vital importance. The policy-procedure as a minimum should include the following elements:

### Purpose and Definitions

- All accidents and incidents to be reported
- Accident types and their investigation

## ACCIDENT RATIO STUDIES

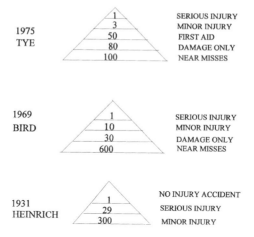

**Figure 5-14** Three of the best-known accident ratio studies.

- Who is to be notified regarding occurrence of accidents, particularly serious events
- Reporting accidents to regulatory agencies.

Apart from regulatory, insurance, and workers' compensation reporting requirements, an employer should establish an internal investigation report format which preferably will be multipurpose (for any type of occupational accident). One exception might be traffic accidents on public highways. This type of accident involves reporting data that are somewhat unique, and frequently the form is provided by your insurance agency.

Principal accident types that should be considered are as follows:

- Employee occupational injuries and illnesses
- Damage-only accidents
- Significant incidents (near misses)
- Consultant/utility/telephone employee accidents on owner's property
- Vendor/contractor injuries and illnesses on owner's property
- Delivery vehicles and their occupants' accidents on owner's property
- Potential third-party liability accidents
- Visitor accidents on owner's property

In the event of third-party liability potential or a serious accident requiring a report to a regulatory agency, a team investigation is the recommended approach. Membership of an accident investigation team will vary with the circumstances. Membership will include

- Immediate supervisor
- Next level of supervision/department head

- Safety and health professional
- Person/s with technical expertise relevant to the accident situation
- Legal counsel
- Insurance representative

Reporting of accidents to regulatory agencies varies with the jurisdiction. In California, employers are required to report any work-connected fatality or serious injury or illness. A "serious injury" is defined in the California Labor Code as:

> ... any injury or illness occurring in a place of employment or in connection with any employment which requires inpatient hospitalization for a period in excess of 24 hours for other than medical observation or in which an employee suffers loss of any member of the body or any degree of permanent disfigurement.

Check with the agencies concerned at your work locations to determine reporting requirements.

## Procedure and Guidelines

The following are general guidelines for accident investigation and reporting.

### Priorities Immediately After an Accident

- Get to the scene as quickly as possible.
- Size up the situation.
- Ensure that emergency services have been notified and render first aid care for the injured.
- Protect others from injury.
- Protect equipment and facilities.
- Secure the site.
- Notify the proper persons concerned.

### Gathering Information

1. Speed is essential. As soon as you are aware of an accident, begin investigation immediately. The longer you delay, the less likely you are to get the facts. People tend to forget details quickly and, if they have time to talk over the accident with others, they may include details that never happened.
2. Go to the scene of the accident and interview all witnesses at the scene.
   - Interview each witness separately.
   - Make it clear at the outset of the interview that the purpose of the interview is not to establish blame but to establish the facts of the accident in order to prevent a reoccurrence.

- Ask each witness to explain in his/her own words what he/she saw. Let the witness tell you the story without interruption; then ask questions for clarification.
- Record notes on all witnesses reports or obtain signed statements. A hint on taking witnesses statements: have the witness describe in his/her own words what he/she saw. Questions to ask include
  - What impressed you?
  - How close did you get?
  - What did you say?
  - What did you do?
  - At any moment did you notice the time?

  Tell the witnesses to write down each point as they recall it occurred. Then as other things occur to them, they should write them down. The sequence is not important. Finally, ask the witness, "Is there anything else you remember that would help us understand the accident?"
- Do not complete the interview until you are satisfied that you have sufficient facts to establish the following:
  - The position of all persons involved
  - The job being done
  - How the job was being done
  - The unexpected happening
  - The result of the unexpected happening
3. At the scene note conditions carefully
   - Conditions of the working surface — wet or dry, oily, greasy, rough or smooth, obstructions, level or sloping, skid marks.
   - Position of valves, switches, equipment, vehicles, tools, etc.
   - Lighting — general level of lighting, lights not working, etc.
   - Any unsafe conditions that may have produced or contributed to the accident; check work area, machines, equipment, tools, materials, electrical, vehicles
4. Make a sketch indicating layout and measurements.
5. Photograph the area. Take both a general view and close-ups of relevant items and equipment.
6. Keep site secured until at least this part of the investigation is complete. In cases of fatality or serious accidents, the site should remain secured until cleared by the law enforcement and regulatory agencies concerned.
7. Secure evidence — In the event of a potential insurance damage claim, third-party liability, or serious accident where there is legal liability potential, it may be prudent to secure certain equipment items and records. Check with legal counsel for guidance as to what should be secured and how it should be secured. Items to consider for securing include
   - Operating and maintenance logs
   - Copies of agency reports
   - Complete accident investigation file with reports, notes, statements, sketches, pictures, etc.
8. Additional Hints for Accident Investigations
   - Always bear in mind that people will not always tell you all the sides of the accident due to their desire to protect themselves or protect their ego.

Most accident victims will rationalize and tell the story differently whether they mean to or not.
- The investigator should always be suspicious if the person changes his/her story.
- Example: On a construction project, two trucks collided. One was backing up and one was going forward. Both drivers said the other was at fault. Investigation revealed that one truck was stationary and the other reversed into it without looking. Conclusion: It is very important to find out what motion took place prior to the accident.
- Example: A truck backing at a dump at night time backed so far it went over the side of the pile, overturned and the driver was killed. The spotter at the dump said he signalled the driver to stop but the driver continued coming back. The investigator asked the following:

Q: "How fast was he coming?"
A: "Pretty fast."
Q: "Did he stop before he went over?"
A: "I think so."
Q: "Any signals?"

At this point, the spotter became confused. On two occasions his story differed. Finally, he broke down and admitted he was not there at the time; he had gone to a nearby cafe for a cup of coffee.

Questions to ask include
- What position were the people in?
- What did they do or not do?
- What made the person lose balance?
- Why did they fall?
- Where were they standing?
- How were they standing?
- Where was he/she supposed to have been at the time?

### Analyzing the Facts

Having done the best possible job of obtaining the facts as to how the accident happened, the investigator is now ready to determine the cause and/or source of the accident. To do a thorough analysis, it is recommended that you use a format as illustrated in Figure 5-15.

Record all unsafe actions/work practices and unsafe conditions which in any way could have contributed toward producing the accident. Keep an open mind at this stage — do not prejudge the accident.

Next determine the reasons why the actions occurred or the conditions existed. This may cause you to go back and reinterview witnesses or reexamine the equipment and is useful toward ensuring that you have established all the facts of the accident. This will also help to reveal underlying causes such as management system failures, inadequate training, improper maintenance, poor engineering design, etc.

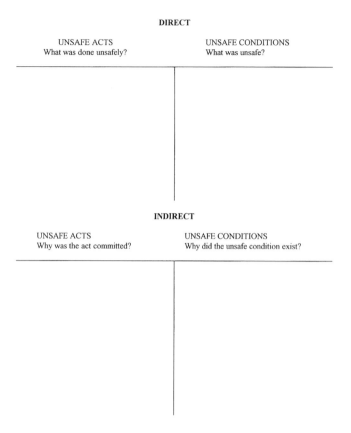

**Figure 5-15**  Accident cause-source analysis.

## *Establishing Controls to Prevent Recurrence*

1. Review your analysis of the accident facts to determine the key causes and/ or sources of the accident.
2. Determine the action necessary to control those key unsafe actions and conditions.
3. Take immediate temporary control measures as necessary to prevent further injury or damage.
4. Establish long-range or permanent actions to control the hazards in the future.
5. Establish other controls, system improvements, etc. that are recommended. Identify which safety action programs may be involved in the correction process, such as job hazard analysis or hazard control.
6. Follow through to ensure that all control actions have been taken.

## *Complete and Distribute an Accident Investigation Report*
Figure 5-16 illustrates a sample format for this report.

**ACCIDENT INVESTIGATION REPORT**

| | | | | |
|---|---|---|---|---|
| **LOCATION** | PLANT OR FACILITY | | LOCATION OF ACCIDENT (THE NAME OR NUMBER OF BLDG. STORE, DEPT., FLOOR, ETC. | |
| **EMPLOYEE** | EMPLOYEE NAME | BADGE NUMBER | DATE AND HOUR OF ACCIDENT | |
| | DEPT. OR OR SECTION IN WHICH EMPLOYEE WORKS | | EMPLOYEE'S JOB OR POSITION | |
| | DESCRIBE THE INJURY AND/OR DAMAGE : | | | ESTIMATED LOSS: |
| **DESCRIPTION** | SEQUENCE OF EVENTS: (A) EMPLOYEE'S POSITION IN RELATION TO SURROUNDINGS; (B) HOW WORK WAS BEING DONE; (C) UNEXPECTED HAPPENING; (D) RESULT OF UNEXPECTED HAPPENING; (E) ADDITIONAL FACTS IF AVAILABLE; (F) ATTACH DRAWINGS AND/OR PHOTOGRAPHS WHENEVER POSSIBLE. | | | |
| **ACCIDENT CAUSE & SOURCE** | UNSAFE ACT (WHAT WAS DONE UNSAFELY) | | UNSAFE CONDITION (WHAT WAS UNSAFE) | |
| | WHY ACT WAS COMMITTED | | WHY CONDITION EXISTED | |
| **HAZARD CONTROL** | HAZARD (CAUSE) CONTROL: IMMEDIATE CONTROL ACTION | | CONDITION/SOURCE) CONTROL: IMMEDIATE CONTROL ACTION | |
| | LONG RANGE | | PERMANENT | |
| **FOLLOW-UP** | RESPONSIBILITY FOR CONTROL ACTION/EST. COMPL. DATE | | RESPONSIBILITY FOR CONTROL ACTION/EST. COMPL. DATE | |
| **JHA** | NO. & TITLE APPLICABLE WRITTEN JOB HAZARD ANALYSIS | | HAZARD IDENTIFIED IN JHA ☐ YES      ☐ NO | JHA NEEDS REVISION   ☐ YES ☐ NO    COMPL. DATE: |
| **CONTROL** | REPORT COMPLETED BY:          DATE: | | REVIEWED BY:      DATE: | REVIEWED BY:       DATE: |

**Figure 5-16** Typical accident investigation report format.

## Evaluation of Accident Reports

A sample quality evaluation format for an accident report is illustrated in Figure 5-17. The immediate supervisor of the investigator should use this as a guide in reviewing accident reports. The evaluation will also assist the reviewer in pinpointing training needs and possible additional actions that may be needed in the investigation process.

| EVALUATION OF ACCIDENT REPORT | | | | |
|---|---|---|---|---|
| **SUPERVISOR** | **DEPARTMENT** | | **DATE** | |
| **REPORT** | Is the report accurate and complete?<br><br>Are all the sections completed, including date, loss estimate, employees involved, and JHA activity? | | 5 | |
| **DESCRIPTION** | Is the description clear?<br>Does the description include the sequence for events, employee's position in relation to surroundings, how was job being done, the unexpected happening and result of the unexpected happening? | | 5 | |
| **ANALYSIS OF THE PROBLEM** | Are all the cause and source factors contributing to the incident identified?<br>Does the analysis include all direct people factors (unsafe acts) and indirect people factors (reasons why the employee acted as they did?)<br>Does the analysis include all direct environment factors (unsafe conditions) and all indirect environment factors (reasons why the unsafe condition existed?) That is, equipment deterioration, improper or inadequate engineering, etc.<br>NOTE: (a) Causes are the actions of people<br>　　　　(b) Sources are the conditions of things | | 30 | |

| | | | | |
|---|---|---|---|---|
| **CONTROL ACTION** | Does the control action protect the employee and prevent reoccurrence of the accident? | IMMEDIATE<br>Remove the employee from the hazard, protect the employee from the hazard by temporary guarding, tell the employee how to do the job safely, etc. | 20 | |
| | | LONG RANGE<br>Remove the hazard, guard the employee from the hazard, tell the employee how to control the hazard (Safety meeting or JHA). | 30 | |
| **FOLLOW-UP** | Does this section specify WHO will do WHAT by WHEN? | | 10 | |
| | | **TOTAL** | **100** | |
| | | | | |

| | |
|---|---|
| Date: | Reviewed by: |

**Figure 5-17**   Example of a quality evaluation for accident/incident investigation reports.

## Accident Investigation Checklist

1. During regulatory agency "special investigation"
   - "Cooperation" — the basic rule
   - Legal counsel readily available
   - Escort at all times
   - Parallel photographs, samples, etc.
   - Representation at interviews
2. Accident investigation kit
   - Chalk, pens, pencils, felt tip markers, China marker, flashlight, magnifying glass

- Graph paper, lined paper, clipboard, tape, adhesive labels, tags with string hangers, ruler, tape measure
- Sample containers — manilla and plastic envelopes, containers with caps
- Camera with flash film and photo log
- Cassette recorder with tape
- Barrier tape, warning signs

3. Witness interviews

**DOs**

- Show concern.
- Be friendly.
- Listen carefully — do not interrupt.
- Ask broad, open-ended questions.
- Preferably, interview at the scene.

**DON'Ts**

- Sarcasm
- Pressuring witness
- Blaming anyone
- Asking leading questions
- Making assumptions

4. Witness' statement tape-recorded

- State the date, time, place, who you are, and with whom you are talking.
- Record the witness' consent: "I am giving this statement of my own free will, and the facts and observations are true to the best of my knowledge and belief."
- At the end of conversation, state the same information.

5. Witness' statements written by witness

- Ask witness to "write in your own words".
- Describe as you saw it.
- What impressed you?
- How close did you get?
- What did you say/see/do?
- Did you notice the time?
- Sign and date your statement and note time.

6. Witness' statements written down for the witness

- Write down the witness' statement in double line spacing.
- Date the statement and state the place in which it was signed.
- Have the witness read statement, initial changes, and initial each page.
- Put a clause at the end indicating the witness' consent.

7. Witness' statement

- The above statement was written down by according to what I have told him/her. I have read through the statement, and it accurately records my account of the facts.
- Signature of witness and date

8. Witness' interviews, investigators notes

- Background information
- Names of other parties at the scene
- Where were the other witnesses/what were they doing?

- Weather/lighting conditions
- Physical limitations (hearing, eyesight)
- Witness' location/observation/actions

9. Photography
- 35-mm camera a must; Polaroid only for quick reference
- Color film
- Take pictures as soon as possible.
- Take general shots of overall accident scene.
- Take closeups.

10. Photo log
- Date/time/where picture taken
- Name of photographer
- What does the photograph show?
- Camera, lens, and film used

Note: Much of this information can be covered in a photo ID board on the first shot of the film roll.

11. Sketches and diagrams
- Scale drawing
- Equipment or facilities involved
- General topography
- Signs
- Witnesses
- Key dimensions

12. Securing evidence — guidelines
- Small items in envelopes; label each item, tag larger items
- Store evidence in controlled area
- Files indexed by injured person's name
- Cross-reference the main file to bulky items

13. Securing evidence — files should contain:
- Witness/in-house statements and reports
- Results of tests
- Documents relating to machinery or equipment
- Operating and staff schedules
- Pictures, sketches, notes, logbook, photo log, list of evidence items, personnel records
- Policies, procedures, regulations
- Other relevant information

## FURTHER READING

### Books

American National Standards Institute. Method of Recording Basic Facts Relating to the Nature and Occurrence of Work Injuries, No. 216.2.
American Society for Industrial Security (ASIS). *Basic Guidelines for Security Investigations.* Arlington, VA: ASIS, 1981.

Bird, F. E. and Germain, G. L. *Practical Loss Control Leadership.* Loganville, GA: International Loss Control Institute, 57–119, 1990.

Boylston, R. P., Jr. *Managing Safety and Health Programs.* New York: Von Nostrand Reinhold, 95–108, 1990.

Colvin, R. J. *The Guidebook to Successful Safety Programming.* Boca Raton, FL: Lewis Publishers, 65–74, 1992.

Ferry, T. S. *Modern Accident Investigation and Analysis.* New York: John Wiley & Sons, 1988.

Ferry, T. S. *Accident Investigation for Supervisors.* Des Plaines, IL: American Society for Safety Engineers, 1988.

Hendrick, K. and Benner, L., Jr. *Investigating Accidents with STEP.* New York: Marcel Dekker, 1987.

Kuhlman, R. L. *Professional Accident Investigation.* Loganville, GA: Institute Press, 1977.

National Safety Council. *Accident Investigation — A New Approach,* 1983.

Ridley, J. et al. *Safety at Work.* London: Butterworths, 180–197, 1983.

Slote, L. et al. *Handbook of Occupational Safety and Health.* New York: John Wiley & Sons, 99–123, 1987.

Vincoli, J. W. *Basic Guide to Accident Investigation and Loss Control.* New York: Van Nostrand Reinhold, 1994.

Bureau of Business Practice. *The BBP Safety Management Handbook.* Waterford, CT: Prentice-Hall, 1986 (or latest edition).

## Other Publications

Smith, S. L. Near misses: safety in the shadows. *Occup. Hazards* 56(9), 33–36, September 1994.

Tiedt, T. and Kindley, R. A lawyer's prospective on accident investigations. *Prof. Saf.* 32(8), 11–17, August 1987.

U.S. Department of Labor, Occupational Safety and Health Administration. OSHA Safety and Health Program Management Guidelines: Issuance of Voluntary Guidelines. *Fed. Regist.* 54(16), January 26, 1989.

# 6 TRAINING, INTEREST, AND MOTIVATION — PRACTICAL APPLICATIONS THAT WORK

**Joan F. Woerner**

## TABLE OF CONTENTS

## INTRODUCTION

Implementation and success with training, interest, and motivation pro-grams for safety can be very challenging to the safety professional. Through many years of practical application, I have found several methods and tech-niques to be extremely beneficial. This chapter will give you specific examples that have worked over and over again and provide a positive direction in making a safety program successful.

## EMPOWER EMPLOYEES TO CORRECT UNSAFE CONDITIONS

Employees generally can recognize hazards, but they do not have either the know-how or the authority to get unsafe conditions corrected. This lack of information creates problems for the safety professional, who gets targeted as the safety fix-it person all too frequently. It also creates an atmosphere of frustration and complaints with employees who feel that safety issues are not being addressed or corrected in a timely manner.

1-56670-054-X/96/$0.00+$.50
© 1996 by CRC Press, Inc.

To solve this problem, the safety professional must first be able to define the process or procedure to get unsafe conditions corrected. This can vary greatly from company to company, so it is important that you are well versed in safety maintenance prioritization and procedures before you take it out to the employees for implementation. To begin the process, the unsafe condition should be evaluated by severity of the hazard and cost to repair. Those conditions that have greater potential to cause severe injury should take precedence over low hazard potential conditions. Also, the cost of repair will make a tremendous difference in how quickly the condition can be corrected. A project requiring capital appropriations and a significant amount of engineering design will take much longer than correcting a low-cost maintenance item. We all have businesses to operate, and safety maintenance must be integrated appropriately into the day-to-day business.

Since every unsafe condition can not be corrected instantaneously, employees must be educated in the process of assessing and prioritizing the repair of unsafe conditions. Each employee should have access either to a system to input work orders for correction of unsafe conditions or to an individual in their immediate operating area who has the responsibility of performing this task. The use of a line or area safety representative to perform this job can also be effective if individual access is impractical.

Once a safety work order is input into the maintenance system, there must be a method to identify and handle unsafe conditions separately from other maintenance repair issues. It helps to put a safety identifier on the work order so it is easily recognized as a safety issue and given proper prioritization. There should also be a method for the person who issued the work order to check the status of the job. This provides feedback to the employee and helps alleviate the feeling that no one is working on the problem. Once completed, safety work orders can then be sent to the safety professional for tracking the progress of the system. Work orders also provide excellent documentation for audits and inspections.

## UTILIZE YOUR IN-HOUSE RESOURCES AS INSTRUCTORS

We often fail to utilize our own in-house resources for safety training purposes. Many of us look to outside consultants to provide expertise. If this is the case in your operation, you need to stop and look at what you are missing. Every employee in your operation has some special skill or interest that could add value to your safety program. The best safety instructors are those with knowledge, operational experience, and an ability to communicate to others. Employees are much more receptive to a safety class taught by a knowledgeable peer than they would be to one taught by an outside expert. There are many organizations which provide train-the-trainer programs that last from three days to several weeks. Identify your own in-house experts and then provide them with teaching skills and an opportunity to write their own program. Once they are trained, they are there at your facility, not only as on-site instructors, but as experts to assist in accident investigations, problem solving, and advising. Some examples of areas in which operators make great instructors include the following: forklift operators and forklift maintenance mechanics are naturals as forklift instructors; employees who have experienced injuries and illnesses due to repetitive motion or have interest in competitive athletics make excellent exercise or ergonomic instructors; emergency responders make great first aid, CPR, confined space rescue, and hazwoper instructors. Make sure that those employees chosen as trainers have respected operating skills, are knowledgeable in the area they are teaching, have a personal commitment or interest in the subject, and have an ability to communicate in front of a group. You do not want to set someone up for failure. If you do it right, you will find a commitment to safety that you have never experienced before.

## MANAGEMENT COMMITMENT AND VISIBILITY = EMPLOYEE AWARENESS AND MOTIVATION

Without true management commitment and visibility throughout the operation, the safety program will never reach its greatest potential. There is nothing more greatly noticed by employees than a plant manager who regularly makes himself visible and accessible by walking through the operation and randomly stopping and talking to employees about safe work practices. When this practice is also carried out by department managers and supervisors it leaves a very strong message to employees that management is committed to safety. Be careful, though, because this can backfire if not done correctly. Make sure your management staff and supervisors are well versed in safety issues and regulations before they actively participate in this type of activity. Sincerity and follow-up are also critical. The average employee is amazingly sensitive to management behavior. If managers do not set the appropriate example and provide reasonable responses to safety questions and comments, they will immediately lose credibility with the workforce. One additional way to get increased management visibility is to have them periodically stand at the facility entrance during shift change and greet employees as they come and go to work with a safety hat, pin, pen, coffee cup, free lunch ticket, etc. The safety giveaways should tie in with a company safety theme or positive safety milestone that the employees have been working to achieve. This gives managers an opportunity to be visible and to personally thank employees for their participation in the safety program.

# 7

## INCIDENT REPORTING AND ACCIDENT PREVENTION

**George E. Swartz**

Every accident is an incident, but every incident is not an accident. Safety professionals know and understand that during the course of a normal workday there are numerous exposures which cause injury to any employee. For years, the Heinrich theory identified a set number of serious or fatal injuries, less serious injuries, and minor injuries in a triangle. This theory was further developed by adding a fourth, larger base to the bottom of the triangle which identified potential exposures or "opportunities" for injury (see Figure 7-1).

There are many that do not agree with the accident numbers originally set forth by Heinrich, but one thing is certain: every individual who has spent any time around workers who have accidents knows that there are definitely hundreds — perhaps thousands — of opportunities for injury or physical harm to employees.

It stands to reason that in the course of a normal workday there will be breakdowns of machines and unsafe situations created by fellow employees that pose a threat. Many go unseen by workers, either because they fail to really notice the danger or because they are unfamiliar with the hazard due to a lack of training and understanding. Yes, there are other hazards or situations that employees know are definitely unsafe, but they either work around them or fail to alert others to the potential perils facing them.

It is not uncommon for employees to take shortcuts in their jobs. They are convinced through past practice that they can run the machine without a guard for a few more parts or walk around the oil spill on the floor. They did it before; nothing happened to them or anyone else. "It's all a part of my job", one employee was heard to have said when confronted by a supervisor. The employee was reaching into a punch press to retrieve parts while the press was running!

1-56670-054-X/96/$0.00+$.50
© 1996 by CRC Press, Inc.

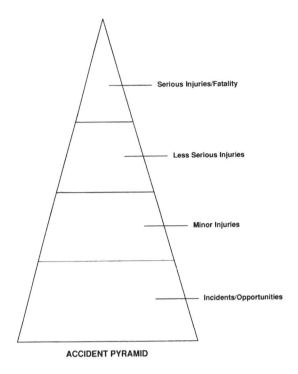

ACCIDENT PYRAMID

**Figure 7-1**  Accident pyramid.

In some cases, the supervisors fail to observe dangerous conditions during the course of the day in their departments. They unknowingly allow employees to face hazardous situations — many of which the employee accepts as being "normal for the job". In another situation, a new employee was being reprimanded for not wearing a face shield while pouring a hazardous chemical. She reminded the supervisor, "You've seen me do this before — even talked to me while I was working — but never reminded me that a face shield was required."

Who is wrong? Both the supervisor and employee are wrong. The employee is wrong for not thinking of her own welfare while being exposed to a hazardous chemical. She could have asked, "What equipment is required?" She could have observed experienced employees performing the same tasks and realized that the chemicals were harmful to the skin. The supervisor was wrong because he failed to properly inform the employee of the job requirements. A job hazard analysis could have been posted for that job, and the employee should have been required to read it before performing a new job. Also, the supervisor must be constantly on the alert for "things that are out of place", like the lack of personal protective equipment.

The situation mentioned above is just another example of an opportunity for injury in a workplace. The employee being reprimanded performed the task dozens of times without a face shield and was never harmed. She could have continued for weeks, even months, and escaped harm each time. Each repetition of the job convinced her that she was okay and that there was nothing out of place. Her bad habit became a fixed habit. The fact remained that her next pouring of the acid could have resulted in disfigurement, or it could have been months before the catastrophe struck.

Opportunities abound for injury in any workplace. Employees many times notice fellow workers performing tasks unsafely. In some cases they remind them of their folly; in other cases they take the position, "Let them find out for themselves — it will teach them a lesson." How many times have you heard the comment, "I knew he was going to get hurt one day by doing that." If you knew your child was going to touch a hot stove and suffer burns to the hand, wouldn't you, as a responsible parent, keep the child from harm's way through education, enforcement, and perhaps moving hot pots to the back burners?

Since many hazards in the workplace cannot be totally removed or guarded, other means of safeguarding must be employed. Employees must be educated as to the inherent dangers in their departments. Safe behavior must be required of every employee and supervisor. Visitors must be required to follow basic safety rules. Enforcement must take place on a regular basis — not just when everyone is reacting because of a bad accident.

A good phrase to follow is "condition corrected — accident prevented". An ongoing effort must be made to utilize a program in which all employees are constantly on the alert for a hazard — be it physical or behavior related. Each time a hazard is corrected the potential for injury or economic loss is reduced or removed. Wipe up the oil spill; do not walk around it. Remove the box from the steps; do not walk around it. Fix the guard on the machine even though the machine is not running. Remind the forklift operator to slow down. Remind employees to wear all required personal protective equipment, etc.

At the beginning of this chapter the Heinrich theory was mentioned. Whether you agree with this concept or not, it is hard to deny the significance of bottom portion or base of the pyramid. There are hundreds to perhaps thousands of opportunities for injury in the workplace on any given day. Management should focus on these day-to-day incidents so that more serious injuries are prevented. Many fail to realize this because most of the time these close calls or "accident" opportunities do not result in injury to anyone. Because of this, there is usually a lack of effort to identify and correct such hazards. Only when a serious injury takes place does someone take notice of the circumstances leading to the event. The employee could have been performing this job in a dangerous manner for years before his injury occurred.

These incidents must be corrected on an ongoing basis. Unsafe conditions or unsafe behavior — regardless of degree — must be corrected to prevent injury or loss. Keep in mind that many times the smallest or most insignificant event could lead to potential disaster. Today's near miss could be tomorrow's fatality.

```
┌─────────────────────────────────────────────────────────────────────────────┐
│  ⟨mïDAS·⟩           NON–INJURY INCIDENT REPORT                                │
│                      PROPERTY DAMAGE REPORT                                    │
│                              AND/OR                                           │
│   A Whitman Company   DISCOVERY OF POTENTIAL HAZARD                            │
├─────────────────────────────────────────────────────────────────────────────┤
│                    (CHECK ALL THOSE THAT APPLY)                                │
│     □  Near Miss    □  Non–Injury    □  Property Damage   □  Incident/Event   □  Fire Loss │
│                                                                               │
│  Market: _____  Shop Location: _____  Date Incident Occurred: _____ │
│  1)  Describe the incident or what occurred:_____  │
│      _____ │
│      _____ │
│      _____ │
│      _____ │
│                                                                               │
│  2)  Machine or equipment involved: _____ │
│  3)  Extent or damage/costs (describe):_____ │
│      _____ │
│      _____ │
│                                                                               │
│  4)  How much lost time or down time involved:_____ │
│  5)  How many employees were involved in incident: _____ │
│  6)  Was anyone  injured:        Yes □     No  □                               │
│  7)  Was first aid administered:  Yes □     No  □                              │
│  8)  Was there a fire:            Yes □     No  □                              │
│  9)  Was employee wearing required personal protective equipment:  Yes □    No □ │
│  10) What steps were taken by you to prevent recurrence:_____  │
│      _____ │
│      _____ │
│                                                                               │
│  11) What steps were taken by others (maintenance, management, outside sources, etc.) to prevent recurrence: │
│      _____ │
│      _____ │
│                                                                               │
│  12) Was a health hazard or exposure involved:   Yes □    No  □               │
│  13) Can this incident or event take place again: Explain on back side.  Yes □    No □ │
│  14) Identify additional factors or comments on back of this report.          │
│  15) Have you drawn a sketch of the details on the back of this report or attached a photo:  Yes □   No □ │
│                                                                               │
│      Signature: _____       Date: _____   │
│  Forward Top Copy to:   Corporate Safety                                      │
│  Middle Copy:           Shop Files                                            │
│  Bottom Copy:           Market/Administrator's File                           │
│                                                                               │
│  3400–64                                                        Revised 11/91 │
└─────────────────────────────────────────────────────────────────────────────┘
```

**Figure 7-2**  Sample form for reporting and correcting hazards.

For a safety program to be truly effective the effort must focus on all close calls or near misses; damaged equipment, property, or products; a fire of any size; or failures in the maintenance program. As an example, a forklift has defective brakes. The operator fails to notify his supervisor. In addition, there is no system in place which requires daily lift truck inspections. The employee fails to slow down at a turn in the factory, the brakes fail, and he crashes into a pallet of electronic parts. No one is injured, but thousands of dollars have been lost through product damage. The operator knew for weeks that the brakes were defective but did not step forward to inform management. Perhaps someone could have been in front of the speeding forklift when the brakes failed, resulting in a serious injury or death.

Management cannot expect employees to step forward and report all hazards. Systems must be in place that require inspection, reporting, and discussion of defective machines, close calls, etc. Organizations should use a simple yet effective form for reporting and correcting incidents (see Figure 7-2). Supervisors should be required to look for these "opportunities" to prevent loss of any kind. Employees should be encouraged to report various incidents to management. The completed forms can serve as material for safety meetings and education. Photos can be taken of various situations so employees can be better informed as to the hazards in their departments. What better way to alert everyone to a nearly fatal accident than to record it, photograph it, discuss it, and post the reminder for everyone to see? The same hazard or situation could be present in another department or on the next shift, and disaster could be avoided by management action.

Supervisors are in the best position to process incident reports. A safer department will be the result of their ongoing efforts. As the program evolves, many unsafe conditions will be corrected, which will help remove many hazards from the workplace. The investigation and correction of incidents on a regular basis will definitely reduce accidents. In addition, economic losses will be reduced significantly by taking early action on all incidents.

## FURTHER READING

### Books

National Safety Council. *Accident Prevention Manual for Industrial Operations,* 10th ed. Itasca, IL: National Safety Council, 1992.

Manuele, F. H. *On the Practice of Safety.* New York: Van Nostrand Reinhold, 1993.

Colvin, R. J. *The Guidebook to Successful Safety Programming.* Boca Raton, FL: Lewis Publishers, 1992.

### Other Publications

Swartz, G. Incident reporting — a vital part of quality safety programs. *Prof. Saf.,* 38(12), 32–44, December 1993.

# 8

# PRINCIPLES, TECHNIQUES, AND OBJECTIVES FOR EFFECTIVE SAFETY MANAGEMENT

**Herman Woessner**

## TABLE OF CONTENTS

## INTRODUCTION

There are six safety management objectives which must be achieved by and incorporated into a safety and health program for it to be effective. This chapter will analyze the objectives and the principles upon which they are based and will describe proven techniques for implementing them.

The techniques described in this chapter by no means exhaust the universe of possibilities. They simply are measures that have proven to be effective over a wide range of public and private organizations with which I have been

1-56670-054-X/96/$0.00+$.50
© 1996 by CRC Press, Inc.

associated in a professional career spanning 25 years. Other safety practitioners, working with other types of organizations, at other times, undoubtedly have developed and will continue to develop different, equally effective methods for accomplishing safety objectives.

Even though many different methods can be employed to accomplish the six safety management objectives, the objectives themselves and the principles upon which they are based are universal. They define and determine the effectiveness of all safety and health programs.

## ROLES AND RESPONSIBILITIES

The fundamental responsibility of a safety and health professional is to facilitate and coordinate the development and implementation of an effective injury and illness prevention program. The key role to be played in fulfilling this responsibility is that of program leader, not hazard controller, which is often the role many practitioners assume and some even relish, particularly those who have not been mentored.

Controlling hazards (i.e., unsafe conditions and unsafe work practices) is primarily the responsibility of employees and supervisors. They are closest to the hazards and have the greatest capability of recognizing and controlling them. There can never be enough safety and health professionals present in the workplace, with enough authority, to effectively control the myriad of hazardous employee exposures that can occur on any given workday. Thus, most of the organizational time and energy of safety and health professionals should be spent on influencing and assisting managers, supervisors, and employees to successfully meet their responsibilities.

To influence the behavior of others, one must be able to effectively communicate to them how the desired behavior supports or enhances their valued interests and why it is more beneficial than competing behaviors. While everyone is interested in "safety", the degree of interest among and between managers, supervisors, and employees varies, for each group and each individual holds values and interests which compete for time, energy, and resources with their safety interests.

To be a successful safety leader, the safety and health professional must first identify and understand the individual and organizational interests and competing values of the person or persons whose behavior she/he is trying to influence and then speak to those interests by demonstrating how the desired safety behavior will positively affect them.

Assume, for example, that a safety and health practitioner for a private company discerns through interactions with a plant manager and his/her peers that cost control is one of the manager's major organizational concerns and that receiving a high performance appraisal from his/her boss is of significant personal interest. Also, assume that the safety and health professional wishes to have the manager demonstrate greater safety leadership to improve the quality of housekeeping and employee training in the plant. How should she/he proceed to convince the manager to lead the recommended safety improvement effort?

The answer lies in providing the manager with valued information that supports his/her organizational and personal interests. In this example, the valued information must relate to cost control and management effectiveness. The safety and health professional should, therefore, obtain and analyze all available records to determine the historical, current, and potential costs of work-related injuries and illnesses at the plant and the affect they have had in the past, and are likely to have in the future, on the manager's budget and cost containment goals. In addition, the records should be analyzed to identify the significant causal factors, which, presumably, will include poor housekeeping and inadequate employee training.

If available, the following experience and cost data should be analyzed, organized, and presented to the manager to help him/her better understand and appreciate the need to take the desired action:

| Loss Experience Data | Cost Data |
| --- | --- |
| Injury and illness cases | Workers compensation cost |
| | Productivity loss |
| Property damage incidents | Property damage cost |
| Environmental impairment incidents | Environmental remediation cost |
| | Public image loss |
| Hazardous materials exposures | Medical cost |
| | Labor/public relations loss |
| Business interruption incidents | Productivity loss |
| | Customer loss |
| Regulatory citations | Penalty cost |
| | Government relations loss |
| Legal cases | Legal cost |
| | Labor/public relations loss |

The above is a win-win approach; it gives positively valued information for valued actions, which is much more effective than threats of regulatory penalties or predictions of dire consequences. Too many safety and health practitioners rely on this "Chicken Little" approach to influence their organizations. They do so because scare tactics require little effort to develop and sometimes produce the desired result faster than would the systematic process of analyzing and marshalling positively valued information. However, the continual use of such tactics can eventually lead to a loss of creditability and effectiveness for the safety and health professional. Over the long run, no one appreciates the bearer of only negative news. Eventually, the messenger will become associated with the message.

The positive approach also is not limited by rank or organizational type, as are the negative tactics. Employees and public agencies, for example, are little influenced by the threat of an OSHA citation, for they are usually exempt from any penalties from failure to comply. On the other hand, they will respond to information that helps them to realize their personal and organizational goals.

The interests of persons and organizations, and the information required to influence them, will vary, but the approach need not. The challenge to the safety and health practitioner is to identify the organizational and personal interests that motivate individuals in his/her organization and provide them with accurate and credible data in a manner that complements those interests. This is easier said than done, for it requires the safety and health professional to have good analytical and communication skills, organizational credibility, and a nonthreatening personality, and it requires management to ensure performance accountability.

To be effective, the safety and health professional must first develop the necessary analytical and communication skills and earn creditability within the organization. Then the professional can help establish accountability criteria for acceptable safety performance by managers, supervisors, and employees.

## SAFETY AND HEALTH PROGRAM

A safety and health program is a network of dynamic policies, goals, responsibilities, subprograms, plans, procedures, and standards designed to guide an organization to proactively and systematically identify and control workplace hazards and risks. The wide scope and complexity of such a system require that it be written down to ensure completeness, clarity, and consistency.

The development of a safety and health program should be planned and designed to accomplish the six universal safety management objectives. It should not just evolve in response to promulgated OSHA regulations or as a reaction to injury/illness incidents experienced by the organization. This reactive approach invariably produces unrelated, inconsistent, duplicative, and ineffective program elements.

A quality control process must be an integral part of every safety and health program, for it is critical to the program's long-term effectiveness. Without a quality control process, any system will become inconsistent and outmoded over time. To prevent this from happening to the safety and health program, the responsibilities and operations comprising the quality control process should be written into the program, either as a separate element or as the quality control section in each safety plan and procedure.

The quality control process is a self-regulating subsystem for ensuring continuous improvement of program performance and content. It should provide for (1) periodic, random observation and focused analysis of required program activities; (2) evaluation of written program elements in light of the performance observations and analyses; (3) feedback of the analytical and evaluation findings to persons with the authority and responsibility for correcting program deficiencies; and (4) observation and analysis of the improvement actions taken and evaluation of the results.

## OBJECTIVES, PRINCIPLES, AND TECHNIQUES

### Safety Leadership

The primary safety management objective that must be achieved by and incorporated into an organization's safety and health program is to have top and middle management actively lead employees toward achievement of the organization's safety goals. Achievement of the other safety management objectives is largely dependent upon the quality and consistency of leadership demonstrated by management.

Leaders convey vision and values to their organizations through interactions and communications with subordinates. Subordinates judge their leaders' commitment to the vision and values by the frequency, consistency, and sincerity of their written and verbal statements and body language.

The significant measures of a leader's commitment to stated policies and objectives are the quantity and quality of the time and resources (human, financial, physical) she\he devotes to achieve them. Thus, managers must devote their time and the organizational resources they control in support of the safety and health program if they expect employees to fully appreciate its value and to enthusiastically work to achieve its goals. Moreover, managers must be consistent, positive, and sincere in their safety interactions and communications with employees if they expect to be believed and consistently followed.

Including a safety objective in the organization's mission statement is an effective first step toward establishing safety as a valued organizational function. It places safety on the same level as the other critical mission objectives, profit and quality, and conveys that the safety and health program is integral to their achievement.

To illustrate how and why a safety objective can be incorporated in an organization's mission statement consider the examples of ABC Company and The Public Agency. The stated mission of ABC Company is "to manufacture the highest quality widget at the lowest cost", while The Public Agency does not have a written mission statement.

ABC Company's mission statement could be revised to read: "The mission of ABC Company is to **safely** manufacture the highest quality widget at the lowest cost." By inserting the word "safely" in the statement, management is committing the organization to the goal of injury/illness prevention. It implies that even if the company succeeds in becoming the high-quality, low-cost leader of the widget manufacturing industry, its mission will not be fully achieved unless the risk and experience of employee injury are low.

A mission statement for The Public Agency could be written to read: "The mission of The Public Agency is to provide public services in a courteous, expeditious, and efficient manner while protecting the health and safety of employees, the public, and the environment." Such a statement defines occupational and environmental protection as important operational objectives and serves as a visual reminder to the organization of management's commitment to employee and public safety and to environmental protection.

Developing a safety policy statement that defines and reinforces the safety objective expressed in the mission statement is an essential next step in institutionalizing the organization's safety management commitment. The safety policy statement should describe the organization's core beliefs, commitments, and responsibilities regarding safety and should connect the success of the safety and health program to the success of the organization's overall mission.

An example of a connective statement would be: "We believe that XYZ Organization has a fundamental responsibility to protect the health and safety of our employees and that fulfillment of that responsibility is critical to the organization's overall mission. We further believe that establishment of an effective safety and health program is essential for meeting our safety responsibilities and for achieving our mission objectives."

The owner and/or top manager(s) in the organization should sign the finished policy statement, and it should be distributed and posted throughout the organization along with the mission statement.

The process of developing and/or revising the organization's mission and policy statements undoubtedly will serve to broaden management's understanding of its safety roles and responsibilities, and the statements themselves will serve as visual reminders of the organization's commitment to safety.

Another technique that management can use to lead the organization to improved safety performance is to develop and communicate behavior-directed and results-directed safety goals and measures. Establishing the objective of a daily supervisory-conducted "toolbox" safety meeting with employees would be an example of a behavior-directed safety goal, while a 10% reduction in a facility's annual injury/illness rate as an objective is an example of a results-directed goal.

By establishing and communicating such goals and visibly monitoring performance toward their achievement, management conveys that prescribed safety behavior and desired injury/illness results are of value to the leadership and the organization.

Management should conduct regularly scheduled reviews of the facility's/company's safety performance versus the established goals to demonstrate its safety vision and interests, as well as to identify and correct any program deficiencies.

Safety and health professionals should seek to influence and assist management's safety leadership efforts. This can be done by using the valued information approach discussed earlier to convince them to adopt the methods covered in this section.

## Employee Participation

A safety management objective closely related to that of having top and middle managers actively and visibly lead the organization's safety efforts is to achieve widespread employee participation in safety program activities. The more employees are involved in the program and the greater the involvement, the more likely they will feel responsible for ensuring its success.

Employees are much more likely to develop a sense of ownership for the safety program and to understand, appreciate, and support its requirements if they help to develop and maintain it as well as be expected to implement it. Management should encourage employees from every department and level of the organization to participate in the development and monitoring of the organization's safety and health policies and procedures, including the mission and safety policy statements. This can be done by establishing a steering committee for each policy/procedure and an implementation monitoring committee in each department.

The committee members must be trained to perform their assignments. Those involved in program development should be informed by management about the purpose, objectives, and format of the proposed policy/procedure and the process for developing and writing it. Those committee members engaged in monitoring policy/procedure implementation activities should be trained in the requirements of the policy/procedure and in methods for behavior monitoring and reporting.

Management should recruit qualified volunteers from each department to serve as safety communicators, problem solvers, and/or trainers. These employees would be trained to assist supervisory personnel in communicating hazard awareness and hazard control information to their workers and in facilitating solutions for program implementation conflicts. Bilingual employees could be chosen as communicators and problem solvers for issues affecting individuals and ethnic groups for whom English is a second language.

There are many ways in which employees may participate in an organization's safety and health program; the more participation the better. Employee participation, like the other safety management objectives, has a much greater probability of being achieved if it is planned. Management must support the plan and ensure that employees are properly trained and their contributions recognized. The likelihood of these requirements being met will be much greater if the "Employee Participation Plan" is institutionalized as an element of the organization's written safety and health program. The organization's safety and health professionals should take the lead in having a written plan drafted and approved as an important element of the safety and health program.

## Knowledgeable Program Participants

Developing knowledgeable program participants is the third safety management objective that is essential to establishing and maintaining an effective safety and health program. Managers, supervisors, workers, safety committee members, program monitors, trainers, safety communicators, problem solvers, and safety department personnel are all program participants. The more they know about how to carry out their program responsibilities, the more success-ful they are likely to be.

A safety training subprogram should be included in every organization's safety and health program to ensure that a sufficient amount of safety training occurs each year. The subprogram should specify the frequency, duration, and type of training and retraining required by OSHA and by other elements of the safety and health program; a minimum amount of safety leadership training for managers and supervisors; training documentation requirements; and the person(s) or department responsible for conducting or arranging the required training and for maintaining the training records. The subprogram should also include a provision for conducting safety training for employees who, for whatever reason, miss a scheduled training session.

The safety and health professional should recommend that the top manag-ers of the organization attend an in-house or external seminar on the principles and techniques of safety leadership and management at least once a year and should seek their approval to include a commitment for such training in the safety training subprogram.

Many organizations hold annual planning or operational review meetings at which either staff or outside consultants make presentations on various topics applicable to the objectives of the organization. This type of setting would be ideal for a session on safety leadership/management. It could serve as the forum for annually educating and motivating managers about their safety leadership roles and responsibilities. To convince top managers of the value of including safety on the agenda of such meetings, the safety and health profes-sional can point out that it will be supportive of and consistent with their expressed commitments in the organization's mission and safety policy statements.

The safety training subprogram, like all elements of the safety and health program, must be regularly monitored to determine if the frequency and quality of required activities are sufficient. The subprogram should outline reporting and follow-up actions to be taken in response to any training and record-keeping deficiencies found by the monitors.

## Hazard Identification and Control

Establishing a proactive system for identifying, evaluating, and control-ling unsafe conditions and unsafe work practices is the fourth essential safety management objective. The occurrence of an injury or illness is largely depen-dent upon the frequency and duration of exposure to uncontrolled hazards. Thus, the longer and/or more frequently an employee is exposed to an unsafe condition or commits an unsafe work practice, the greater the likelihood that a harmful encounter will occur. The severity of that encounter depends on the amount of energy or toxicity that can be imparted to the exposed employee by the hazardous condition or unsafe work practice.

## RISK ASSESMENT GUIDE

| AVERAGE NUMBER OF EMPLOYEES EXPOSED TO HAZARD EACH DAY | TOTAL DURATION OF DAILY EXPOSURE TO Hazard | AVERAGE DAILY FREQUENCY OF EXPOSURE TO HAZARD | TOTAL DAILY & YEARLY EXPOSURE TO HAZARDS IN HOURS | LIKELIHOOD OF AN INJURY OR ILLNESS OCCURRING FROM THE TOTAL DAILY AND YEARLY EXPOSURE TO THE HAZARD |
|---|---|---|---|---|
| A | B | C | $D = A \times B \times C$<br>$Y = D \times 256$ days | |
| Example:<br><br>10 exposed workers | Example:<br><br>1.5 hours of exposure | Example:<br><br>2 exposures per day | $D = 10$ employees x 1.5 exposure hours x 2 exposures per day<br><br>$Y = 30$ hours of worker exposure per day x 256 days in year<br><br>$Y = 7,680$ annual exposure hours | (a) Very Likely<br>(b) Likely<br>(c) Somewhat Likely<br>(d) Unlikely<br>(e) Very Unlikely |

| MAXIMUM AMOUNT OF ENERGY AND/OR TOXICITY LIKELY TO BE TRANSFERRED FROM/BY THE HAZARD TO THE EXPOSED EMPLOYEE(s) | LIKELY HARM TO THE MOST EXPOSED EMPLOYEE(s) FROM THE MAXIMUM AMOUNT OF ENERGY/TOXICITY TRANSFERRABLE FROM/BY THE HAZARD | R I S k | (a) | (b) | (c) | (d) | (e) |
|---|---|---|---|---|---|---|---|
| | (1) | | CLASS 1 RISK | CLASS 1 RISK | CLASS 1 RISK | CLASS 1 RISK | CLASS 2 RISK |
| ----------------> | A S S E S S | (2) | CLASS 1 RISK | CLASS 1 RISK | CLASS 2 RISK | CLASS 2 RISK | CLASS 3 RISK |
| Very Large<br>Large<br>Moderate<br>Low<br>Negligible | (1) Death<br>(2) Permanent Impairment<br>(3) Temporary Impairment<br>(4) Injury but no Impairment<br>(5) Pain but no Impairment | M E N T (3) | CLASS 1 RISK | CLASS 2 RISK | CLASS 3 RISK | CLASS 3 RISK | CLASS 4 RISK |
| | ----------------> | M (4) | CLASS 2 RISK | CLASS 3 RISK | CLASS 4 RISK | CLASS 4 RISK | CLASS 4 RISK |
| Note: harmful energy may be in the form of electrical current, mechanical movement, chemical reaction, radiation, gravitation, heat, etc.<br><br>Toxicity may be in the form of airborne gases, vapors, particulates, fungal spores, bacteria etc., as well as corrosive and flammable liquids/solids. | Note: Impairment is Dismemberment, Disfigurement, or Functional Disability. | A T R I X (5) | CLASS 4 RISK | CLASS 4 RISK | CLASS 4 RISK | CLASS 5 RISK | CLASS 5 RISK |

| RISK CLASSES | HAZARD CONTROL PRIORITIES: |
|---|---|
| Class 1: Great Risk<br>Class 2: High Risk<br>Class 3: Moderate Risk<br>Class 4: Low Risk<br>Class 5: Negligible Risk | Priority 1: Control Class 1 Risks immediately<br>Priority 2: Control Class 2 Risks as soon as possible within 30 days<br>Priority 3: Control Class 3 Risks as soon as possible within 90 days<br>Priority 4: Control Class 4 Risks as soon as possible within 1 year<br>Priority 5: Control Class 5 Risks when practical |

The key to an effective hazard identification, evaluation, and control system is to prepare and require those closest to the hazards (i.e., supervisors and employees) to regularly and systematically conduct workplace inspections. The preparation is accomplished through training that focuses on teaching supervisors and employees to carefully observe and analyze job conditions and activities and to assess the injury/illness risk they present to exposed workers.

Risk assessment is an evaluation of the probability of an injury/illness occurrence based upon the frequency and duration of employee exposure to a hazard and an evaluation of the likely magnitude of the harm from such an occurrence based on an assessment of the amount of toxicity or energy that would likely be imparted from the hazard to an affected employee. The highest risk hazards have high energy/toxicity potential and frequent employee exposure. Identification and control of these risks and the hazards should be the first priority of the hazard control system.

In addition to requiring hazard recognition and control training for supervisors and employees, the system must establish minimum standards for inspection frequency and for the timeliness of implementing hazard control measures. The inspection frequency standard should be no less than monthly, and the time requirement for controlling identified hazards should be based on the risk.

## Performance Goals, Standards, and Measures

The fifth safety management objective that the safety and health program must achieve is the creation of behavior-directed and results-directed safety performance goals, standards, and measures. Behavior-directed goals seek to focus the organization's attention and efforts on desired safety behavior itself, rather than on the expected consequence of that behavior. Results-directed goals, on the other hand, focus attention and efforts on the expected consequences (i.e., lower injury/illness numbers, rates, and costs) in order to encourage the desired safe behavior. A goal to have supervisors and employees conduct a weekly safety inspection of their work areas is an example of a behavior-directed safety management goal. It directly encourages specific supervisory and employee behavior known to be effective in the identification and control of hazards. A goal to have supervisors and employees reduce the number of injuries and illnesses in their work units by 10% is an example of a results-directed goal. It exhorts supervisors and employees to take unspecified actions to prevent and control the causes of injuries and illnesses.

Performance standards establish the preferred ways and means for achieving performance goals and, together with the goals, serve as the measures of performance effectiveness. A performance standard for the above example of a behavior-directed goal could be to have a team consisting of the unit supervisor and at least two unit employees conduct the weekly safety inspection using a checklist of safe work practices and good housekeeping requirements. The same performance standard could be used in support of the results-directed goal.

By comparing and evaluating actual behaviors and comparing results to performance goals and standards, an organization can measure the effectiveness of its safety and health program and the individuals involved. The frequency, timeliness, and quality of prescribed behaviors are the principle measures of performance relative to behavior-directed goals and standards, while injury/illness experience rates and associated costs are the key measures of results-directed goals.

The following performance indicators are some of the important measures that can be used to evaluate the overall effectiveness of the safety and health program and to point out areas in need of improvement:

| Results Measures | Behavior Measures |
|---|---|
| Total case rate: number of injuries and illnesses × 200,000 hours ÷ number of hours worked | Frequency and quality of program-required workplace inspections |
| Lost time case rate: number of lost time cases × 200,000 hours ÷ number of hours worked | Frequency and quality of program-required job observations |
| Lost and restricted case rate: number of lost time and restricted work cases × 200,000 hours ÷ number of hours worked | Frequency and quality of program-required training |
| Lost and restricted day rate: number of days lost and restricted × 200,000 hours ÷ number of hours worked | Frequency and quality of program-required audits |
| Total workers' compensation reserves: the amount reserved by the insurance company to pay for current year's injuries/illnesses | Frequency and quality of program-required safety meetings |
| Experience modification (Ex-Mod) factor: the multiplier, based on injury/illness experience, used by insurance companies to determine workers' compensation premiums; an ex-mod factor greater than one is high | Frequency and quality of program-required safety documentation |
| Property damage and business interruption costs: the direct and indirect costs resulting from accidents involving property damage and/or business interruption | Frequency and quality of program-required safety performance appraisals |
| Vehicle accident incidence rate: the number of vehicle accidents × 25,000, 100,000, or 1,000,000 miles ÷ number of miles driven in a year | Number and frequency of employee safety suggestions |
| Total vehicle accident costs: the direct and indirect costs to pay for accident and to restore vehicles | Timeliness of program-required responses to employee safety suggestions |
| Annual number of OSHA citations: the total number of citations received in a year | Timeliness and quality of accident/incident investigation |
| Cost of OSHA penalties: annual cost of penalties for OSHA citations | Number of employees properly wearing program-required personal protection equipment |

## Safety Performance Accountability

Accountability for performing to program standards is the sixth essential safety management objective that must be achieved by and incorporated into a safety and health program. It provides program participants the necessary impetus and reinforcement to consistently strive to achieve program goals and objectives.

Accountability involves a supervisor's evaluation of the actual performance of a subordinate over a specific period of time as measured against established performance goals and standards. The supervisor reviews his/her findings with the subordinate and takes steps to reinforce acceptable behavior/results and to modify unacceptable behavior/results. The supervisor has to acknowledge with praise, or some other form of recognition, subordinate performance that has met or exceeded standards and must provide encouragement and support to improve performance that is not up to standard. This support can take many forms, ranging from verbal encouragement, to support for additional training, to providing needed tools and equipment, to discipline.

Discipline should only be used when it is necessary to correct persistently substandard performance. It should be employed as a last resort, only after positive accountability methods have failed to achieve acceptable performance.

Supervisors communicate value to their subordinates through the accountability process. Whatever responsibilities supervisors hold their subordinates accountable for will be a priority for subordinates because of their obvious importance. Supervisors would not spend the time and effort to hold subordinates accountable for achieving them if they were not important. Conversely, responsibilities for which subordinates are not held accountable by their superiors, or for which accountability is sporadically or inconsistently administered, will be recognized by subordinates as being of less value than their accountable responsibilities.

Accountability cannot be an effective guarantor of performance, however, unless the means for achieving the desired goals and standards are adequate and available to the responsible persons. Employees cannot be expected to work safely and avoid injury, for example, unless they are trained to recognize workplace hazards and are provided the means to control them. Thus, the key responsibility for which all superiors in an organization should be held accountable is the responsibility of providing their subordinates with sufficient information and resources to meet their safety responsibilities.

The resources needed to achieve safety performance goals, and the goals themselves, should be the product of a carefully thought-out safety management plan, one which is based on an evaluation of past safety program accomplishments and failures and which is supported by budgeted funds. An action plan unsupported by budgeted funds is only a wish list.

The process for holding safety program participants accountable and the process for establishing a safety plan and budget should be incorporated as separate elements in the written safety and health program. This would help to ensure their consistent implementation.

The accountability process element should describe the actions to be taken by supervisors and subordinates to determine the appropriate behavior-directed and results-directed goals and standards against which the subordinate's performance will be measured and should set the frequency, format, and documentation requirements for conducting performance evaluations. The program element also should describe the actions to be taken by supervisors and subordinates in response to the evaluation findings.

One action that must be taken to ensure the effectiveness of the accountability process is to tie the evaluation findings to the organization's pay-for-performance system. While the prospect of gaining or losing a monetary reward alone cannot guarantee desired performance results, it does convey value to the affected individuals in the organization and should serve as a powerful performance motivator.

The program element describing the safety planning and budgeting process would cover the required actions to (1) collect performance and costs information for identifying program strengths and weaknesses; (2) obtain input from affected and knowledgeable employees, supervisors, and managers on existing safety problems; (3) formulate an improvement plan based on the data collected and input received; (4) identify the financial, human, and material resources needed to accomplish the plan; (5) create organizational and departmental budgets to pay for planned activities; and (6) establish plan responsibilities, completion dates, and an evaluation process.

## CONCLUSION

The effectiveness of an organization's safety and health program is defined and determined by the degree to which it achieves the six safety management objectives discussed in this chapter. Achievement of the objectives can be greatly facilitated by including them in the written safety and health program along with the policies, procedures, and processes required to achieve them. The primary responsibility of safety and health professionals is to facilitate and coordinate the development and implementation of the program to achieve the stated objectives.

## FURTHER READING

Bird, F. E. and Germain, G. L. *Practical Loss Control Leadership.* Loganville, GA: International Loss Control Institute, 1990.

Drucker, P. *The Practice of Management.* New York: Harper & Row, 1985.

Grimaldi, J. V. and Simonds, R. H. *Safety Management,* 4th ed. Homewood, IL: Richard D. Irwin, 1984.

Krause, T. R., Hidley, J. H., and Hodson, S. J. *The Behavior-Based Safety Process.* New York: Van Nostrand Reinhold, 1990.

Petersen, D. *Techniques of Safety Management, a Systems Approach.* Goshen, NY: Aloray, 1989.

Roland, H. E. and Moriarty, B. *System Safety Engineering and Management.* New York: John Wiley & Sons, 1990.

# 9 PROCESS SAFETY MANAGEMENT

**Richard C. Nugent and Mark D. Hansen**

The Occupational Safety and Health Administration (OSHA) has implemented a standard designed to protect our processes, employees, and the public from accidents involving the use and manufacturing of hazardous chemicals. The standard, OSHA CFR 1910.119, "Process Safety of Highly Hazardous Materials Standard", published in February 1992, requires companies to conduct a formal hazards analysis and risk assessment on all applicable processes. Once the analysis is complete, the company must eliminate the hazard or establish controls aimed at reducing the risk of catastrophe due to fire, explosion, failure, or release. This is to be done in the design phase, as well as being performed on existing operations. In addition to the U.S., similar regulations have been enacted in the U.K., Canada, Australia, The Netherlands, Mexico, and Italy. In an effort to provide their subscribers with other standards/guides, industry associations in the U.S. have developed the standards listed below:

1. American Petroleum Institute, "Recommended Practice RP-750, Management of Process Hazards"
2. American Institute of Chemical Engineers, Center for Chemical Process Safety, "Guidelines for Technical Management of Chemical Process Safety"
3. Chemical Manufacturer's Association, "Responsible Care Program"

The U.S. Environmental Protection Agency (EPA) also published similar requirements as a part of the Clean Air Act Amendment, section 112(r), published in November 1990. This section establishes the requirements for the development of a risk management plan.

Over the past 10 years, it has been estimated that failures in processes involving hazardous chemicals have cost industry more than $2 billion. Figure 9-1 indicates that the most serious losses were incurred in 1989 (Krembs and Connolly, 1990).

| Incident | Estimated Loss (in Millions) |
|---|---|
| Richmond, CA (4/89)<br>$H_2$ gas fire | $90 |
| Morris, IL (6/89)<br>Vapor cloud explosion | $41 |
| Martinez, CA (9/89)<br>$H_2$ and hydrocarbon release | $50 |
| St. Croix, Virgin Islands (9/89)<br>Hurricane Hugo | $60 |
| Pasadena, TX (10/89)<br>Vapor cloud explosion | $725 |
| Baton Rouge, LA (12/89)<br>Pipeline rupture | $43 |
| Total | $1009 |

**Figure 9-1**  Serious Industrial Losses, 1989.

In addition to the financial loss, many lives were lost or affected due to these accidents. Based on this information, it would seem that it makes plain good business sense to perform the analysis and protect the process against failure. Whether you are applying good business practices or you are complying with this standard, the application of sound management practices will provide the necessary means to ensure success.

As in the management of any project, program, or process, you must first establish a plan to determine your course of action. The plan to comply with the requirements of "process safety" is as follows:

1. Determine if the standard applies to your operation.
2. Familiarize yourself with the required elements of the standard.
3. Determine what resources are needed for compliance.
4. Draft a written plan or procedure to state how you will comply.
5. Implement the specific plan.
6. Develop and complete recommendations or action items, resulting from assessments/analyses.
7. Document, document, document.

It should be stated that obtaining management support for this undertaking is necessary in each of the listed steps. They must buy in to this effort if it is to succeed. A discussion of each of the above listed steps follows.

1. **Determine if the standard applies to your operation.** The standard requires compliance if your operation uses hazardous, flammable, explosive, or toxic chemicals above a listed minimum threshold. Exceptions are written in the rule, such as flammables used for fuel. Know the chemicals and quantities used in your process.

2. **Familiarize yourself with the required elements of the program.** The required elements of the standard are

- **Process safety information** — This element establishes the means by which the hazards of a chemical process or a mechanical or electrical operation are identified and understood. It consists of hazards of materials used, process and mechanical design of equipment or systems, and documentation such as engineering drawings, equipment specifications, materials of construction, relief system design, etc.
- **Process hazard analysis** — This consists of the systematic identification, evaluation, and elimination and/or control of process-related hazards to prevent catastrophic incidents. Various hazard analysis methodologies are suggested, i.e., "what if" analysis, hazard and operability study (HAZOP), fault tree analysis, and failure mode and effect analysis (FMEA). Figure 9-2 provides a list of considerations/questions for use when conducting the process hazard analysis (PHA).
- **Management of change** — Written procedures stating how you will manage changes in technology, facilities, and equipment must be developed. These changes, usually "not in kind", must be properly managed to avoid serious process incidents. The basis for change, safety and process impacts, consequences of deviations, engineering, and administrative control of changes must be considered and documented. All affected personnel must be advised of any changes.
- **Operating procedures** — Written procedures must be developed. Included in these procedures are operating limits, safety, health and environmental considerations, how to correct or avoid deviations, and clear logical instructions for the safe operation and maintenance of the process.
- **Safe work practices** — Written procedures must be developed for the following activities: lockout/tagout, confined space entry, line breaking, hot work or burning and welding, and any other procedure specific to your operation.
- **Process safety reviews** — This element is similar to that of element #2, except this element requires the reviews to be conducted at various stages. The stages are design, construction, installation, commissioning, start-up, operation, maintenance, and demolition. These reviews require the use of a review team consisting of members representing various backgrounds of expertise such as chemical/technical, safety, electrical, mechanical, operations, etc.
- **Training** — This is often one of the most overlooked elements and is perhaps one of the most important. Operating and maintenance personnel must be informed concerning the proper and safe operation of the process. Chemical and process hazards must be stressed.
- **Contractors** — Provisions for instructions, including safety rules for contractors working on or near hazardous processes, must be provided to contractors. The contractor must be informed about chemical hazards, safety procedures, emergency procedures, and other requirements specific to your operation. Requirements for information to be provided to you by the contractor are also part of this element.
- **Emergency response** — A comprehensive emergency response program must be developed and maintained. Included are the following:

**PROCESS HAZARDS CHECKLIST**

| | CATEGORY | SUBJECTS TO BE INVESTIGATED |
|---|---|---|
| | Transition Periods | Startup, Standby, Shutdown, Upset Situations, Abnormal Operation |
| SECTION I GENERAL | Procedures | Startup, Normal, Shutdown, Emergency Sequence Checklists |
| | Loss of Utilities | Electric, Heating, Cooling, Air, Inerts, Agitation |
| | Location | Plant Public Exposures |
| | Storage Tanks | Design, Separation, Inerting |
| | Dikes | Capacity, Drainage |
| | Emergency Valves | Remote Control - Hazardous Materials |
| SECTION II STORAGE | Inspections | Flash Arresters, Relief Devices Vents |
| | Procedures | Contamination Prevention, Analysis |
| | Specifications | Chemical, Physical, Quality, Stability |
| | Limitations | Temperature, Time, Quality |
| | Pumps | Relief, Reverse Rotation, Identification |
| | Ducts | Explosion Relief, Fire Protection |
| SECTION III MATERIAL HANDLING | Conveyors, Mills | Stop Devices, Coasting, Guards |
| | Procedures | Spills, Leaks, Decontamination |
| | Piping | Ratings, Codes, Cross-Connections, Leaks |

**Figure 9-2** Process hazards checklist.

emergency plans, accountabilities, drills, and periodic updates. Emergency equipment and procedures must be inspected/audited on an ongoing basis. A "most catastrophic" and "most probable" scenario(s) should be developed and addressed in planning and mock exercises.

- **Incident investigation** — Any incident, from near miss to catastrophic failure, must be investigated, cause(s) determined, and recommendations developed to prevent recurrence of the incident.
- **Audits** — There must be a documented plan for the periodic examination of the process to determine compliance with process safety elements and provide feedback and recommendations where improvements are needed.
- **Employee participation** — A written plan must indicate how employees will participate in the process hazard analyses and in the development of other elements. Operations and maintenance employees must be

| CATEGORY | SUBJECTS TO BE INVESTIGATED |
|---|---|
| Vessels | Design, Materials, Codes, Access |
| Identification | Vessels, Piping, Switches, Valves |
| Relief Devices | Reactors, Exchangers, Piping |
| Review of Incidents | Plant, Company, Industry |
| Inspections, Tests | Vessels, Relief Devices, Corrosion |
| Electrical | Area Classification, Conformance, Purging, Grounding |
| Operating Ranges | Temperature, Pressure, Flows, Ratios, Concentrations, Densities, Levels, Time, Sequence, Records of Critical Process Variables |
| Ignition Source | Peroxides, Acetylides, Friction, Static Electricity, Heaters, Chemical Reaction |
| Compatibility | Heating Media, Lubricants, Flushes, Packing |

SECTION IV
PROCESS

| CATEGORY | SUBJECTS TO BE INVESTIGATED |
|---|---|
| Controls | Ranges, Redundancy Warranted, Fail-Safe Calibration, Inspection Frequency, Adequacy, Backup during Repairs, Sample Conditioning |
| Alarms | Adequacy, Limits, Fire |
| Interlocks | Tests, By-Pass Procedures |
| Relief Devices | Adequacy, Vent Size, Discharge |
| Process Isolation | Fire-Safe Valves, Remote-Operated Valves, Purging |
| Instruments | Air Quality, Time Lag, Reset, Wind-Up |
| Hazards | Fires, Runaways, Vapor Clouds |
| Computer | Exposure Malfunction Effects, Protection |

SECTION V
INSTRUMENTATION
& EMERG. DEVICES

**Figure 9-2 (continued)**

involved. Employees and their representatives must have free access to all process information.

- Trade secrets — Employers must make information necessary for compliance with the regulation available to all employees involved in the implementation of the elements, without regard to the possible trade secret status of the information. Employers may require confidentiality agreements with personnel receiving trade secret information.

Figure 9-3 provides a comparison of process safety management (PSM) elements as they apply to each standard, OSHA and EPA.

| CATEGORY | SUBJECTS TO BE INVESTIGATED |
|---|---|
| **SECTION VI** | |
| Ditches | Flame Traps, Reactions, Exposures |
| **WASTE DISPOSAL/** | |
| Vents | Discharge, Radiation, Mists, Location |
| **SPILLS & RELEASES** | |

| | CATEGORY | SUBJECTS TO BE INVESTIGATED |
|---|---|---|
| | Location | Accessibility, Control, Exposure |
| **SECTION VII** | Procedures | Normal, Abnormal |
| **SAMPLING** | Equipment | Personnel Protection, Containers, Storage, Disposal |

| | CATEGORY | SUBJECTS TO BE INVESTIGATED |
|---|---|---|
| | Fixed Protection | Sprinklers, Halon, Monitors |
| | Portable Protection | Extinguishers, Fire Hoses |
| **SECTION VIII** | Flammable Vapor/$O_2$ Detectors | Reliability, Applications |
| **FIRE PROTECTION** | Fire Walls | Adequacy, Condition, Doors, Ducts |
| | Drainage | Slope, Drain Rate, Location |

**Figure 9-2 (continued)**

3. **Determine what resources are needed for compliance.** The resources needed to comply with the standards fall mainly in the areas of people, time, and money. You must consider the education and experience of the employees you will need to perform the PHAs. As indicated in required element #6, process safety reviews, personnel with expertise in chemical/technical, safety, electrical, mechanical, operations, environmental, or clerical procedures, etc., depending on the type of process, will be required to participate in the reviews and in incident investigations. Financial considerations will determine whether to use existing personnel or if outside consultants should be brought in to augment existing personnel or to bring the facility into 100% compliance. Finances should also be considered when developing the list of recommendations as to what is needed to bring the facility into compliance, where deficiencies are noted. Depending on the size and type of processes, time requirements can range from 200 man-hours for small processes to 10–20 man-years for larger chemical facilities.

4. **Draft a written plan or procedure to state how you will comply.** This document is probably the most important element in determining your level of compliance. The document will, if it is well thought out, serve as your "action blueprint". All of the required elements must be addressed: processes to be identified, activities to be performed, time limits, documentation strategies, and assignment of responsibilities. Please note that the PSM standard is performance based and, therefore, does not require "black and white" remedies for compliance. Be innovative in the use of existing resources and do not reinvent the wheel. If you have other applicable procedures, such as contractor safety, line breaking, etc., you do not have to rewrite these

| TITLE | Guidelines for Management of Process Hazards API RP750 — Section | Process Safety Management of Highly Hazardous Chemicals OSHA 1910.119 — Paragraph | Guidelines for Technology Management of Chemical Process Safety AICHE, CCPS — Chapter | Process Safety Code of Management Practices CMA Code for Responsible Care — Tabs | 1990 Clean Air Act Amendment OSHA-CAAA S 1630 Air Toxics — Section |
|---|---|---|---|---|---|
| General | 1 | a,b,c | 1,2,3 | 1,2,3 | 301,304 |
| Process Safety Information | 2 | d | 4 | 5.5 | 301 Regs, 304 std |
| Process Hazard Analysis | 3 | e | 4,6 | 5.7 | 301 Regs, 304 std |
| Management of Change | 4 | l | 7 | 5,10.8 | 304 Standard |
| Operating Procedures | 5 | f | 4 | | 301 Regs, 304 std |
| Safe Work Practices | 6 | k | 4 | 7.4 | 301 Regs, 304 std |
| Training | 7 | g | 10 | 7.7 | 301 Regs, 304 std |
| Assuring Quality & Mechanical Integrity of Critical Equipment | 8 | j | 8 | 6.10 | 301 Regs, 304 std |
| Pre-Start-up Safety Review | 9 | i | 5 | | 304 Standard |
| Emergency Response & Control | 10 | n | | 6.16 | 301 Regs, 304 std |
| Investigation of Process Related Incidents | 11 | m | 11 | 4.10 | 301 Regs, 304 std |
| Audit of Process Hazards Mgt Systems | 12 | o | 13 | 6.7 | 301 Regs, 304 std |
| Contractors | | h | | 7.13 | 304 Standard |
| Capital Projects | | | 5 | | |
| Human Factors | | | 9 | 7.9 | |
| Standards, Codes & Regulations | | | 12 | 6.4 | 304 Standard |
| Waste Management | | | | 2.5-2.7 | |
| References | 13 | Appendices A,B,D | 14 | 9 | |

**Figure 9-3** Process safety management: Comparison of basic program elements. (From DeHart, R. E. and Gremillion, E. J., *Process Safety — A New Culture*, Occupational Safety and Health Summit, February 1992.)

SAFETY PROCEDURE                    _____

                                                        MANAGER, SAFETY

**PROCESS REVIEW**                  _____

                                                            MANAGEMENT

I.    PURPOSE

      The purpose of the Process Review Procedure is to establish the method by which all
      installations, modifications, additions, or other changes are made to plant buildings or
      equipment.

      Additionally, this document will establish the means by which a formal review of all
      applicable "processes" will be conducted.

II.   A.    For any new installation or modification, the Engineer developing the project
            shall inform the Safety Department of the project.

            1.    The Safety Department shall review the scope of the project and
                  determine to what extent Safety Design Criteria is needed.

            2.    All concerns, questions, or comments shall be noted on the comment
                  sheet.  The sheet shall then be signed and returned to the listed
                  Engineer.

                  2.1    Should the project be of such magnitude or of a complex nature,
                         then the Safety Department shall develop, in writing, a Safety
                         Design Criteria Package.  This will be provided to the Engineer
                         during the planning phase of the project.

                  2.2    The Safety Design Criteria document shall become a permanent
                         part of the project file.

                  2.3    The Safety Design Criteria shall stipulate any and all inclusions
                         into the project so that any state, local, or federal regulations,
                         corporate policy or standards, applicable codes, and good
                         engineering practices are adhered to.

                  2.4    Safety Design Criteria shall be approved by Corporate Safety.

                  2.5    Prior to the final approval of any project request, the Safety
                         Department shall review and sign off on the project, by review of
                         the Approval Transmittal.

**Figure 9-4**  Safety procedure.

procedures in your written plan. A simple cross-reference statement in each
of the procedures is sufficient for compliance. Finally, having the highest
level of management applicable to the organization sign the written plan is
an excellent method of demonstrating management's support for the program. An example of a written procedure is found in Figure 9-4.

5. **Implement the specific plan.** This is probably the second most important
   requirement because you must now actually implement the plan developed
   in step #4. The first step is to assemble the team chosen to conduct the
   reviews of each of the elements. These personnel must be educated in the
   requirements of the standard and educated in the applicability of their
   individual areas of expertise. Next, present your plan to complete your

II.    A.    2.    2.6    During construction/installation, the Engineer in charge of the project shall make frequent inspections to insure the adherence to the Safety Design Criteria. The Safety Department shall also conduct design reviews as needed.

2.7    Prior to startup or use of new installations, the Safety Department shall inspect/audit the facility/equipment. This inspection shall be noted on "Safety Checklist for New Installation or Major Equipment Modifications", Form No. _____. This form shall be placed into the project file and a copy shall be maintained by the Safety Department.

2.8    Prior to startup of any new equipment, procedures and JSA's will be written and posted. All employees involved with the equipment shall be trained in the correct and safe operation of the equipment.

Employee training for processes involving toxic, explosive, or flammable materials shall, at a minimum, consist of the following information:

| | |
|---|---|
| Chemical Hazards | Process Chemical Capability Grid |
| Toxicities | Material Safety Data Sheets |
| Permissible Exposure Limits | Process Chemistry |
| Physical Data | Process Flow Diagrams |
| Reactivity Data | Maximum Inventory |
| Corrosivity Data | Process Limitations & |
| Thermal & Chemical Stability | Consequences of Deviations |
| Electrical Classification | Process Protection including |
| Code Requirements | safety systems, detection, |
| Equipment and Process Materials | monitoring, alarms, pressure relief |

All training is to be documented.

2.9    Consideration shall be given to conducting meetings, for information purposes, with residents in the immediate area of plant property and the local Emergency Planning Committee shall be informed whenever the new process involves any flammable, toxic, or explosive material.

2.10   Contractors working in areas as noted in 2.8, above, shall receive training and information as listed, as applicable to the work being performed. Specific procedures are noted in the Contractor Safety Procedure.

**Figure 9-4  (continued)**

review(s) and modify the plan as team members contribute to the effort. Upon assigning responsibilities, set intermediate completion dates and hold update meetings as necessary until the project is completed. Written minutes of your update meetings should be maintained, or status reports compiled from each member or subteam.

6.  **Develop and complete recommendations or action items resulting from assessments/analyses.** Included in this step are two components. The first is to consider the necessary actions involved in compliance with the requirements of the standard. The second involves recommendations made as the result of completion of process hazard analyses. The first component simply requires that all of the necessary elements be addressed and a simple

II.    B.    For smaller installations or modifications such as those initiated through the plant Work Order System, it shall be the responsibility of the Maintenance Planner to notify the Safety Department of any concern which may involve Safety.

NOTE:        Where any changes are made to existing equipment or new equipment installed, all applicable drawings and procedures shall be updated.  JSA's shall be updated as necessary.

       C.    For existing processes where there is a greater than usual hazard potential, such as those involving toxic, explosive, or flammable chemicals, or where there are processes that, when failed, significantly impact the operation of the plant, the following shall apply:

             1.    There shall be a Process Review Team formed and the team shall consist of representatives from the following areas:

                   1.1    Operations

                   1.2    Engineering

                   1.3    Safety

                   1.4    Maintenance

             2.    The purpose of this team will be to review the particular process to determine consequences of various scenarios applicable to the safe operation of the process.

             3.    The review technique shall be determined by the team; however, an organized approach is vital to the effectiveness of the review.  Review techniques shall be of the following types:

                   3.1    "What If" Analysis

                          A "What If" Analysis is a technique aimed at forecasting various scenarios and evaluating the hazard of the consequences. Experienced individuals ask pertinent questions relating to every phase of the process from raw material unloading and storage to shipment of finished product.  This is a straightforward tool and, when properly executed, does not require a great deal of time or sophistication.  A checklist guide and worksheet to document findings can be a useful tool (see Attachment A).

**Figure 9-4  (continued)**

statement must be included indicating how you are, or intend to be, in compliance. The second component is more involved since it relates to the recommendations made to improve your process and it involves the application of a risk assessment. Risk, or what could happen, and the probability of occurrence, can be evaluated qualitatively and/or quantitatively. Quantitative risk analysis is the most commonly used method of evaluation of an episodic risk. A common approach to qualitatively analyzing hazards to determine potential risk is the use of a risk matrix method as demonstrated in Figure 9-5 and Figure 9-6. These techniques, although based on intuitive reasoning and the analyst's experience, can be very effective in determining the relative risk of each incident scenario and in identifying risk reduction

II.    C.    3.    3.1    The "What If" Analysis can be very helpful in identifying serious potential hazards. For those portions of a process identified as a high potential hazard, a more sophisticated method, such as HAZOP or Fault Tree Analysis, may be required to more accurately identify all undesirable deviations leading to the potential hazard. As an example, the "What If" approach might ask: "What if nitric acid were to mix with ethylene oxide in the reactor?" The Consequence/Hazard would be a potential uncontrolled runaway reaction. The "Conclusion" may be to perform a Hazard and Operability (HAZOP) Study (see 3.2) to identify all deviations leading to the mixing of nitric acid and ethylene oxide.

3.2    Hazard and Operability Study (HAZOP)

A HAZOP study is a method for systematically examining a part of a process to discover how deviations from intended design can occur and how these deviations can give rise to hazards.

3.3    Fault Tree Analysis

When a small portion of a process is to be studied, a "Fault Tree Analysis" can also be used to determine the combinations of component or human failures and the sequences of failure events which can lead to occurrence of a specific undesired event.

4.    Regardless of the type review used, the following shall be included:

4.1    Characteristics and hazards of raw materials, intermediates, and finished products.

4.2    Process chemistry including principal reactions and side reactions.

4.3    Review up-to-date P & I Diagrams for modes of failure.

4.4    Critical equipment.

4.5    Operating Procedures, concentrating on startup, shutdown, and emergencies.

4.6    Preventive Maintenance Program, including inspection and testing of safety devices.

NOTE:    Additional considerations are listed in Attachment A.

**Figure 9-4  (continued)**

methods. The objective of the risk matrix is to determine how the severity and/or frequency of the identified hazards can be reduced through some mitigating action. Quantitative risk analyses are performed using engineering techniques and computer models to analyze incident scenarios to determine their impact — more specifically, what damage may occur from a fire, explosion, and/or toxic release and how frequently it will occur. The potential impacts are systematically estimated to determine the magnitude and severity of the incident. Upon completion of the development of your recommendations, it is very important that, where you assess the risk to be acceptable for a specific situation, you state "why" the risk is acceptable and what is in place to maintain the risk at an acceptable level. All other

II.    C.    5.    Written reports shall be maintained of the review and updates of action items shall be provided until all are completed.

      6.    Processes and review frequencies for the plant are as follows:

|      |       |          | Frequency | Next Due |
|------|-------|----------|-----------|----------|
| 6.1  | _____ | Process  | 3 Years   | 1996     |
| 6.2  | _____ | Handling | 5 Years   | 1997     |
| 6.3  | _____ | Units    | 5 Years   | 1997     |
| 6.4  | _____ | Process  | 5 Years   | 1997     |

The base year for _____ Process shall be 1990.  The base year for the remaining processes shall be 1992.

III.   MANAGEMENT OF CHANGE

      A.    Where any change or modification is made, other than "replacement in kind", whether through Engineering projects or Maintenance work orders, a "Management of Change Approval" form, #_____, Attachment B, shall be executed.

            1.    "Replacement in kind" does not require use of the approval form. "Replacement in kind" is described as follows:

                  1.1    Equipment replaced within the parameters of the original or existing design criteria.

                  1.2    Ingredients, previously used in approved recipes, where hazards or risks are known.

                         NOTE:    For materials added where process air emissions will be affected, the Environmental Department shall approve the use of the material, prior to the change.

            2.    Examples of changes which will require the use of the form include:

                  2.1    Any change in safety alarm settings, interlocks or process tripping devices.

                         NOTE:    Bypassing these types of devices is covered by Safety Procedure #_____, Safety Control Bypass Permit.

**Figure 9-4  (continued)**

recommendations requiring action must be documented, responsibilities established, and completion dates assigned. This process must be followed through until all recommendations are completed.

7. **Document, document, document.** In each and every step of compliance with the process safety standard, documentation is required. This documentation not only serves as a guide, but it also serves as proof of your intentions or efforts should you be subject to an inspection by OSHA. Where actions are required of other employees, such as attending training sessions, participating in reviews, or mock emergency scenarios, have the employees sign their name, indicating attendance. All documentation should be retained for a period not less than the life cycle of the applicable process.

III.   A.   2.   2.2   Physical changes in process equipment, piping, instrumentation or electrical components, " not in kind".

2.3   Use of a new chemical or reintroduction of a chemical after an absence of 2 or more years.

2.4   Any new installation.

2.5   Changes in procedures or process parameters affecting pressures, flow, temperature or rates, not within the intended design specifications.

2.6   Changes in process control computer software not within intended design specifications.

B.   The origination of the "Management of Change Approval" form shall be the responsibility of the Engineer or Maintenance Associate responsible for planning the project or the work order. For changes involving chemicals, the origination of the form shall be the responsibility of the supervisor of the Department where the chemical will be used. Should the change arise from the Technical Department, Science & Technology, or Process Support, then it shall be the responsibility of the particular department employee responsible for the particular product being changed. NOTE: Where RFT's are written for polymer changes or formulation changes, the RFT form shall be approved by the Manager, Safety and Environmental prior to beginning the trial.

1.   The originator initiating the form is to complete the sections for Area, Project Title or Work order number and give a brief description of the change, in the space provided.

2.   The form shall then be forwarded to the Safety Department where the "Requirements" check list will be completed and the form signed.

3.   The form shall then be forwarded to the supervisor of the affected area, for approval. Once approved by the supervisor, the requested change may take place. The form shall be retained in the Area until all noted requirements are completed. NOTE: In the case of Request for Trials (RFT's), where hazardous chemicals, increased emissions or Industrial Hygiene exposure is not affected, the RFT will be returned to the appropriate Technical or S & T person.

4.   Upon completion of the requirements, the form shall be returned to the Safety Department for retention.

III.   B.   5.   Other approval signatures may be obtained as warranted by the work to be performed.

NOTE:   Approval of the form, by the Safety Department, shall take no longer than 48 hours. The time limit begins upon receipt of the form by the Safety Department.

6.   The Safety Department shall maintain a "Management of Change Approval" log, to track status of each form initiation.

7.   In the absence of the Safety Department, the Plant Manager shall designate appropriate personnel for review.

**Figure 9-4   (continued)**

**WHAT-IF WORKSHEET**

LOCATION _____

DATE _____

TEAM _____

DESCRIPTION _____

| WHAT-IF | CONSEQUENCE/HAZARD | CONCLUSION/RECOMMENDATION |
|---------|--------------------|---------------------------|
|         |                    |                           |
|         |                    |                           |
|         |                    |                           |
|         |                    |                           |
|         |                    |                           |
|         |                    |                           |

**Figure 9-4  (continued)**

Management commitment is critical to the success of this undertaking. The support needed does not stop with top management. It also involves the commitment of all levels of employees involved with the process. Resources must be allocated and priorities defined to ensure compliance and possibly prevent your company or facility from becoming a statistic. We all have the resources available to us to perform the required steps of this standard. Following the guidelines presented here will provide you with a sound basis for compliance.

## MANAGEMENT CHANGE APPROVAL

AREA: _____     PROJECT OR WORK ORDER NUMBER: _____

| REQUIREMENTS | COMPLETION (initials/date) | ORIGINATOR |
|---|---|---|
| | | **DESCRIPTION OF CHANGE** |
| __ Before __ After __ NA  MSDS | ___/___ | |
| __ Before __ After __ NA  PROCESS CHEMISTRY | ___/___ | |
| __ Before __ After __ NA  PROCESS SAFETY INFO | ___/___ | |
| __ Before __ After __ NA  PFD'S,P&ID'S, AND EFD'S | ___/___ | |
| __ Before __ After __ NA  CRITICAL EQUIP.IN PM | ___/___ | |
| __ Before __ After __ NA  MAINTENANCE PROC. | ___/___ | |
| __ Before __ After __ NA  INVENTORY LIMITS | ___/___ | |
| __ Before __ After __ NA  SOP | ___/___ | |
| __ Before __ After __ NA  EMERGENCY PLAN | ___/___ | |
| __ Before __ After __ NA  VENTILATION INFO | ___/___ | |
| __ Before __ After __ NA  OPERATOR TRAINING | ___/___ | |
| __ Before __ After __ NA  MAINTENANCE(Elect&Mech) | ___/___ | |
| __ Before __ After __ NA  HAZARD ANALYSIS | ___/___ | |
| __ Before __ After __ NA  ELECTRICAL CLASS MAPS | ___/___ | |
| __ Before __ After __ NA  SAFETY DESIGN CRITERIA | ___/___ | |
| __ Before __ After __ NA  PRESTARTUP REVIEW | ___/___ | |

**FINAL STATUS OF CHANGE**

CHANGE MADE: ___/_____

ALL DOCUMENTATION COMPLETE: ___/_____

| APPROVALS | NOTES & INSTRUCTIONS |
|---|---|
| | |
| Originator _____ (Date) | |
| Safety _____ (Date) | |
| Area Supervisor _____ (Date) | |
| Others _____ (Date) | |
| _____ (Date) | |
| _____ (Date) | |
| _____ (Date) | |

FORWARD IN TURN:

SAFETY _____

AREA _____

SAFETY _____

## Figure 9-4   (continued)

| SEVERITY CATEGORY | FREQUENCY CATEGORY | A - FREQUENT Will occur twice or more in system lifecycle | B - PROBABLE Will occur at least once in system lifecycle | C - REMOTE Possible, but unlikely to occur in system lifecycle | D - IMPROBABLE A credible accident event cannot be established |
|---|---|---|---|---|---|
| I. CATASTROPHIC PERSONNEL: Death FACILITIES/EQUIPMENT/VEHICLES: System loss, repair impractical, requires salvage or replacement | | IA | IB | IC | ID |
| II. CRITICAL PERSONNEL: Severe injury/occupational illness. Requires admission to a health care facility. FACILITIES/EQUIPMENT/VEHICLES: Major system damage. Damage greater than $1,000,000 or which causes loss of primary mission capability | | IIA | IIB | IIC | IID |
| III. MARGINAL PERSONNEL: Minor injury/occupational illness. Lost time accident of more than one day that does not require admission to a health care facility. FACILITIES/EQUIPMENT/VEHICLES: Loss of any non-primary mission capability, or damage more than $200,000 but less than $1,000,000 with no direct impact on mission capability. | | IIIA | IIIB | IIIC | IIID |
| IV. NEGLIGIBLE PERSONNEL: Less than minor injury/occupational illness. May or may not require first aid but lost time is less than one day. FACILITIES/EQUIPMENT/VEHICLES: Damage equal to or less than $200,000 | | IVA | IVB | IVC | IVD |

**Figure 9-5**   An example of a risk matrix index. (From DeHart, R. E. and Gremillion, E. J., *Process Safety — A New Culture*, Occupational Safety and Health Summit, February 1992.)

| Probability of Success / Category | A. Frequent (1) Event Likely to Occur Once or More Per Year | B. Probable (2) Event Likely to Occur Once Every Several Years | C. Occasional (3) Event Likely to Occur Once In Lifetime of Facility | D. Unlikely (4) Event Unikely, But Not Impossible |
|---|---|---|---|---|
| I. Catastrophic (1) Personnel-Life Threatening Environment-Large, Uncontrolled Release Equipment-Major Damage Resulting in Loss of Unit | 1 | 2 | 3 | 4 |
| II. Critical (2) Personnel-Severe Injury Environment-Moderate, Uncontrolled Release Equipment-Moderate Resulting in Unit Downtime | 2 | 4 | 6 | 8 |
| III. Marginal (3) Personnel-Lost Time Injury Environment-Small, Uncontrolled Release Equipment-Minor Damage Resulting in Unit Slowdown | 3 | 6 | 9 | 12 |
| IV. Negligible (4) Personnel-Minor Injury Environment-Small, Controlled Release Equipment-Negligible Damage | 4 | 8 | 12 | 16 |

High: Requires Action    Moderate: Further Study Required    Low: Investigate As Time Permits

**Figure 9-6** An example of a risk matrix. (From Hansen, M.D., *CSOC Integrated System Safety Program Plan,* United States Air Force Space Command, Loral Command and Control Systems, Colorado Springs Division, December 31, 1993.)

## REFERENCES

DeHart, R. E. and Gremillion, E. J., *Process Safety — A New Culture,* Occupational Safety and Health Summit, February 1992.

Hansen, M. D., *CSOC Integrated System Safety Program Plan,* United States Air Force Space Command, Loral Command and Control Systems, Colorado Springs Division, Colorado Springs, CO, December 31, 1993.

Krembs, J. A. and Connolly, J. M., *Analysis of Large Property Losses in the Hydrocarbon and Chemical Industries,* 1990 NPRA Refinery and Petrochemical Plant Maintenance Conference, May 23–25, 1990, San Antonio, TX.

M & M Consultants, *Large Property Damage Losses in Hydrocarbon-Chemical Industries — A Thirty Year Review,* 13th ed.

## FURTHER READING

### Books

American Institute of Chemical Engineers, *Guidelines for Hazardous Evaluation Procedures,* 2nd ed., New York, American Institute of Chemical Engineers, 1992.

American Institute of Chemical Engineers, *Guidelines for Technical Management of Chemical Process Safety,* Center for Chemical Process Safety, New York, American Institute of Chemical Engineers, 1989.

Bird, F. E. and Germain, G. E., *Practical Loss Control Leadership,* Atlanta, GA, Institute of Publishing, 1990.

Lees, F. P., *Loss Prevention in the Process Industries,* London, 1980.

NUS Training Corporation, *OSHA's Process Safety Management of Highly Hazardous Chemicals (29 CFR 1910.119) in Plain English,* NUS Training Corporation, undated.

## Other

American Petroleum Association, *Management of Process Hazards, API Recommended Practice (RP) 750,* 1st ed., American Petroleum Association, January 1990.

Environmental Protection Agency, Chemical Process Safety Management, Section 304, 1990 Clean Air Act Amendment, *Fed. Regist.*

Environmental Protection Agency, Prevention of Accidental Releases, Section 301, 1990 Clean Air Act Amendment, *Fed. Regist.,* November 1990.

Hawks, J. L. and Mirian, J. L., Create a good PSM system, *Hydrocarbon Processing,* August 1991.

Occupational Safety and Health Administration, Process Safety Management of Highly Hazardous Chemicals, (OSHA) 1910.119, *Fed. Regist.,* August 1992.

Occupational Safety and Health Administration, Process Safety Management, OSHA, 3, 32, 1993.

Occupational Safety and Health Administration, Process Safety Management Guidelines for Compliance, OSHA, 3.32, 1983.

# 10 ENVIRONMENTAL AND HAZARDOUS MATERIALS MANAGEMENT ASPECTS

Robin Spencer and Marc Bowman

## TABLE OF CONTENTS

## INTRODUCTION

In most companies today, any discussion of safety and health management must consider the several aspects of handling hazardous materials or hazardous wastes in day-to-day operations. Successful companies promote an environmental management program that has been developed to establish and follow corporate environmental philosophy, policies, and procedures to promote compliance and minimize future liabilities. This chapter is intended to provide an overview of environmental management aspects when operating with hazardous materials. The environmental management program can often parallel corporate safety and health programs.

## THE IMPORTANCE OF REGULATORY COMPLIANCE

It should be clearly understood by all employees at a facility that compliance with the laws, regulations, and procedures is not only company policy, but that it is expected to be incorporated into every aspect of doing business on a daily basis. The environmental laws and regulations have been established to ensure employee health and safety as well as to protect public human health and the environment. In addition, it makes good business sense. The costs of remediation can be enormous from the acute effect of spills and releases or the chronic effect of long-term mismanagement of hazardous wastes and hazardous materials.

If these reasons are not enough, every employee should understand the enforcement provisions for not complying with laws and regulations. While the actual assessment of fines and penalties is a complicated process, as discussed briefly below, they may be very significant. For example, the federal hazardous waste law includes civil penalties of up to $25,000 per day for each violation if the Environmental Protection Agency (EPA) determines that a person has violated or is violating any provision of the law. In addition, individuals may be subject to fines of up to $50,000 per day for each violation and imprisonment for up to 2 years.

The Federal Guidelines for Environmental Compliance were proposed in 1993 and outline seven basic components for courts to weigh when evaluating a sentence for an environmental crime:

1. Line management has to be routinely involved in the environmental compliance program.
2. Environmental policies, standards, and procedures must be integrated throughout the organization (up and down the hierarchy).
3. An independent auditing, monitoring, reporting, and tracking system that uses terms familiar to all employees should be in place.
4. The organization must have regulatory expertise.
5. The organization has a system of incentives or rewards for environmental compliance consistent with programs for sales or production.
6. The organization has consistently enforced corporate environmental policies using proper disciplinary procedures.
7. The organization has a way to measure continuing improvement in its efforts to achieve environmental excellence. Both internal and external audits as part of an overall environmental management program can measure compliance.

It is clear that a coordinated proactive environmental management program is highly desirable, if not essential.

## ENVIRONMENTAL MASTER PLANNING

An environmental management program is based on an approach and concept called environmental master planning. It can be applied to a large organization, a single facility, or even an operation within the facility. The overall objective of environmental master planning is to establish and implement the necessary management systems in order to ensure compliance with environmental laws and regulations and to identify, reduce, and minimize environmental risks and liabilities. Many significant environmental liabilities found in companies today are not the result of noncompliance with laws, but rather a result of poor management practices.

Environmental master planning generally has four components:

1. **Establish and follow environmental philosophy, policies, and proce-dures to promote compliance and minimize future liabilities.** Policies establish the principles and standards under which facilities will operate and give the authority to enforce the policies. Procedures are the "regulations" under which a corporation or facility will operate to ensure compliance with the established policies.
2. **Establish databases to allow for the tracking of environmental compli-ance and to monitor operations which have a high environmental risk.** Databases bring key environmental information necessary for management uses into a usable format. The databases can be either electronic or in a paper file format.
3. **Establish and implement an internal inspection and audit program.** The audit program will identify noncompliance with laws, regulations, policies, and procedures; management practices which pose unnecessary environ-mental risk; and existing liabilities.
4. **Establish and implement an environmental risk management system.** The environmental risk management system provides for the measurement of the risk; prioritizes issues; identifies corrective action measures and the measurement of progress toward those measures; utilizes a cost/benefit analysis; gives highest priority to those environmental issues posing the greatest threat; and becomes an integral part of the existing management system, e.g., capital spending plans, operating budgets, etc.

The environmental master planning process should be reviewed after the system has been tested, e.g., after an internal compliance audit. It also should be reviewed following a change in applicable federal, state, or local laws or regulations, or after a regulatory agency inspection or Notice of Violation. Hazardous material or hazardous waste usage at an organization or business determines the amount of attention which should be spent on the environmental master planning.

## DEFINITION OF HAZARDOUS MATERIALS AND HAZARDOUS WASTES

Hazardous materials, substances, chemicals, and wastes are legally defined terms in laws such as the Occupational Safety and Health Act (OSHA), the Resource Conservation and Recovery Act (RCRA), the Toxic Substances Control Act (TSCA), the Transportation Safety Act, the Comprehensive Environmental Response, Compensation, and Liability Act (CERCLA), and the Emergency Planning and Community Right-to-Know Act (EPCRA). Generally, a hazardous material is any substance or mixture of substances having properties capable of producing adverse acute or chronic effects on human health and safety or the environment. Rule-of-thumb guidance is that if a material has a Material Safety Data Sheet (MSDS), it is probably a hazardous material.

Hazardous wastes are typically hazardous materials which no longer have a beneficial use. According to the Resource Conservation and Recovery Act the law which regulates solid and hazardous wastes (the laws regulating hazardous wastes are specifically defined in RCRA), for a waste to be a hazardous waste, it must first be determined to be a solid waste. Solid wastes include solids, semisolids, liquids, and containerized gases. Generally, solid wastes are materials which are discarded, abandoned, disposed of, disposed of by burning or incineration, accumulated in certain ways, recycled or reclaimed in certain ways, or are "inherently waste-like".

Once it is established that a material is a solid waste, a hazardous waste determination must be made and, in fact, is required by law. Solid wastes are determined to be hazardous wastes by checking to see if the material is on a list of hazardous wastes and/or checking to determine if the waste exhibits certain characteristics. These characteristics include toxicity, corrosivity, ignitability, and reactivity.

## CORPORATE ENVIRONMENTAL MANAGEMENT PROGRAM

A corporate environmental management program will consider which federal, state, or local regulations apply to its operations. The program should be headed by a person with the authority to make decisions and to commit both human and financial resources. To be truly successful, the program must have top management support and commitment.

In addition to the environmental policies, procedures, and databases described above, the environmental management program should consist of routine compliance audits (with follow-up), community involvement, em-

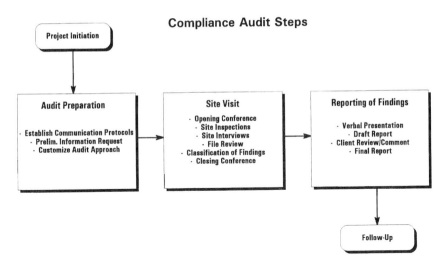

**Figure 10-1** Compliance audit steps.

ployee training, periodic regulations reviews, environmental review for new projects, and professional activities. Each of these components is described below.

## Compliance Audits

Environmental compliance audits are tools used by most companies to assess the status of compliance with environmental laws and company policies and procedures. An audit is a "snapshot" of conditions at the facility at the time of the audit and provides a starting point for activities to bring the facility into full environmental compliance if exceptions to compliance are identified. There are three key components of a compliance audit: (1) audit preparation, (2) the site visit, and (3) management of the audit findings. These key components are shown in Figure 10-1 and are discussed below.

Audit preparation is critical for an effective audit. Prior to visiting the site, the audit team should establish lines of communication with the facility, corporate staff, and legal council (if involved). The audit team should then submit to the facility a preaudit questionnaire, which will help in the compilation of all the information needed to conduct the audit. Some information is sent to the audit team in advance of the site visit, while other information is made ready for inspection during the site visit. Based on a review of the completed preaudit questionnaire, the audit team should then customize the audit approach in preparation for the site visit.

While at the facility, the audit team will hold an opening conference with site managers, conduct site inspections, review environmental files and records, interview key facility personnel, identify exceptions to compliance, and verbally discuss preliminary audit findings during an exit interview with site managers. The site visit portion of the compliance audit is usually conducted using audit checklists. These checklists may be simple or very lengthy and detailed, depending on the complexity of the issues expected to be encountered at the site.

Management of audit findings is a critical component of the auditing process which involves both the audit team as well as the facility and corporate personnel responsible for environmental compliance. Audit findings can be classified into four issues:

- Compliance issues — those issues which, in the opinion of the auditor, are out of compliance with existing, applicable laws, regulations, or codes
- Management issues — those issues (which are not compliance issues) that do not reflect the current best management practices in similar industries for management of environmental issues
- Significant liability issues — those issues (which are not compliance issues) which, in the opinion of the auditor, pose significant potential risk to the facility, either as significant environmental liabilities, significant risks to public health, and/or significant threats to process equipment or production capability
- Punchlist issues — those minor issues which can be readily corrected with minimal resources (usually that day or the next)

Any written report of findings should be objective and without recommendations and should have limited distribution to prevent access by unauthorized persons.

How the audit findings are reported varies widely depending on the needs of the company as well as advice from legal counsel. Of primary importance, however, is the development of a corrective action plan to address all of the audit findings. For each audit issue, the corrective action plan should identify the proposed corrective actions to mitigate the audit finding, the person or team responsible for implementing the corrective action, and the proposed completion date for the corrective action. The corrective action plan should be reviewed at least quarterly thereafter to track the progress of the corrective actions. The status and completion of the corrective actions should be documented and filed with the audit report.

## Community Right-to-Know

Community involvement in and knowledge of a facility's operations and the hazardous materials managed there is not only a good business practice; it is required by law. Under EPCRA, the Community Right-to-Know provisions of the Superfund Amendments and Reauthorization Act (SARA) Title III, four types of reporting of hazardous materials may be required. These are

- Emergency Release Reporting (Section 304)
- Material Safety Data Sheet (MSDS) Reporting (Section 311)
- Inventory Reporting (Section 312)
- Toxic Chemical Release Reporting (Section 313)

SARA Section 304 reporting is triggered when a spill occurs of an extremely hazardous substance in quantities greater than the designated reportable quantity and the spill enters the environment. Extremely hazardous substances are defined in the SARA law and are included in the Extremely Hazardous Substances List of the Comprehensive Environmental Response, Compensation, and Liability Act (CERCLA).

SARA Section 311 reporting is required for any hazardous substance for which an MSDS is required and which is stored on site in quantities greater than 10,000 pounds (or the threshold quantity for extremely hazardous substances). Under this reporting requirement, copies of MSDSs (or a listing of these MSDSs) for each of these hazardous materials stored in quantities in excess of 10,000 pounds must be submitted to the state emergency response commission, the local emergency planning committee, and the local fire department. The basic purpose is to provide emergency response personnel with information about what is stored on site. This is a one-time reporting which must be updated within 90 days upon a change of any hazardous material inventory which triggers reporting.

SARA Section 312 is an annual report which is submitted on Tier I or Tier II forms and provides an inventory reporting for each hazardous material. The reporting requirements are triggered by the same determination for the same hazardous materials as for SARA Section 311 reporting. The purpose of this reporting is to provide emergency response personnel with information about the storage locations and quantities of hazardous materials.

SARA Section 313 is an annual report of emissions and discharges of certain toxic chemicals. These emissions are reported on Form R. The purpose of this report is to inform the community of releases of hazardous materials into the environment.

It is also prudent (and required by law for certain hazardous waste activities) to include community leaders and emergency response organizations in emergency planning at a facility. This could include sharing copies of the emergency management plans, familiarization with the site by emergency response organizations and hospitals, participating in community emergency drills, and establishing links of communication with local media.

## Training

Federal and most state laws require employees who handle hazardous materials and hazardous wastes to be trained in their safe handling and their hazards. Typically, the training is provided upon job assignment, then annually thereafter. At times, hazardous materials training overlaps with safety and health training.

## Regulations Review

The environmental compliance manager should be aware of new laws and regulations which affect the regulated community. The manager can subscribe to any number of environmental or trade journals which cover hazardous materials. There are update services for regulations and agency policies. Weekly or monthly newsletters keep the manager up to date on what is happening in the EPA and elsewhere. If the budget allows, the business can even subscribe to the daily *Federal Register.* Regulatory agencies offer classes or public hearings for new or proposed regulations. The manager could attend the periodic meetings of local chapters of professional organizations both to gain new technical knowledge and to develop contacts with other professionals with similar interests. One can also attend training classes, seminars, conferences, or university-level classes to receive training on regulations.

## Environmental Review for New Projects

Changes are a normal part of daily operations at most facilities. While most of these changes do not have environmental or regulatory consequences, it should be company policy to ensure that all new projects are in compliance with the law, minimize the release of pollutants to the environment, minimize the generation of wastes, minimize the use of toxic substances, and minimize health and safety risks associated with operations. In order to meet this policy, proposed significant changes in any facility should be evaluated to ensure that environmental regulatory compliance will be maintained despite the result of the change. Due to the complexity of existing environmental regulations, no one person can be expected to understand all applicable rules and regulations. Therefore, a "management of change procedure" is typically used. The purpose of this procedure is to ensure that proposed significant changes go through an internal environmental review process. As with all management decisions, discretion must be used in applying this procedure, keeping in mind that this procedure is intended to protect the public, neighbors, employees, and operations. Typically, "significant changes" which should be reviewed per this procedure include any and all proposed changes which may have an impact (positive or negative) on

- Employee health and well-being
- Public or resident health and well-being
- The environment, including air, surface water, groundwater, soil, etc.

## Environmental Review Form

| Location: | Project #: | | | |
|---|---|---|---|---|
| **Brief Project Description:** | | | | |

| **Air Environmental Review** | Yes | No | N/A |
|---|---|---|---|
| Will the project create any new point source or fugitive emissions? | | | |
| Will the project change (increase or decrease) existing point source or fugitive emissions? | | | |
| Will there be any dust or VOC emissions from construction or demolition activities? | | | |
| Are there any air permitting issues associated with this project? | | | |

| **Water Environmental Review** | Yes | No | N/A |
|---|---|---|---|
| Will the project create any new or different water discharges to the sewer (POTW)? | | | |
| Will the project create any new or different water discharges to surface waters or the land? | | | |
| Will the project create any new or different waste water discharges to site septic systems? | | | |
| Will the project create any new or different waste water discharges to storm drains? | | | |
| Will the project create the potential for soil or groundwater contamination? | | | |
| Are there any NPDES, storm water, or POTW permitting issues associated with this project? | | | |

| **Hazardous Material/Solid Waste Environmental Review** | Yes | No | N/A |
|---|---|---|---|
| Have outside contractors' wastes/hazardous materials been addressed? | | | |
| Will there be any new hazardous materials used or stored on site resulting from the project? | | | |
| Will any haz. materials or wastes remain on-site after completion of the project? | | | |
| Will there be a change in the inventories of any existing hazardous materials? | | | |
| Will any existing hazardous material storage plant/project be closed as result of this project? | | | |
| Will the change trigger changes to SARA 311, 312, or 313 Right-To-Know reporting? | | | |
| Have plans for responding to spills of hazardous materials been adequately addressed? | | | |
| Will there be any new or different wastes resulting from the project? | | | |
| Will there be any wastes generated during construction or demolition phases of the project? | | | |
| Will any wastes associated with the project be hazardous wastes? | | | |
| Have provisions for storage and disposal of hazardous wastes been addressed? | | | |
| Has a reduction in the generation of new or existing hazardous wastes been considered? | | | |
| Will the project involve any treatment, storage or disposal of hazardous waste on site? | | | |

| **Asbestos/PCB Environmental Review** | Yes | No | N/A |
|---|---|---|---|
| Will the project involve the removal or demolition of Asbestos containing materials? | | | |
| Will all asbestos removal be completed by state-certified, asbestos-trained personnel? | | | |
| Have provisions been made for proper packaging, transporting and disposing the asbestos? | | | |
| Will the project involve the removal of any PCB or PCB-contaminated equipment? | | | |
| Have provisions been made for proper packaging, transporting and disposing of the PCBs? | | | |

**Review Signatures:**

_____    _____    _____    _____
Plant/project Environmental Coordinator    Date    Plant/project Manager    Date

**Figure 10-2**  Sample environmental review form.

- Permits or memoranda of understanding with governmental agencies
- Any air, water, or waste discharges or emissions from any plant/project

At the initiation of a proposed change, an environmental review of the proposed change is performed. This review should be documented using a form similar to Figure 10-2.

## Professional Activities

The environmental manager and others interested in the field have the opportunity to increase and share their knowledge through participation in workshops, seminars, and professional meetings and as instructors for classes. There are many professional environmental organizations to which one may belong or in which one may participate: Registered Environmental Professionals, Institute of Certified Hazardous Materials Managers, Groundwater Resources Association, Water Well Association, and others. Many states, such as California and Nevada, have environmental registration programs. Most of the professional organizations have local chapters which offer periodic meetings, seminars, and national conferences.

The knowledge gained by a professional's experience is invaluable for students trying to figure out career paths; if the environmental manager has the opportunity, he should try to be a guest speaker at elementary or high schools or at colleges. Whether corporate or academic, sharing knowledge is the responsibility of the environmental manager or professional. This way, the environmental manager and business can benefit from enhanced public understanding of hazardous materials management.

## CONCLUSION

Environmental management of hazardous materials or hazardous wastes can be a full-time job. It is essential that top management supports and commits to making effective environmental policies. An environmental management program will reduce the risks and liabilities associated with facilities operating with hazardous materials or hazardous wastes. Employees increasingly understand that environmental protection is in the best interest of themselves and their community. Combined with training and a general awareness of environmental compliance goals, employees are generally cooperative with management to achieve environmental compliance since they know that in order to have a job, the company must stay in business. Also, a proactive approach to environmental compliance is appreciated by regulatory agencies.

## FURTHER READING

### Books

Blakeslee, H. W. and Grabowski, T. M. *A Practical Guide to Plant Environmental Audits.* New York: Van Nostrand Reinhold, 1985.

Cahill, L. B., Ed., with Kane, R. W. *Environmental Audits,* 5th ed. Rockville, MD: Government Institutes, 1987.

Carson, H. T. and Cox D. B., Eds. *Handbook on Hazardous Materials Management,* 5th
    ed. Rockville, MD: Institute of Hazardous Materials Management, 1995.
Saunders, T. *The Bottom Line of Green is Black.* San Francisco, CA: Harper & Row,
    1994.
Seldner, B. J., Cothrel, J. P. et al. *Environmental Decision Making for Engineering and
    Business Managers.* New York: McGraw-Hill, 1994.

# Section III
# Health Program Management Aspects

# OVERVIEW OF THE KEY ELEMENTS OF AN EFFECTIVE OCCUPATIONAL HEALTH MANAGEMENT PROGRAM, PART 1

# 11A

Barbara K. Cooper and Ellen E. Dehr

## TABLE OF CONTENTS

## INTRODUCTION

This chapter provides practical information needed to plan, organize, and implement an effective health management program. As a manager of occupational health for a small company that has never had any health programs, where do you begin? What programs should receive priority? What constitutes

1-56670-054-X/96/$0.00+$.50

the best use of your resources? These questions will be addressed and the information presented here can also be used to evaluate and improve already existing health programs.

The manager should always take a comprehensive approach to the promotion of health while resisting extreme pressure toward fragmentation. Emphasis on prevention of disease must permeate the goals and mission of the entire organization at all levels. You will, therefore, be the consummate politician, as well as the great communicator. This chapter discusses the steps involved in developing and implementing occupational health programs and how to:

- Conduct a needs analysis.
- Prioritize the health needs of your organization.
- Determine which regulations apply.
- Prepare a plan of action.
- Gain support for your programs.

## DEFINITION OF HEALTH PROGRAMS AND HEALTH PROGRAM MANAGEMENT

An occupational health program has many of the same goals as a safety program, but has a different emphasis. The profession of industrial hygiene involves the anticipation, evaluation, and control of environmental factors arising in or from the workplace that may result in injury, illness or impairment or affect the well-being of workers and members of the community. The industrial hygienist analyzes operations and materials and recommends procedures to protect human health. The approach emphasizes prevention of disease arising from the following: noise and vibration, hazardous materials such as lead and asbestos, ionizing and nonionizing radiation, bloodborne pathogens, psychological stress, heat and cold stress, poorly designed workstations or tools, and other occupational factors.

The key to the prevention of health problems lies in the development and implementation of a comprehensive employee health program. Such a program includes a written plan of action which outlines a schedule of activities and procedures to be followed. These health programs, because of the preventive approach, often involve extensive health education. Other program elements include hazard identification and evaluation, hazard communication, and employee compliance.

Why do health professionals recommend the program approach, and why should these programs be comprehensive in nature? Health professionals often receive direction from those who may not have a clear understanding of health and safety management. Sometimes environmental regulatory compliance constitutes a higher priority for other managers than occupational health issues and employee wellness. Many health professionals find themselves coping with the crisis of the moment.

Having an organized plan for accomplishing the important tasks relating to employee health will help protect you from the changing priorities of all levels within your organization. In addition, the presence of such a program helps demonstrate progress toward regulatory compliance. Finally, a program approach facilitates program documentation and follow-up.

However, health programs, plans, and policies will fail if developed without the direct input of others. You cannot know the health priorities of an organization unless you do your homework. Strive to gain support from key individuals while providing education about industrial and occupational health issues. Your programs will succeed if everyone has a stake in the prevention of employee health problems.

Comprehensive programs avoid duplication of effort and competition among different departments. Roles, responsibilities, and duties are outlined in detail, eliminating confusion and reducing misunderstandings. Health programs often have similar elements: health hazard evaluation, exposure monitoring, employee training, record-keeping, medical surveillance. Grouping such programs streamlines such difficult tasks as scheduling training sessions or medical appointments. Routine monitoring of the work environment can include measurement of a variety of environmental stressors at the same time.

In addition to these practical considerations, the program approach will relieve you of the pressure of having to cope with a variety of conflicting expectations. Determine your initial goals using a consensus approach. Work with your managers to determine their priorities and concerns. Incorporate these ideas into an overall strategy for health programs. Make each goal realistic. Consider how your organization works as a unit to accomplish tasks.

While the plan is always subject to review and change, you will feel more in control. Those with other priorities involving the use of your time and resources will often respond more favorably once they understand that your programs are part of an overall plan to improve health among your workers and comply with occupational health regulations.

Health, safety, and environmental management programs should report to the highest management level. Executive management must approve and actively work toward the goals of the health management program in order for it to succeed. Otherwise, not all mid-level managers will embrace the changes necessary to promote a healthy workforce. Health professionals should have the authority and the executive mandate to carry out their function. If the health programs report to the operating units themselves, a conflict of interest arises.

Grouping industrial hygiene with other technical functions such as engineering may prove more successful than placement of this function in human resources or some other administrative unit, since employees often view those in scientific disciplines as resources rather than adversaries. In addition, scientific units often have better access to laboratory or research facilities.

Obviously, you may have little control over where your unit reports within the organization. After educating managers about occupational and environmental compliance, you may successfully negotiate a "dotted-line" reporting relationship to the highest levels of management to ensure communication and cooperation within the organization.

## AVAILABLE RESOURCES

1. Conduct a needs analysis. Beginning a health program in any organization can prove to be a daunting task. Evaluate past efforts. Technical functions such as engineering may have performed analyses of ventilation equipment or plant design. Often human resources divisions investigate accidents or illnesses and oversee workers' compensation programs. Legal counsel can recommend specific wording for policies, procedures, and directives involving health matters. Operating units responsible for production or maintenance often have already established safe work practices.

2. Interview those within your organization who have the greatest knowledge about employee health. Review the records kept on occupational injuries and illnesses. Compare the injury and illness rates to those within your industry. Make note of the most common types of illnesses and injuries. Review in depth a representative sample of the incidents listed, including the workers' compensation files and medical records.

3. Visit production and maintenance areas and talk to employees and first-line supervisors. What are their concerns? Discuss the injury and illness statistics with them. Conduct brief walk-throughs of production, research, and maintenance operations and make flowcharts summarizing the processes and materials involved. How many employees work in each area? What types of materials and equipment do they use?

4. Try to visit all plants and locations for which you are responsible. Every plant has its own culture and idiosyncrasies. You may want to conduct more formal interviews or audits during this information-gathering stage if your area of jurisdiction is wide.

5. Contact labor union representatives to discuss health and safety concerns. What existing agreements include health and safety requirements? What procedures are recommended for specific operations? Often unions have general operating procedures for the trades they represent. Your human resources department should be able to provide you with a list of the unions representing the employees of your organization.

6. Interview key managers for their perspective. Do not neglect those from the legal, human resources, and finance departments. These managers may budget for health and safety equipment, supplies, and training and may manage projects which involve hazardous materials, such as asbestos in buildings or chemical spills.

7. Build support for your programs by appearing to be organized and confident; keep in mind that soon you will be presenting a plan of action for management approval. A few managers will see health programs as your total responsibility. In addition, some may appear defensive about their illness or injury rates. Always present yourself as a resource to these individuals. The function of the health professional involves making

recommendations to improve working conditions. The actual implementation of such recommendations and other health policy matters remains a line function.

8. Prioritize the health needs of your organization. Review the information you collected during your needs analysis. Which operations present the greatest health risk to your employees? Noise-induced hearing loss may threaten some workers due to their work with power tools. Others may use organic solvents without proper personal protective equipment and without adequate ventilation. New chemicals or procedures may present unknown risks to your workforce due to the lack of a hazard communication program. Increased workers' compensation costs are often a result of an increase in the severity or number of injuries among employees.

9. Determine which regulations apply to your operations. Visit the Occupational Safety and Health Administration (OSHA) offices in your area and research your organization's health and safety track record. Introduce yourself to the industrial hygiene and safety compliance officers and consultants. Explain your efforts and discuss any past regulatory violations or incidents. These individuals can provide a perspective on your industry and can provide additional resources.

10. Obtain and review the relevant regulations. Read carefully the sections pertaining to your priority health risks. Evaluate, to the extent possible, the regulatory status of your operations and make a realistic analysis of the actions required to achieve compliance. State and federal OSHA offices often have resources to assist businesses.

11. Locate model programs within your industry and summarize their approach. Check trade and professional journals for articles describing health risks to those in your industry. Contact the authors of these studies for more information. Obtain sample policies and programs. Find out what works.

12. Become active in business and professional associations. Consider serving on a committee or as an officer. You will make friends, improve your support system, and strengthen your technical knowledge.

## PLAN OF ACTION

1. Conduct long-term planning. Prepare a business plan or work plan listing your overall program priorities for the next 1 to 3 years. Make the document conform as much as possible to the style used within your organization for reports, audits, and plans. List your program goals, in order of overall priority, at the beginning of the plan in a one-page summary. Next outline the objectives or intermediate steps needed to achieve the goals. Finally, list for each objective the activities needed to carry out the objective. Include a time line or give dates of completion for the objectives.

2. Carefully consider your staffing requirements and overall budget. Given your resources, what can you accomplish in the time period indicated by your plan? Make conservative estimates of the funding and time needed to complete these tasks; overestimate the amount of time by a factor of three.

PROFESSIONAL OCCUPATIONAL HEALTH ORGANIZATIONS

American Association of
Occupational Health Nurses
50 Lenox Pointe
Atlanta, GA 30324
(404) 262-1162

American Board of
Industrial Hygiene
4600 W Saginaw, Ste. 101
Lansing, MI 48917-2737
(517) 321-2638

American Conference of Governmental Industrial Hygienists
Kemper Woods Center
1330 Kemper Meadow Drive
Cincinnati, OH 45240
(513) 742-2020

American Industrial Hygiene
Association
2700 Prosperity Ave.
Suite 250
Fairfax, VA 22031
(703) 849-8888

American Public Health
Association
1015 Fifteenth St., NW
Suite 300
Washington, DC   20005
(202) 789-5600

Institute of Hazardous
Materials Management
11900 Parklawn Dr.
Suite 450
Rockville, MD 20852
(301) 984-8969

National Hearing Conservation
Association
431 East Locust St., Suite 202
Des Moines, IA 50309
(515) 243-1558

National Safety Council
P.O. Box 558
Itasca, IL 60143-0558
(708) 285-1121

For more associations and other sources of help, see Appendix
A in Fundamentals of Industrial Hygiene (see FURTHER READING).

**Figure 11-1**  List of professional occupational health organizations.

SAMPLE INDUSTRIAL HYGIENE PROGRAM POLICY

It is the policy of [organization] to provide a safe and
healthful workplace.  Line management will be responsible for
the recognition and assessment of potential health hazards in
the work environment and for the implementation of necessary
control procedures and practices to eliminate these hazards or
reduce them to the lowest reasonably achievable level.
Operating management will also be responsible for ensuring
compliance with all government regulations and standards
relating to employee health from factors arising from the
workplace.  The industrial hygiene staff will provide
professional assistance and guidance to operating management
in achieving these objectives.

**Figure 11-2**  Sample industrial hygiene program policy. (From Ross, D. M., *Appl. Ind. Hyg.*, 3(5):F30–F34, 1988. With permission.)

Your initial program goals must succeed within the time frame and budget
indicated.
3. Gain organizational consensus for your overall plan. Include key staff
members whom you contacted during the needs analysis in the planning

SAMPLE WORK PLAN FORMAT

I. Summary
     (List all of the primary goals)

II.   Goal 1: Survey of Paint Shop
       To perform a complete industrial hygiene survey of the
       Paint Shop and to address the need for a Respiratory
       Protection Program for these employees.

       Objective 1:
       Complete an Industrial Hygiene survey of the Paint Shop.

       Activities:
       1.   Review all operations and materials used with shop
            supervisor.  Observe work practices.
       2.   Perform air monitoring of all operations involving
            paint which contains lead.
       3.   Perform additional air monitoring of painting,
            cleaning and abrasive blasting operations involving
            toxic materials.
       4.   Review sampling results and prepare technical
            report.  Include recommendations for improved
            working conditions.

     Et Cetera

**Figure 11-3** Sample work plan format.

process. Route a draft copy of the overall health plan to these individuals and obtain their responses. Present a working version of the plan at senior staff meetings. Begin your presentation by discussing the results of the needs analysis. Why is this program needed? Show how these programs will benefit the company as a whole. Include a discussion of the costs of these programs and be prepared for extensive questions regarding this subject. Conclude by presenting concrete benefits related to health programs, such as reduced numbers of injuries, illnesses, and workers' compensation cases and increased employee morale and productivity.

4. Prepare general health policy statements for each of the proposed health programs listed in your work plan. Such policy statements define the scope of the program and list the responsibilities of those involved. Have these policies officially endorsed and announced by executive management before beginning your work.

What common problems cause health programs to fail? Many health professionals respond that they have too much work, not enough resources, and not enough time to complete projects. However, occupational health programs sometimes suffer from a lack of planning and organization. Time spent planning and positioning yourself within the organization often results in the most efficient implementation of employee health programs. Your goals become the organization's goals. When important health and safety matters arise, you will communicate your ideas effectively.

Unless your programs enjoy the support of executive management, they cannot succeed. These managers control the resources available to the organization and set the tone for attitudes about worker health. First-line supervisors form the critical link between health policy as it appears in training sessions and on paper and the actual working conditions. Concentrate on providing quality training and access to technical information to your managers and supervisors. As your employees and supervisors mature in their careers and move into management themselves, they take with them their new attitudes and knowledge about employee health and wellness, making future programs easier to develop and implement.

You and your staff need to devote time to gaining and maintaining the professional skills required to carry out health management programs. Support your staff in achieving their career goals and in developing themselves as professionals. Frequent attendance at educational seminars, forums, and conferences will help these individuals stay current in a rapidly changing field. Subscribe to information services, legislative reviews, and professional journals. Require your staff to spend 4 to 5 hours each week reading such literature, and encourage them to take a speed-reading course.

Finally, know your limitations. If your budget permits, make use of contract employees or consultants to perform technical evaluations or training. As a manager, you must constantly find ways to accomplish your goals through the work of others. Focus on your own area of technical expertise and always request the assistance of others when necessary. You will develop professionally and serve your organization at the same time.

## SPECIFIC PROGRAMS

Incorporate specific programs into an overall health plan which conforms to legal and organizational requirements. Federal regulations will soon mirror the California Code of Regulations, which requires employers to have an injury and illness prevention program (summarized below):

1. A written program outlining the scope of the program and which designates the person(s) responsible for implementation
2. Hazard identification through periodic inspection of the workplace
3. Hazard correction procedures for correcting health hazards identified during the inspections
4. Hazard communication through use of material safety data sheets, labeling, and training of employees regarding occupational hazards
5. Investigation of occupational injuries and illnesses
6. Employee compliance with health and safety regulations
7. Health and safety training which includes both general and specific information regarding occupational health hazards

INDUSTRIAL HYGIENE TECHNICAL REPORT

I.  INTRODUCTION
    Describe the purpose of the survey: baseline, routine
    monitoring, health hazard evaluation or emergency
    response activity.  Identify work area or shop and
    manager, supervisor or foreman.  Provide names and titles
    of inspection team.

II.  DESCRIPTION OF OPERATIONS
    In tabular form, list employee job titles, work locations
    and personal protective equipment.  Include duration of
    all activities and a description of equipment and tools.
    Describe routine and extraordinary tasks.  Describe
    supervision and training.  Itemize all potential health
    hazards (such as noise, solvent vapors, or dust).

III.  SURVEY METHODS
    For each potential health hazard, describe method of
    evaluation in detail.  If industrial hygiene sampling was
    performed, list the sampling strategy (how you determined
    which employees to monitor).  Include the make and model
    of all  monitoring equipment, as well as methods of
    calibration.  Describe how others have surveyed similar
    operations, and cite journal articles or other references
    relevant to your methodology.

IV.  SURVEY RESULTS
    Tables of monitoring results should include the following
    information:  Name and job title of employee sampled;
    date; duration of sample in minutes or hours; contaminant
    or material of interest; volume of air sampled (if
    applicable); time-weighted average of exposure; 8-hour
    time-weighted average; Permissible Exposure Limit (PEL)
    and Threshold Limit Value (TLV) for material of interest.
    List all health and safety hazards noted during the
    survey.

V.  DISCUSSION
    Compare the results of the monitoring to the recommended
    exposure limits.  Are employees overexposed to airborne
    contaminants or to excessive levels of noise, radiation,
    heat or cold?  Describe any technical difficulties which
    could affect the validity of your results.  Briefly
    compare your results to those obtained by others.

VI.  RECOMMENDATIONS
    Describe changes in operations, materials used, personal
    protective equipment, or equipment that may reduce
    employee exposures and improve working conditions.
    Engineering controls may provide the best protection for
    employees.

VII.  APPENDICES
    Include sampling sheets, laboratory results, and copies
    of appropriate regulations.

**Figure 11-4** Sample industrial hygiene technical report.

A general program which includes these elements can form the foundation for
all occupational health and safety programs within your organization. The
program should also include procedures for medical surveillance, exposure
monitoring, and record-keeping. Today's occupational health regulations

emphasize this comprehensive approach as opposed to the somewhat fragmented standards of the past. This approach also helps document each step as your organization achieves compliance.

Make your program easy to reference. Put your written programs together in a binder for your supervisors and managers. Report on program implementation at staff meetings. Track illness and injury statistics. Consider printing an employee wellness newsletter, or contribute regularly to already existing publications.

Listed below you will find a summary of each of the specific health and safety programs. Consider this information a guide and not a comprehensive listing of the required elements of each program. Carefully review the relevant regulations, which may be specific for your geographic area or industry and which may not be discussed in this summary chapter.

## Hearing Conservation

Protecting the hearing of your employees involves physical evaluation of occupational noise exposures as well as monitoring the actual hearing acuity of those exposed to noisy environments. Perform noise evaluations of all work sites, following the guidelines set forth in the federal and state regulations. Determine which employees are exposed to noise in excess of 85 decibels, measured on the A scale (dBA), as a time-weighted average (TWA). Include such employees in a medical surveillance program. Each should receive a baseline audiogram and should have his/her hearing acuity measured at regular intervals. The hearing conservation program (HCP) seeks to prevent significant shifts in the hearing acuity of employees.

Shifts in hearing acuity at certain frequencies require the employer to remove this employee from the noisy environment or reduce his or her exposure to noise. Such shifts should also cause the employer to evaluate the effectiveness of the rest of the hearing conservation program. Please see Chapter 11B for specific information on medical surveillance.

In order to evaluate employee noise exposures, obtain the basic industrial hygiene equipment needed for this task: sound-level meters (type 2) for direct measurements of noise sources and several sets of noise dosimeters. During your walk-throughs of employee work areas, use the sound-level meter to take preliminary readings of the noise levels. Evaluate individual pieces of equipment in the operator's hearing zone. Determine the length of time each piece of equipment is used. Note the approximate amount of time employees receive exposure to noise levels in excess of 80 dBA.

Return to work areas where employees may be exposed to more than 80 dBA (TWA) of noise and conduct noise dosimetry on a representative number of employees over the course of several full working days. Determine if you are sampling usual working conditions; return on another day if necessary.

Focus your efforts on the reduction of noise exposures in the work environments where employees are exposed to noise in excess of 80 dBA. Is the equipment properly installed? Have noise-abating covers been removed to facilitate operations? Can the noise source be insulated from the work environment through the use of engineering controls? Perhaps employees can perform their work from inside a noise-proof booth or at a greater distance from the noise source. Can the company replace very noisy equipment? Provide hearing protection while these changes are being implemented and for all operations involving even brief exposures to high noise levels, such as the use of hand tools.

Provide hearing protection for all employees exposed to noise in excess of 85 dBA. Permit employees to choose from several different varieties of hearing protection. Most hearing protectors must be fitted by a qualified individual due to anatomical variations in the size of the ear canal. Fitting for these earplugs can take place during the annual hearing test (audiogram). Other types of earplugs, such as expandable foam plugs, require no fitting and provide excellent protection for some individuals; however, these plugs will not fit everyone. Earmuffs also provide excellent noise attenuation and can be worn in conjunction with earplugs, increasing the amount of noise attenuation. Those performing very noisy operations for any length of time should wear both ear plugs and ear muffs.

The noise attenuation factors printed on the packaging materials for hearing protectors often overestimate the actual ability of the device to protect employees from excessive noise. These attenuation factors result from simulation of working conditions in a laboratory setting. Many factors contribute to the actual protection factors provided by hearing protection. The seal on earmuffs, for example, may be broken by the employees' use of eye protection. The muffs themselves may become cracked or hard from age or exposure to heat or light. Ear plugs may not be inserted properly, or they may be dirty or in poor repair. Consequently, employees need training in the use of hearing protection in order to protect their hearing.

Conduct employee training in hearing conservation. Describe the program in detail, emphasizing the importance of prevention of hearing loss. Often employees think hearing can be easily restored, as simply as they can improve their vision by wearing a pair of glasses. Ask an older employee to discuss the difficulties and isolation associated with noise-induced hearing loss, particularly in situations involving large groups of people, such as a party. Explain that hearing aids, while useful, often cannot restore hearing acuity to previous levels. Those with hearing loss will experience a decrease in quality of life.

Because of the extreme complexity of human hearing, avoid focusing at length on the anatomy and physiology of the ear, except to stress that noise-induced hearing loss involves nerve damage, which is irreversible. Discuss the importance of reducing the daily dose of noise through the use of engineering and administrative controls and hearing protection.

## Asbestos Safety

A confusing plethora of federal, state, and local regulations governs the handling of this hazardous substance. Extensive media coverage of the health problems caused by exposures to high levels of airborne asbestos fibers has created extreme concern among employees and the general public. High-profile litigation involves many manufacturers of asbestos-containing products and employers of those engaged in asbestos operations. How can you protect your employees from a known human carcinogen but diffuse unnecessary fear of asbestos-related disease? How can you keep abreast of all asbestos regulations but still have time to focus on other health concerns and programs?

Despite the controversy, asbestos exposure remains an important environmental and occupational issue. The health risks relating to occupational exposures to airborne fibers of asbestos have been well documented. Modern environmental and occupational regulations involving asbestos-containing materials include extensive controls to protect worker health and the environment. OSHA promulgated the first asbestos standard in 1971 and has revised the standard several times as new information has become available. The permissible exposure limit (PEL) set for this substance by OSHA in 1971 was 12 fibers per cubic centimeter of air (f/cc). On October 11, 1994, a revised PEL of 0.1 f/cc went into effect for all asbestos workers.

In addition to the revised PEL, this standard creates four categories of asbestos work for those involved in the construction or shipyard industries. Employers must use the work practices outlined for each category of asbestos work, regardless of the exposure levels involved. This approach differs from previous standards, which relied upon an action level (AL) of exposure to asbestos. Employers were required to determine if an operation exposed workers to more than the specified AL before utilizing appropriate work practices and controls.

The current OSHA standard incorporates many of the provisions of the Asbestos Hazard Emergency Response Act (AHERA) and the Asbestos School Hazard Abatement Reauthorization Act (ASHARA), enacted by Congress in 1986 and 1990, respectively. This legislation established a program to address the problem of asbestos in schools. AHERA mandated inspections of school buildings for asbestos and required schools to prepare plans to manage the material if found. AHERA also created extensive training requirements for those involved in performing asbestos work in schools. ASHARA extended the training requirements to persons performing such work in public and commercial buildings. In April 1994, the Environmental Protection Agency (EPA) Asbestos Model Accreditation Plan (MAP), which clarifies and expands some of these training requirements, went into effect.

Many state governments have established their own asbestos regulations, including accreditation programs for those involved in performing asbestos-related work. Many of these regulations are much more stringent than the federal standard. Other regulations involving asbestos include the EPA's National

Emission Standards for Hazardous Air Pollutants (NESHAPS), which require the removal of asbestos prior to demolition or renovation. The EPA and the Department of Transportation regulate the removal and disposal of asbestos as a hazardous waste.

As with other health programs, begin by identifying the hazardous material involved (in this case asbestos). Building surveys also permit the evaluation of asbestos-containing materials. These surveys must be performed prior to renovation or demolition (under NESHAPS), but should also be scheduled for all buildings. Protocols for such surveys are outlined in the AHERA legislation and should be followed. Management plans for asbestos incorporate the results of such surveys and permit the abatement of the most serious exposure hazards first. All asbestos-containing materials should be clearly labeled, and employees should not disturb these materials unless they are trained and have the proper protective equipment.

Focus on educating your organization about the correct manner of dealing with asbestos, particularly asbestos in buildings. An independent industrial hygienist who holds EPA certification in project design and supervision and is also a Certified Industrial Hygienist (CIH) should plan and oversee all abatement work. This practice is directly analogous to an architect preparing plans prior to the commencement of building construction. Individuals performing building surveys for asbestos should also have an industrial hygiene background and should have attended an EPA-approved course in building inspection. Similarly, maintenance and custodial employees need to know how to identify potential asbestos-containing materials. Occupants of buildings containing asbestos materials should receive awareness training. Therefore, your asbestos training program may involve nearly all employees.

During training sessions, provide information in a matter-of-fact way. Your employees may think of asbestos as a dangerous, human-engineered chemical, although asbestos minerals are mined from the ground like gold or silver. Although many who worked with asbestos have contracted lung cancer or asbestosis, most of these workers experienced exposure levels many times higher than those permitted by today's asbestos regulations. The actual risk of contracting asbestos-related disease decreases with level of exposure.

Discuss the latency period associated with asbestos disease and the synergistic effect of smoking and asbestos exposures. Those working with asbestos should not smoke!

Establish a method for reviewing all planned building renovations. Compare these plans with the building surveys to determine if asbestos-containing materials will be involved. In many cases, more extensive surveys of building materials will be required. Removal of materials in occupied buildings, particularly sprayed-on fireproofing which contains asbestos, requires expert planning and oversight. If your organization leases property, you need to ensure that these tenants adhere to your asbestos policy and program when planning and conducting building renovation or demolition.

While the recent regulatory emphasis involves asbestos in buildings, you should not overlook asbestos materials in your manufacturing or maintenance operations. Review product material safety data sheets (MSDSs) to identify any asbestos-containing products in use by your workers. Often roofing materials, flooring, adhesives, gaskets, and other products contain asbestos. Find effective substitutes for these products. Phase out the use of asbestos-containing products, such as brake shoes, during periodic maintenance operations by replacing them with asbestos-free materials.

Complying with the asbestos regulations will require an enormous commitment of time and effort from your staff. You may decide to contract out some or all of these responsibilities while retaining oversight capacity.

## Respiratory Protection

One of the greatest problems with respirators is that they are used haphazardly and in ways that may be detrimental to the health of employees. You will need to take control of respirator use by determining when they should be worn, what types should be issued, and how they are to be used and maintained. OSHA requires that you formalize your program in written form. Your program should address the selection of proper respirators, training in their use and limitations, cleaning, disinfection, storage, inspection, evaluation of work conditions and program, medical surveillance, and approved respirators.

First, determine the operations requiring respirator use. Conduct air monitoring of operations, starting with those that you feel may be problematic. If the exposures are high, determine the best engineering or administrative control measures, such as product substitution or ventilation. OSHA mandates that you institute other controls before personal protective equipment (PPE) because PPE puts much of the burden of protection on the employees. A respiratory protection program is also more costly than it might seem, and it will take a lot of effort to ensure that the program is being followed. While control measures are being implemented, you should have employees wear respirators. After you implement control measures you should conduct more air monitoring. You may find that the control measures are not completely effective and that the use of respirators should continue. In some cases, such as when working with asbestos, the general practice is to always wear some level of respiratory protection, regardless of air monitoring. This is because you want to keep exposure as low as possible and because air concentrations are neither uniform nor entirely predictable.

Obtain copies of the respiratory protection regulations and standards. The American National Standards Institute (ANSI) has published many standards on respiratory protection, including ANSI Z88.2, which is referenced in the OSHA regulations.

Put your program in writing and get it approved by management. First you must decide who will administer the program. This person will be responsible for purchasing, training, fit testing, and the other elements of your program and will probably be you. Some of the duties could be given to others in the organization, but make sure that everyone with responsibilities has the proper training or professional background, including yourself. If you do not have all the necessary skills, you could contract parts out or attend classes. Work with management and your human resources department to decide what you will do if employees are unable to wear respirators. What efforts will be made to accommodate them? You should also work with the applicable employee unions.

Determine the types of respirators required for your operations. In selecting respirators consider the contaminant, its form, its concentration, the exposure limit, the possibility of an environment immediately dangerous to life or health (IDLH), flammability, warning properties, and eye or face protection needed. Make sure you familiarize yourself with the uses and limitations of various types of respirators.

Schedule medical examinations for all the employees who work (or could work) in the areas requiring respirators. Tell the physician what types of respirators you would consider having the employee wear because, depending on the type, the examinations may vary slightly. The physical exam is required because some medical problems may prohibit an employee from wearing a respirator. Respirators can cause physiological stresses due to factors such as restricted air flow and greater weight. However, because being restricted from wearing a respirator can affect the livelihood of an employee, efforts should be made to accommodate physical problems by selecting a different type of respirator. You should work with the physician on this problem. If this fails, you will need to rely on the policy you developed earlier.

Employees who are qualified to wear respirators must be assigned respirators that fit them. This is especially important for negative air-purifying respirators. Fit testing should be carried out at least annually and should be performed in accordance with standards and regulations.

There are two types of fit testing: qualitative and quantitative. Quantitative fit testing is the preferred method because it actually measures the amount of contaminant getting inside the respirator. However, it can be expensive to set up a system and may not be practical if you only have a small number of employees wearing respirators. Traditionally, quantitative fit testing was performed in a booth with a particle generator and counter. However, there are now portable models which are based on counting the particles in ambient air that may make quantitative fit testing more reasonable for your situation. There are also outside consulting and training firms that can perform fit testing for you.

If quantitative fit testing is not feasible, you can perform qualitative fit testing. This involves finding out if a respirator wearer can taste or smell a test agent. You must follow the protocol carefully for the test to be valid.

Whichever form of fit testing you decide on, you will need to order respirators from various manufacturers and in different sizes so that employees have a selection. One face piece size is not going to be adequate for everyone in your company. Various sizes from just one manufacturer may not be adequate either. Read through the respirator fit testing protocol and obtain the equipment you will need.

Respirators will only protect employees if they are properly maintained. If possible, employees should be issued a personal respirator rather than have to share one with others. All respirators must be cleaned and disinfected regularly (right after wearing if shared). There are disposable towelettes for cleaning and disinfection, but it is better to wash the respirator in soap and water when possible. There are disinfectant soaps that can be dissolved in water. If the work area does not have a clean sink for this purpose, you could supply clean buckets. The respirator should be dried and then stored in a clean place. One good way to do this is to put the respirator in a clean plastic container on the top shelf of a clean locker (where it will not get crushed). Respirators should not be left on a hook in the work area, on the dashboard of a truck, or anywhere else they may become dirty or damaged.

Before use, respirators should be inspected to verify that all the parts are in place and functioning properly. If a part is missing or broken, you will need to order a replacement part from the manufacturer for the particular model of respirator. Respirators are approved as a whole unit, and if you use a different part (for example, use a safety pin to hold a strap in place) you are no longer using an approved respirator.

Training for employees and supervisors is a critical part of your program. The success of your program depends upon employees wearing respirators correctly and maintaining them correctly. Since you cannot check on all work areas all of the time, supervisors should ensure compliance with the program, so both employees and supervisors should be trained. You should cover the basics of the types of respirators, their uses and limitations, and all of the aspects of your program. Try to make the training as hands-on as possible. Have employees bring in respirators and actually clean them, inspect them, try them on, etc. After your official training, you should try to visit the work areas and check on respirator usage and storage. Talking individually with employees in their work areas and addressing their specific questions or deficiencies can be a very good form of training. Training must be refreshed at least yearly.

Continue to do air monitoring periodically to make sure conditions have not changed. Some regulations for specific chemical hazards specify when monitoring should be done.

Audit the program to be sure that it is being followed. Talk to respirator wearers and inspect the condition of respirators. Evaluate your program and change written procedures if necessary.

## Hazard Communication

An effective hazard communication program involves a careful review of the hazardous materials purchased and used by your employees. Spend time with the purchasing department and first-line supervisors to find out how hazardous materials enter the workplace. You may discover that salespersons leave products, with or without MSDSs, for your employees to use on a trial basis. In addition, supervisors or employees themselves may purchase hazardous materials in small quantities for urgent or unplanned activities. This may result in improper use of these materials by your employees.

Try to establish the most practical point in the purchasing process for review of the materials used. Approve the use of a material *prior* to purchase. Manufacturers of hazardous materials will always provide a MSDS for each of their products and can send this information by facsimile if necessary. Review the ingredients section of the MSDS and determine whether the product contains any materials considered to be very toxic or carcinogenic. If so, you will want to work with the supervisor and possibly the supplier to find a substitute material for the job.

Next, review the MSDS for completeness using the regulatory requirements in checklist form (see Figure 11-5). Contact the manufacturer regarding an inadequate MSDS. Provide a copy of your checklist, indicating the missing information. Sections often deficient in required information include (1) chemical ingredients, (2) toxicity, (3) personal protective equipment, and (4) date the MSDS was prepared. Occasionally, a MSDS may not have the most basic information, such as the manufacturer's name, address, and telephone number. Do not approve a material for use without having the information needed to recommend safe work practices. By requesting this information, manufacturers will improve their MSDS or risk losing business.

Give employees ready access to a file containing the MSDSs after you have approved each one. Each material used by the employee should have an available MSDS. Generic MSDSs generally do not meet the requirements of the Hazard Communication Standard. Different brands of the same product may contain slightly dissimilar chemical ingredients. There are some computerized MSDS systems available, but make sure the system is available to every worker at every work site. Every product should have a label clearly identifying the material and allowing the employee to reference the relevant MSDS. Each label should contain a brief description of the major health effects associated with the chemical ingredients, as well as information concerning appropriate handling and use. Contact manufacturers regarding inadequate labeling. Most safety equipment suppliers sell labels or label-making kits for any products you have that lack manufacturer's labels.

What about materials already in use? Follow the procedures described above for new materials after obtaining a detailed inventory from each work area. You will have to provide guidelines for employees responsible for providing such inventories. Provide a brief training session defining hazardous materials and explaining the purpose of the inventory. Provide a format for the inventories. Make use of already existing inventories of stock if the materials listed constitute the actual ones in use. Alternatively, your staff can manually

Material Safety Data Sheet Adequacy Checklist

Material Name _____ Product # _____

Manufacturer _____

MSDS Date of Preparation or Revision _____

| REQUIRED INFORMATION | 29 CFR 1910.1200(g)(2) | CHECK IF INADEQUATE |
|---|---|---|
| I.   Product Identification, including 1, 2 or 3: | | |
|     1. If single substance: | (i)(A) | |
|       a. chemical name(s), | | |
|       b. common name(s), | | _____ |
|     2. If mixture: | (i)(C) | _____ |
|       a. chemical name(s) of ingredients, | | |
|       b. common name(s) of ingredients, | | _____ |
|     3. Common name(s) of mixture, | | _____ |
|       if tested as a whole. | (i)(B) | _____ |
| II.  Physical and Chemical Characteristics, such as vapor pressure and flash point | (ii) | _____ |
| III. Physical Hazards, such as: potential for fire, explosion, reactivity | (iii) | _____ |
| IV.  Health Hazard Data, including: | (iv) | |
|     1. Signs and symptoms of exposure, | | _____ |
|     2. Medical conditions aggravated, | | _____ |
|     3. Primary routes of entry, | (v) | _____ |
|     4. OSHA PEL, ACGIH TLV, and any other recommended exposure limit, | (vi) | |
|     5. Whether listed as a NTP, IARC, or OSHA carcinogen. | (vii) | _____ |
| V.   Safe Handling and Use, including: cleanup of spills and leaks | (viii) | _____ |
| VI.  Control Measures, such as: Engineering controls, PPE, work practices. | (ix) | _____ |
| VII. Emergency and First-Aid Procedures | (x) | _____ |
| VIII. Date of MSDS Preparation or Revision | (xi) | _____ |
| IX.  Name, Address, and Telephone No. of Responsible Party | (xii) | _____ |

Reviewed by:_____ Date:_____

**Figure 11-5** Sample material safety data sheet checklist.

list hazardous materials during a walk-through inspection of each work area. All inventories should note the manufacturer's name, the product name (and number if available), common name, size and type of container, and quantity of material stored. The inventories should also list how these materials are stored and used. Periodic updates of hazardous materials inventories will keep your information current.

Even small organizations will find their inventories to be extensive. Assembling corresponding files of MSDSs will require extensive clerical support. Cross-reference your file of MSDSs by manufacturer, product name, and

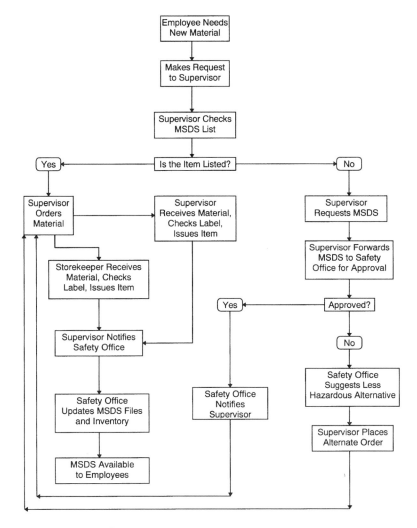

**Figure 11-6** Sample flowchart showing hazard communication system.

common name. Using a computer database is helpful. Make it as easy as possible for your employees to find the material, especially during an emergency.

Develop a flowchart illustrating the system for hazard communication within your organization (see Figure 11-6). Use this chart during general training sessions. Present the basic concepts of hazard communication, including definitions of hazardous materials, MSDSs, the use of protective equipment, and basic toxicology. All employees should receive this general training, perhaps during regularly scheduled safety or staff meetings.

Training in the use of specific materials may present the most challenging portion of the hazard communication program. Make use of a number of resources for this training; do not rely on the MSDSs alone. Develop standard operating procedures (SOP) by working with the supervisors and employees. Include detailed instructions for safely using and handling hazardous materials. Glossaries of the terms used by toxicologists and industrial hygienists may make it easier to present technical information. Group together materials with similar chemical properties, such as organic solvents. Schedule a number of brief sessions; do not try to present too much information at one time.

Some organizations effectively use the "train-the-trainer" approach. The occupational health staff trains first-line supervisors and provides them with the curricula, audiovisual materials, and other supplies needed. Since employees tend to attach greater significance to information learned from their supervisors, this method often works well. Supervisors may need extensive support from your group, however, in order to feel comfortable presenting technical information.

## Hazardous Materials

Many environmental regulations concerning hazardous materials go beyond the stipulations of the Hazard Communication Standard; hazard communication forms a subset of hazardous materials programs. A partial list of hazardous materials regulations appears in Figure 11-7. Many of these overlap.

Include in the hazardous materials inventory the information required for public disclosure for each product, including the chemical ingredients (and percentage contained), toxicity, storage provisions, and emergency procedures. Include detailed maps showing the location(s) where hazardous materials are stored and used. During a fire, earthquake, chemical spill, or other emergency situation, this information, as well as the MSDSs for all products, should be readily available to emergency responders.

Public interest and environmental action groups will have access to the above information. In addition, your organization must provide the aggregate amount of hazardous materials released to the environment. Substitution of less toxic materials, as well as reduction of the overall use of these materials, will reduce environmental and occupational health liability.

Storage requirements are often not clearly spelled out by regulation. General storage requirements are also difficult to apply to specific situations. However, pay particular attention to the storage and handling of the following types of hazardous materials:

1. Flammable and combustible liquids, such as gasoline, diesel fuel, solvents, paints, and paint removers
2. Corrosives, including acids and bases
3. Oxidizers and reactive materials, such as peroxides, chlorine gas, and acetylene
4. Poisons and toxic materials: inorganic lead and other metals, pesticides, etc.
5. Compressed gases

SELECTED FEDERAL HAZARDOUS MATERIALS REGULATIONS

Occupational Safety and Health Administration (OSHA)
  Hazard Communication: Title 29, Code of Federal Regulations,
    Section 1910.1200 (29 CFR 1910.1200)
  Emergency Action Plan: 29 CFR 1910.38(a)
  Fire Prevention Plan: 29 CFR 1910.38(b)
  Hazardous Waste & Emergency Response: 29 CFR 1910.120
  Chemical Process Safety: 29 CFR 1910.119
  Laboratories: 29 CFR 1910.1450
  Respiratory Protection: 29 CFR 1910.134
  Bloodborne Pathogens: 29 CFR 1910.1030

Environmental Protection Agency (EPA)
  Emergency Planning & Community Right-To-Know (SARA Title
    III): 40 CFR 355, 370, 372
  Underground Tanks: 40 CFR 280, 281
  Pesticides: 40 CFR 152-186
  Hazardous Waste: 40 CFR 240 et seq.

Department of Transportation (DOT)
  Transportation of Hazardous Materials and Waste: 49 CFR 107,
    171-197

See Also:
  Uniform Building Code
  Uniform Fire Code

(Please note that this is not a complete list.   Contact
agencies listed and their state and local counterparts for
more information.)

**Figure 11-7** Examples of federal hazardous materials regulations.

Storage of large quantities of flammable materials requires special containers or buildings equipped with fire suppression systems, containment features, ventilation, and explosion-proof lighting. Smaller quantities of materials will require flammable-liquid storage cabinets or special storage containers. Follow the Uniform Fire Code, National Fire Protection Association (NFPA) standards, ANSI standards, and local fire code for your specific situation.

Combustible and reactive materials should be stored separately from flammable materials since they may provide fuel for a fire. Oxidizers require separate storage. Corrosives require careful handling because these chemicals can cause severe burns, and they also should be stored apart from other materials.

Conduct periodic walk-throughs of operations involving hazardous materials. Particularly in small operations, you may find incompatible materials stored together or a complete lack of storage lockers for flammables. Containment of these materials in the case of a spill is critical, and secondary containment should always be used. Berms and containment dikes separate large tanks of these materials and provide a collection point in case of spills. Even simple pallets that can contain liquids can serve this purpose in smaller operations.

Many employees will not have an intuitive understanding of the properties of chemicals or the consequences of improper storage of incompatible materials, even after extensive training. Provide detailed procedures for these employees. Begin by grouping materials used in each work area by chemical properties, as listed above. For each group, choose the materials used most often or which are the most representative of these materials as a whole. Next, go to the MSDSs for these materials and check under the sections on storage, handling, and emergency procedures. If necessary, contact the manufacturer for more detailed information.

Go through work areas or check your inspection records for examples of how these materials are presently being used and stored. Break each operation down into steps. Specify the use of personal protective equipment (especially chemical splash goggles and face shields), safety cans for flammable liquids, and a laboratory hood or other ventilation during mixing or transfer. Include provisions for spills by providing spill kits for quick cleanups of small spills of less toxic materials (be sure your training describes which spills employees may clean up themselves). All work sites (including mobile or temporary ones) where hazardous materials are used should have plumbed eye washes and showers or other provisions for flushing the eyes or skin with water following an accidental spill.

Employees and supervisors should check the procedures for completeness. To save time, have them prepare the breakdown of the operation into steps and assist with the incorporation of safe handling procedures. When complete, post the procedures in the workplace and conduct training sessions.

## Emergency Planning

Many regulations require the employer to prepare emergency plans, including procedures to follow in case of fire, flood, earthquake, or other disaster. Hazardous materials require special consideration. All employees who routinely work with hazardous materials should know how to cope with small spills of these materials, as described in the previous section. Your emergency plan should include this information, as well as procedures for more catastrophic situations. This section lists the most common elements of emergency planning.

A release of an unknown or very toxic material, or of any hazardous material in an unventilated space or any other uncontrolled situations, may require immediate evacuation. Employees should contact the emergency response team. Hazardous materials response often is performed by specially trained units of the fire service. However, your company may have internal response units, especially if your operations take place in remote areas. Clearly post the telephone number of the appropriate emergency response team in all work areas. In many cases, employees should call 911.

Before disaster strikes, prepare a hazardous materials inventory as described in the previous section. Include maps indicating the locations of all hazardous materials storage areas. Briefly describe the operations which utilize the hazardous materials.

The hazardous materials team will act to contain the immediate emergency situation — for example, a spill involving about 500 gallons of diesel fuel on a street near your facility's cafeteria. The responders will contain the spill using absorbents and keep the material from entering the storm drains. The response team will find and eliminate sources of ignition and estimate the wind speed and direction to determine the extent of any vapor generation. They may order evacuation of the cafeteria and other nearby buildings. After stabilizing the situation, the responders will probably expect you to have a contract hazardous materials responder perform the actual cleanup of the spill.

Some organizations will train their own emergency response team, but most of the time it makes sense to contract with a hazardous materials/waste company for these services. Have several on contract so that at least one is available at all times. Screen these companies carefully. What is their general reputation in the environmental community? Review their internal operating procedures and their hazardous waste disposal procedures. Your company, and possibly you personally, will be held liable for their actions. Poorly handled hazardous materials spills cost corporations a great deal, both in terms of actual cleanup costs and in credibility with the public.

Emergency planning should also include employees and operations where hazardous materials are not in use. All employees need to know what to do in case of medical emergencies. Exit routes from all buildings and lists of emergency phone numbers may be included in brief emergency action plans. Some organizations will also have in place more extensive plans to deal with the response to a major disaster.

Communication forms a major element of both emergency action and response planning. How can your employees communicate if telephone lines do not work? Are existing radio communication systems compatible? Upgrade such systems gradually, if necessary. You will need an integrated, comprehensive communication system.

Who will be in charge during an emergency? How will you account for all employees? What are the roles of those designated as emergency response coordinators? What should employees do if an emergency occurs outside of regular business hours?

After developing a workable program incorporating these points, conduct training and host mock emergencies. Share your plan with the emergency responders and include them in the exercises. Everyone should know what to do if an emergency occurs.

The hazardous materials inventory forms the backbone of the hazardous materials program. You must know what materials are used, the quantities used, and how they are stored. In the process of building the chemical inventory, develop storage, handling, and emergency procedures. Include these in the emergency response plan.

## Lead Safety

Currently, the regulations and standards regarding lead are changing. There has been an increase in public concern and policy making about environmental lead, and this is affecting occupational lead programs as well. Congress passed the Housing and Community Development Act of 1992 (commonly referred to as "Title X") to address lead poisoning, and the law is being implemented in stages by various agencies. OSHA has a new standard for the construction industry. The Housing and Urban Development Administration (HUD) and the EPA are developing new guidelines and laws for environmental lead exposure. Many states are also enacting occupational and environmental lead regulations. For example, the lead paint abatement industry is becoming increasingly regulated.

Since the regulations are changing quickly, when you are first beginning a lead program you will need to gather information on current regulations and the regulatory trends in your region. You should check with OSHA, EPA, HUD (if you deal with housing), and local agencies responsible for environmental and occupational safety and health. You will have to ascertain which regulations apply to your operations. Environmental regulations will probably apply to anyone dealing with lead; they will be most applicable to lead abatement.

In terms of occupational health regulations, you will need to decide if you are covered by the general industry standard (29 CFR 1910.1025) or the new construction industry standard (29 CFR 1926.62). You may need to check with OSHA, but generally, if a product is made (for example, a battery plant where exposure is fairly consistent), you are covered by the general industry standard. If lead work involves construction, demolition, or repair (such as a painting contractor), you are covered by the construction standard. Maintenance activities can be a gray area, and you will have to find out how the standards are being enforced or follow the most stringent ones.

As with any program, you need to start by getting management support. You will have to explain the need for the program. You should be prepared to discuss the budgetary impact of the program and the consequences (both regulatory and health-based) of less than full commitment. If you are dealing with lead in buildings, you may want to relate your lead program to your asbestos program, since they have many similarities.

You will need to evaluate the work which involves lead. You should do air monitoring for the various activities that involve lead. Evaluate the work practices. This is a good time to get the employees involved in the program. For example, you can have them help you evaluate improved work practices.

When you know where your lead exposures are, look at the ways you control the hazards. As with all programs, start by looking at engineering and administrative controls. Some of these controls may be spelled out in the regulations. Reevaluate the work areas where lead exposures occur to validate that the controls you have chosen are working. By now you should have your program in writing and have full management support.

Employees, supervisors, and line managers should all get training on the hazards of lead and the program you have put in place to address them. Check with the regulators in your area because certified training providers may be required (particularly for abatement). Send employees for medical exams and blood lead monitoring, and set up a program for regular medical surveillance.

## ACKNOWLEDGMENTS

The authors wish to thank the following persons, who are current or former employees of the City and County of San Francisco: Richard Lack, Richard Lee, Neva Nishio Petersen, Pamela Reitman, Vickie Wells, and Karen Yu. These individuals provided invaluable assistance during the development of health programs for the Port of San Francisco. Their support and friendship are greatly appreciated.

## REFERENCES

### Books

American Conference of Governmental Industrial Hygienists (ACGIH). *Industrial Hygiene Program Management.* Cincinnati, OH: ACGIH, 1988–1989.

Berger, E. H., Ward, W. D., Morrill, J. C., and Royster, L. H., Ed. *Noise and Hearing Conservation Manual,* 4th ed. Fairfax, VA: American Industrial Hygiene Association, 1986.

Clayton, G. and Clayton, F., Eds. *Patty's Industrial Hygiene and Toxicology.* New York: John Wiley & Sons, 1991.

Garrett, J., Cralley, L. J., and Cralley, L. V., Eds. *Industrial Hygiene Management.* New York: John Wiley & Sons, 1988.

Plog, B. A., Ed. *Fundamentals of Industrial Hygiene,* 3rd ed. Chicago: National Safety Council, 1988.

### Other Publications

American Industrial Hygiene Association Management Committee. Health and safety implications of European Community 1992 (EC92). *Am. Ind. Hyg. Assoc. J.* 53(11):736–741, 1992.

American Journal of Public Health (Editorial). Workers' health and safety (WHS) in cross-national perspective. *Am. J. Public Health* 78(7):769–771, 1988.

Andersen, G. H., Smith, A. C., and Daigle, L. L., An approach to occupational health risk management for a diversified international corporation. *Am. Ind. Hyg. Assoc. J.* 50(4):224–228, 1989.

Bardsley, C. A., Lichtenstein, M. E., and Nusbaum, V. A. Industrial hygiene career planning. *Am. Ind. Hyg. Assoc. J.* 44(3):229–233, 1983.

Belk, H. D. Implementing continuous quality improvement in occupational health programs. II. The role of the corporate medical department — present and future. *J. Occup. Med.* 32(12):1184–1188, 1990.

Bosch, W. W. and Novak, J. J. Strategies for developing comprehensive occupational health programs. *Water Eng. Manage.* April: 26–30, 1993.

Brandt-Rauf, P. W., Brandt-Rauf, S. I., and Fallon, L. F. Management of ethical issues in the practice of occupational medicine. *Occup. Med. State of the Art Rev.* 4(1):171–176, 1989.

Bridge, D. P. Developing and implementing an industrial hygiene and safety program in industry. *Am. Ind. Hyg. Assoc. J.* 40(4):255–263, 1979.

Brief, R. S. and Lynch, J. Applying industrial hygiene to chemical and petroleum projects. *Am. Ind. Hyg. Assoc. J.* 41(11):832–835, 1980.

Burdorf, A. and Heederik, D. Guest editorial. Occupational and environmental hygiene; more than just a name game. *Am. Ind. Hyg. Assoc. J.* 53(10):A484–A490, 1992.

Colton, C. E., Birkner, L. R., and Brosseau, L. M., Eds., Respiratory Protection: A Manual and Guideline, 2nd ed., Fairfax, VA, American Industrial Hygiene Association, 1991.

Committee report. Scope of occupational and environmental health programs and practice. *J. Occup. Med.* 34(4):436–440, 1992.

Cooper, W. C. and Zavon, M. R. Health surveillance programs in industry. In: *Patty's Industrial Hygiene and Toxicology,* 3rd ed., Vol. 3, Part A. Harris, R. L., Cralley, L. J., and Cralley, L. V., Eds. New York: John Wiley & Sons, 1994.

Corn, M. and Lees, P. S. J. The industrial hygiene audit: purposes and implementation. *Am. Ind. Hyg. Assoc. J.* 44(2):135–141, 1983.

Dalton, B. A. and Harris, J. S. A comprehensive approach to corporate health management. *J. Occup. Med.* 33(3):338–347, 1991.

Dwyer, T. Industrial safety engineering — challenges of the future. *Accid. Anal. Prev.* 24(3):265–273, 1992.

Fallon, L. F. Organizational behavior. *Occup. Med. State of the Art Rev.* 4(1):93–104, 1989.

Fallon, L. F. An overview of management. *Occup. Med. State of the Art Rev.* 4(1):1–10, 1989.

Feitshans, I. L. Potential liability of industrial hygienists under USA law. II. Prescription for reducing potential liability. *Appl. Occup. Environ. Hyg.* 6(3):205–214, 1991.

Geiser, K. Protecting reproductive health and the environment: toxics use reduction. *Environ. Health Perspect. Suppl.* 101 (Suppl. 2):221–225, 1993.

Gough, M. PCIH lecture: Zero Risk or Acceptable Risk. *Am. Ind. Hyg. Assoc. J.* 52(10):A556–A560, 1991.

Hazzard, L., Mautz, J., and Wrightsman, D. Job rotation cuts cumulative trauma cases. *Personnel J.* February: 29–32, 1992.

Henry, B. J. and Schaper, K. L. PPG's safety and health index system: a 10-year update of an in-plant hazardous materials identification system and its relationship to finished product labeling, industrial hygiene, and medical programs. *Am. Ind. Hyg. Assoc. J.* 51(9):475–484, 1990.

Hughes, J. T. An assessment of training needs for worker safety and health programs: hazardous waste operations and emergency response. *Appl. Occup. Environ. Hyg.* 6(2):114–118, 1991.

Jacobs, H. C. Managing safety. *CHEMTECH* July:400–402, 1989.

Johnson, S. L. Manufacturing ergonomics: a historical perspective. *Appl. Ind. Hyg.* 4(9):F24–F29, 1989.

Lichtenstein, M. E. Guest editorial: megatrends and the occupational safety and health professional. *Am. Ind. Hyg. Assoc. J.* 45(11):A17–A18, 1984.

Lichtenstein, M. E., Buchanan, L. G., and Nohrden, J. C. Developing and managing an industrial hygiene program. *Am. Ind. Hyg. Assoc. J.* 44(4):256–262, 1983.

Minerva Education Institute. Management's safety and health imperative: eight essential steps to improving the work environment. *Am. Ind. Hyg. Assoc. J.* 52(4):A218–A221, 1991.

Molyneux, M. K. and Wilson, H. G. E. An organized approach to the control of hazards to health at work. *Ann. Occup. Hyg.* 34(2):177–188, 1990.

Moser, R., Meservy, D., Lee, J. S., Johns, R. E., and Bloswick, D. S. Education in management aspects of occupational and environmental health and safety programs. *J. Occup. Med.* 31(3):251–256, 1989.

Moser, R. Quality management in occupational and environmental health programs — benefit or disaster? *J. Occup. Med.* 35(11):1103–1105, 1993.

Novak, J. J. and Bosch, W. W. Strategies for developing comprehensive occupational health programs for the paper industry. *Tappi J.* 76(9):87–95, 1993.

Occupational Medical Practice Committee of the American College of Occupational and Environmental Medicine (ACOEM). Scope of occupational and environmental health programs and practice. *J. Occup. Med.* 34(4):436–440, 1992.

Reynolds, J. L., Royster, L. H., Royster, L. J., and Pearson, R. G. Hearing conservation programs (HCPs): the effectiveness of one company's HCP in a 12-hr work shift environment. *Am. Ind. Hyg. Assoc. J.* 51(8):437–446, 1990.

Robins, T. G., Hugentobler, M. K., Kaminski, M., and Klitzman, S. Implementation of the federal Hazard Communication Standard: does training work? *J. Occup. Med.* 32(11):1133–1140, 1990.

Roughton, J. Integrating quality into safety and health management. *Ind. Eng.* July: 35–40, 1993.

Roughton, J. Safety and health management through bar code technology. *Appl. Ind. Hyg.* 4(6):F29–F31, 1989.

Sass, R. The implications of work organization for occupational health policy: the case of Canada. *Int. J. Health Serv.* 19(1):157–173, 1989.

Snyder, T. B., Himmelstein, J., Pransky, G., and Beavers, J. D. Business analysis in occupational health and safety consultations. *J. Occup. Med.* 33(10):1040–1045, 1991.

Stanevich, R. S. and Stanevich, R. L. Guidelines for an occupational safety and health program. *Am. Assoc. Occup. Health Nurs. J.* 37(6):205–214, 1989.

Tuskes, P. M. and Key, M. M. Potential hazards in small business — a gap in OSHA protection. *Appl. Ind. Hyg.* 3(2):55–57, 1988.

White, K. Managerial style and health promotion programs. *Soc. Sci. Med.* 36(3):227–235, 1993.

Yarborough, C. M. System for quality management. *J. Occup. Med.* 35(11):1096–1102, 1993.

# OVERVIEW OF THE KEY ELEMENTS OF AN EFFECTIVE OCCUPATIONAL HEALTH MANAGEMENT PROGRAM, PART 2

# 11B

Mary E. O'Connell

## TABLE OF CONTENTS

## OCCUPATIONAL EXPOSURE TO BLOODBORNE PATHOGENS

### Definition of Bloodborne Pathogen

Bloodborne pathogens are disease-causing germs carried by blood. More precisely, bloodborne pathogens are pathogenic microorganisms that are present

1-56670-054-X/96/$0.00+$.50
© 1996 by CRC Press, Inc.

in human blood and can cause disease in humans. These pathogens include, but are not limited to, hepatitis B virus (HBV) and human immunodeficiency virus (HIV), which cause two of the most serious diseases: respectively, hepatitis (acute or chronic infection of liver) and autoimmune deficiency syndrome (AIDS). Both diseases can be transmitted through occupational exposures. The purpose of this section is to review guidelines for preventing HBV and HIV exposures and subsequent illness or death.

Occupational exposure means any reasonably anticipated skin, eye, mucous membrane, or parenteral contact with blood or other potentially infectious materials (OPIMs) that may result from the reasonably anticipated performance of an employee's duties. OPIMs include the following:

- Parenteral: piercing of mucous membranes or skin barrier through such incidents as needle sticks, human bites, cuts, or abrasions
- Human body fluids: semen, vaginal secretions, cerebrospinal fluid, synovial fluid, pleural fluid, pericardial fluid, peritoneal fluid, amniotic fluid, saliva in dental procedures, all body fluids visibly contaminated with blood, and all body fluids in situations where it is difficult or impossible to differentiate between body fluids
- Any unfixed tissue or human organ (other than intact skin) from a human

## Scope of Concern

In the world community, health care workers have the greatest exposure and risk of contracting HBV or HIV from infected source individuals. Although safety and health legislation worldwide was written primarily to protect health care workers, the legislation applies to all employers and almost all workers who, depending on job responsibilities, written or implied, have potential exposures to bloodborne pathogens. While this category of worker is obvious in some cases (public safety, emergency response personnel), it is likely that in your own organization you will have categories of workers and/ or individuals who, by the nature of their jobs and your knowledge of what they do or could be asked to do, have the potential for exposure to bloodborne pathogens (e.g., hotel/motel housekeepers, municipal workers who clean city parks and restrooms, security personnel). The reality is that *anyone,* including you and me, can have an occupational exposure to blood or OPIMs and is at risk.

Therefore, an effective exposure control plan should cover everyone at some point. For instance, everyone should receive basic information and follow-up if they are exposed, but not everyone needs to be vaccinated against hepatitis B. If your organization has more than one work site, you need a separate plan for each location based on the potential exposures and the groups with potential for exposure at each. Keep in mind that the performance-orientation of your written plan is in your identification of exposed workers, their training, and your documentation of their training.

Also, when writing your exposure control plan, you need to take into consideration your legal liabilities under tort law, health and safety legislation, and workers' compensation.

## Management Aspects

1. Determine who is at risk. Make a list of groups of employees who have job tasks that involve exposure to blood, body fluids, or tissues. You may need to look at job descriptions, evaluate the essential functions/duties of the position, conduct task analyses, and/or look at collateral duties or duties secondary to the primary job, workers' compensation data, first-aid logs, or incident reports. Look at a wide variety of data to help you understand the scope of the potential exposures. It may be smaller or larger than you anticipated. (See item 2.)
2. Determine who else is at risk. Make a list of tasks and procedures (other than those involving exposure to blood, body fluids, or tissues) in which occupational exposure occurs at your workplace.
3. Determine how you are going to prevent/control exposures. What types of controls will you use? Engineering? Work practices? Personal protective equipment and clothing? All or some?
4. Determine the need for a voluntary, employer-paid hepatitis B vaccination program. Who will be included? Who will administer the vaccinations? Who will do post-exposure evaluations and follow-up?
5. Develop procedures to manage exposure incidents: reporting, follow-up, post-exposure treatment and counseling, record keeping.
6. Write the exposure control plan. Update it annually using knowledge gained during the previous year.
7. Write the training program. Be sure to include legally required and company-specific information. Training for emergency responders should focus on the practical application of handling and cleaning up potentially contaminated materials. Provide facts about how HBV and HIV are transmitted. Dispel myths. Provide facts about personal behaviors which put people at risk and how to prevent them. Provide facts about the occupational transmission of HBV and HIV. When writing your program, remember that anyone who has occupational exposure (obvious or implied) must be included.
8. Develop training records. How will you document it? Who is going to keep the records? Develop an implementation schedule for your exposure control plan.
9. Purchase personal protective equipment and have a ready supply on hand.
10. Develop housekeeping procedures for cleaning and decontaminating surfaces prior to reuse.
11. Use only an approved disinfectant. Any disinfectant that will kill HBV, HIV, and *Mycobacterium tuberculosis* is acceptable.
12. Consider the medical requirements of your program.
13. Obtain puncture-resistant, leak-proof sharps disposal containers and biomedical labels/bags for contaminated materials.

## Management Controls

Your written exposure control plan must include your procedures for preventing exposures. The primary methods are

- **Universal Precautions** — This means treating all others as if their blood and OPIMs are infected with HBV, HIV, or other infectious organisms. Universal precautions is a general work practice control that is extremely critical to the success of your program. Everyone must follow them. It is also a mind-set that can mean the difference between an exposure incident resulting in or not resulting in infection.
- **Engineering Controls** — Engineering controls are used to eliminate the hazard (e.g., puncture-resistant disposable sharps containers, tongs to pick up contaminated materials, self-sheathing needles or needless i.v. systems). You should evaluate your engineering controls periodically to determine if they are effective or need to be improved.
- **Work Practice Controls** — Work practice controls focus on safe behavior (e.g., hand-washing techniques that minimize the likelihood of contamination). If soap and water are not available, them antiseptic hand cleaners (e.g., towelettes) must be readily available. Your policies and procedures must contain workplace practices that prohibit exposures to blood or OPIMs, based on the nature of your tasks and operations — for instance, no eating in a laboratory; procedures to minimize splashing, spraying, or splattering of blood or OPIMS; no food or drinks allowed in areas where blood or OPIMs are present.
- **Personal Protective Equipment** — Personal protective equipment (PPE) must be used when engineering and work practice controls do not eliminate the exposure. PPE is then required to prevent exposures through clothes, undergarments, skin, eyes, mouth, or other mucous membranes. PPE must be provided to all employees who need it, free of charge. You must factor these costs, as well as the costs of cleaning, repairing, and maintaining it, into your safety program budget. When you require PPE to be worn, you must monitor and enforce its use. PPE includes appropriate gloves, mask and/or eye protection, gowns, aprons, surgical caps and hoods, and/or shoe coverings. Procedures for removing it when contaminated and removing and disposing of it prior to leaving work areas must be part of your program.
- **Housekeeping** — *decontamination:* All fixed and nonfixed work sites that could become contaminated with blood or OPIMs must be included as part of your routine housekeeping schedule to ensure that they are kept clean and sanitary. Employees or others responsible for housekeeping must be trained in your procedures. Any disinfectant capable of killing HIV, hepatitis B, and *M. tuberculosis* is appropriate. *Regulated waste:* Regulated waste must be disposed of in accordance with applicable state, local, or national laws. You must research these regulations to ensure that your program is in compliance. *Laundry:* Special procedures are necessary to prevent exposure due to contaminated laundry. They include bagging or placement in containers at the location where it is used.

## Program Monitoring

Periodic review of compliance with universal precautions engineering controls, infection control rounds, and observance of work practices while administering first aid or medical care are all elements of program monitoring. What you do and how frequently you do it will determine your monitoring schedule. Regardless, you will need an evaluation tool designed to address your particular workplace. This can be a separate tool (e.g., hospitals) or an integrated tool (e.g., safety inspection form with a place to note safety practices). The goal is to adjust your program by reviewing your practices and making improvements as indicated.

## MEDICAL SURVEILLANCE GUIDELINES

### Definition and Purpose

The purpose of occupational medical surveillance is to prevent disease and disability through the early identification of occupation-related conditions in individuals or groups of workers. This is done in communities at large or in specific workplaces. Medical surveillance is either active or passive. Active surveillance means that targeted workers (e.g., those who work with lead) are selected and examined. Passive surveillance means that workers present themselves at a medical provider with symptoms that may be work related. Active surveillance is more reliable in that the goal is to establish preexposure baselines for comparison to future test results. Passive surveillance is less reliable because it detects symptoms and is less likely to detect conditions in their early stages.

### Management Aspects

The nature of your work environment will determine the type of program you will need. Key management aspects for program design, control, and monitoring include the following:

1. Conduct a walk-through health and safety survey and industrial hygiene evaluations to determine baseline conditions in the workplace.
2. List the potential exposures. Determine appropriate baselines and periodic tests for each exposure. Some of these will be based on regulatory requirements and some will be based on the recommendations of occupational health practitioners.
3. Select workers for testing. Those workers selected for medical surveillance will be those at high risk for injury (e.g., audiograms for possible hearing loss) or disease (e.g., pulmonary function tests for asbestos and blood tests for lead) or those whose jobs impact public safety issues (e.g., drug testing for drivers of commercial vehicles).

4. Assess workers' exposures prior to testing. Obtain work history, past exposures, history copies of past test results, or other information that will provide raw data.
5. Determine if the selected tests are likely to detect adverse health effects.
6. Determine the side effects of the selected tests (if any). Will the test provide more benefit than harm?
7. Identify specific tests by job category.
8. Develop a medical removal program and policy for workers who need temporary or permanent removal (legislative or otherwise). Remember that it is better to prevent the hazard than to subject employees to medical removal, which could result in dislocation and/or wage loss.
9. Provide feedback to employees about the results of testing, what the results mean, and any health implications.
10. Arrange for counseling if appropriate.
11. Review the program periodically and make adjustments as needed.

## SHIFT-WORK STRESS CONTROL

It would be nice to provide a tidy "how to" guide in this section that health and safety managers could use to control the effects of stress on shift workers. However, because shift work and the variables which affect it are so complex, we still lack reliable data in many areas of research, particularly the relationship between shift work and injuries, which is inconclusive. If this relationship is inconclusive, how do we justify programs to control the effects of shift work on our employees?

First, shift work is inherently an organizational stressor and as such can contribute to occupational stress, a related health and safety concern. Our efforts to control the effects of shift work preclude an understanding of stress in general, our organization and how it is managed, and who is at risk. While shift-work research conducted over the past 25 years is in general inconclusive, it has provided us with information about the types of variables that have an impact on shift work and/or the shift worker's ability to adapt to changing work schedules. Some variables are controllable, some are uncontrollable, and some may be mitigated through workplace interventions and enlightened management approaches. Variables which impact shift work/shift workers include

- Ability to adapt to shift work
- Level of physical exercise
- Use of bright light
- Noise
- Use of caffeine
- Ability to sleep
- Gastrointestinal functioning
- Social factors
- General mood and attitude

- Shift rotation
- Shift assignment
- Shift tenure
- Clockwise rotation
- Counterclockwise rotation

When examining your organization and your program needs, keep in mind that both internal and external environments impact how an employee adapts to shift work. Management controls that may be helpful include

1. Education of managers and supervisors on the known effects of shift work on workers
2. Replacement screening for shift workers to assess adaptation to shift work using a validated adaptation index tool; candidates with low scores may be poor choices for shift work
3. Regular work breaks, especially during the first week of shift rotation
4. Wellness classes at the work site on the effects of circadian rhythm disruption and effective coping mechanisms
5. One-to-one counseling of shift workers regarding sleeping patterns, or discussions regarding the continuance of shift work if appropriate
6. Transfer of employee from shift work to a more normal work routine
7. Education of shift workers about the positive benefits of the use of caffeine, a regular exercise program, and use of bright light to manage disrupted circadian rhythms
8. Assessment of overtime and management's expectation of shift workers to meet production deadlines, quotas, or work schedule
9. Where shift work is unavoidable, improve conditions by arranging work hours more naturally
10. Consider eliminating night work, which has shown in studies to improve sleep, mood, gastrointestinal functioning, social well-being, and attitude toward work

Clearly, additional research is needed to assist us in managing the occupational health and safety implications of shift work in our employee populations.

## MENTAL STRESS MANAGEMENT

### Scope and Definition

Simply stated, stress is a complex response to a series of life experiences. Stress is a universal problem and concern, so much has been written about it. By their very nature, all organizations have stressors which exact a high toll through direct and indirect costs. Yet, with our collective knowledge, occupational stress has taken on a new life of its own in our modern, fast-paced world.

When perceived stress at work produces mental injury and subsequent disability, the stress claim is termed mental-mental because there is no physical injury or stimulus. These are very difficult cases to manage, primarily because there is no objective measurement of mental illness. Consequently, employees may not report this type of injury for fear of ridicule, while others who demonstrate stress related to physical injuries or mental stress that results in physical injuries are more easily believed and therefore more likely to report it.

Trying to manage mental stress without assessing, preventing, and controlling personal and work stress would not be appropriate. Like it or not, we are all vulnerable to all types of stress. How we or our employees deal with it depends on our own understanding of our responses to life events, as well as the resources available to help us.

Our dilemma as occupational health and safety managers is how best to assist our employers and employees in balancing the effects of stress in their daily lives. This becomes even more challenging because the causes and manifestations of organizational and employee stressors, while providing behavior clues, are often obscure and unknown. Managing stress means looking at our companies, and our employees, and then helping both to manage stress properly, to take control of it, to make it the focus for positive and effective change.

## Management Aspects: Identification and Assessment of Stress

The initial step for workplace management of stress is identification of the problem. Because occupational stress is so difficult to identify, it, more than many other types of occupational diseases, requires a thorough analysis to achieve desired results. It is best to obtain and use commercially available assessment tools which have already been validated for use in organizations. Typically, professional assessment tools focus on job or task characteristics, organizational factors, physical conditions related to the organization, and employees' stress levels and perception of workplace stress.

1. Organizational stressors — The assessment tool used should address your company's definition of stress as well as stressors in and from your organizational structure, physical environment, managers and supervisors, and employees. Organizational stressors include policies, procedures, people interactions, employee turnover rates, workload, attitudes about mediocre work, absenteeism rates, decreased use of vacation time, job dissatisfaction, increased accident frequency, decrease in work quality (errors), low creativity, reduced productivity, disloyalty, thefts, disability payments, increased sick leave, damage and waste, and sabotage.
2. Employee stressors — Identification and assessment of stress in individual employees or groups of employees is also essential for controlling mental stress. The first type assesses the employee's own assessment of his/her current stress level or status, based on the number of major life events that have occurred over the previous year. One frequently used assessment tool

is the classic one developed by Homes and Rache, or a modified version of it. The second type of tool asks employees to rate work-related statements that represent stress-producing situations. Even without formal assessment tools, what we often see in employees are the behavioral manifestations of stress. Behavioral clues are related to how employees feel, their attitudes, and their physical responses, and are all important to consider.

## PREVENTING, CONTROLLING, AND MANAGING STRESS

Following identification, the emphasis should be on the prevention and control of stress and, where appropriate, the elimination of organizational stressors. Stress is part of our lives, is universal, and therefore can be the focus for change. Successful programs will focus on prevention and control rather than the complete elimination of stress or stressors. This may require a paradigm shift about what stress is and is not, as well as an understanding that the highest challenge is in overcoming our own and our organizations' prejudices to achieve results and realize the potential of both. It also means being able to see things in a different way, being open to new solutions. Managing the program must include developing recommendations from your assessments, an action plan to implement them, and periodic evaluation of the results so that adjustments can be made. Management aspects for preventing, controlling, and managing stress include the following:

- By the organization
    1. Positive management attitude toward stress prevention (formal and informal) and unwanted occupational stress
    2. Strong communication and performance feedback systems
    3. Commitment of time and money to the program
    4. Timely response to stress-related problems
    5. Education of management, supervisors, and employees on the signs, causes, effects, and control of work-related psychological disorders
    6. Provision of voluntary and confidential employee assistance program (mental health services)
    7. Integration of wellness programs and mental health services with occupational health and safety programs
    8. Reasonable workloads and meaningful work
    9. Comfortable work climates
    10. Ergonomic design/redesign of workstations and work areas (e.g., well-designed jobs)
    11. Job training and retraining programs
    12. On-site fitness centers or subsidized corporate memberships to health clubs
    13. Smoke-free and drug-free workplace
    14. Strong disability management program and return-to-work program
    15. Early case management of all stress-related claims (workers' compensation and group health)

16. Periodic program assessments to monitor burnout indicators versus evidence of healthy behaviors
17. Continual identification of organizational stressors
18. Continual correction of organizational stressors
19. Ongoing compliance with health and safety standards
20. Following treatment for mental stress problems: a treatment monitoring plan, support once returned to work, cooperation with aftercare programs, and strong social support

- By the employee
  1. Talk about problems with a trusted friend or colleague
  2. Participate in worksite wellness, health, and safety programs
  3. Get regular exercise (walking, swimming, etc.)
  4. Eat a healthy diet
  5. Participate in active stress reduction activities (progressive relaxation techniques, biofeedback, and behavior modification)
  6. Participate in community activities
  7. Use the Employee Assistance Program (EAP) when problems arise

## FURTHER READING

### Books

Keita, G. and Sauter, S., *Work and Well Being: An Agenda for the 1990s,* Washington, D.C.: American Psychological Association, 1992.

Petersen, D., *Managing Employee Stress,* Goshen, New York: Aloray, 1990.

Waldron, H. A., *Occupational Health Practice, Occupational Health Department, St. Mary's Hospital, London,* London: Butterworth and Company, 1989.

### Other Publications

Agnew, J., Management of job stress at the organizational level, *AAOHN Update Ser.* 3(25), 4, 1989.

Begley, J., An occupational health nurse manages mental health cost at the worksite: a case study, *AAOHN Update Ser.* 4(17), 1991.

deCarteret, J., Occupational stress claims: effects on workers' compensation, *AAOHN J.* 42(10), 494-498, 1994.

Federal Code of Regulations, Chapter 29, Section 1910.1030, Occupational Exposure to Bloodborne Pathogens.

Gilbert, B., Employee assistance programs: history and program description, *AAOHN J.* 42(10), 488-493, 1994.

Hunter, E., Managing mental health costs at the worksite, *AAOHN Update Ser.* 4(16), 1991.

Marks, L., Stress. I. Assessment of stressors in the workplace, *AAOHN Update Ser.* 1(9), 1984.

Marks, L., Stress. II. The challenge of making stress a positive force, *AAOHN Update Ser.* 1(10), 1984.

OSHA, *Bloodborne Facts, Reporting Exposure Incidents,* Washington, D.C.: OSHA Publications.

OSHA, *Bloodborne Facts, Personal Protective Equipment Cuts Risks,* Washington, D.C.: OSHA Publications.

OSHA, *Bloodborne Facts, Holding the Line on Contamination,* Washington, D.C.: OSHA Publications.

OSHA, *Bloodborne Facts, Protect Yourself When Handling Sharps,* Washington, D.C.: OSHA Publications.

OSHA, *Bloodborne Facts, Hepatitis B Vaccination — Protection for You,* Washington, D.C.: OSHA Publications.

Phillips, J. A. and Brown, K. C., Industrial workers on a rotating shift pattern: adaptation and injury status, *AAOHN J.* 40(10), 468–476, 1992.

# 12     ERGONOMICS FOR MANAGERS — A PRACTITIONER'S PERSPECTIVE

**Donald L. Morelli**

## TABLE OF CONTENTS

## INTRODUCTION

Ergonomics has been considered primarily a tool to improve workplace safety and health. While these are important factors, the potential for proper application of ergonomic principles to enhance operational and organizational efficiency and effectiveness also should be recognized. An understanding of

1-56670-054-X/96/$0.00+$.50
© 1996 by CRC Press, Inc.

the basic philosophy of ergonomics, along with readily usable approaches to apply it, will allow the engineer or manager to make ergonomics an integral part of any plant layout, job design, or business decision.

Regardless of the title of a manager — Safety, Environment, Production, Human Resources — many forces shape the decisions that must be made every day. These forces include operational efficiency and effectiveness, compliance with laws and regulations, judicious use of limited — and often sparse — resources, and, not least of which, acting in accordance with professional and personal ethics.

In light of these pressures, consider the following hypothesis for managers: *As professionals responsible for making decisions that affect people and organizations in a world of rapid and unavoidable change, we should be committed to*

- *The development and application of sound approaches for all elements of work design*
- *Selecting the best workers for each task and then training and developing those workers*
- *Developing a spirit of cooperation between management and workers*
- *Dividing work equally between management and workers, each doing the part for which they are best fitted*

Many management professionals would agree with all four ideas listed above. After all, managing has long been defined as those activities necessary for planning, staffing, organizing, directing, and controlling the actions of groups of people to some desired outcome. However, some may take issue with the notions of a real cooperative spirit and equal division of work between management and workers as simplistic approaches to complex problems.

Granted, these four elements are not really very close to gaining total acceptance in our workplaces now, nor were they when they were first written in the 1920s. These are, paraphrased somewhat, the four principles of scientific management of Frederick W. Taylor. They were a bit revolutionary for the time and to a degree remain so today.[1]

If these ideas were revolutionary, the ideas and forces that are shaping tomorrow are many times more so. Addressing the work world of the future will require concepts and approaches that will reshape not only our ideas but the nature of work and the workplace. As John Naisbitt suggested back in the 1980s in *Megatrends,* the future work world will require a balance of "high touch" with "high tech."[2] The need for such a balance could not have been foreseen in Taylor's time, but his principles consider it indirectly in advocating a "balance" of the division of work. Tomorrow will require an enhancement of Taylor's principles and the management practices that they spawned — it requires a proper application of ergonomics.

## ERGONOMIC PERSPECTIVE

To understand ergonomics and its applications, one must appreciate its basic approach. This is vital because ergonomics is more than an area of interest or field of study. It is not just a science in terms of a collection of rules and data. Ergonomics more closely resembles a philosophy to understand the world and how it functions. Although once considered a derivative of the industrial engineering and industrial psychology disciplines, it continues to evolve and now represents a further development of, an enhancement to, these disciplines.

The ergonomic perspective considers the industrial setting to be comprised of four components:

- Human — the person or group of people engaged in a purposive activity
- Task — the series of actions necessary for the human(s) to accomplish the activity
- Machinery/equipment — the hardware and devices provided to the human(s) to assist in the performance of the activity
- Environment — the overall arena in which the purposive activity takes place, to include not only the physical factors of temperature, lighting, noise, etc., but the organizational and psychological factors that can affect human performance

Any ergonomic evaluation considers the interaction of these four components as a "human/task/machinery/environment" system. The primary objective of ergonomics is to attain and maintain balance *in* the system and thereby maximize the performance *of* the system. Analysis of this system seeks to determine if imbalances exist between the elements of the system and to identify the sources of the imbalance. This is done by examining whether the demands of any of the components of the system require performance beyond the capabilities of any *other* component.

In most cases of physical work, and where the risk of worker injuries is a focus, the analysis tends to focus on whether the demands of the **task** require capabilities the **human** worker either cannot perform (resulting in a traumatic incident) or cannot *sustain* (a cumulative incident) without undesirable results. This imbalance will lead to a "system failure" which can manifest itself as injuries, reduced productivity, quality problems, and/or, more likely, a combination of all three.

In some instances, the **equipment** provided to aid in performing a task may be the source of demands that can also lead to such system failures. Tool handles may be too large for smaller workers to use effectively. Complex systems may require extensive training and experience to operate properly, limiting the people who should be allowed to work with these systems but also limiting the number of people available for these jobs.

The work **environment** may present demands beyond worker capabilities, such as "downsized" organizations expecting fewer workers to do the work of the entire previous staff, with no improvements in work design or methods. In these ways, the problems that are routinely faced in business are looked upon not as single issue events, i.e., "human error", but as the result of the interaction of the components of the workplace.

To return to safety issues, ergonomic analysis considers the basic elements of the individual tasks that must be accomplished and identifies the various demands of each work element. These demands are then compared to known human characteristics and capabilities to determine if system imbalances exist. Corrective measures can then be devised. If risks of physical injury are found, the corrective measures generally focus on

1. Improved postures
2. Reduced force demands
3. Reduced frequency of activity[3]

The objective of this systems approach is to maximize overall system performance. A corollary to this is to fit the task to the human worker; i.e., do not require something the human either cannot perform or cannot sustain.

Design criteria to accomplish this generally seek to set task requirements so that they are acceptable to the largest practical population segment. This usually translates to attempting to design for the 5th percentile worker, either at the small end of the range in reach or at the large end for clearance dimensions. This approach of "disadvantaging" the fewest people is especially important in strength or force requirements. In the case of muscular-skeletal injuries and task demands, application experience suggests that requirements set such that at least 75% of the theoretical worker population can perform them can reduce the risk of injury some 30 to 50%.[4]

## HUMAN CHARACTERISTICS

Once the basic "systems" approach of ergonomics is accepted, it is necessary to understand human capabilities and limitations — human characteristics — in order to apply the systems analysis technique. In other words, fitting the task to the worker can only be done if the characteristics of the worker are known so that they can be compared to the task demands.

Many sources for information on human characteristics exist. Several excellent texts are listed in the references. Discussing all the human characteristics pertinent to safety considerations would be beyond the scope of this entire text. However, human characteristics can be considered in terms of the type of task demands required in any specific work element.

Again considering the more common situations of physical tasks, the nature of task demands can be broken down into three main areas: force, frequency, and posture. Simply, these terms relate to the following:

- Force — the requirement to exert strength to perform the task
- Frequency — the number of repetitions of the activity
- Posture — the body positions assumed to perform the work element

## Force

The strength required to perform a task is dependent on several factors; among the major ones are

- The size or bulk of an object that is handled or held
- The posture or body position assumed by the worker
- The size or configuration of hand tools used
- The frequency of the exertions

It should be apparent that force, frequency, and posture are interrelated factors, as suggested above. To illustrate this, consider a task demand to pick up a cassette of silicon wafers out of a dip tank with one hand, hold it over the tank to drain, and then move and insert it into a water rinse tank. The weight of the cassette obviously has to be such that the worker can lift it. However, the arm position — determined by the horizontal reach to the tank — is also vital. The functional forward reach of the 5th percentile worker is about 27 in., so the tank must be located no further than that from the operator.

However, this is not the full solution. If the tank *is* placed such that a 27-in. reach is required, that will mean the small worker will have to lift the cassette with a fully extended arm. The available strength of the 5th percentile worker in this posture is roughly only half of the maximum strength for such a task demand. Therefore, the cassette will have to be very light in weight, or the tank should be moved closer to the worker to permit a more desirable arm posture in which maximum upward force can be generated.

How frequently the cassette is dipped and how long it is held to drain are also important. If the holding time and the repetitiveness of the activity do not allow for adequate recovery time for the muscles involved, overall performance will suffer.

## Frequency

The frequency of activity determines the factors that must be analyzed. For tasks performed infrequently, strength limitations, as outlined above, generally take precedence. In tasks that are performed continuously or for relatively long time periods, e.g., unstacking a full pallet of items all at one time, energy demands and appropriateness of recovery times become most important. Several resources addressing energy demands and recovery time calculation are provided in the references.[5,6]

## Posture

Posture during manual work, especially material handling, is the subject of continuing debate. However, a universally agreed upon risk factor is the horizontal distance from the worker's lower back to the point where a load is grasped. The guiding principle is to keep this distance as small as possible.

This is where the "classic" squat lift can be advantageous, if the load handled is small enough to fit close to the body or between the knees. If the load is large or cannot be grasped tightly to the body, then virtually all of the research and analytical tools suggest that the overall body posture assumed is not so important — the task will probably present a high risk of injury. These large "moment arms" associated with material handling of awkward loads result in high compressive loads in the lower back, which have been correlated to incidence of back injury.[4]

Also a postural concern is hand and wrist position during gripping activities. Here again, flexed or extended wrist positions, as well as awkward or "pinching"-type grips, reduce the maximum available force a worker can generate. Therefore, to perform what may appear to be a simple task may actually require a very high percentage of the worker's available hand strength. This will lead to high forces and loads in the connective tissues of the hand, wrist, and arm (the tendons and ligaments) and more rapid fatigue of the muscles involved. The high loads in the tendons and ligaments increase the risk of sprain or repetitive motion disorder (if the frequency is high), and the muscle fatigue will diminish performance.

## REDESIGN REQUIREMENTS

Once system imbalances have been identified, it is often too easy to jump to a "fix" to get a problem solved. This temptation should be avoided. Rather, the information regarding what may be out of balance should be turned around into "redesign requirements" which state clearly, but in broad terms, what factors must be addressed and how. At this point, input should be solicited from every viable source to construct an intervention, improvement, design change, or process upgrade. Workers, engineers, operating supervisors and managers, and your equipment venders all should have input to the decision process. Some general redesign requirements are given below upon which to model your future thought processes.[3]

- General redesign requirements
  Reduce forces required
  Reduce frequency of repetitive activities
  Improve postures
- Redesign requirements — force
  Reduce the weights handled
  Reduce the force exerted in awkward postures
  Reduce the length of time the force is exerted
  Reduce the gripping force when the wrist is bent
  Reduce the length of time an item or body part is held unsupported
- Redesign requirements — frequency
  Reduce the number of times an item is handled or lifted
  Reduce the occurrences of extreme body positions
  Provide adequate recovery time for repetitive work
  Reduce the repetition of identical work elements
- Redesign requirements — posture
  Reduce the deviations from "neutral" body positions
  Reduce torso inclinations, especially while lifting
  Reduce shoulder abduction and adduction
  Reduce forward arm reach to grasp, hold, or dispose of material
  Reduce wrist deviations while gripping

## A WORKPLACE GUIDE

To aid your future efforts to identify and eliminate system imbalances and the risks they create, the following tools have been developed. The first is a list of typical work situations and task demands that present "invitations" to improve workplace ergonomics. These items can help you identify potential problems **before** undesirable events take place. By identifying the associated force, frequency, or posture demands, you can apply the redesign requirements provided or devise your own to address the situation.

The second is a guide to the basic ergonomic imbalances linked to four general and common "business problems". Although the items presented focus somewhat **after** "events" have happened, they should help in directing efforts to identify the root causes of these system failures and eliminate or minimize their reoccurrence.

Third is a collection of ergonomic fundamentals and principles against which to compare the job designs and task demands that characterize your workplace. These aids are not intended to be all-inclusive of every ergonomic principle, but do provide a framework against which to assess a wide variety of typical work situations.

### Invitations to Improve Workplace Ergonomics

- Is absenteeism high on a particular job?
- Is turnover higher than with similar jobs?

- Is production efficiency lower than expected/predicted?
- Is product quality unacceptable?
- Does the process result in too much material waste?
- Is there high equipment or tool damage on specific jobs?
- Is the worker frequently away from the workplace?
- Is the work pace rapid and beyond the worker's control?
- Does the job require frequent use or manipulation of hand tools?
- Is the worker required to maintain any single posture for long time periods?
- Do workers sit on the front edge of their chairs?
- Are work chairs modified by the use of additional cushions?
- Are the workers required to hand-hold parts that could be positioned in jigs, clamps, or fixtures?
- Are dials, controls, or displays difficult to read and identify?
- Does the job require special lighting?
- Do hand tools or other equipment transmit vibrations to the worker's hands, arms, or whole body?

## Ergonomic Target Guide

If your operational problems include

1. Product damage/high scrap rates
   Most likely worker injuries:
   Strains/sprains of the back and shoulders

   Look for
   - Static or awkward postures created by workplace design
   - Excessive material handling
   - Manual material handling over a large range of motion

2. Poor product assembly/high rework
   Most likely worker injuries:
   Carpal tunnel syndrome
   Finger cuts and lacerations
   Shoulder "complaints"

   Look for
   - Worker modifications to chairs or tools
   - Work surfaces at inappropriate heights
   - Extreme positions of the wrists (flexion, extension, and deviation)
   - Repetitive pinching/gripping requirements

3. Production upsets
   Most likely worker injuries:
   Strains/sprains of the back and shoulders
   Cuts and contusions to the fingers

   Look for
   - Improper queuing of parts

  • Excessive walking along assembly lines
  • Absence of line-stop authority
  • Manual material handling over a large range of motion

4. Warehousing/order-filling deficiencies
    Most likely worker injuries:
        Strains/sprains of the back
        Contusions to the upper extremities

  Look for
  • High turnover of workers
  • Inadequate recovery time in job cycle
  • Excessive rehandling (palletizing/depalletizing)
  • Poor scheme of assigning warehouse addresses

## Fundamental Ergonomic Principles

  • People vary in size and work capacity.
  • Variance among humans is considered to be normally distributed.
  • Reach dimensions must be designed to accommodate at least the 5th percentile of the potential worker population.
  • Clearance dimensions must be designed to accommodate at least the 95th percentile of the potential worker population.
  • Strength demands must accommodate at least the 75th percentile for infrequent tasks and should accommodate the smallest percentile human worker possible for high-frequency tasks.

## Design Specifications

| | |
|---|---|
| Whole body force — lifting | ≤250 kg compressive load in the lower back (posture and range of motion dependent)[4,7] |
| Arm strength — lifting | ≤25 lbs |
| Arm strength — pulling | ≤20 lbs |
| Gripping force | ≤25 lbs (actual force generated in the hand, not the load lifted by the hand) |
| Static holding force | ≤40% of the maximum force available in the working posture |
| Static holding time | ≤20% of the maximum possible holding time for the working load and posture |
| Forward reach for tasks | |
|   Repetitive | 18 in. |
|   Occasional | 24 in. |
| Dimensions for keyboard interface | |
|   Height (standing) | Range 39 to 44 in. |
|     Fixed | 41 in. |
|   Height (seated) | Range 25 to 28 in. |
|     Fixed | 26 in. (if thigh clearance permits) |
|   Height adjustable chairs | Range 15 to 20 in. |
|   Height of CRT screen | Approximately 16 to 18 in. above the keyboard height |

## Selection Preferences and Characteristics — People and Machines

| People | Machines |
| --- | --- |
| Perform best when doing a variety of motions and tasks | Can do highly repetitive motions with high force and speed |
| Are creative; can plan and invent | Can perform accurately in endless repeated tasks |
| Can interpret data, but at a finite rate | Can process high volumes of data |
| Can become bored and distracted | Can detect small variations in routine tasks |
| Are highly affected by environment | Can perform in environments hostile to humans |
| Have predictable physical and mental limits of performance | Have, by comparison, almost limitless design possibilities |
| Can repair themselves when given adequate rest and recovery time | Need external repair and maintenance |

## JOB STRESS

Job stress is an increasing source of concern and cost for many organizations. A consistent theme in the research literature suggests that job satisfaction and job stress are inversely related.

Job satisfaction can be linked to

- Greater degrees of worker control
- Autonomy
- Input and involvement in decision making

Indicators of job stress are associated with

- Lack of control or input
- Repetitive and monotonous jobs
- Environments from which the worker gains little or no respect
- Jobs that have high levels of physical or mental workload

Performance studies of typical quality control/inspection tasks suggest that very low and very high levels of stimuli (e.g., nonconforming products to be detected) both correlate to decreased performance.

Research in job satisfaction versus job complexity suggests a similar pattern of effect: low complexity and high complexity are both related to low satisfaction and reduced efficiency. Both low and high complexity tasks also show the biochemical waste product signatures associated with increased levels of stress.

## TOMORROW'S CHALLENGES

The basic concepts and applications of ergonomics can help in dealing with problems faced by virtually every business or organization today. However, the forces shaping tomorrow's work world are even farther-reaching than those that gave birth to Taylor's principles.

The growing interest in the total quality of work life, reducing stress while increasing job satisfaction, can be viewed as a variation of the concept of maximizing our "system" performance, with the performance measure being a person's overall well-being. If we do not learn to address our system failures in the workplace, how will we begin to deal with them on a social, or societal, basis?

To design workplaces and manage workers in the future, we will face greater differences among our workers and in the work environment. How will we specify our tasks and equipment (the two elements of the system we can readily affect) to minimize injuries and attain maximum performance?

We are facing a workforce that is constantly increasing in average age. What capabilities will our workers have as they work well into their 60s and beyond?

The U.S. continues as a land of opportunity to many from other countries. As the demographics of the workforce evolves, do we understand the demands placed on an Asian woman to use tools designed for a large North American male? Are we willing to find out and make the necessary changes to assure her a safe work situation, one that may enhance her performance?

The apparently increasing gap between skilled and unskilled workers may present a requirement to design jobs and equipment very differently for each group. Are we willing to do this so the less skilled can have jobs, support themselves, and contribute to society?

The ergonomic system approach provides a framework to enhance performance today and avoid problems tomorrow. Ergonomics simply must be an integral part of the overall skills and decision approaches of every person involved in creating safe places to work.

## REFERENCES

1. Taylor, F. W., *The Principles of Scientific Management,* Harper & Brothers, New York, 1929.
2. Naisbitt, J., *Megatrends, Ten New Directions Transforming our Lives,* Warner Books, New York, 1982.
3. Drury, C., *Ergonomics in Manufacturing,* Taylor & Francis, Philadelphia, 1987.
4. National Institute of Occupational Safety and Health (NIOSH), *A Work Practices Guide for Manual Lifting,* Tech. Rep. No. 81–122, NIOSH, Cincinnati, 1981.
5. Hutchingson, R. D., *New Horizons for Human Factors in Design,* McGraw-Hill, New York, 1981.
6. Rodgers, S. H., Recovery time needs for repetitive work, *Semin. Occup. Med.,* 2(1), March 1987.
7. Waters, T. R. et al., Revised NIOSH equation for the design and evaluation of manual lifting tasks, *Ergonomics,* 36(7), 749–776, 1993.

# 13 PRINCIPLES OF RISK COMMUNICATION

Michael L. Fischman and Lucy O. Reinke

## TABLE OF CONTENTS

## INTRODUCTION

Given the increased level of concern expressed by employees and the public in recent years about occupational and environmental hazards, risk communication has become an increasingly important topic for industry and policy makers. After many failures reflecting poor or no communication about environmental and occupational hazards, industry and government agencies are looking to other approaches. There has been recognition of the need to address not only the scientific data, but also the concerns and perceptions of employees and the public, in part by adopting a more democratic process in which members of these groups serve as active participants in the collection and communication of information and in the decision-making process. Risk communication is fundamentally similar whether addressed to a community resident or an employee, with the exceptions that occupational exposures tend to be higher and more prolonged than environmental exposures and that employees are generally more familiar with the processes, hazards, controls, and safety philosophy at a facility.

1-56670-054-X/96/$0.00+$.50
© 1996 by CRC Press, Inc.

While risk communication efforts attempt to convey scientific information to groups of concerned individuals, it is clear that risk communication itself is not yet a science. Literature searches in this area do not identify articles which have scientifically evaluated the impact or effectiveness of different approaches to communicating risk information. Certain authors, most notably Sandman and Chess, have developed some very useful but nonetheless theoretical constructs by which we can try to understand the layperson's reactions to hazards and thereby adjust our approach to communicating information about these hazards. They support their publications with numerous instructive examples, but there are no controlled studies to cite in this field. Apart from an understanding of this body of work and thorough technical knowledge about the hazard, one cannot be successful in risk communication without possessing the skills of any good communicator — good listening skills, common sense, sincerity, and sensitivity to the audience.

## THE ROLE OF RISK ASSESSMENT AND RISK MANAGEMENT

Before one can communicate risk, one must understand it. The process of risk assessment must precede risk communication, although a good risk assessment does not ensure good risk communication. Risk assessment may be a qualitative or a quantitative process. While a full discussion of risk assessment methodology is beyond the scope of this chapter, risk assessment basically involves four steps — hazard identification, dose-response assessment, exposure assessment, and risk characterization. Hazard identification involves two steps — identification of exposures and risk identification (identifying the adverse impacts of an exposure, as observed in epidemiologic studies or animal toxicology tests). The dose-response assessment attempts to determine the doses required to elicit particular adverse effects and the no observed adverse effect level (again from epidemiologic studies or animal toxicology tests). Exposure assessment involves a determination of the exposures to the group of interest, e.g., by industrial hygiene monitoring. Risk characterization integrates the above information in order to make an assessment as to whether significant risk exists.

Management of identified risks is also a process separate from risk communication, though risk communication will likely not be well received without a plan for managing identified risks. If the risk assessment process identifies a likely hazard, efforts should include reduction of exposure (by industrial hygiene or environmental control measures) and consultation with the occupational medicine physician as to the likely impact of prior exposures. In the workplace setting, employees should be notified of their exposure and its significance. In some cases, it may be necessary to consider recommending the removal of the employee from exposure, allowing voluntary transfers, or modifying duties. In these situations, it is best to involve the employee(s) in the decision-making process.

## DIFFERING PERCEPTIONS OF RISK

There is often a polarization in attitudes toward health risks associated with environmental and occupational exposures. Individuals often react angrily toward these risks as they perceive them. Many individuals subscribe to a notion that private industry puts profits ahead of safety. From the point of view of many industries, the public is thought to consist of "lay persons" who are unable to understand science, do not want to listen, and are afraid of the "wrong" risks. While the public has a tendency to ignore some serious risks, it frequently gets alarmed at significantly less serious ones. The layperson's response to risk not surprisingly reflects his or her perception of that risk, not necessarily the actual magnitude of the risk. As part of a risk communication effort, it is essential to accept these different mind-sets and to recognize that it may not be possible to change them.

Sandman explains differences between "experts" and the public. To the experts, Sandman points out, risk is the death rate (or a disease or injury rate). The public, however, has a broader view. Sandman has defined the death rate as the "hazard" and the other factors as "outrage". He then defines risk as the sum of hazard and outrage. While the public does not pay much attention to the hazard, industry or agencies do not pay much attention to the outrage. Thus we have two entities in dialogue about two fundamentally different issues. Ignoring the issues that concern the public just because the data do not support their concerns will only multiply their outrage and at the same time increase their distrust.

## THE PERSPECTIVE OF THE AUDIENCE

Whether from enlightenment, regulation, or fear of liability, there is a growing tendency for companies to involve affected citizens, employees, and/or agencies in environmental and occupational health decisions. Although there are different ways of doing this, it is important to understand who your audience is and how they view or perceive risk. In "Improving Dialogue with Communities: A Short Guide for Government Risk Communication", Chess outlines, as quoted below, some reasons why public risk perceptions may not correlate with the scientific evidence. While focusing on community concerns, these factors apply equally well to employee concerns:

- *Voluntary risks are accepted more readily than those that are imposed.* When people do not have choices, they become angry. Similarly, when communities feel coerced into accepting risks, they tend to feel furious about the coercion. As a result, they focus on government's process and pay far less attention to substantive risk issues; ultimately, they come to see the risk as more risky.
- *Risks under individual control are accepted more readily than those under government control.* Most people feel safer with risks under their own

control. For example, most of us feel safer driving than riding as a passenger. Our feeling has nothing to do with the data — our driving record versus the driving record of others. Similarly, people tend to feel more comfortable with environmental risks they can do something about themselves rather than having to rely on government to protect them.

- *Risks that seem fair are more acceptable than those that seem unfair.* A coerced risk will always seem unfair. In addition, a community that feels it has been stuck with the risk and little of the benefit will find the risk unfair — and thus more serious. This factor explains, in part, why communities that depend on a particular industry for jobs sometimes see pollution from that industry as less risky.
- *Risk information that comes from trustworthy sources is more readily believed than information from untrustworthy sources.* If a mechanic with whom you have quarreled in the past suggests he cannot find a problem with a car that seems faulty to you, you will respond quite differently than you would if a friend delivers the same news. You are more apt to demand justification rather than ask neutral questions of the mechanic. Unfortunately, ongoing battles with communities erode trust and make the agency message far less believable.
- *Risks that seem ethically objectionable will seem more risky than those that do not.* To many people, pollution is morally wrong. As former Environmental Protection Agency (EPA) Assistant Administrator Milton Russell put it, speaking to some people about an acceptable level of pollution is like talking about an acceptable number of child molesters.
- *Natural risks seem more acceptable than artificial risks.* Natural risks provide no focus for anger; a risk caused by God is more acceptable than one caused by people. For example, consider the difference between reactions to naturally occurring radon in homes and the reactions to high radon levels caused by uranium mine tailings or industrial sources.
- *Exotic risks seem more risky than familiar risks.* A cabinet full of household cleansers, for example, seems much less risky than a high-tech chemical facility that makes the cleansers.
- *Risks that are associated with other memorable events are considered more risky.* Risks that bring to mind Bhopal or Love Canal, for example, are more likely to be feared than those that lack such associations.

Employee and community groups deserve credit for being capable of understanding the scientific aspects of risk assessment. Many professionals who are part of these groups work as physicians, environmental engineers, safety engineers, chemists, lawyers, etc. Acknowledgment of their capabilities and, in some cases, using them as intermediaries for the communication of information lays down a foundation for trust and effective interaction.

Understanding the audience, including its demographics, employment, education, and prior experiences with health risks, is a crucial part of this process. A lecture on the risk of a hazardous gas leak will differ significantly when addressed to the employees of a facility versus members of the surrounding community. Similarly, the lecture will differ significantly when the audience is primarily scientists from the EPA versus a group of concerned citizens. Once one establishes the nature of the audience, it is important to gather information as to what the audience already knows, in addition to assessing their needs and concerns. In order to do this, one may need to do some research by making phone calls or interviewing representative members of the audience. Make sure your presentation reflects their needs as well as your needs.

## PRESENTING INFORMATION

It is then important to establish those crucial pieces of information that are most important to communicate. Often, the issues in these situations are quite complex; from this, one must distill out the main messages which must be conveyed, in order to avoid overwhelming the audience. Identify main ideas and gather relevant data and supporting details. Throughout the presentation, be attentive to signals from the audience which may indicate confusion or frustration. The audience may merely want a definition of a term or an explanation of a more complicated issue. Expect that there may be questions that you will not be able to address or answer. If there are other speakers who will address this issue later, let the audience know it. If you cannot answer a question, let the questioner know that you do not know the answer at present but that you will research it and provide a response in a timely manner. It is helpful to allow questions throughout the presentation and, at the same time, to permit time for a question and answer period at the end.

The use of a team of professionals from appropriate disciplines to present information and serve as a panel permits each team member to present information that he or she has gathered or analyzed (and with which he or she is most familiar). The team may consist of an industrial hygienist, a safety or environmental engineer, a facilities manager, an occupational health nurse, and/or an occupational medicine physician, as indicated. Relevant industrial hygiene or environmental monitoring data should be presented to support subsequent exposure assessment conclusions. It is important for one team member, often the occupational physician, to integrate the medical, toxicologic, and exposure assessment information into one coherent picture. Since no one professional is likely to be able to address all questions, the presence of the team strengthens the presentation and inspires greater confidence in the audience. It is critical that someone on the team be very familiar with the relevant literature (e.g., epidemiologic data, animal toxicology studies) in order to respond promptly to concerns and to appear to be competent to provide information and reassurance. It is not difficult for an audience to recognize that a speaker has not prepared for his or her presentation, which may further any feelings of mistrust. Many individuals attending these meetings have researched the subject, in some cases using resource and reference materials similar to those that were available to the speakers. It is prudent to never underestimate your audience.

The language of the presentation should be such that it can be understood by lay-people but should not be condescending. Practicing the presentation by rehearsing it prior to the meeting will aid in the delivery. If rehearsed in front of a small group, the ultimate presentation can reflect their constructive feedback and prepare the speaker for possible questions. During the presentation, it is important to acknowledge any concerns or symptoms experienced by audience members, giving credence to them even if not accepting that they are connected to the exposure. One should respond directly to the concerns expressed by attendees. When possible, be definitive; do not "waffle". If the data support a conclusion with a high degree of probability, one should not convey an equivocal answer because there remains some very small amount of uncertainty. If you are not definitive, you will not be successful in reassuring your audience, even though the data indicate that they should feel reassured.

It is very important to put the concerns into perspective with regard to severity, the potential for long-term consequences, and the limits of uncertainty. For example, with an indoor air quality problem, symptoms may be quite uncomfortable and may even be temporarily disabling. However, a statement supported by scientific literature that such symptoms have been commonly observed in problem buildings, universally resolve after leaving the building, and do not result in permanent disability should help to ease some of these concerns. With regard to the limits of uncertainty, individuals will often state to an occupational health professional something to this effect: "Science doesn't know everything there is to know about that chemical; therefore, how do you know that this exposure won't cause cancer?" While the degree of toxicologic and epidemiologic data for different chemicals varies considerably, it is most often possible to address these concerns by discussing the low dose, the transient nature of the exposure, the "requirements" for chemical carcinogenesis, and the existing knowledge base. Another common misconception may have to be dispelled: "They didn't think that asbestos was harmful in the past; how do we know that what you're assuring us now won't be proven wrong in a few years?" Indicating that the health risks from asbestos were known but concealed in the past and focusing again on the limits of uncertainty should help to dispel this notion.

One should avoid characterizing symptoms as being psychological in origin, even if it is clear that psychological factors are contributing to the development of symptoms. Unfortunately, many individuals would perceive such characterizations as refined ways of indicating that you think they are "crazy". You will lose your credibility and increase the intensity of the outrage factors. However, there are nontoxicologic phenomena which may account for symptom development in some situations. Neutra et al. and Shusterman et al. from the California Department of Health Services have demonstrated, in a series of epidemiologic studies conducted near hazardous waste sites, that odor and "environmental worry" are powerful risk factors in the induction of common symptoms such as headache and nausea:

Retrospective symptom prevalence data, collected from over 2000 adult respondents living near three different hazardous waste sites, were analyzed with respect to both self-reported "environmental worry" and frequency of perceiving environmental (particularly petrochemical) odors. Significant positive relationships were observed between the prevalence of several symptoms (headache, nausea, eye and throat irritation) and both frequency of odor perception and degree of worry. Headaches, for example, showed a prevalence odds ratio of 5.0 comparing respondents who reported noticing environmental odors frequently versus those noticing no such odors and 10.8 comparing those who described themselves as "very worried" versus "not worried" about environmental conditions in their neighborhood. Elimination of respondents who ascribed their environmental worry to illness in themselves or in family members did not materially affect the strength of the observed association. In addition to their independent effects, odor perception and environmental worry exhibited positive interaction as determinants of symptom prevalence, as evidenced by a prevalence odds ratio of 38.1 comparing headache among the high worry/frequent-odor group and the no-worry/no-odor group. In comparison neighborhoods with no nearby waste sites, environmental worry has been found to be associated with symptom occurrence as well. Potential explanations for these observations are presented, including the possibility that odors serve as a sensory cue for the manifestation of stress-related illness (or heightened awareness of underlying symptoms) among individuals concerned about the quality of their neighborhood environment. (Shusterman et al., p. 30)

There was no excess of serious health effects observed in residents near these waste sites, nor did air measurements document any toxicologically significant exposures. Experimental studies document headaches, irritation, and other symptoms in volunteers exposed to the odors of solvents present in very low, toxicologically insignificant concentrations (Otto et al.). In addition to the physiologic effects induced by odors, e.g., nausea, environmental odors may serve as a sensory cue for the manifestation of autonomic or stress-related symptoms (e.g., headache and nausea). According to Shusterman et al., several studies document a 4 to 10 percentage point increase in the reporting of symptoms associated with stress. These studies document a mechanism for symptom induction that is not mediated by toxicologic effects or psychological factors. Under some circumstances, it may be appropriate to convey the findings of these studies to employees or community residents to help in allaying their concerns about symptoms.

## CONCLUSION

Despite the fact that occupational and environmental exposures are, in general, much better controlled now, the level of concern about such real or perceived hazards has increased dramatically in recent years, coincident with increased media attention and the aftermath of a number of well-publicized environmental disasters. The principal tasks of environmental, health, and safety staff remain in hazard reduction. However, effective communication of actual risk is essential in order to avert fear and anger associated with unjustified concerns, to maintain good employee and community relations, and to avoid unnecessary disruptions in production or other key activities. Risk communication is not and probably never will be amenable to rigorous scientific methodology, even if the informational content of the message is scientifically demonstrable. We need to approach our audience, whether employees or public citizens, with concern, sensitivity, honesty, and a willingness to understand their point of view. The combination of a well-conceived risk assessment and appropriate and sensitive risk communication can, at least in many cases, permit resolution of vexing employee and community concerns to the satisfaction of all parties.

## REFERENCES AND FURTHER READING

### Books

Hance, B. J., Chess, C., and Sandman, P. *Industry Risk Communication Manual: Improving Dialogue with Communities.* Lewis Publishers, Boca Raton, FL, 1990.

### Other Publications

Chess, C. et al. *Improving Dialogue with Communities: A Short Guide for Government Risk Communication.* Environmental Communication Research Program, Division of Science and Research, New Jersey Department of Environmental Protection, Trenton, NJ, 1988.

Fisher, A. et al. One agency's use of risk assessment and risk communication. *Risk Anal.,* 14(2):207–212, 1994.

Neutra, R., Lipscomb, J., Satin, K., and Shusterman, D. Hypotheses to explain the higher symptom rates observed around hazardous waste sites. *Environ. Health Perspect.,* 94:31–38, 1991.

Otto, D. et al. Neurobehavioral and sensory irritant effects of controlled exposure to a complex mixture of volatile organic compounds. *Neurotoxicol. Terat.,* 12(6):649–652, 1990.

Sandman, P. *Explaining Environmental Risk.* Office of Toxic Substances, United States Environmental Protection Agency, Washington, D.C., 1986.

Shusterman, D., Lipscomb, J., Neutra, R., and Satin, K. Symptom prevalence and odor-worry interaction near hazardous waste sites. *Environ. Health Perspect.,* 94:25–30, 1991.

Sparks, P. J. and Cooper, M. Risk characterization, risk communication and risk management. The role of the occupational and environmental medicine physician. *J. Occup. Med.,* 35(1):13–17, 1993.

# PRACTICAL APPLICATION OF OCCUPATIONAL HEALTH PROGRAMS

**Neva Nishio Petersen**

In this section, I plan to address some of the obstacles that you may encounter as you plan and implement your occupational health programs and to discuss some ideas to help keep your program "on track".

First, let us look at some of the problems that you may encounter — this is by no means an exhaustive list, but hopefully it will provide some insight into "real world" issues that you may be forced to deal with.

- There is not enough time or staff to adequately address projects and/or complete a needs assessment. This is a problem most of us are already familiar with, and one that will probably continue indefinitely as corporations and governments try to control costs.
- Your planned programs meet resistance due to the cost. This includes the cost of the equipment and the cost of personnel to attend training or implement the safety procedures. Often, compliance with safety and health programs will mean that employees need additional time to complete the job. Job preparation and setup may take longer (as in the case of asbestos work); employees may find it slower to work with personal protective equipment; additional paperwork (such as obtaining a confined space permit) may take time; and cleanup after a job may also add additional time (time to clean and store safety equipment such as respirators).
- Apathy may exist on the part of managers and supervisors. Often safety is perceived as the safety department's problem, and managers and supervisors may look to you to fix the problem rather than take ownership themselves.
- Manager and supervisors try to reset your priorities based on their perception of the issues or on information from the other sources.
- Employees are enthusiastic about the new programs, but interest wanes and the program dies.
- A cluster of occupational injuries or illness resets your priorities.

So, how can you overcome these and other obstacles that you may run into?

One of your most valuable tools is your occupational injury and illness record, your OSHA 200 log. By tracking your injury and illness frequency and severity rates, you can track your progress and success. By breaking down your log into categories — how many of each type of injury you have had — you can identify the areas and programs that should be a priority for attention.

For example, if you look over that last 3 years and you discover that you have had several chemical exposures, but only one cumulative trauma injury, you should prioritize your respiratory protection and hazard communication programs, and ergonomics may be lower on your list. This is a "quick and dirty" way to set priorities if you do not have time to complete a full needs assessment. (Please keep in mind, however, that a needs assessment would be preferable if possible.)

Tracking injury and illnesses may also help justify the priorities you have set and can help to convince other managers and supervisors that your priorities are the right occupational health priorities for your organization. However, as you set these priorities, you must keep legal compliance in mind — if you have a high rate of cumulative trauma injuries, you will still need to ensure that your organization meets respiratory protection and hazard communication regulations. Often, your programs will be driven by both factors.

The cost of industrial injuries and illnesses and the cost of regulatory noncompliance (fines and liability) can help demonstrate the *cost effectiveness* of occupational health and safety programs. Workers' compensation carriers often can provide the direct costs of your injuries and illnesses — medical cost and disability — and you can estimate the cost of lost time or lost productivity due to injuries. The total injury cost can then be compared to the estimated cost of the occupations health program — necessary equipment, training time, and any additional job time. Usually the cost of compliance is lower than the cost of the injury.

However, once you start tracking injuries and illnesses, do not forget to look beyond the statistics at the root cause of the incident. Your policy and program should be designed to address the problem so that the injuries and illnesses decrease. Also, feedback to each department is a critical step in giving the department "ownership" of any issue. Departments need to know of any problems so that they can correct them. They must also be involved in identifying the "fix" for the problem, and they must be a part of implementing the fix. This way, they will readily accept ownership of the program, and if they have ownership they will follow through and continue with the program.

For example, you begin implementing a respiratory protection program in response to all your chemical exposures — if the department is aware of the problem and is involved in developing the program (Where do they want to store respirators? When is training most practical for them? Who in the department is best suited to ensure compliance in the department?), they are more likely to implement the program and follow through to ensure continued compliance. And without continued compliance, you will not see any reduction in chemical exposures.

To help ensure that your programs continue, you should also provide continued support and feedback — let managers and supervisors know about their successes in implementing a program or reducing injuries. Alternatively, if necessary, provide continued information on problem areas and continue to solicit ideas for improvement as well as suggest your own ideas to correct problems.

Finally, be flexible. Your priorities and your programs will need to change and evolve. As new regulations are passed, you will need to update your programs and develop new ones. A cluster of occupational injuries or illnesses may reset your direction. However, if you keep your focus on your overall goals, you will be able to successfully develop, implement, and maintain an effective occupational health program.

## FURTHER READING

Plog, B. A. et al., *Fundamentals of Industrial Hygiene,* National Safety Council, Itasca, IL, 1988.

# Section IV
# Safety Program Management –
# Regulatory Compliance Aspects

# 15 SAFETY AND HEALTH MANAGEMENT — REGULATORY COMPLIANCE ASPECTS

**Gabriel J. Gillotti**

## TABLE OF CONTENTS

## INTRODUCTION

The cost of preventable workplace injury, illness, and death in the United States is almost incomprehensible to the typical person on the street. In economic terms, it is in the billion dollar range every year, a magnitude that the average worker whose well-being is at stake cannot relate to in any terms.

Data gathered from state workers' compensation agencies and the National Safety Council (NSC) are very revealing with respect to the costs associated with the prevention of deaths, injuries, and illnesses. Workers' compensation figures indicate that in 1983 approximately $17.6 billion was paid out to employees, whereas in 1993 the figure was $50.2 billion. These are payouts to employees — actual costs to employers were about 30% higher, or $83 billion.

The NSC claims that the true cost to the nation for work-related deaths and injuries is greater than the cost of compensation alone. Their estimate as to the

1-56670-054-X/96/$0.00+$.50
© 1996 by CRC Press, Inc.

cost for 1993 was $112 billion. In other words, each worker in the country must produce an additional $94 in goods and services to offset the $112 billion.

Every human, compassionate being is for improved safety and health — whether a worker or employer, union or nonunion, Republican or Democrat. The mission of the Occupational Safety and Health Administration (OSHA) is the same in 1995 as it was at the time of the passage of the enabling legislation in 1970: to save lives and to prevent deaths, injuries, and illnesses in order to protect the American working man and woman.

OSHA's task is difficult in large part because of changes in the work environment, including the associated technology, worker-employer relationships, impact of priorities set by society, marketplace conditions, and the impact of government priorities; all have varied enormously over the 25-year life of OSHA. It has been extremely difficult for OSHA to maintain its focus on this moving target; the changing workplace.

Unavoidably, OSHA's effectiveness at meeting its mandate is frequently challenged. One or more measures are used to argue the point, e.g., the number of inspections conducted annually, the level of monetary penalties, the negative impact on the small-business community, the negative impact on specific industries such as agriculture or chemical processing, and others.

The available data bear out the fact, however, that OSHA's enforcement of its standards has resulted in a fulfillment of its mission, at least for 18 of its nearly 25-year existence. According to the Office of Technology Assessment of the federal government, the rules on lead and cotton dust, as examples, have dramatically reduced workers' exposures and substantially reduced illness rates. Also, the federal Bureau of Labor Statistics data indicate that in 1975, 9.4 workers per 10,000 were killed on the job, while in 1993 the number was 5.5 per 10,000 workers. In 1973 there were 11 injuries or illnesses per 100 full-time workers, while in 1993 the number had fallen to 8.5.

For a variety of reasons, both political and administrative, OSHA's approach has varied dramatically as well. The results of one approach, i.e., strong commitment to and emphasis on enforcement, have been measured at least once. After studying illness/injury data from 6842 manufacturing plants, Wayne Gray and John Scholze concluded in 1991 that when OSHA inspects and imposes penalties for violations, there is a measurable injury reduction in those workplaces following the inspection. It is nearly impossible to measure the other way (i.e., training and consultation) that prevention is achieved, either directly or indirectly, as a result of OSHA action.

Since the law was interpreted to have the Congressional *intent* that prohibits federal OSHA from providing direct on-site consultation and assistance, a few years after the law's passage contracts with state and other agencies to deliver such services were initiated. The results of the efforts of those OSHA agents have not been measured in any credible manner in the past 18 or so years. To what extent the current reductions are attributable to that approach and, in what proportion consultation and enforcement go hand in hand to prevent injuries and illnesses, are indeterminate. How does one measure the number of accidents prevented, lives saved, etc., with any confidence? The Bureau of Labor Statistics' OSHA data are but a snapshot of the universe of workplaces and are not a measure of "success".

I believe that the efforts of responsible employers, particularly those who never have the occasion to be face to face with an OSHA representative, are difficult to measure also. The enlightened owner and manager who view OSHA's standards as a minimum and who have the compassion and concern for the well-being of their workforce are responsible for a significant portion of the reduction. Their efforts are essential and must continue if the numbers are to continue to improve.

Having commented on roles and on some of the results, let us address the extent to which relationships between employers, workers, and government bring about success, or the lack thereof.

## NEW ERA FOR LABOR-MANAGEMENT RELATIONSHIPS

In December 1994, a report was issued on workers' opinions about their jobs. The year-long study, coordinated by a Harvard economist and a University of Wisconsin law professor, polled 2048 workers about how they felt about job satisfaction, loyalty, and particularly about participation. Although a single study does not make a case, it is timely to this discussion because labor-management committees are viewed as the new era mechanism, essential to the establishment of effective safety and health programs in all workplaces.

What did the study conclude? Over half of the workers polled want participation and influence in decision making and nearly the same number feel loyalty toward their employers. That combination suggests that labor-management joint committees in all probability could be successful in addressing safety and health issues. The axiom "who knows better what hazards exist in a given work situation than the exposed worker" is self-evident and certainly worthy enough for management to solicit opinions from that worker.

The 1994 attempt at a reform of the OSHA law by the Democrat-controlled Congress prompted extensive debate and controversy on a provision which mandated labor-management safety and health committees. Currently, 10 of the 24 states that administer the OSHA program require workplace accident prevention programs, and 7 of the 10 require joint committees. The effectiveness of this provision in those seven states has not been measured such that a convincing case for adoption by other states and federal OSHA has been made.

There are, however, at least two NLRB cases, the Electromation and Dupont cases, which clearly "dampen" the efforts of employees, employers, and government to encourage the formation of joint committees. There are those who believe that it is virtually impossible to establish committees for any purpose, including safety and health, without violating the National Labor Relations Act (NLRA). Although in the two cited cases — Electromation, a company which has no unions, and Dupont, which is a union workplace — the fact situations are different and are specific to that workplace, the decisions are being badly interpreted to disallow committees in any workplace.

An example of such an interpretation is that, if a committee is established and becomes effective, the employees will consider themselves a "labor organization" and not attempt to exercise their rights to join a union. The NLRB views such a situation as an effort by the two parties, employer and employee, to "deal with one another via bilateral discussions that include an exchange of ideas". That is what a labor union engages in under the NLRA, particularly if the safety and health committee gets into issues that affect manning and production, which is often the case.

Until the NLRB gives a "green light" to all parties, employers risk violating the law (NRLA), however constructive their intent. OSHA and unions are equally frustrated because they see committees as a potentially new and progressive mechanism for empowering workers and preventing injuries and illnesses.

Currently, there are efforts in Congress to amend the NLRA to lessen the effect of the recent decisions and to encourage the creation of committees.

In spite of what has been discussed to this point regarding committees and programs and the hurdles that impede the effort, OSHA has been working with companies for about 15 years in establishing about 200 voluntary protection programs (VPPs). The VPPs include "committees" and accident prevention programs and are proving to be very successful based on site-specific data. I will elaborate on the VPP effort further on in the text.

## UNION AND NONUNION WORKPLACES

Clearly, the driving force behind the initial enactment of the OSHA law was the organized labor establishment. Without their persistent effort to represent their members by bringing their voice to the agency, other interests may have prevented the agency from staying the course and protecting workers.

One truly appreciates labor's efforts when one realizes that about 85% of the country's workers have no representation and that what labor does for one of its members benefits about six other workers who have no voice but their own. We all know how difficult it is for a lone voice to be heard when addressing the government rule makers on issues that affect all of us.

Are there any real differences between union and nonunion workforces when it comes to safety and health programs, accident prevention, problem resolution, or whatever else necessitates positive interaction between employers and workers? If I am an employee whose supervisor practices a style of management that relies on intimidation and absolutely control, I will have at some time a compelling need for a union to represent and protect me using their knowledge, the contract, labor laws, and any legitimate means available. The nonunion employee has no such protection, although he/she may have the good fortune of having an enlightened, compassionate manager who does not put productivity before safety.

The labor unions have been very aggressive in their support of the OSHA law reform provisions for labor-management committees as well as OSHA's efforts to promulgate a proposed accident prevention program enforceable standard, presently scheduled for release in late 1995. Other interactions with OSHA include participation on advisory committees, at public hearings on standards, as commenters on proposed rules, etc. Again, the direct voice of the majority, the nonunion workplace, is virtually silent but for the effort of the unions. In the end, the forces of both will prevail if the concept of joint labor-management problem resolution is accepted as a valued method.

As for conflicts in unionized workplaces over what is subject to collective bargaining and what is a safety and health issue, they will arise and must be addressed quickly to avoid compromising either the negotiated contract or the joint committees' efforts. All too often, conflicts have been labeled as safety or health matters when they have not been, and care and effort should be made to separate the "agendas" unless and until labor and management choose to include the safety and health program and procedures as an integral part of the contract. Many unions have chosen this approach, and in many cases it works very well.

## BEHAVIOR VS. MECHANICAL FIX

Over the years, veterans in the safety profession have repeated an old axiom, "80% of hazards can be changed only by changing human behavior, 20% by instituting mechanical fixes." It is not evident when one reviews federal or state OSHA standards that drafters of standards have sufficiently believed that axiom. They still "specify" the mechanical solution to the exclusion of any consistent mention of the need to direct employees toward assuming a greater share of responsibility for themselves, and their fellow worker, while carrying out their daily tasks.

Leaders of government programs in many foreign nations have for years put greater emphasis on efforts to modify employee behavior to become more safety conscious. Some nations make it the focus of their efforts through extensive training programs and by strongly supporting labor-management cooperative efforts. Workers who have a stake in determining and instituting a preventive measure will surely change behavior to benefit co-workers and themselves.

Psychologist E. Scott Geller, Ph.D., has spent over 20 years lecturing about the need to have employees not only involved in safety; he stresses that they must become a force behind making safety a positive, proactive practice. People care about the well-being of family members and may develop interpersonal relationships at work that can be nearly as strong. His emphasis is on teaching workers a "learned optimism" so that they expect things to be safe and improving rather than believing that, no matter what they do, accidents will happen.

Dr. Geller also points out that the same classic management principles that apply to safety apply to other workplace issues, i.e., seeking employee input and leadership on key issues, keeping people informed, creating feedback systems, auditing progress, setting short- and long-term goals, and rewarding employees for safe behaviors, something they can control. As for control, the reward and acknowledgment system should be designed to recognize employees for that which they can control, not for those changes about which they have little or no individual impact or control.

The federal government's effort to affect employee behavior is contained largely in numerous required training standards. There are approximately 158 specific federal standards that call for training on both narrow and broad topics. Of the total, 83 apply to general industry, 41 to construction, 32 to the maritime industry, and 2 to agriculture. The topics range from personal protective equipment (PPE) to electrical hazards. Some training programs require a minimum number of hours, while others are more performance oriented and the employer is expected to design and deliver the information. For the most part, the training is designed to sensitize the employee to the hazard and to alert him/her to the proper work procedure, work practice, or equipment safeguard.

As is true in most cases, there is no follow-up to determine whether the "lessons were learned", the safe practices were applied or injuries/illnesses decreased because of training.

The standard that specifically calls for employee training to be conducted in the construction industry by the employer is 29 CFR 1926.21(b)(2), which reads as follows:

> The employer shall instruct each employee in the recognition and avoidance
> of unsafe conditions and the regulations applicable to his work environment
> to control or eliminate any hazards or other exposure to illness or injury.

Over a 5-year period (1990 to 1995) 6821 inspections addressed this standard, and 82% of the time the employer was cited, with an average penalty of $800. Obviously, the 41 training requirements applicable to the construction industry are not being adhered to if the 82% level is indicative of employer effort.

## ACCIDENT PREVENTION PROGRAMS

OSHA's direct experience with the subjects of programs and committees has been primarily in three areas, the voluntary protection programs (VPP) initiative, the issuance of the Safety and Health Program Management Voluntary Guidelines in January 1989, and enforcement of 29 CFR 1926.20(b), *Accident Prevention Responsibilities,* in the construction industry.

The VPP concept can be described as follows:

- The concept of the voluntary protection programs is based on the realization that workplace compliance with OSHA standards *alone* will never completely accomplish the goals of the Occupational Safety and Health Act. Good workplace safety and health programs which go well beyond OSHA standards can provide a system of rules which can be set and enforced quickly, rewards for positive action, and safety and health improvements in ways simply not available to OSHA.
- Voluntary protection programs are intended to supplement OSHA's enforcement effort by identifying employers committed to protecting workers through internal systems so that limited enforcement resources can be directed to workplaces where the most serious hazards exist; they are also intended to encourage voluntary protection systems.
- All approved participants in voluntary protection programs are required to have implemented safety programs and meet all relevant OSHA standards. Participation will not diminish either employer or employee rights or responsibilities under the Act.
- OSHA will verify qualifications; remove approved participants from routine scheduled inspection lists; provide any necessary technical support; investigate formal, valid complaints and serious accidents; and evaluate the programs.

Employee participation is an essential and required element in these programs, and many of the participating companies have very successful and effective committees. Not only have employees clearly bought into being "part of the solution and not part of the problem", but the enthusiasm and commitment is very evident. OSHA, the government partner, creates an incentive for companies to apply for approval by exempting the workplace from scheduled inspections. Both the employers and employees have made safety and health concerns as much a part of day-to-day operations as production, maintenance, etc.

OSHA continues to encourage companies to apply for VPP recognition and approval, not only because it results in the institutionalization of the program in their company over the long term, but because it also enables OSHA to direct its resources to other high-hazard establishments.

Although there are no particularly quantifiable measures of VPP labor-management committee contributions other than reduced rates, which may be a consequence of a number of factors, one has only to meet and discuss the important intangible consequences of employee involvement to know how employee behavioral changes have occurred. The workers themselves clearly change their behavior because they have a stake in the success and well-being of the entire group.

The Safety and Health Program Management Voluntary Guidelines of 1989 have proven to be one of the most useful documents published by OSHA since its inception. These guidelines are particularly useful to small employers who have neither the in-house experts nor the resources to hire their services but do want to have an effective program. As stated earlier, responsible and enlightened employers do want to protect their workers and to reduce their compensation and insurance costs, and they need the guidance that this document provides. The Guidelines will be discussed in more detail later in this chapter.

The enforcement of the 29 CFR 1926.20 standard in construction has produced interesting results. The standard reads as follows:

*Accident Prevention Responsibilities.* It shall be the responsibility of the employer to initiate and maintain such programs as may be necessary to comply with this part.

Such programs shall provide for frequent and regular inspections of the job sites, materials, and equipment to be made by competent persons designated by the employers.

The use of any machinery, tool, material, or equipment which is not in compliance with any applicable requirement of this part is prohibited. Such machine, tool, material, or equipment shall either be identified as unsafe by tagging or locking the controls to render them inoperable or shall be physically removed from its place of operation.

The employer shall permit only those employees qualified by training or experience to operate equipment and machinery.

Over the past 5 years (1990 to 1995) there have been 4764 inspections that involved the identification of an accident prevention program as a significant issue at a workplace. In 63% of the cases, the lack of the program resulted in serious violation citations, with an average penalty of $609. That clearly indicates that this long-standing requirement has not become an integral part of the construction industry's effort to comply with the standard and recognize the importance of the intent and purpose of the standard.

It should be mentioned here that employees who participate in these programs, as committee members or in any other respect, are protected from recrimination by employers for that participation. The drafters of OSHA's law foresaw the need to protect those workers who do exercise their rights to be committee persons, to file complaints, to report hazards, etc. by including the 11(c) provision. OSHA expends considerable resources and effort investigating employees' allegations of recrimination, in many cases where unions are nonexistent and grievance procedures are not available to the employee. Many employees have been "made whole" as a result of OSHA's 11(c) action.

OSHA has attempted to learn more about the value of accident prevention programs as part of this effort to promulgate both the 1989 Guidelines and the enforceable standard presently being drafted.

In an attempt to get beyond typical anecdotal evidence which tends to focus on larger employers, federal OSHA had a report prepared by outside experts on the subject of the benefit to construction companies who institute effective worker protection programs. The evidence in this report released on June 9, 1994 demonstrates major gains in injury reduction when individual firms undertake safety and health programs. The report also estimated that a net savings to the entire construction industry, if programs were instituted, could be as high as $16 billion annually due to the prevention of deaths and injuries.

The report cited the successful, "nonanecdotal" U.S. Army Corps of Engineers program to reduce construction injuries and fatalities. Results have been compiled to indicate that, between 1984 and 1988, contractors working for the Corps registered an average lost workday case rate of 1.34 to 1.54 per 100 full-time workers compared with the nationwide construction industry average of 6.8 to 6.9 per 100 full-time workers. For that period and since, the Corps has required a written safety and health program and has required the contractor to conduct work-site hazard analyses, to implement hazard prevention and control measures, and to provide safety and health training.

In 1994, the U.S. General Accounting Office (GAO) released a report entitled "Occupational Safety and Health: Differences between Programs in the U.S. and Canada" (GAO/HRD-94–15S). As a result of this study comparing the U.S. as a whole with the three most populous subdivisions of Canada: Ontario, Quebec, and British Columbia because the authority rests with provinces in Canada, GAO found differences in three major areas. These were: operations and funding of safety and health programs, the extent of worker involvement in policy making and accountability, and enforcement.

With respect to programs and worker involvement, in Canada there must be a joint worker/employer health and safety committee at all sites with 20 or more workers, with a least one half of its members being chosen by workers. Workers can refuse to do hazardous work until the condition is corrected, and joint committees can shut down a hazardous work process until the hazard is abated (Ontario). The emphasis is placed upon the internal responsibility of the joint committee rather than on the number of provincial government inspectors whose presence is all that determines the level of protection. Unlike the U.S. statute, workers can be held accountable and fined by the government as well.

Unfortunately, there are few data to demonstrate the success of the provincial programs and to compare with the U.S. data. On the other hand, there are few critics of the philosophy and approach taken by the provincial government according to the GAO report.

## FEDERAL OSHA GUIDELINES

As recently as July 1994 OSHA hosted a "stakeholder" meeting of industry, labor, state agencies, and others to discuss the OSHA agenda for the future. Predictably, a "major issue" with broad support was the matter of OSHA stimulating effective safety and health programs with top management commitment and meaningful employee involvement. The workplace-level, bottom-up solution to problems was the notion promoted by attendees provided that employee empowerment should be accompanied by stronger OSHA 11(c) whistleblower protection for workers who agree to committee involvement. Top-down effort is essential also, and OSHA's burden was to develop strategies for enhancing awareness on the part of our nation's business leaders and then holding them accountable. The goals of safety and of high performance or productivity are compatible, and top management must be convinced of this.

The stakeholders also shared their ideas as to what elements must be integrated into a safety and health program. The bottom line to OSHA was that OSHA model programs and self-help materials and other "stimuli" were necessary to get the employers engaged.

Even prior to the stakeholder session, as far back as the institution of the VPP policies in the early 1980s and at various OSHA top level management meetings prior to 1989, the need for published guidelines was discussed. On January 26, 1989, OSHA published a document entitled *Safety and Health Program Management Guidelines; Issuance of Voluntary Guidelines.* As a result of extensive staff work and the input of 67 commenters from a broad spectrum of individuals, labor, trade associations, professional associations, etc., the guidelines were published. This document is limited in its approach to all industries except construction because of the uniqueness of construction work sites.

Convinced that there were a number of basic principles that underlie an effective program, OSHA began its Guideline language by stating that (1) the program must provide for systematic policies, procedures, and practices that are adequate to recognize and protect employees from hazards; (2) systematic identification, evaluation, and prevention or control of general, specific, and potential hazards must be included; (3) the employer's effort must go beyond the minimum the law requires, must prevent injuries and illnesses regardless of whether OSHA has an enforceable standard to force compliance; and (4) effectiveness is the measure, not the fact that it is in writing.

Given those principles, OSHA identifies four elements fundamental to an effective program: (1) management commitment and employee involvement, (2) work-site analysis, (3) hazard prevention and control, and (4) safety and health training.

Let us expand and clarify those elements:

**Management commitment and employee involvement** — it must be clearly articulated in writing so that *all* personnel with *any* responsibility on site, or in a support function elsewhere, understand the priority of safety and health in relation to other organizational values. By making their own management practices visible to the entire organization, employees will get the "priority" message.

As is true with the setting of production goals, a goal for the safety and health program which sets forth the results desired and the measures of effectiveness is a crucial first step.

Employees will commit to the program only if they buy into the responsibility that comes from involvement in decision making. Realizing, as stated earlier, that employees have the most intimate knowledge of the job they perform and the hazards presented, the commitment is there if management values and respects the employee's perspective.

The employee's role can take various shapes, including, but not limited to, being involved in actual inspections for hazards and recommending abatement methods, being engaged in analyzing job hazards, providing input to the drafting of the work rules, training co-workers when his/her expertise is relevant, being active at safety meetings, and assisting in accident investigations. Performance of these tasks as an employee can be in the context of joint committees, labor committees, rotational assignments, or volunteer opportunities presented by the employer.

In place in successful organizations is the clear understanding of assignments and the constant practice of communicating responsibility to managers and employees. With that must be the delegation of adequate authority and resources so that the assigned responsibility can be met.

Accountability, after the above actions are in place, must be instituted so everyone knows he/she must carry his/her share of the load. By conducting comprehensive reviews and audits to confirm success, note surface deficiencies, and institute program improvements, the loop will be closed on what is intended by this first element. This is not constructive, however, if done with an eye toward punishment, but it may be necessary to institute major changes in roles and responsibilities, which can be healthy for the program.

**Work site analysis** — the hazards, actual and potential, cannot be acted upon unless comprehensive workplace surveys are conducted at least annually or as necessary. When major changes occur, such as the installation of new equipment or processes or the purchase of new material such as chemicals, a hazard analysis, however small in scale, must be conducted.

The policy should provide for a continuous flow of information on perceived hazards from employees, with the understanding that a timely and thorough response will be provided by management.

Two other forms of "analyses" that must occur as part of the program are (1) investigations of accidents, and of "near misses", focusing on accident cause; and (2) the periodic review of injury/illness data for identifying patterns and causes.

**Hazard prevention and control** — engineering control is the place to start when determining how to prevent an incident. Designing safe equipment and equipment features is the most effective control. Additionally, there must be safe work procedures and provisions for purchasing and providing appropriate PPE. Administrative controls, such as reducing the hours an employee may be exposed, are another acceptable means if engineering the prevention is not achievable. A combination of two or all three of the engineering, administrative, and PPE controls may be appropriate in some cases.

A comprehensive facility and equipment maintenance program will prevent many accidents. All too often it is not preventative, but is reactive, which can be costly in terms of injuries, illnesses, and equipment.

Every workplace must exercise control with respect to emergency situations by training employees how to act quickly and avoid harm's way.

If an incident occurs, medical response — in order to minimize and control harm — should be adequate by virtue of a well-instituted medical program.

**Safety and health training** — all employees must be trained to understand how they may be exposed while doing their particular job and while working with others. That way they will understand and implement the programs's policies, procedures, and the established protection. Training that enables an employee to recognize hazards, particularly if he/she is a member of a safety and health committee that conducts walk-through inspections, is crucial.

Managers must be trained how to analyze the work under their supervision in order to identify hazards, how to carry out their roles within their assigned managerial function, including, but not limited to, instituting and maintaining controls.

Often it is more effective to incorporate the safety and health training within job practices training so that the employee has an even better understanding of how safety and health are an integral part of this work.

The need that is most often overlooked is the need to ensure that those trained comprehend the material. Various approaches such as observing their work, oral questioning, and even formal testing should be undertaken. Traditionally, workplace training is conducted such that background and comprehension skills are overlooked and all employees are treated as equals in these regards. The verification of the understanding of the message will ensure, to some degree, a measure of success.

Our discussion is now complete — the government's challenge is to create the momentum necessary to shift the exercise of responsibility from government to employers, but it has not yet been met. In these times of lesser government intervention into the business community due to Congressional mandates, the roles of the partners in injury/illness prevention — namely, the government, employers, and employees — is shifting in terms of their relative magnitude and importance. We therefore, in a constructive fashion, must each fulfill our role effectively, however differently. Accident prevention programs may well be the vehicle that enables that to happen. In fact, we may learn that it more closely meets the intent of the OSHA law than our past practice. As well, it may result in a greater appreciation by the industrial community of what can be achieved in reducing the loss of life and the injuring of their workers, without government involvement.

## FURTHER READING

1. "Worker Representation and Participation Survey" published by Gannett News Service and authored by R. Freeman and J. Rogers.
2. OSHA 29 CFR 1910 published July 1, 1994, OSHA Regulations applicable to General Industry.
3. *Occupational Hazards Magazine,* January 94, Article entitled, "Activating Your Greatest Safety Resource", by Gregg LaBar.
4. OSHA publication 2254 (Revised 1992) entitled, "Training Requirements in OSHA Standards and Training Guidelines".
5. GAO Report (GAO/HRD-94–155), dated 1994, entitled "Occupational Safety and Health; Differences between Programs in the U.S. and Canada".
6. OSHA Guidelines dated January 26, 1989, entitled, "Safety and Health Program Management Guidelines; Issuance of Voluntary Guidelines; Notice".

# 16

# THE REGULATORY VIEWPOINT — CALIFORNIA STANDARDS: A REGULATOR'S PERSPECTIVE ON SAFETY AND HEALTH MANAGEMENT

**Thomas A. Hanley**

California Senate Bill 198 required California employers to comply with the most stringent and far-reaching legally mandated occupational safety and health program in the United States. California employers were required to develop, implement, and maintain an effective injury and illness prevention program (IIPP) with seven mandated, integrated components. Employers responded to the requirement with varying motivational perspectives: from recognizing the value of the concept for loss control to apprehension of substantial penalties following a Cal/OSHA inspection. Those employers, irrespective of motive, who in good faith developed, implemented, and maintained an effective IIPP are a real asset to California and its workers.

Real direct benefit has come to these responsible employers as well. Not only have these employers experienced the bottom-line benefits of loss control, but they have also found that effective injury and illness prevention tends to place an employer in an advantageous posture with respect to Cal/OSHA by minimizing both the likelihood of a Cal/OSHA inspection and the impact of any Cal/OSHA inspection that does occur. Furthermore, the process of maintaining an effective IIPP tends to make an employer familiar with the organization and structure of Cal/OSHA, as well as the services available through Cal/OSHA and the employer's rights under the Cal/OSHA program. From the perspective of one Cal/OSHA regional manager, I will develop this position.

Those employers who have developed, implemented, and maintained effective IIPPs, based on internal motivations or in response to concern about potential sanctions following a Cal/OSHA inspection, tend to be the subject of fewer Cal/OSHA inspections than other employers. In spite of the recent establishment in California of a High Hazard Compliance Inspection unit, the predominant focus of Cal/OSHA compliance assets is on employee complaints of unsafe conditions and reports of serious occupational injury and illness. Cal/OSHA, unlike federal OSHA, is largely complaint driven. Employee complaints of unsafe conditions and mandatory employer reports of serious injury accidents and occupational illness still motivate most Cal/OSHA inspections.

Employers with effective injury and illness prevention programs have given serious attention to establishing and committing resources to the required communication component of the IIPP. Responsible employers have established one or more means to communicate with employees and to facilitate employee communication with safety and health management. Whether these employers have selected postings of written material, distribution of written material, meetings of safety and health committees, or some combination thereof to communicate expectations and requirements of their safety and health program is not as important as the fact that some form of effective communication does take place.

The communication component, in addition to providing and reinforcing a general awareness of the structure and organization of the company's program and safety rules, invites and encourages employees to communicate their concerns about safety and health matters to the employer. For these responsible employers, a valuable employee is one who reports his or her concerns about potentially unsafe and unhealthful working conditions. In implementing the two-way communication component, the responsible employer clearly delineates one or more routes by which employees can relay their concerns. Often this includes directing employees to report their concerns to the line supervisors, an identified safety and health committee person, or the safety and health coordinator. Frequently, the employer will provide an anonymous route by which concerned employees may express their concerns without any fear of discrimination.

The required sanctioning component in the responsible employer's IIPP ensures that employees are motivated to report their concerns through one of several reporting channels. Most responsible employers not only institute a system of graduated discipline in compliance with the requirement that they maintain a system to ensure conformance to the requirements of the company's IIPP, but also have a system of positive inducements. The responsible employer in California is confident that concerned employees know how to report, know what to report, and are motivated to report.

The responsible employer, through supervisory staff, the safety and health office, or other means, investigates employee-reported problems and commits resources to control and abate the hazard when it is substantiated. Even when the employee-reported condition is determined not to represent a hazard, employees are briefed and informed as to the investigation and results. The reporting employees, in observing or learning of their employer's response to their concerns, are motivated to continue the practice and are confident that the employer wants to maintain a safe and healthful workplace and can accomplish such a goal with its own assets. The safety and health effort receives information on potential hazards not only from employees, under the communication component, but also from its own in-house inspection program, under the hazard identification component, and by isolating the causes of in-house accidents under the IIPP requirement to investigate accidents.

When employees have confidence in their employer's own system of complaint reporting, as a result of the employer's response to employee complaints, employees continue to report their concerns in-house and do not complain to Cal/OSHA. Since the Cal/OSHA program is largely complaint driven, the employer who has maintained an effective IIPP is subject to fewer Cal/OSHA inspections. An effective IIPP solicits, accepts, and resolves complaints in-house.

Apart from other economic benefits that result from effective loss-control measures, employers with effective IIPPs who solicit, investigate, and resolve employee safety and health complaints demonstrate that they are good corporate citizens who are sensitive to the limited resources of government, especially the regulatory arm of Cal/OSHA. To the extent that Cal/OSHA field compliance resources are not assigned to inspect a business with an effective IIPP, they can be focused on hazardous settings that deserve attention. The responsible employer, as a good corporate citizen, is preserving both governmental and societal assets.

In addition to requiring inspections in response to employee complaints, the California Labor Code requires that all industrial accidents resulting in serious injury or illness be investigated, contrary to the mandate of federal OSHA, which limits its focus to fatal and catastrophic events. The requirement to investigate all serious injury accidents and illnesses is the other primary motivator which brings the Cal/OSHA inspector to an employer's door in California.

An effective IIPP prevents and limits the severity of destructive and injurious health and safety incidents and again limits the likelihood of Cal/OSHA having to respond to the responsible employer's place of business. The net result of the efficient identification of safety and health hazards, and their subsequent control or elimination, is fewer severe hazards to cause accidents or illnesses. To the extent that the implemented IIPP of the responsible employer motivates the investigation of near-miss events, minor accidents, serious accidents, and incidents of serious occupational illness, the likelihood of serious injury or illness in the future is reduced along with the likelihood of a visit from Cal/OSHA.

The responsible employer assigns a competent and qualified employee or supervisor to investigate the accident or near-miss event with an eye to identifying causes, particularly preventable causes. The accident causes are shared through report and discussion with the safety director, coordinator, safety and health committee, and line supervisors. Such discussion tends to serve as a quality control mechanism for the in-house investigation. Through such a process there is a greater likelihood that all relevant factors and causes are considered and the abatement actions, appropriately selected, are prioritized and implemented.

For the responsible employer, such a process decreases the frequency of occurrence of accidents and occupational illnesses and generally reduces the severity of such incidents that do occur. Reducing the occurrence of serious injury and illness-producing events simultaneously reduces the frequency of Cal/OSHA visits. Accidents are also reduced by the general inspection process required of the IIPP. The mandated periodic inspections conducted by responsible employers are not just *pro forma*. The inspection results are reported and discussed; the hazards are identified, prioritized, and corrected under the IIPP's abatement requirement.

Since Cal/OSHA enforcement scheduling is so sensitive to employee complaints and the requirement to investigate serious occupational injury and illness events, effective employer activity, which results in preventing accidents and motivates employees to report their complaints and concerns in-house, tends to minimize or prevent Cal/OSHA inspections and investigations. Fewer Cal/OSHA inspections means less time disruption to an employer's workplace and less potential for citations and civil penalties.

Although keeping the Cal/OSHA enforcement arm away from the employer's door is one of the positive results of an effective safety and health management and genuine IIPP compliance, the same responsible employers tend to be the employers who acquaint themselves with the structure and organization of Cal/OSHA and the Division of Occupational Safety and Health, the employer's rights under the Cal/OSHA program, the benefits and services available through the program, and Cal/OSHA inspection procedures. With this knowledge, responsible employers can avail themselves of opportunities to further reduce the chances for an inspection, to influence the outcomes of any inspections that do occur, and to impact the development and amendment of California's occupational safety and health standards.

Whereas California State Assembly Bill 110 (Workers' Compensation reform legislation) has mandated that Cal/OSHA refer problematic employers with high experience modifications to the loss control services of their workers' compensation carrier or the expanded Cal/OSHA Consultation Service/Unit, responsible employers have already discovered the benefits and services of the Cal/OSHA Consultation Service and have used the Consultation Service as another resource in their overall loss control effort. It is paradoxical that the employers who need it the least are the ones inclined to utilize the Cal/OSHA Consultation Service more often. The responsible employer tends to be the employer who has learned about the Cal/OSHA Consultation Service and has utilized the service to receive technical guidance and information, publications, consultative inspections, and safety and health videos from the video library.

The process of developing, implementing, and maintaining an effective IIPP leads an employer, through its own efforts or those of its consultants, to secure the Cal/OSHA orientation publications: "The Guide to Cal/OSHA", Guides to developing IIPPs, Model IIPPs, Cal/OSHA Policy and Procedures, and contact information for the local Cal/OSHA Compliance Office and Consultation Service office. The responsible employer knows how to "plug in" to Cal/OSHA and knows how to contact the local Cal/OSHA supervisors and managers. These same employers are the ones who request Cal/OSHA representatives to deliver speeches and make presentations (to the extent Cal/OSHA resources permit). These employers have the opportunity to bring personnel responsible for IIPP support activities to such presentations and, in so doing, stay abreast of changes, trends, and Cal/OSHA policy interpretations. The responsible employer stays ahead of the loss-control power curve by maintaining an effective IIPP, by minimizing Cal/OSHA inspections, and by embracing available resources and knowledge available through Cal/OSHA on a proactive basis.

Responsible California employers are finding not only that the resources expended to maintain an operative IIPP are cost-effective, but that an effective IIPP acts as a force multiplier. The number of personnel actively or partly, but meaningfully, focused on safety and health issues increases because all supervisors and employers are mindful of, and encouraged to do, their part in accord with the IIPP. The responsible employer includes consideration of compliance with the IIPP and loss control efforts in performance appraisals of employees, supervisors, and managers. The performance evaluation is a suitable vehicle for motivating employees, supervisors, and managers to do their part under the IIPP. When efforts in support of a company's IIPP are recognized in performance evaluations, occupational safety and health becomes as locally significant as production. Even the language of IIPP, when used by senior supervisors and managers, motivates subordinate supervisors to understand that continued success and advancement within the company requires effort and support of the IIPP.

The IIPP acts as a force multiplier which literally increases the number of people focused on safety and health issues. The more eyes, ears, and noses focused on safe and healthy work operations, the better. The more eyes, ears, and noses involved, the sooner potential hazards, whether they be physical, chemical, biological, or human, are reported, investigated, and controlled. Effective safety and health management mobilizes more than just a company's safety and health personnel. Nearly everyone is mobilized into the effort. IIPP effectiveness spawns additional resources to build on its own effectiveness.

It is my experience that effective IIPPs tend to increase overall employee morale and, consequently, production. Effective IIPPs create a culture of respect for the knowledge, opinions, and experience of a company's employees. Employees are encouraged to report their ideas and concerns about safety and health. Demonstrated employer response reinforces the employee's belief in the employer's respect for the employee's individual contribution to safety and health and the overall company mission. From explaining how procedures, methods, tools, and processes can or should be modified to achieve safer operations, it is but a short distance to employee suggestions as to how to improve the quantity and quality of production. Mutual respect creates loyalty and high morale. High morale feeds self-respect and pride, which in turn feeds efficient and improved production. Employers with effective IIPPs tend to be productive, competitive companies.

The IIPP can serve as an umbrella or integration mechanism for other specialized safety and health program requirements driven by particular safety and health standards relative to the nature of an employer's operation. The emergency action plan, fire prevention plan, hazard communication program, respiratory protection plan, bloodborne pathogen program, and other plans and programs are incorporated under the general IIPP as specialized components of the plan. By incorporation into the IIPP, the particular substance and requirements of a specialized plan are strengthened by the various components of the IIPP. So, the emergency action plan is the subject of the communication component, compliance component, hazard identification component, and training component of the IIPP, which enhances its effectiveness and efficiency. The same is true of the other programs. The language of the IIPP standard reinforces the language of the specific plans. The most effective and vital IIPPs reference and incorporate subordinate plans and programs.

The effective IIPP will tend to minimize the occurrence of those events which motivate Cal/OSHA inspections — complaints and accidents. But even the most sophisticated employer with a very effective IIPP cannot keep all employees happy all the time. The disgruntled employee can generally find some issues to raise in a call to the local Cal/OSHA office to generate a valid complaint, which will be assigned for field inspection. However, the responsible employer with an effective IIPP will consistently minimize any adverse impact of a Cal/OSHA inspection.

Responsible California employers with an effective IIPP typically know Cal/OSHA inspection procedures and will react proactively to facilitate the inspection in their favor. When the Cal/OSHA field compliance person arrives in the reception area of the responsible employer, the receptionist will know which office or person to call. The safety and health coordinator or director will know in advance which person to assign to greet the inspector and represent management. The plan is already in place. The Cal/OSHA inspector will be greeted positively and graciously in a way that acknowledges that Cal/OSHA cannot make appointments for initial responses and must conduct unannounced inspections. The representative for the responsible employer takes a tack similar to, "How do you do, Mr. or Ms. Jones. I am Herb Gibson, Assistant Safety and Health Coordinator here at Business, Inc. If you will show me your identification, we can step into my office and discuss the parameters of the opening conference and inspection." Contrast this with, "Mr. Callanan is the person responsible for safety and health here, and he is on vacation. He will be back next week; why don't you make an appointment then?"

The assistant safety and health coordinator of the responsible employer with an effective IIPP and knowledge of the Cal/OSHA program knows that Cal/OSHA is not mandatorily required to conduct a comprehensive wall-to-wall inspection in response to every complaint. Cal/OSHA policy provides sufficient latitude to conduct partial inspections. District Office scheduling and individual workload often limit inspections to partial inspections.

An assistant safety and health coordinator for the responsible employer knows that the Cal/OSHA inspection is often a response to a confidential complaint. The assistant safety and health coordinator knows that the Cal/OSHA inspector cannot provide, in addition to the confidential complainant's name, the specific and exact nature of the complaint, out of caution that the specific nature of the complaint may tend to identify the complainant relative to the number of workers/employees associated with specific processes, tools, or parts of the operation.

Our model assistant safety and health coordinator typically proceeds to sketch and delineate the units, departments, and sections with their associated missions and the approximate number of employees engaged in each. The assistant safety and health coordinator from our model company knows that this approach will have a reasonable chance of generating the following response from the Cal/OSHA inspector: "As you know, I must always preserve the right to conduct a comprehensive inspection, but at present I intend to inspect the warehouse, including forklift operations and the spray coating operation. When I conclude my inspection of those areas, I will determine which, if any, additional areas to inspect."

During the opening conference, our model assistant safety and health coordinator takes advantage of opportunities to demonstrate the effectiveness of the employer's IIPP and the good faith of the employer with respect to safety and health issues generally. During the opening conference, the Cal/OSHA inspector provides evidence and input going to judgment about the effectiveness of the IIPP and the good faith of the employer, including the IIPP being in writing, the presence of the necessary supplemental plans and programs, the active presence of the required IIPP components in the written plan, the maintenance of training and inspection records, the impressive demonstration of supervisory knowledge of the IIPP, the expression of full cooperation with the pending inspection, the stated inclination to improve the program and correct any hazards that may be found, and the Log 200 suggesting no serious unreported accidents or unmitigated trends. Even before the inspection has started, the responsible employer with an effective IIPP, through the assistant safety and health coordinator, has gone a long way toward limiting the impact of the Cal/OSHA inspection. After the opening conference, only two other criteria will be considered in the Cal/OSHA inspector's final determination of IIPP effectiveness — the overall results of the inspection in terms of the nature and number of hazards identified, and the impact of the IIPP on employees.

Assuming that the Cal/OSHA inspector does not find an excessive number of violative conditions or plain-view serious violations and that employee contacts reveal employee knowledge and involvement with the IIPP, the inspector will find the IIPP fully effective and determine a high degree of good faith on the part of the employer. A high good faith rating significantly reduces any civil penalties that may be subsequently issued. Furthermore, the Cal/OSHA inspector will likely conclude that expanding the inspection beyond the partial inspection already conducted, in view of the employer's very effective IIPP and sophisticated assistant safety and health coordinator, is not an effective utilization of the Cal/OSHA inspector's time, given his or her backlog of pending assignments. An effective IIPP communicated by a knowledgeable safety and health representative can save time, money, and production resources.

The sophisticated employer also knows an employer's rights under the Cal/OSHA program and readily takes advantage of such rights. In the current scenario, if the Cal/OSHA inspector has issued any citations and proposed any civil penalties, the employer has the right to an informal conference by appointment with the Cal/OSHA district manager and the Cal/OSHA inspector at the local Cal/OSHA office. The district manager has the authority to amend the citations and civil penalties in response to the employer's presentation of new credible evidence, not yet considered, or in response to supported and reasonable interpretations of existing evidence.

In the event that the sophisticated employer does not prevail at the informal conference, but still considers its position persuasive, that employer can initiate the Cal/OSHA appeal process with a phone call and subsequent completion of a simple form within 15 days of citation service. The sophisticated employer also knows that abatement of the cited violative conditions and payment of the associated civil penalties are stayed, pending appeal. Further, an appeal hearing will be held locally, before an administrative law judge. Although the process is sufficiently formal, to the extent that a record is maintained and general rules of evidence apply, the sophisticated employer knows that it is not necessary to retain an attorney.

It has been my experience that occupational safety and health directors, coordinators, and production supervisors and managers who are capable managers of IIPP activity within their purview are very effective advocates at Cal/OSHA Occupational Safety and Health Appeals Board hearings. They understand the evidence burden Cal/OSHA must present to sustain any citations and civil penalties. They know how to present contrary evidence with respect to the significant elements, particularly with respect to employer knowledge of the violative conditions cited and employee exposure to the violative conditions. Even after the inspection, the responsible employer with an effective IIPP can continue to minimize the overall impact of a Cal/OSHA inspection.

The Division of Occupational Safety and Health, as the unit of California state government exclusively responsible for the enforcement of occupational safety and health, recognizes the vital role of employers who take occupational safety and health seriously and develop effective IIPPs. These employers are doing their part to protect the safety and health of California workers. As a Cal/OSHA regional manager, I salute the efforts of responsible California employers and offer my congratulations for their concurrent rewards of reduced losses, high morale, competitive production, and minimized adverse Cal/OSHA impact.

## FURTHER READING

### Books

National Safety Council. *Accident Prevention Manual for Industrial Operations: Administration and Programs,* 9th ed., National Safety Council, Chicago, IL, 1988.

### Other Publications

*1994 California Labor Code,* Parker Publications, Carlsbad, CA, 1993.
*Guide to Cal/OSHA,* State of California, Division of Occupational Safety and Health, San Francisco, 1993.

# 17

# THE EMPLOYER'S VIEWPOINT — STANDARDS AND COMPLIANCE

**Peter B. Rice**

## TABLE OF CONTENTS

## INTRODUCTION

Occupational safety and industrial health issues have become increasingly prominent in recent years. Employers' viewpoints are changing. This is an exciting time in which employers are abandoning the belief that programs like safety and health are hindrances to production. Many companies have recognized the importance of safety and health; however, for many reasons, mostly financial, there are thousands of employers who are just now recognizing their importance.

In response to the directives of the Occupational Safety and Health Administration (OSHA) and various state OSHA plans, employers are recognizing the importance of strong, effective occupational safety and health programs and regulatory compliance. Most employers recognize that the primary purpose of complying with OSHA regulations is to prevent occupational injuries and illnesses and to provide a safe and healthy workplace. Employers are increasingly becoming aware of other important reasons to recognize and comply with OSHA standards.

The progressive employer recognizes that OSHA compliance actually allows him/her to be more competitive in the marketplace, with a strong impact on productivity and profitability. With OSHA compliance, how can this make an employer more competitive, more productive, more profitable? The first issue to consider is the impact of an injury/illness on the bottom line. Injuries/ illnesses and noncompliance mean more workers' compensation claims. Workers' compensation claims translate into direct dollar losses. Noncompliance with the regulations can also lead to more frequent OSHA inspections and the potential for thousands of dollars in penalties.

The astute employer should be aware of not only the consequences of an OSHA inspection but, in cases of intentional negligence, the potential for criminal charges as well.

Failure to comply with OSHA regulations can also have a big impact on other, more intangible factors such as employee morale. What employee is going to be motivated to work for an employer who has little or no regard for safety and health in the workplace? On the other hand, when employees see that an employer is genuinely concerned with compliance with the regulations, setting the example for safety, employees recognize this and are also positively affected.

Public and community relations are increasingly becoming important to employers. Failure to recognize the impact of public relations can produce dramatic negative effects.

Employers also need to maintain strong client and customer relationships. Many clients will choose not to purchase goods or solicit services from a company with a poor track record in safety and health. OSHA is currently drafting regulations to prevent employers with poor safety records, based largely on workers' compensation claims, from bidding on federal construction projects. It is only a matter of time before other goods and services come under the same scrutiny.

## OSHA + EMPLOYER = A PARTNERSHIP

Maintaining good relations with the OSHA Compliance Office and its representatives makes good sense and is strongly encouraged. OSHA may appear bureaucratic in trying to enforce the hundreds, if not thousands, of safety and health rules and regulations currently in place. There is, however, a strong positive advantage of working with OSHA. OSHA is also made up of knowledgeable, often highly educated professionals with varied backgrounds such as engineering, safety management, chemistry, physics, industrial hygiene, construction, and public administration. The vast majority of OSHA compliance personnel are strongly motivated to establish solid working relationships with employers. They recognize that cooperation, rather than enforcement, is much more effective in creating these relationships. The result has been such programs as the Small Employers Voluntary Compliance Program (SEVCP) and the Cooperative Self-Inspection Program (CSIP), which includes the STAR and REACH programs.

Whether or not an employer participates in any of these voluntary programs, developing an effective partnership with OSHA is still possible and very much encouraged. Employers should establish contact with the local OSHA office, meet the area manager, and tour their facilities. OSHA is generally very receptive to this approach and is usually eager to review an employer's existing safety and health program to evaluate the employer's level of compliance. Other means for strengthening employer/OSHA relationships include (1) meetings or conferences featuring OSHA representatives as speakers or attendees, (2) panel discussions with OSHA representatives, and (3) direct telephone contact with representatives with regard to OSHA compliance questions.

Making use of an OSHA consultation, a free service funded by the U.S. Department of Labor to assist employers in their safety and health programs, can help in the development of a partnership between OSHA and the employer. The addresses and phone numbers of the OSHA consultation programs are listed in Table 17-1.

## EMPLOYERS' OBLIGATIONS

As specified in the Occupational Safety and Health (OSH) Act, an employer is responsible for providing a safe and healthy workplace. This implies that the employer not only must comply with OSHA requirements (which really are minimum standards), but also must know which ones apply to their situation. Employers and their representatives (supervisors and managers) are required to ensure compliance with safety and health standards in facility and process design, equipment selection and setup, and normal operations. It is important to recognize that all safety and health hazards need to be identified, evaluated, and controlled, regardless of whether or not there is a corresponding OSHA regulation. General safety and health hazards are listed by category in Table 17-2.

Management is required to maintain records and conduct employee training with regard to general and specific hazards and safe work practices in the

**Table 17-1   State OSHA Consultation Project Directory**

| State | Telephone | State | Telephone |
|-------|-----------|-------|-----------|
| Alabama | (205) 348-3033 | Montana | (406) 444-6401 |
| Alaska | (907) 451-2888 | Nebraska | (402) 471-4717 |
| Arizona | (602) 542-5795 | Nevada | (702) 688-1380 |
| Arkansas | (501) 682-4522 | New Hampshire | (603) 271-2024 |
| California | (415) 972-8515 | New Jersey | (609) 984-3507 |
| Colorado | (303) 491-6151 | New Mexico | (505) 827-2877 |
| Connecticut | (203) 566-4550 | New York | (518) 457-2810 |
| Delaware | (302) 577-2889 | North Carolina | (919) 733-3900 |
| District of Columbia | (202) 219-6091 | North Dakota | (701) 250-4521 |
| Florida | (904) 922-8955 | Ohio | (614) 644-2631 |
| Georgia | (404) 894-8274 | Oklahoma | (405) 528-1500 |
| Guam 9-011 | (671) 646-9246 | Oregon | (503) 378-3272 |
| Hawaii | (808) 548-7510 | Pennsylvania | (800) 382-1241 |
| Idaho | (208) 385-3283 | | (Toll-free in state) |
| Illinois | (312) 353-2220 | | (412) 357-2561 |
| Indiana | (317) 232-2688 | Puerto Rico | (809) 754-2134-2171 |
| Iowa | (515) 281-5352 | Rhode Island | (401) 277-2438 |
| Kansas | (913) 296-4386 | South Carolina | (803) 734-9599 |
| Kentucky | (502) 564-6895 | South Dakota | (605) 688-4101 |
| Louisiana | (504) 342-9601 | Tennessee | (615) 594-6180 |
| Maine | (207) 624-6460 | Texas | (512) 482-5783 |
| Maryland | (410) 333-4196 | Utah | (801) 530-6901 |
| Massachusetts | (617) 727-3463 | Vermont | (802) 828-2765 |
| Michigan | (517) 335-8250 (H)[a] | Virginia | (804) 464-7774 |
| | (517) 322-1814 (S)[b] | Virgin Islands | (809) 772-1315 |
| Minnesota | (612) 297-2393 | Washington | (206) 956-5439 |
| Mississippi | (601) 987-3981 | West Virginia | (304) 347-5937 |
| Missouri | (314) 751-7954 | Wisconsin | (608) 266-8579 (H) |
| | | | (414) 521-5063 (S) |
| | | Wyoming | (307) 777-7786 |

*Note:* Consultation programs provide free services to employers who request help in identifying and correcting specific hazards, want to improve their safety and health programs, and/or need further assistance in training and education.

[a] H — Health.

[b] S — Safety.

workplace. Employers are required to provide safety and health services and resources, including personal protective equipment, first aid supplies, and fire protection equipment. An employer's program must be documented with clearly defined policies and procedures since OSHA regulations are in part generic and are written for a wide range of industries or businesses. The employer needs to examine these regulations and apply them to the specific needs of their facilities, operations, and processes. The employer also needs to establish a hierarchy for implementing and enforcing OSHA regulations within their company.

## UNDERSTANDING OSHA

Occupational safety and health issues have been a concern to employers for more than 2000 years. Hippocrates recognized and documented the hazards

**Table 17-2   Example Classification of Occupational Safety and Health Hazards**

| Chemical | Physical | Biological | Ergonomic | General Safety |
|---|---|---|---|---|
| Fumes[a] | Cold stress | Insects (e.g., ants, bees, scorpions, spiders) | Circadian rhythm (e.g., shift work/rest cycles) | Construction[a] |
| Gases[a] | Heat stress | Microbes (e.g., bacteria [tuberculosis[b]], parasites, viruses [hepatitis B,[a] HIV,[a] others]) | Fatigue (extended work hours) | Maintenance[a] |
| Liquids[a] | Ionizing radiation[a](e.g., alpha-, beta-, gamma-, X-rays) | | Hand tools[a] | Electrical[a] |
| Mists[a] | | | | Emergencies[a] |
| Particulates/dusts[a] | Noise[a] | Toxic plants (e.g., poison oak) | Manual material handling[b] (e.g., biomechanics, lifting, pushing, pulling, carrying) | Environmental conditions[a] |
| Vapors[a] | Nonionizing radiation[a] (e.g., lasers, radio frequencies, microwaves, ultraviolet light) | Reptiles (e.g., snakes) | Mental task overload | Fires/explosions[a] |
| | Pressure | Sanitation[a] (e.g., drinking water, hygiene facilities) | Stress (occupational and nonoccupational) | Mechanical and machinery systems[a] |
| | Vibration | Small mammals (e.g., dogs, rodents, skunks) | Substance abuse | Motorized equipment[a] |
| | | | | Pressurized systems[a] |
| | $O_2$ deficiency | Potentially violent people[b] | Work station design[b] (e.g., dials, controls, signals, labeling, office [computer] workstations) Repetitive physiological stress | Fall protection[b] |
| | | | | Motor vehicle occupant safety[b] |

*Note:* All hazards must be addressed in employer's overall safety and health program if present at work site.

[a] Employee safety and health regulations adopted.

[b] Employee safety and health regulation in development (1995).

of lead to miners. Laws to enforce worker protection surfaced in the 18th and 19th centuries. The implementation of these laws coincided with the start of the industrial revolution. In the United States, the first workers' compensation law was passed in 1911 in Wisconsin, and today all states have various forms of workers' compensation and OSHA-type regulations.

Approximately half of the states in the United States have their own federally approved OSHA programs. The remaining states rely on federal OSHA for both regulations and enforcement. All of the OSHA-type programs consist of (1) standards generated from a standards development process, (2) enforcement activities generally conducted by OSHA enforcement officers, and (3) an appeals process. Employers may appeal any or part of an OSHA enforcement action that results in a penalty or citation.

## OSHA INSPECTIONS AND THE EMPLOYER'S RESPONSE

Inspections generally are the result of (1) imminent danger of serious violation, (2) catastrophes and fatal accidents, (3) employee complaints, (4) programmed inspections (generally of high-hazard industries), and (5) reinspections.

Although OSHA inspections cannot be totally avoided, they certainly can be minimized by having a strong safety and health program. It is worth noting that the number one cause of an OSHA inspection is an employee complaint. While it is illegal to prevent an employee from complaining to OSHA, minimizing these complaints can be achieved by implementing an internal mechanism whereby employees can air their complaints and grievances without fear of employer reprisals. Once a complaint is received, the employer should investigate the situation and take the necessary control and other follow-up actions. Above all, take action and let the employee(s) know what has been done.

## OSHA INSPECTIONS — THE EMPLOYER'S ROLE

OSHA compliance officers generally have free access to investigate and inspect any workplace during regular working hours and at other reasonable times if necessary to enforce safety and health compliance.

What should the employer expect and even require?

- The compliance officer must present appropriate credentials to the employer.
- The investigation or inspection must be carried out within reasonable limits and in a reasonable manner.
- During the course of any investigation or inspection, the OSHA compliance officer may obtain any statistics, information, or physical materials in the possession of the employer that are directly related to the purpose of the investigation or inspection.

- The OSHA compliance officer may conduct any test necessary to the investigation or inspection.
- The OSHA compliance officer may also take photographs. Any photograph taken during the course of an investigation or inspection is to be considered confidential information. Also, it is important to note that any employer information that might contain or reveal a trade secret is protected.
- Since there are requirements that prevent OSHA from giving advance warning or notice of an inspection, except under specific conditions, it is important for an employer to have a plan as to what to do when OSHA does come to inspect the facility.

The employer should have a standing policy and procedure plan or manual, such as "How to Prepare for a Compliance Inspection". Employees should be informed of the elements of the policy and procedure, which should be readily available to all those who may be in the position of greeting an inspector (e.g., receptionist, foreman, safety officer).

If the employer is suspicious of the inspector's credentials, the employer should contact the inspector's supervisor or manager for verification. Under most circumstances, it is prudent to allow the inspector entry. However, there are cases where an employer may refuse entry. Of course, the OSHA inspector may, with probable cause, request and obtain a search warrant. An employer may be able to effectively narrow the scope of an inspection by refusing entry and requiring a search warrant. An inspection warrant generally cannot permit an OSHA compliance officer to search beyond the scope created by the declarations upon which the warrant is based. Once admitted to a workplace by the employer, the OSHA compliance officer may not inspect beyond the scope of the warrant.

Some tips for handling an inspection include:

- Consent to any inspection can be given by any person with "ostensible authority". The employer's plan to deal with an inspection should specify who or what position (i.e., safety director) should allow entry. Having a knowledgeable person involved at the time of the OSHA compliance officer's visit is desirable. An employer's representatives should be familiar with the processes, hazards, safe work practices, and overall safety and health program.
- The employer's representatives need to be knowledgeable and careful in expressing the employer's position, since anything said during such discussions can be used against the employer at an appeals hearing, assuming an appeal is filed.
- The employer's representatives should keep careful and detailed notes of such meetings to document both what they say and what the OSHA compliance officer(s) say, since their comments can also be used at an appeals hearing.
- An employer may be able to avoid a citation altogether if the employer is able to convince the OSHA compliance officer that no violation, and no hazard, exists at the time of the inspection.

- At the closing conference, required by law, an employer may be able to convince the OSHA compliance officer that the proposed citation is erroneous and should be dropped or modified.
- An employer can request an informal conference to discuss the matter with other OSHA management personnel (e.g., the district manager).

## Practical Considerations of an OSHA Inspection

- In deciding whether to require an inspection warrant, employers should consider the impact on the company's relationship with OSHA. Apart from easing the inspection and penalty process, maintaining a good relationship with OSHA is beneficial from the standpoint of the wealth of information resources available from the agency.
- If the employer is not prepared to grant the OSHA inspector access to its facility, and it is expected that the OSHA compliance officer may request a warrant, the employer may notify the appropriate court and agency officials issuing the warrant. The presiding judges of the courts in which the employer is located should be advised of the employer's desire to be present in the event that the OSHA compliance officer applies for an inspection warrant. In most cases, an employer does have the right to participate in the warrant issuing process and may have the opportunity to have a warrant declared invalid.
- The employer should designate an individual to act as the key person for OSHA inspections (e.g., safety director). Upon arrival, the OSHA compliance officer should be referred to the key person. If that key person is unavailable, the compliance officer should be informed that the company program requires the presence of that person and that they are currently unavailable. Also, the inspector should be asked to wait until the key person is available. Generally speaking, OSHA compliance officers will wait for a reasonable period of time.
- The key person should understand the nature of an inspection and the rights of both OSHA and the employer. If the key person is unsure of the employer's rights, he or she should ask the compliance officer.
- Nearly all OSHA inspections will entail an evaluation of the employer's compliance with occupational safety and health standards that apply to all employers. These standards, to which all employers must comply, include
  - An emergency action plan to address emergency response capabilities and procedures to address anticipated emergencies; types of emergencies would include those that are "reasonably foreseeable" (e.g., chemical spills, power failures, adverse weather conditions)
  - A fire prevention plan that addresses fire hazards and proper procedures for responding to fire emergencies
  - An employee exposure and medical records plan that provides for the access and maintenance of such records
  - A hazard communication plan to address the labeling of hazardous substances in the workplace, employee training, and access to material safety data sheets
  - Compliance with general requirements for workplaces and work surfaces

- An injury and illness prevention plan — required in some states, a plan to identify, evaluate, and control occupational safety and health hazards, including employee training, accident investigation, identified responsible person(s), and record keeping
  - During the inspection, the employer's representative should be present and take notes throughout. If the OSHA inspector takes a photograph, the employer's representative should also take a photograph. The employer's representative should be able to identify any trade secret information and so inform the OSHA representative. Details of tests (e.g., air monitoring) or samples collected by an OSHA inspector should be noted. Also, the employer may want to collect their own "companion" samples or tests.
  - Inspection of any records other than those specifically described in a warrant (assuming a warrant was used to gain entry), or beyond those records which employers are required to maintain under the OSH Act (such as the log and summary of occupational injuries and illnesses, the supplemental record, and the annual summary), should not be permitted.
  - If possible, to minimize disruption, try to schedule the inspection outside of working hours or during a lunch break. Generally speaking, if these concerns are made known to an OSHA compliance officer, he/she will be open to rescheduling the inspection.

Since it is the objective of the OSHA compliance officer to evaluate the employer's safety and health program, it is important for the employer's representative to share with the inspector all that the employer is doing to promote safety and health — in a sense, to "sell" their program.

Finally, all permits, licenses, and previous reports from compliance agencies should be maintained on site or be readily available.

## CITATIONS AND PENALTIES

Following an inspection, the OSHA compliance officer and his/her management will make a determination as to what citations, if any, will be issued and what penalties, if any, will be imposed. If a citation is issued, the employer will receive it by certified mail. The employer also has the obligation to post a copy of each citation at or near the place where the violation occurred for 3 days or until the violation is abated, whichever is longer.

Although the types of citations and penalties may differ between state plan programs and the federal OSHA program, they both follow the same pattern and include the categories Other than Serious Violations, Serious Violations, Willful Violations, and Repeat Violations. All types may carry financial penalties. The willful and repeat violations, because of their nature, are assessed at a much higher rate.

In additional to financial penalties, criminal actions may be considered and taken against employers and employer representatives for violations of OSHA regulations. Generally these actions are taken against employers that are intentionally negligent in carrying out their responsibilities of providing a safe and healthy workplace for employees. Criminal prosecution is becoming a more widely used tool by local law enforcement agencies to hold employers accountable.

## APPEALING A CITATION

When issued a citation or notice of a proposed penalty, the employer has the right to an appeal. The employer may appeal either the citation itself, the violation type, or the penalty.

Why appeal a citation? There are several reasons to consider appealing a citation. However, before an appeal is submitted, the employer should request an informal meeting with OSHA management responsible for the citation. As more and more employers are appealing citations, to avoid legal disputes and prolonging the abatement and legal process, OSHA management is oftentimes motivated to settlement agreements that revise citations and penalties.

Should you appeal? That depends. The employer may consider the following:

- *The possibility of repeat citations.* Having a repeat citation down the road for the same condition is possible. As noted earlier, a repeat citation carries with it much greater penalties.
- *Other legal ramifications.* There may be legal ramifications other than OSHA to consider. These include the potential impact of civil litigation, workers' compensation, and criminal prosecution.
- *The potential for setting a precedent.* For example, an employer does not want to become an "easy target" for OSHA compliance activity by acquiescing to every citation that is issued regardless of its merits.
- *Cost of abatement.* The cost of abatement may be more than the employer can afford at the moment. Having to abate the violation may outweigh the cost and hassles of the appeal.
- *Cost of appeal.* Depending upon the type of citation and whether an attorney is put on your appeal team, appeal costs could be substantial.

Should the employer be represented by an attorney in the appeal process? Generally yes, because OSHA compliance representatives have usually handled dozens or hundreds of appeal cases, and without the representation of an experienced OSHA attorney the employer would be at a serious disadvantage.

Potential defenses to a citation that an attorney familiar with OSHA law might use are

- The evidence does not support a violation.
- The evidence does not support the characterization of the citation.
- The wrong safety order was cited.
- An independent employee action occurred.

If the employer decides to contest either the citation, the time set for abatement, or the proposed penalty, they have 15 working days from the time the citation and proposed penalty are received in which to notify the OSHA area office in writing. An orally expressed disagreement will not suffice. This written notification is called a "Notice to Contest", and forms can be obtained by calling the local OSHA area office.

Once received, the OSHA area director will forward the case to the Occupational Safety and Health Review Commission (OSHRC) for action. The OSHRC will assign the case to an administrative law judge for review. More often than not a hearing is scheduled, in which the employer and even the employees have the right to participate. The OSHRC does not require that the parties be represented by an attorney; however, depending upon the case, because it is a legal hearing, an attorney familiar with the process can be instrumental in defense of the employer and the employer's rights.

States with their own occupational safety and health programs have a state system for review and appeal of citations, penalties, and abatement periods. The procedures are generally similar to federal OSHA's, but cases are heard by a state review board or equivalent.

## RECORD KEEPING AND REPORTING

Safety and health records must be maintained since they are required in some cases and can be very meaningful over the long term. These records may serve as compliance and may be admissible in legal proceedings. Also, records are essential for an effective safety and health program. They aid in identifying and evaluating problem areas and accident causes and assist management in measuring overall safety and health performance. Certain records are required by law (i.e., Material Safety Data Sheets, OSHA log of occupational injuries and illnesses, employer's accident records) and are retained for specified periods. It is advisable to keep these required records indefinitely.

Most OSHA-related records are available to the OSHA inspector during an inspection. The types of records that the employer must generate and maintain are many. What records are kept is in some cases dependent upon the size of the company and its type.

What type of records are generally required to be developed and maintained by the employer to satisfy the OSHA compliance?

- Injury and illness records
- Log of occupational injuries and illnesses
- Reporting a work-related death or serious injury
- Medical and exposure records
- Safety and health inspection checklists

## Injury and Illness Records

Injury and illness records differ by state, and the OSHA office with jurisdiction should be contacted for identification of the necessary records and their requirements. In general, injury and illness records include

- Employee's claim for workers' compensation benefits
- Employer's report of occupational injury or illness
- Doctor's first report of occupational injury of illness

## Log of Occupational Injuries and Illnesses

With limited exceptions, every employer must complete the "Log and Summary of Occupational Injuries and Illnesses", also known as the OSHA log. When an employee in the course of his/her employment suffers an injury or illness requiring medical treatment (generally other than basic first aid), that injury or illness must be recorded on the log. Information to be included on the log includes the employee's name, occupation, department, description of the injury or illness, and number of days of restricted work activity or lost workdays. The log is kept for a calendar year and posted during the month of February in the year following the one for which the log was maintained.

With limited exceptions, the employer maintains this log, even if he/she has no recordable incidents. In the event of an OSHA inspection, odds are the Log will be requested for inspection by the OSHA compliance officer.

The logs must be maintained for 5 years in most cases (it is recommended that the logs be indefinitely maintained as they serve as a good source of record). In some cases the employer may be exempt from keeping the OSHA Log. The employer should check with OSHA to determine whether they meet the small employer exception or other exception by industry type (certain industries are exempt based upon their standard industry classification).

## Reporting a Death or Serious Injury

With limited differences, OSHA regulations require an employer to submit a report to the nearest OSHA compliance office if an employee is seriously injured on the job or in connection with the job, suffers a serious job-related illness, or dies on the job or in connection with it. The report must be made as soon as practically possible, but not longer than 8 hours (federal OSHA) after the employer knows or otherwise becomes aware of the death, serious injury, or illness.

Although at this time there are differences in reporting requirements between certain state plan states and federal OSHA, the employer should check with the local OSHA compliance office and identify the applicable requirements. Incorporate those requirements into your program and make sure that those with safety and health program responsibilities are informed of the reporting requirements.

## Medical and Exposure Records

Records that the employer has acquired for employees pertaining to medical conditions (e.g., physician and employment questionnaires or histories, laboratory results, medical opinions, diagnoses, treatments, etc.) and exposure records (e.g., air monitoring data) must be

- Maintained by employer
- Preserved by the employer
- Made available to employees and in some cases to OSHA

They can also be a wonderful source of helpful safety and health information for the employer.

## CONCLUSION

Compliance with the OSHA regulations has advantages far exceeding simply the avoidance of OSHA citations and penalties. Also, developing a positive relationship with OSHA and its representatives can bring rewards to the employer. In the experience of this author, putting a little energy and effort into a safety program can oftentimes bring the employer significant rewards in the form of fewer injuries and illnesses, fewer workers' compensation claims, improved production and morale, fewer regulatory and legal hassles, and in some cases more clients and an overall improved product and image.

## FURTHER READING

### Books

*Accident Prevention Manual for Business and Industry: Administration and Programs,* National Safety Council, Itasca, IL, 1992.

*Supervisors Safety Manual,* National Safety Council, Itasca, IL, 8th ed., 1993.

Blosser, F., *Primer on Occupational Safety and Health,* Bureau of National Affairs, Washington, D.C., 1992.

### Other Publications

Cohen, J. M. and Peterson, R. D., *The Cal/OSHA Handbook,* California Chamber of Commerce, 1994.

*All About OSHA,* U.S. Department of Labor, Occupational Safety and Health Administration, OSHA 2056, 1995.

# Section V
# Legal Aspects

# 18

# AN OVERVIEW OF REGULATORY COMPLIANCE AND LEGAL LIABILITY ISSUES

**James R. Arnold, Roberta V. Romberg, Susan Bade Hull, and Peter C. Lyon**

## TABLE OF CONTENTS

## INTRODUCTION

This chapter provides an overview of some of the key laws and regulations governing employers whose workers may be exposed to hazardous conditions or toxic substances in the workplace. It also surveys several major environmental laws that cover employers in their capacity as potential polluters of the environment.

Within the last 20 years, there has been dramatic and rapid growth and change in both areas surveyed by this chapter. Occupational safety and health laws and environmental protection statutes have been enacted at both the federal and state levels. Federal and state agencies have been established to enforce those laws and to promulgate regulations under them. These laws and regulations are enforced simultaneously, with federal law establishing in effect minimum compliance standards and each state's laws requiring potentially greater standards.

This increased regulation of the workplace and the environment creates complex legal problems and risks for employers, their managers, and their employees. The laws described below illustrate the myriad occupational safety and health regulations now imposed on employers and the environmental protection restrictions imposed on potential polluters.

## LABOR CODE

### Occupational Safety and Health Requirements

The federal[1] Occupational Safety and Health Act imposes two duties on employers: (1) a duty to comply with specific safety standards promulgated by both the federal and state agencies, and (2) a general duty to provide workers with a safe and healthful working environment, free from recognized hazards likely to cause serious harm.

Each state is authorized to adopt occupational safety and health legislation so long as it is at least as protective as federal law. So far 21 states[2] have adopted mini-OSHA statutes. California's mini-OSHA, enacted in 1973, is a good example of these state statutes.[3]

Government agencies are the primary enforcers of occupational safety and health requirements. They may initiate enforcement action on their own, triggered by an accident perhaps or through a routine audit, as well as through worker complaints.

#### Specific Safety Standards

The federal Occupational Safety and Health Administration (OSHA) has promulgated numerous regulations establishing specific safety standards for a wide array of potential workplace hazards, from airborne grain dust to ladders and from trenches to noise. State agencies have followed suit, issuing similar specific safety standards. We will focus on one set of safety standards, those dealing with toxic substances in the workplace, as a general example of specific safety standards.

Generally speaking, there are two types of specific safety standards for toxic substances. Some of the standards limit how much a worker can be exposed to a toxic substance in a set period of time. Other standards set such limits but, in addition, require the employer to implement an extensive protective program including medical examinations, training, and record-keeping. OSHA generally prefers that employers use reasonable engineering controls to reduce exposure below the prescribed standard. If engineering controls do not sufficiently reduce exposure, personal protection equipment must be utilized.

There are currently 25 toxic substances for which OSHA has promulgated *specific comprehensive standards*. They include arsenic, asbestos, benzene, formaldehyde, lead, vinyl chloride, etc. OSHA has also set *exposure limits* for more than 375 additional toxic substances, including acetone, acetylsalicylic acid (aspirin), aluminum, carbon dioxide, chlorine, chromium, copper, cotton dust, ethanol, nicotine, silica, starch, wood dust, zinc, etc.

#### Example of Comprehensive Standards

Comprehensive standards, for lead by way of example, require that employers monitor air lead levels to determine if they exceed permissible exposure limits over time.[4] If workers are exposed to an 8-hour time-weighted average of 30 µg or more of lead per cubic meter of air (action level), certain monitoring, medical surveillance, training, and education requirements are triggered. If the exposure limit over 8 hours equals or exceeds 50 µg/m$^3$ of air (permissible exposure limit), the employer must take additional corrective action, including providing employees with respirators of an approved design and training the employees in their proper use. Air monitoring must continue every 3 months until exposure drops below the permissible exposure limit and every 6 months until exposure drops below the action level for two consecutive measurements taken 2 weeks apart.

In addition, the employer must test the blood of employees working with or near lead every 6 months. If any individual employee tests at 40 µg of lead per deciliter of blood, the employer must retest the employee within 2 months. If (a) one blood test reveals more than 60 µg of lead per deciliter of blood or (b) an employee's tests average more than 50 µg per deciliter of blood over 3 months, the employer must remove the employee from the area of exposure (with no loss of salary or benefits) until the employee's blood lead level drops below 40 µg of lead per deciliter of blood.

The comprehensive lead standard also requires warning signs in work areas, and employers must maintain detailed monitoring records, provide employees access to monitoring results, and specifically notify employees of results above exposure limits. Also, employees working with lead must be given free annual medical examinations.

### Record-Keeping

As the standard for lead illustrates, many of OSHA's specific safety standards impose specific record-keeping requirements on employers. In addition, OSHA has promulgated general record-keeping rules governing all employers. These rules require, among other things, that the employer (1) maintain a log of all workplace injuries, including type of injury and number of days lost or restricted; (2) post the log annually; (3) retain and make the log available for employee and OSHA inspection for 5 years; (4) promptly report certain types of injuries, including those involving fatalities, to the local OSHA office; and (5) maintain medical and exposure records and retain them for 30 years from the time the employee leaves the workplace. An employer's failure to maintain records properly is a separate statutory violation.

### Hazardous Substances

In 1983, OSHA promulgated the Hazard Communication Standard[5] requiring employers to give specific information and training to employees who handle chemicals considered to be "hazardous" (e.g., combustible, explosive, flammable, unstable, or water reactive). The key to compliance is maintaining complete and updated Material Safety Data Sheets (MSDSs), developing a written hazardous materials communication program, and giving all workers required training.

Employers who handle hazardous substances must also be alert to OSHA's regulations for hazardous waste operations and emergency response procedures. Employers regulated under the federal environmental statutes discussed below must develop emergency response plans, site characterization and analysis, site control, training, medical surveillance, engineering controls and personal protection monitoring, information programs, and decontamination procedures. In addition, facilities using certain listed "highly hazardous" toxic and reactive materials must meet additional standards designed to minimize the risk of explosions and other catastrophic events.

### Inspections

OSHA officials may inspect a workplace either as part of a routine administrative program of inspection or in response to a reported accident or to an employee complaint. Employers may find themselves inadvertently triggering an inspection by the notice they are obligated by law to give OSHA whenever there is an accident resulting in a fatality or the hospitalization of five or more workers.

Ordinarily OSHA will arrive to conduct an inspection without advance notice. If OSHA specifically wants an employer representative to be present or wants to conduct the inspection after business hours, it may give advance notice, but in no event longer than 24 hours. Once on the premises, OSHA inspectors may test equipment and interview employees to document plainly observable violations. If an employer initially resists a site audit, OSHA is authorized to obtain search warrants and to issue administrative subpoenas to compel interviews with employees. While employers generally cannot prevent an audit and are unwise to try, usually the employer is permitted to and should accompany the OSHA inspector during the inspection; an employee designee may do so as well.[6] Prudent employers ensure that whoever represents them during an inspection is knowledgeable about OSHA rules and regulations.[7]

As an example of the ultimate "catch-22", some courts have now ruled that OSHA may even subpoena an employer's internal self-audit of its own compliance with OSHA standards.[8] Some employers seek to shield these internal audit reports by having attorneys perform them and seeking protection through the attorney-client privilege and/or attorney-work product doctrines. At this writing (see also the section, Protecting Self-Evaluations and Audits from Disclosure, this chapter) the law is not settled on how successful employers may be in protecting these reports.[9] Therefore, employers who take the management initiative to conduct compliance audits should be prepared to remedy any deficiencies they may find or they will be loading the proverbial gun to shoot themselves in the foot.

### General Duty

In addition to the specific safety standards described above, both federal and state occupational and health laws impose upon employers a "general" duty to provide a safe and healthful environment. To date, there is no definition of what that means in any specific instance and there are no minimum general requirements to meet, except on an "after-the-fact" case-by-case review. That is, of course, both good news and bad news.

### Basis for Recent Regulation — Smoking

Within the last several years, regulators at both the federal and state levels have eyed the "general duty clause" as a possible basis for launching both specific enforcement and new regulations regarding some of the current "hot" topics, such as indoor air quality, smoking, and workplace violence. In April 1994, for example, federal OSHA (fed/OSHA) for comment proposed regulations which would require employers across the nation in office buildings, health care establishments, and other nonindustrial settings to establish and implement indoor air quality plans to control indoor air contaminants. The effect of the proposed rules would be to require virtually all of the nation's employers to ban smoking in the workplace. OSHA received many comments and held hearings on the proposed rules in September 1994. The hearing rooms were filled beyond capacity. While final rules are not realistically expected for a considerable period, there is concern that such proposed regulations, regardless of whether they are ultimately promulgated, will nonetheless encourage litigation against employers regarding workplace smoking on the grounds that they violated the OSHA "general duty" standard (i.e., that the employer recklessly disregarded a known risk).

California, for example, has amended the general duty clause of the Cal/OSHA Act, effective January 1, 1995. All employers in the state are now prohibited from knowingly permitting tobacco smoking in enclosed spaces in the workplace.[10] Violations draw increasing monetary fines, with Cal/OSHA being obligated to respond to complaints after the third violation.

### Basis for Recent Enforcement — Workplace Violence

Homicide was the leading cause of workplace death in Texas and California in 1993, eclipsing traffic accidents, fires, and construction falls. That statistic has gotten the attention of employers, employees, and the director of Cal/OSHA, which is currently drawing up enforcement guidelines on workplace violence. Cal/OSHA circulated draft guidelines internally in July 1994. During 1994, Cal/OSHA also issued several citations for workplace violence, all based on the "general duty" clause. During the same year, fed/OSHA issued a citation based only on the "general duty" clause to a Chicago hospital after several psychiatric patients seriously injured hospital workers. While working in a psychiatric hospital presumably includes a certain level of risk, fed/OSHA asserted that the hospital could have taken substantially greater precautions to protect its workers in an inherently dangerous situation. Fed/OHSA and the hospital resolved this particular dispute by the hospital paying a small monetary fine and establishing specific preventive measures.

### Potential Penalties

*Administrative Penalties — Civil and Criminal*

Federal and state OSHA agencies are authorized to issue citations based on their respective compliance officers' reports and findings of alleged violations of either specific or general compliance standards. Citations include a statement of the alleged violation and the proposed length of time to set for abatement. The employer is obligated to post a copy of the citation at or near the place the violation occurred for 3 days or until the violation is abated, whichever is longer.

Additionally, agencies are authorized to initiate civil as well as criminal penalties, depending on the type of violation allegedly found. There are five possible categories of violation in the federal system:

1. General violation (has a direct relationship to job safety and health, but not likely to cause death or serious physical harm); up to $7,000 per violation per day. Discretionary with agency, and may be reduced depending on employer's good faith, history of compliance, and size of business.
2. Serious violation (substantial probability that death or serious physical harm could result and employer knew or should have known of hazard); mandatory penalty of up to $7,000 per day per violation. May be reduced depending on employer's good faith, history of compliance, gravity of alleged violation, and size of business.
3. Willful violation (intentional and knowing); mandatory penalty of up to $70,000 per day per violation, with a minimum penalty of $5,000. May be reduced by business size and compliance history but not good faith. Willful violation may also trigger criminal prosecution, with potential for criminal penalty of up to $500,000 per violation for a corporation and $250,000 per violation for an individual and/or 6 months of imprisonment.
4. Repeat violation (upon reinspection following final citation, the agency finds a substantially similar violation); penalty of up to $70,000 per day per violation.
5. Failure to correct prior violation; civil penalty of up to $7,000 per day for the period the violation continues beyond the prescribed abatement date.

The administrative agency is authorized to impose civil penalties. Criminal penalties, including imprisonment, are imposed through the courts.

Even civil penalties can be overwhelming in particularly egregious cases or where the agency concludes that the employer has not moved swiftly or extensively enough to abate the violation. For example, in September 1993 Hercules Inc., a New Jersey gunpowder manufacturer, was hit with more than $6 million in proposed fines by federal OSHA for allegedly failing to abate safety and health hazards discovered following a 1989 explosion at the company processing plant which seriously injured three workers. The explosion and fire triggered an OSHA inspection resulting in 71 citations, many for serious violations and willful violations. The company contested many of those violations and eventually settled them. Then in 1993 OSHA issued new failure to abate notices alleging failure to abate five of the 1989 violations and failure to maintain appropriate records regarding illnesses and injuries. Most of the total 1993 fine ($6,287,500) related to the five hazards that the agency claimed the company had failed to abate since 1989.

### Appeal Process

As seen from Hercules' experience, there is a process through which employers can appeal OSHA citations, although it can be very time-consuming and expensive. First, employers may request an informal meeting with OSHA's area director, who is authorized to execute settlement agreements revising citations or penalties. Second, employers receiving citations must correct the cited hazard within the prescribed time period or contest the citation or the abatement date by filing a written Notice of Contest within 15 working days of receiving the citation. There is no specific format for the notice, but it must clearly identify the citation, the proposed penalty, the abatement period, and the employer's basis for contesting. Employers must give a copy of the notice to the employees' authorized representative or post it in a prominent place.

The area director then forwards the matter to the Occupational Safety and Health Review Commission (OSHRC), where the case is heard by an administrative law judge. OSHRC is an independent federal agency, not included within OSHA or the Department of Labor. The administrative law judge reviews the contest and then sets the matter for public hearing, where both employer and employees may participate. Any party may request that any ruling of the administrative law judge be further reviewed by the full Commission. Commission rulings may be appealed to the appropriate U.S. Court of Appeals.

### Criminal Penalties

Every corporate employer as well as every individual manager responsible in any manner for health or safety issues should also be aware of the potential criminal liability for health or safety violations — specifically including individual criminal liability. First is the ever-present possibility of manslaughter charges whenever there is a workplace death resulting from negligent or willful conduct. For example, in what is believed to be the harshest judgment ever handed out at the federal level for workplace safety violations, a 20-year prison sentence followed by a 30-year suspended sentence was imposed in 1992 on the 65-year-old owner and president of a chicken processing plant where 25 people died in a fire; the doors had been padlocked. The owner claimed he had padlocked the doors to keep them from being propped open by employees. There are also a handful of state manslaughter cases, almost all in California, which have resulted in somewhat lighter prison terms.

Second, California's Penal Code was amended in 1991 to include provisions making it a crime for corporations as well as individual managers to fail to warn affected employees in writing, and to fail to notify the state agency, immediately when they know of a serious concealed danger in the workplace or their products involving imminent risk of great bodily harm or death. The unsettling part of the mandate (affectionately termed "Be a manager — go to jail") is that the term "knowledge" used in this statute means not only actual knowledge, but also a manager having information that would convince a reasonable person that a serious concealed danger exists. That is a negligence standard which does not require actual knowledge at all, but can nonetheless be used as a basis for criminal prosecution.

## Workers' Compensation

Workers' compensation is exclusively a creature of state law, familiar to almost every employer. While states may vary in the specifics, every state's workers' compensation system has been devised, as a matter of public policy, as a compromise between employer and employee. The employee receives a relatively quick, although limited, award to meet his/her needs for injuries arising from employment. In exchange, the employer reduces the overall risk of monetary damages to which (s)he might otherwise be exposed. "Fault" is irrelevant, except in the most extreme circumstances involving either employer or employee behavior. The response is relatively quick, and the employee gets paid according to a statutory scheme. In exchange, employees and their families surrender their common law right to sue employers and others for damages. Employers end up liable for many more awards, although each is much smaller than it might otherwise be.

Familiarity with the workers' compensation scheme is essential to any safety management professional, both in terms of how it functions and how best to manage it, given the tremendous costs involved. Insurers, insureds, and regulators are all seeking ways to contain spiralling workers' compensation costs. One approach has been to require workers' compensation insurers to provide loss control services to their insureds (and to target employers with the greatest compensation losses). California, for example, requires Cal/OSHA to establish a program for targeting high-hazard employers, including having the employer submit plans designed to reduce hazards.[11] Safety managers can benefit from such programs by drawing on their experience to reduce overall safety risks (and costs).

## Substance Abuse: Drugs and Alcohol

It is undeniable that employee drug and/or alcohol use substantially increases the risk of both accidents and disease, with concomitant increases in employers' costs. Some employers have responded by instituting drug and/or alcohol testing programs for applicants and employees, either uniformly, randomly, or based on "reasonable suspicion". The courts carefully scrutinize such testing programs from the perspective of the employee's constitutional privacy rights, particularly in California. Therefore they must be designed and administered to avoid conflicts with constitutional privacy rights. On the other hand, the federal Departments of Transportation and Defense, as well as the Nuclear Regulatory Agency, have each issued regulations requiring covered employers to develop and implement programs to reduce drug and alcohol use. The regulations specifically require drug and alcohol testing of employees in safety-sensitive positions. Employers covered by these regulations must randomly test half of their employees on an annual basis and promptly test specific employees following accidents or upon reasonable suspicion.

Safety managers who pay specific attention to issues of drug and alcohol use can avoid disruption and save money and lives.

## ENVIRONMENTAL LAWS

Congress, federal agencies, state legislatures, state agencies, and county and city governments have issued hundreds of thousands of pages setting out the requirements of the environmental laws and regulations.[12] Company managers are hard pressed to understand what requirements are enforced by which agency. In addition, the courts have developed rules of liability and, in some states, popular referendums add more standards for environmental compliance.

Safety professionals recognize that the government's trend in both environmental and industrial safety regulatory schemes is away from compliance counseling and toward penalties.[13] Furthermore, environmental regulatory programs assess "no-fault" penalties which become more severe the longer or the greater the violation. Both the safety and the environmental laws contain criminal liability provisions which can be used by prosecutors to impose felony penalties for careless or unwitting violations of environmental protection and worker safety laws.[14] In some circumstances, prosecutors can even use evidence of attempts by managers to shield themselves from information as proof of violations.[15]

These myriad environmental requirements apply to[16]

- Discharges of wastewaters to surface waters and sewer systems
- Emissions to the air from nonmobile ("stationary") sources, including permits and inventorying and reporting of toxic or hazardous air contaminants
- The handling of hazardous materials, managing hazardous wastes (including characterization as such), meeting storage and waste minimization requirements, transporting and disposing of wastes, and obtaining and following permits for on-site storage and treatment
- Hazard communication, community right-to-know laws, and other required warnings

The great majority of the environmental requirements which affect day-to-day management of businesses are found in nonfederal rules and regulations. In addition to state laws, the federal government has delegated its authority to the states for many of the major federal programs. For example, permits authorized by the federal Clean Water Act[17] are issued and administrated by state agencies.[18] In addition, the states (particularly the large industrial states such as Illinois, California, New York, Pennsylvania, and New Jersey) have enacted comprehensive environmental protection laws for water, air, hazardous materials, etc. which generally parallel federal programs.[19]

The following discussion can only describe some of the broad outlines of the major programs. Their goals are to enhance the quality of the natural and human environment, as well as require businesses to operate in an environmentally benign manner. The initial level of compliance and reporting falls most often on the shoulders of safety managers and environmental compliance specialists.

## Discharges and Emissions

Major regulatory efforts are directed toward facilities that discharge waste products from their manufacturing processes into surface waters, sewers (both sanitary and storm), and the air. This section summarizes the relevant laws and regulations. These permitting systems are designed to reduce and eventually eliminate discharges and emissions.[20]

### Discharges to Water: NPDES Permits, Storm Water Discharges, POTW Pretreatment Standards

#### National Pollutant Discharge Elimination System Permits

The Clean Water Act (CWA) and similar state laws require extensive controls whenever a facility discharges materials or waste into surface waters (lakes, rivers, the ocean, etc.).[21] The CWA's primary regulatory program for discharges to surface waters is called the National Pollutant Discharge Elimination System (NPDES) and is enforced by the U.S. Environmental Protection Agency (USEPA) or delegated state agencies. The statute and USEPA regulations require the NPDES permits to include limits on concentrations of organic and inorganic pollutants, flow volumes, and standard water quality measurements, such as temperature, pH, and turbidity. Permits must also include minimum standard conditions.[22] These conditions control all aspects of discharges — and thus directly affect the operation of the source business or plant.[23]

Breach of any permit condition is also a violation of the CWA itself. Additionally, discharges of specific toxics have particular requirements under NPDES permits, and dischargers in a wide variety of industries have industry-specific NPDES permit standards.[24] Violations are generally prosecuted on the basis of strict liability, or liability without a specific intent (what the law calls a *mens rea*). Even managers of publically operated sewer treatment plants (POTWs) have been prosecuted for felony violations of NPDES permits.[25]

### Storm Water Permits

Control of storm water discharges and their pollution load has lagged behind the NPDES "point source"[26] program. Beginning in 1987, industrial and municipal dischargers of storm water were required to obtain NPDES permits. However, the USEPA also authorized states to issue "general permits" to regulate storm water discharges.[27] Compliance with the terms of such "general permits" was deemed a substitute for site-specific NPDES permits. These permits cover rainwater and snowmelt runoff from areas such as roofs, parking lots, streets, corporation yards, plant grounds, etc. The runoff can carry significant amounts of contaminants into receiving waters.

For example, under the USEPA's delegated authority, the California Water Resources Control Board has issued a statewide "general permit" applicable to all industrial storm water discharges, with several basic requirements.[28]

Safety managers have often been assigned the responsibility for complying with such programs. How significant is the threat of prosecution? In early years of these programs, prosecutions (and administrative fines) were generally limited to serious chemical spills which affected health or fishing resources. However, more and more when prosecutions occur they are not limited to companies and they include individual managers.

### Pretreatment Regulations and Discharges to POTWs

Many plants discharge to publically operated treatment works (POTWs). POTWs generally have NPDES permits, issued by the USEPA or state agencies. As a result, industrial sanitary sewer discharges are closely regulated by POTWs, which promulgate and enforce regulatory permit programs similar to the NPDES scheme. A company's history of noncompliance (poor practices, late reporting, obstructing inspectors), coupled with a significant "upset" (or "exceedance or "slug") of concentrations sufficient to qualify as hazardous waste discharges, can result in notorious prosecutions of companies and managers.

## Emissions to Air: The Clean Air Act (CAA) and Toxic and Hazardous Air Contaminants

"Stationary sources" in industry are smokestacks, vents, and cooling towers. They are major sources of air pollution, emitting "criteria" pollutants (carbon monoxide, sulfur oxides, particulates, photochemical oxidants, nitrogen dioxide, and lead) and toxic pollutants (metals and organic chemicals). The 1990 federal Clean Air Act Amendments (CAA 90), particularly Title V, significantly affect industry.[29]

In 1970, as the result of a history of failure in improving air quality, Congress enacted national standards for cleaning the nation's air in "nonattainment areas" and maintaining air quality in "attainment areas". Today, there are two general levels of implementation — the federal government has developed air quality criteria, and the states use the criteria to regulate sources of air pollutants. The states generally use local and regional agencies, often with fee-based permit systems to fund permit programs, research, and enforcement, to carry out their roles.[30]

The CAA requires every state to develop "state implementation plans" (SIPs). These plans describe how the state's regulatory and enforcement program will reduce and eventually eliminate pollutant emissions to the air. In "nonattainment" areas (generally found in states with large concentrations of vehicles and/or manufacturing facilities), the government imposes controls on new or modified sources. Such sources must use the best available emission controls, and the permit-issuing agency must determine that net emissions in the area will not increase. Thus, plant managers may have to acquire "offsets" by reducing emissions at other processes at their plant or by buying them from other companies.

A major expansion of the federal government's control of air emissions occurred in the federal Clean Air Act Amendments of 1990 (CAA 90).[31] Title V of this legislation created a federal operating permit program.[32] Congress thus established a "one facility, one permit" program, which superseded the "source permit" programs.[33]

Hazardous air pollutants (HAPs) were added to the federal regulatory scheme by Title III of CAA 90.[34] Title III applies to quantities of toxic air emissions above several tons per year. State and local agencies have broad authority to include additional substances and set lower regulatory thresholds. This control program includes toxic emissions inventories and submission of plans by businesses,[35] upon which the regulating agency issues a priority rating (and may require the company to prepare a risk assessment).[36] The agency then decides if the risk warrants requiring the company to give warnings to the community and/or install additional abatement equipment (or reduce emissions at their source).

Managers who are responsible for meeting the requirements of air quality control permit laws will find themselves doing a number of jobs (which may not have been described when they were hired). Safety and environmental compliance managers can be assigned the duty of obtaining permits, measuring emissions, controlling concentrations in source chemicals (e.g., volatile organic compounds in paints), working with process engineers on both manufacturing processes and air pollution control equipment, making periodic compliance reports (under penalty of perjury), installing new equipment, obtaining variances, and negotiating for offsets.

The USEPA and state agencies enforce the clean air laws. They can issue civil fines and issue abatement orders to individuals and companies.[37] The U.S. Department of Justice and state prosecutors enforce violations which involve serious hazards, such as improper handling of asbestos (which has resulted in felony convictions of building owners and demolition contractors).[38]

## Hazardous Waste Storage, Handling, and Disposal

### The Resource Conservation and Recovery Act (RCRA)[39]

Most laws controlling hazardous wastes are like the federal Resource Conservation and Recovery Act (RCRA).[40] The RCRA and similar state laws comprise a "cradle-to-grave" program for tracking generation, storage, transportation, and disposal of hazardous wastes. The states' programs are based on both separate state laws and delegations of authority by the USEPA.[41]

The RCRA and state "cradle-to-grave" statutory programs provide the USEPA and state toxics agencies (collectively "RCRA" and "EPA") with the authority to administer a program which, like the CAA and the CWA, relies on businesses to characterize their wastes, maintain records, and periodically report to government agencies. These programs also include inspection and monitoring by the EPA. There are additional specific requirements for generators, transporters, and facilities that treat, store, or dispose of hazardous wastes (TSDFs).[42] RCRA authorizes the EPA to obtain information as to the location of storage and disposal sites, the amount and toxicity of wastes, the owner of the site, the methods of treatment or disposal employed, and the current status of the site.

The EPA, by administrative order, can require monitoring, testing, analysis, and reporting of any site that is believed to present an imminent and substantial hazard. Therefore, the EPA can use its RCRA authority to investigate suspected inactive hazardous waste sites outside the conventional RCRA "regulatory net". Also, the EPA can require RCRA permits for cleanups (e.g., "corrective action") conducted under CERCLA and state laws. Orders can be issued to owners of sites operated in the past by others, as well as to owners of sites not presently known to contain hazardous wastes.

Professional managers recognize that if their company generates wastes, it must (a) determine if the wastes are hazardous,[43] (b) safely handle them,[44] and (c) dispose of them at proper disposal facilities. Managers should also be aware that some state laws are more stringent and/or include more wastes than the federal law.

Federal penalties under the RCRA parallel those under the CWA and CAA, including civil penalties of up to $25,000 per day for each day of noncompliance and criminal felony penalties of up to $50,000 per day and federal imprisonment of up to 5 years for "knowing" RCRA violations. Criminal felony penalties of up to $250,000 for individuals and $1,000,000 for organizations, and imprisonment of up to 15 years for "knowing" endangerment violations, are also included.[45] Penalties and prosecutions are increasing.

### RCRA: A Case Study of a Prosecution

In June 1992, two 9-year-old boys climbed into a dumpster near a manufacturing plant in Tampa, Florida. They were overcome by fumes in the dumpster and died. An investigation revealed that Durex Industries had been using the dumpster to illegally dispose of toluene waste from the manufacturing of rubber rollers for the printing industry.

Durex Industries became the first company to plead no contest to a charge of "knowing endangerment" under the RCRA in July 1994. The company was assessed a $1.5 million fine. *U.S. v. William Recht Co.*, No. 94-70-CRT-17B (M.D. Fla., Jan. 3, 1995). The plant manager and his brother, the plant foreman, were convicted by a jury of illegal disposal of 5 to 7 gallons of toluene without a RCRA permit (although the jury found them not guilty of a "knowing endangerment" charge). The judge sentenced the brothers to 27 months in prison, followed by 2 years of supervised release. *U.S. v. William C. Whittman, et al.*, No. 94-70-CRT-17B (M.D. Fla., Oct. 14, 1994).

## Comprehensive Environmental Response, Compensation, and Liability Act (CERCLA)

The abandoned hazardous waste site cleanup laws, Comprehensive Environmental Response, Compensation, and Liability Act (CERCLA) or "Superfund" (and state "Superfunds," collectively "CERCLA"), are closely related to RCRA-type hazardous waste control laws. CERCLA also prohibits releases of hazardous "substances" (as opposed to hazardous "wastes" regulated under RCRA) into the environment.[46] CERCLA-type laws impose liability and provide funding for cleanup of abandoned hazardous waste sites. However, Superfund laws do not relate to day-to-day compliance by operating facilities and so are not usually directly relevant for safety professionals and other managers. However, safety professionals and managers should be aware that these laws impose cleanup liability on a strict (without fault) as well as a joint and several basis. (These laws also contain spill reporting requirements, discussed in the next section.)

### Community Right-to-Know Laws and Other Required Warnings

The tragic and catastrophic gas release in 1984 in Bhopal, India which killed or injured thousands of people living near a chemical plant led to state and federal community right-to-know laws.[47] The federal law is Title III of the Superfund Amendments of 1986 (or the Environmental Protection and Community Right-to-Know Act — "EPCRA"). These laws are designed to allow public oversight of government agencies and the agencies oversight of businesses that handle hazardous materials (whether wastes, chemicals, products, etc.) above certain quantities.

Managers should understand the basic requirements of the community right-to-know programs. They include immediate reporting of releases of hazardous materials above reportable quantities ("RQs"); preparing and maintaining a business plan and inventory; notifying local agencies of the presence and handling of more than threshold quantities of acutely hazardous materials;[48]

submitting annually the "Toxic Chemical Release Reports" (the so-called "Form Rs") if threshold quantities of toxic chemicals are manufactured, processed, or used; and meeting fed/OSHA and state and local agency requirements for storing hazardous materials.

Like RCRA discussed earlier, the community right-to-know laws are applied through a weblike scheme of federal, state, and local agencies. Managers should remember that the laws require reporting to emergency response officials in order that immediate responses be taken for public safety and health. Violations of the strict reporting requirements, even when seemingly innocuous or merely careless or inconsequential, have resulted in significant fines.[49]

## CIVIL RIGHTS LAWS

Over the next decade, safety managers will have to deal with additional statutory programs under the civil rights laws. We will discuss two areas that are presently emerging for companies and their managers — disability rights and environmental justice.

### Disability and Safety Requirements

Employers may seek to be more selective in hiring in order to reduce the risk of injury, with its increased costs of workers' compensation, medical care, and employee absences. However, employers must consider federal and state prohibitions against employment discrimination based on disabilities. For example, it may seem medically justifiable to refuse to hire cigarette smokers to work with asbestos (and other inhalatory toxins) due to the increased risk of contracting cancer. Similarly, employees with prior back injuries may be medically more likely to suffer further injury from certain types of labor.

These are logical conclusions, but they are contrary to disability antidiscrimination laws. The federal Americans with Disabilities Act (ADA) of 1990,[50] which as of 1994 applies to private and public employers of 15 or more employees, protects the employment (Title I) and accessibility (Title III) rights of the physically and mentally disabled. Title I of the ADA[51] prohibits discrimination against qualified individuals (a) with disabilities, (b) with a record of impairment, or (c) who are regarded as having such an impairment and who can perform the essential elements of the particular job (with or without reasonable accommodation). The ADA applies to all aspects of employment, such as recruitment, hiring, and fringe benefits (including insurance), and to all employment-related activities.

A disability is a "physical or mental impairment" that "substantially limits" one or more "major life activities". A "reasonable accommodation" is a modification or adjustment to a job or work environment that would allow the qualified disabled person to perform the essential job functions. It can include restructuring a job, modifying work schedules, acquiring or modifying equipment, providing readers or interpreters, and the like. However, the duty imposed by the ADA to provide "reasonable accommodation" is limited when it causes an "undue hardship" on the operation of the employer's business. Whether or not the "undue hardship" exception applies is determined on a case-by-case basis, taking into account the nature and cost of the accommodation, the financial resources of the employer, and the type of operation.

Employers are allowed by the ADA to require that workers not pose a direct threat or significant risk to the health or safety of themselves or others. If the threat or risk can be eliminated or lowered by "reasonable accommodation" to an acceptable level, the employer must do so.[52]

## Environmental Justice

Many activists believe that environmental laws have been administered and enforced to the disadvantage of minorities and low-income groups. Early in 1994, the Clinton administration issued an executive order which required federal agencies to develop strategies to prevent disproportionate effects on minority and low-income populations.[53] There are additional initiatives in Congress and through the courts to link the siting of facilities with toxics emissions with the civil rights laws. Disparate impacts of chemical exposures on non-English-speaking communities have already been the basis of lawsuits against a refinery in Louisiana and toxics incinerator projects in California. Managers should anticipate eventually dealing with operational issues (e.g., communications, exposures, risk assessments, cultural, etc.) linked to "environmental justice" concerns.

## OPERATIONAL COMPLIANCE MANAGEMENT

### Safety and Environmental Compliance Auditing

As already noted, government agencies seem to be relying more and more on reporting and enforcement and less on "compliance counseling".[54] The largest companies, particularly in dangerous businesses, have long integrated into their operations a significant level of compliance with safety, occupational health, and environmental protection laws. Some observers believe that large computer chip manufacturing plants, as well as modern steel mills, are far less likely to violate such laws than electroplaters, food processors, and chemical reformulators.

All modern businesses are integrating compliance programs for safety, occupational health, and environmental protection into their operations. The heart of compliance programs are "self-audits" (periodic management reviews of compliance). Compliance auditing can include checking whether a business is fulfilling its duties under worker safety and environmental laws,[55] its permits for discharges and emissions, its labor and employment agreements, and any agreements with government agencies. An audit or self-evaluation can also identify ways a company's management system can promote compliance and reduce risk.

## Multimedia Inspections

Environmental protection and OSHA agencies are turning to "multimedia" inspections. Such inspections review record-keeping for employee training, discharges and emissions, and disposal of wastes. The inspectors, who may be cross-trained (USEPA often assigns trained investigators to state-led teams), also look for physical evidence of violations, such as spills, stains, faulty secondary containment, and the inability of workers to articulate safety and environmental compliance procedures and policies.

Plant managers are responding with training, equipment maintenance, and other types of compliance programs. The most successful programs are those which integrate the sources of information and the reporting which must be done.[56]

## Protecting Self-Evaluations and Audits from Disclosure: The Law

Compliance auditing often reveals violations of laws which, if promptly reported and remedied, can be resolved with governmental agencies. But for companies the written product of such audits, particularly the thought processes and analyses of company managers and their counselors, are highly sensitive and confidential information. Companies feel the need to manage the product as carefully as they do such information as trade secrets and employee medical records.

One way such information can be protected is by understanding the law. If compliance auditing is conducted so that the company can receive legal advice, the company may be able to preserve the confidentiality of the information. Preserving confidentiality can be very important in defending against government investigations and third-party lawsuits ("citizen" or "private attorney general" lawsuits, and those by competitors).

If the company's attorney provides legal counseling based on the information developed by a safety professional and/or the environmental consultant, a company may be able to invoke three "evidentiary" privileges to preserve confidentiality of the information. These privileges include the attorney-client confidential communication privilege, the attorney work product doctrine, and a developing "self-evaluation" privilege. In addition, several states have enacted laws which recognize an environmental audit privilege.[57]

The attorney-client privilege is similar to other "communication" privileges developed by the courts and included in state and federal codes of evidence law.[58] Information can be protected if it is compiled "for the purpose of securing an opinion of law".[59] The attorney work product doctrine protects information gained and communications in anticipation of litigation. Routine audits or self-evaluations may not qualify.[60]

Under a self-evaluation or "self-critical analysis" privilege, it is possible for a business to assess its compliance with regulatory requirements without creating evidence that may be used against it in future litigation.[61] However, the history of the confidentiality of such self-evaluations in the face of government prosecutions is not so positive.[62] The state legislation, mentioned earlier, creates a form of self-evaluation privilege. However, the statutes generally only provide a qualified privilege; for instance, the company must promptly remedy any violations revealed in the audit. The USEPA is presently studying the issue, and federal legislation is being considered which would not allow a privilege, but which would allow mitigation of penalties if an audit had been conducted.

### Watching Out for the Government If Your Company Does an Audit

The USEPA has had a policy for almost a decade which encourages companies to develop, implement, and periodically upgrade environmental auditing programs.[63] However, its policy also includes its position that audit reports are subject to disclosure to the USEPA. Likewise, the U.S. Justice Department's guidance on investigating for and prosecuting environmental crimes offers no safe harbor.[64]

## CONCLUSION

A facility manager must deal with a complex legal regime for health, safety, and environmental compliance. The last two decades have seen an explosion in the duties of traditional safety managers and facility engineers. The broad requirements of these laws must be matched by vigilance from the environmental, health, and safety manager.

There are many places to slip up. Certain basic rules will go a long way: a solid grasp of the regulations, good communications internally and with the regulatory community, a commitment of management to compliance, and common sense. With these tools, the manager should be able to succeed in today's difficult regulatory environment.

## REFERENCES

1. 29 U.S.C. § 654(a).
2. Alaska, Arizona, California, Hawaii, Indiana, Iowa, Kentucky, Maryland, Michigan, Minnesota, Nevada, New Mexico, North Carolina, Oregon, South Carolina, Tennessee, Utah, Vermont, Virginia, Washington, and Wyoming. In addition, New York and Connecticut have approved mini-OSHA statutes for public employees.

3. California Labor Code § 6300 *et seq.*
4. 29 C.F.R. § 1910.1025, App B (1994).
5. 29 C.F.R. § 1910.1200 (1994).
6. 29 C.F.R. § 1903.8 (1994).
7. For a more complete description of the inspection process, see Chapter 17.
8. *Secretary of Labor v. Hammermill Paper,* 796 F. Supp. 1474 (S.D. Ala. 1992).
9. *Secretary of Labor v. Hercules, Inc.,* 857 F. Supp. 367 (D.N.J. 1994).
10. California Labor Code § 6404.5.
11. California Labor Code § 6314.1.
12. There are detailed and strict environmental regulations for certain products and businesses — more regulations than can be covered in one chapter of a book. A partial list would include regulations for chemical manufacturing, importation, and exportation; distribution of hazardous chemicals; manufacturing and distributing pesticides, fertilizers, and radiation materials and equipment; noise-producing activities and products; etc.
13. This trend generally results in internalization for a business of the costs of technical and engineering services to comply with regulatory requirements.
14. Such unlawful practices in businesses include concealing serious dangers in business practices or products, unfair business practices (i.e., penalizing competitive advantages from violations of environmental laws), misrepresenting products as "environmentally safe", and failing to disclose in transactions information about hazardous substances that have come to be located on or beneath property.
15. For instance, in the Clean Air Act there are felony penalties for "knowing releases" of hazardous air contaminants that place victims in imminent danger of death or serious bodily injury. 42 U.S.C. § 7413(c). Prosecutors can use "circumstantial evidence," *including evidence that the defendant took affirmative steps to be shielded from relevant information.* 42 U.S.C. § 7413(c)(5)(B). The Clean Water Act and the hazardous waste management law, the Resource Conservation and Recovery Act, have similar provisions. See Section 309 of the Clean Water Act, 33 U.S.C. § 1319 (c) (B) (i) and Section 3008 of RCRA, 42 U.S.C. § 6928 (f) (2) (RCRA). Thus an employee's knowledge can be imputed to a manager and the head of a company, potentially resulting in criminal liability for the employee, the manager, the head executive, and the company.
16. See "Environmental Requirements: By Subject," *California Environmental Compliance Handbook,* p. 3 (California Chamber of Commerce, 1992).
17. 33 U.S.C. § 1251 *et seq.*
18. For example, these permits ("National Pollutant Discharge Elimination System" or "NPDES" permits) are issued and enforced in California by appropriate California Regional Water Quality Control Boards.
19. Some of the state laws were precursors or models for federal laws. Other laws are amended by state legislatures because of federal requirements for delegation of federal authority (and providing federal funding for administration and enforcement).
20. Surface impoundments, injection wells, evaporative systems, and the like have specialized requirements from the CWA, RCRA (described below), the federal Safe Drinking Water Act, 33 U.S.C. § 300f *et seq.,* other federal laws, and various laws of the individual states.

21. The CWA prohibits the discharge of such wastes ("pollutants") as mine drainage, logging debris, agricultural runoff, cooling water from power plants, alteration of stream flows by sand or gravel mining, silt, etc., 33 U.S.C. § 1311. In addition, Section 404 of the Rivers and Harbors Act (a part of the CWA), 33 U.S.C. § 1344, is administered jointly by the U.S. Army Corps of Engineers and the USEPA and requires a permit to discharge any "fill materials" into wetland areas. An example of a comparable state law is § 5650 of the California Fish and Game Code, which prohibits the "placement" of "any substance or material deleterious to fish, plant life, or bird life" where they can "pass into the waters of this State".

22. 40 C.F.R. § 122.41 (1994).

23. They include, by way of example, properly operating and maintaining all facilities and systems of treatment and control; mitigating and reporting any discharge in violation of the permit; stopping or reducing plant activity to maintain compliance with the limits in the permit; a right of inspection and entry by the agency (city, sewer authority, state, and USEPA) which issued the permit to determine compliance status; monitoring discharges, keeping records, and reporting — generally required to be signed by plant managers under penalty of perjury; formally notifying the issuing agency of planned physical alterations to treatment or conveyance facilities; and periodically reapplying for a permit.

24. 40 C.F.R. Part 122, Appendix A (1994).

25. In *U.S. v. Weitzenhoff,* 1 F 3d 1523 (9th Cir. 1993), *cert. denied,* a federal Court of Appeals for the largest federal circuit applied a "criminal negligence" standard (which had only been applied before to violations of RCRA, *infra*). In 1988, the manager and assistant manager of the East Honolulu Community Services Sewage Treatment Plant instructed two employees to periodically dispose of waste-activated sludge by pumping it directly into the ocean. (The evidence at trial was that, over 14 months, 40 separate pumping events occurred, mainly at night.) When nearby beach users complained, regulators inspected the plant, and the managers repeatedly denied any problem. The federal trial court ruled that a "knowing" violation of the CWA required only that the managers were aware that they were discharging the pollutants, and not that they knew they were violating the terms of the CWA or the plant's NPDES permit. After a guilty verdict, the managers were sentenced to a federal pentitentiary for 21 and 33 months, respectively. Their conviction was upheld on appeal. The potential impact is that discharging *de minimis* of ordinary materials, which are prohibited by state and federal regulations under the Clean Water Act, can be prosecuted as felony violations. See *From* Matthews *to* Weitzenhoff: *How Can California Businesses "Steer a Course Out of Harms Way"?, Calif. Environ. Law Rep.* 13, 1994 (J. R. Arnold with L. A. Callaghan).

26. "Point source" is any natural or artificial conduit which directly discharges into waters of the United States. See 33 U.S.C. § 1362 (12), (14), (16).

27. 40 C.F.R. § 122.41 (1994).

28. The California "General Permit" is Water Quality Order 91-13 DWQ, as amended by 92-12-DWQ. The amended General Permit was issued on September 17, 1992. An applicant must:

   1. File a notice of intent with the State Water Resources Control Board.
   2. Prepare a storm water pollution prevention plan.
   3. Prepare a visual monitoring plan, including dry and wet season monitoring and an annual site inspection.

    4. Sample and analyze storm water discharges.

    5. Submit an annual report to the appropriate California Regional Water Quality Control Board.

    6. Eliminate non-storm-water discharges.

    7. Implement best management practices (including pretreatment, containment, waste minimization, etc., where necessary).

29. 42 U.S.C. § 7661 *et seq.;* 40 C.F.R. Part 70.

30. California, for instance, enacted its own comprehensive Clean Air Act in 1988. California Health and Safety Code § 39000 *et seq.* The California Air Resources Board, to the extent not preempted by federal law, sets ambient air quality standards and regulates fuels, aerosols, and indirect sources (by requiring land use controls such as commuter controls, etc.).

31. 1990 Clean Air Act Amendments, Publ. L. 101-549, 104 Stat. 2467 (1990).

32. 42 U.S.C. §§ 7661–7671; 40 C.F.R. Part 70 (1994).

33. Title V authorized all states to implement these federal operating permit programs, subject to federal review and veto. Title V is intended to specifically define requirements for each regulated "source" while maintaining a company's operational flexibility. The result should be a "one facility, one permit" program. The USEPA has identified three objectives for the Title V program, each of which is substantially different from past air quality regulation. First, USEPA is committed to "market-based principles", or the trading of emission set-offs. Second, USEPA seeks to implement "cross media coordination", to avoid the "media transfer" problem (e.g., groundwater contamination is pumped and treated so the contamination is emitted to the air). Third, USEPA will support companies and state agencies to reduce air pollution through schemes for avoidance, rather than "command and control".

34. 42 U.S.C. § 7412. Congress specified that 189 substances would be regulated as "hazardous air pollutants", including such common chemicals as chlorine, methanol, ethylene oxide, perchloroethylene, diesel exhaust, etc.; 33 U.S.C. § 7412 (r) (3).

35. Such plans describe the equipment which emits the toxics emissions, including stacks, fugitive emissions, and air cleaning devices and processes; they also describe how the business will measure or estimate the volumes emitted.

36. A risk assessment calculates, with detailed quantities, how toxics emitted from the plant are dispersed and how they impact the surrounding human population.

37. The federal environmental protection laws also contain provisions protecting employees from retaliation by their employers. See Section 7001 of RCRA, 42 U.S.C. § 6971, Section 507 of the CWA, 33 U.S.C. § 1367, and Section 322 of the CAA, 42 U.S.C. § 7622.

38. See *U.S. v. Adamo Wrecking Co.,* 545 F.2d 1 (6th Cir. 1975), *rev'd on other grounds,* 434 U.S. 275 (1978). The Title V program is expected to have the most effect in states with rudimentary "nuisance and odor" based permit systems. In industrial states with elaborate operating permit programs, such as the one administered by the South Coast Air Quality Management District in the Los Angeles air basin, the federal operating permit will supplement existing permits. However, the most dramatic change will be enforcement by USEPA and the U.S. Department of Justice (and district attorneys when "cross-deputized" as assistant U.S. attorneys). The CAA 90 authorizes USEPA to veto any air emission permit issued by a state and to issue, notify, or revoke many others.

39. 42 U.S.C. § 6901 *et seq.*

40. Some states' statutory schemes were the pattern for RCRA and the other major federal programs. Today, the federal government seems to be trending toward more selectivity in what programs it adopts from the states.

41. California's Hazardous Wastes Control Law (HWCL), California Health and Safety Code § 25100 *et seq.,* is such a comprehensive program. However, unlike RCRA, the HWCL does not exempt small quantity generators, farmers who generate pesticide wastes, etc.

42. See 40 C.F.R. Parts 260, 261.

43. Many wastes are listed as hazardous. See 40 C.F.R. Part 261. Other wastes will be considered hazardous if they exhibit hazardous characteristics, including ignitability, corrosivity, reactivity, or toxicity.

44. Hazardous wastes must be contained, with proper labelling, and cannot be stored on site without an EPA permit for longer than 90 days. 40 C.F.R. Part 264.

45. "Knowing endangerment" violations involve violations that place another person in imminent danger of death or serious bodily injury. "Knowing" violations include omitting information or providing false information in reports or filings.

46. 42 U.S.C. §§ 9601–9675.

47. Safety managers should also be familiar with hazardous communication rules, or "worker right-to-know" programs. These programs require companies to inventory and label hazardous chemicals in the workplace and to inform and train workers about the chemical hazards they may encounter on the job. These programs generally are "performance oriented"; namely, communication and training goals are set, but it is the duty of the employer to meet the goals.

48. Acutely hazardous materials include such common chemicals as ammonia, tetraethyllead, aniline dyes, sulfuric acid, etc.

49. Managers should also be aware of unique state laws in this area. For example, California's Proposition 65, California Health and Safety Code § 25249.5 *et seq.,* was directly enacted by California voters through a referendum. It compels businesses to warn people (including employees) before exposing them to chemicals included in a state list. If businesses do not provide "clear and reasonable warnings", they must prove that the exposure posed "no significant risk".

50. 42 U.S.C. § 12101 *et seq.*

51. 42 U.S.C. § 12111.

52. For instance, employers may not ask, on an application form or during an interview, whether and to what extent an individual is disabled. The employer can only describe a job and ask the applicant if he or she can perform the essential elements of the job. If the applicant's disability is obvious, or is known to the employer, the employer may ask the applicant to describe or demonstrate how (with or without "reasonable accomodation") the applicant can do the job. Also, the ADA prohibits preemployment physical screening until an offer has been extended. If a disability is discovered, the employer cannot withdraw its offer until it meets the "reasonable accommodation" standard (i.e., a modification to the job that allows the disabled applicant to perform the essential job functions).

53. Executive Order No. 12898, issued Feb. 11, 1994, 59 Fed. Reg. 7629, directs federal agencies to collect, analyze, and assess data on environmental health risks in disadvantaged communities. The order is beginning to have a substantial impact on the USEPA's regulatory programs for clean water, clean air, community right to know, and hazardous wastes control — as well as the exercise of its prosecutorial discretion. It has resulted in lawsuits against expansion of landfills and incinerator projects.

54. One of the most difficult issues for government is funding for safety and environmental compliance programs. The political strength of imposing more and higher levels of penalties (particularly when the public is outraged by a spill or release) is simply not matched by a commitment to fund the resulting regulatory schemes. The growing popularity of "private attorney general" or "citizen" lawsuits exacerbates the mismatch and raises the level of uncertainty as to compliance efforts.

55. And including the civil rights laws, e.g., the Americans with Disabilities Act, *supra.*

56. For instance, some companies review MSDS sheets which they receive from their suppliers and reformulate them into a standard format, both for their workers' comprehension in worker right-to-know programs and for meeting requirements for reporting to regulatory agnecies.

57. S.B. 94-139, Colo. Rev. Stat. § 13-25-126.5 *et seq.;* P.L. No. 16, Ind. Stat. Ann. 13-10-3; Ky. Rev. Stat. Ann. 224.01-040; Or. Rev. Stat. § 468.963 (1). Also, on January 24, 1995 Illinois enacted a similar audit privilege (SB 1724).

58. These include such protected communications as those between spouses, doctors and patients, clergymen and penitents, domestic violence victims and couselors, etc. See Calif. Evid. Code, Ch. 4.

59. See, for example, *Olen Properties Corp. v. Sheldahl Inc.,* No. 91-6446-WDK-JWM, 38 ERC 1887, 1994 U.S. Dist. Lexis 71725 (C.D. Calif., April 12, 1994). A supervisor testified in an affidavit that he prepared environmental audit memoranda to gather information for company in-house attorneys to assist the company in evaluating compliance with environmental laws.

60. The attorney work product doctrine can be overcome if the opposing party demonstrates substantial need and undue hardship and expense in obtaining the same information from other sources.

61. This privilege was originally recognized in *Bredice v. Doctor's Hospital, Inc.,* 50 F.R.D. 249 (D.D.C. 1970), *aff'd without opin.,* 479 F.2d 920 (D.C. Cir. 1973), where doctors' peer review records were protected from disclosure to survivors of a deceased patient in a medical malpractice case. The court held that such peer review was necessary to improve health care and would not be conducted if discussions were discoverable. In *Reichhold Chemicals, Inc. v. Textron, Inc.,* No. 92-30393-RV, 1994 U.S. Dist. Lexis 13806 (N.D. Fla., Sept. 20, 1994), documents prepared by a company during its investigation of groundwater contamination are immune from discovery because of the public interest in companies candidly assessing their compliance with past regulations.

62. See *U.S. v. MacDonald & Watson Waste Oil Co.,* No. CR-88-0032 (D.R.I., 1990), *aff'd in part, vacated in part,* 933 F.2d 35 (1st Cir. 1991) (internal environmental audits can be used to prove that violations were "knowing"); *U.S. v. Hammermill Paper, Inc.,* 796 F. Supp. 1474 (S.D. Ala. 1992) (OSHA subpoena of voluntary safety audits upheld, against a claim of self-evaluative privilege).

63. *Environmental Auditing Policy Statement,* 51 F.R. 25004 (July 9, 1986). USEPA's view of an effective environmental compliance auditing program contains seven elements: (1) top management support and commitment to follow-up; (2) independence; (3) adequate staffing and training; (4) explicit objectives, work, commitment of resources, and regular audits; (5) sufficiency of the process; (6) candid and clear written reports of findings, corrective actions, and schedules for implementation; and (7) quality assurance procedures, training, and accountability.

64. *Factors in Decisions on Criminal Prosecutions for Environmental Violations in the Context of Significant Voluntary Compliance or Disclosure Efforts by the Violator,* July 1, 1991. In fact, some of U.S. Department of Justice "factors", such as immediately turning over to the government information about violations and internal investigatory reports, as well as disciplining and surrendering the names of all involved individuals to the prosecuting agency will be regarded by companies as an unwarranted intrusion into corporate management.

# 19 AN OVERVIEW OF PRODUCT LIABILITY

**Kenneth M. Colonna and Steven M. McConnell**

## TABLE OF CONTENTS

## FOREWORD

The writings in this chapter do not focus on the legal aspects of product liability. Discussions of the legal mechanics of product liability are reserved for texts which focus on business law and are beyond the expertise of the authors.

This chapter does present the basis for product risk assessment within the framework of a product safety management program. It is intended to provide the readers guidance for assessing and reducing risk in the design, manufacture, and distribution of their company's products.

1-56670-054-X/96/$0.00+$.50

## CASE STUDY AND INTRODUCTION

The following is a typical case study of product liability. Judy worked for your customer, a large food-packaging company, for over 16 years. In all her time on the job she always worked on the can-packaging line and was an experienced operator who knew each piece of equipment on the line. Her normal job at the time of the accident was label operator. In her capacity as label operator, she ran the packaging equipment that applied labels to finished sealed containers of her company's product. Traditional practices dictated that Judy and her co-workers rotate to each piece of equipment during a shift for relief of the normal operator during breaks and as a means of training. At the time of her accident, Judy was operating the can-filling machine. The can filler, a large rotating piece of equipment capable of filling 10-ounce cans of food products at a rate of 1100 per minute, is not difficult to operate but requires certain skills acquired only from experience. It was common for operators to make minute adjustments to the filler during the day. This practice is known as "fine tuning" the run. At the time of her accident Judy was fine tuning the filler, which had become necessary because cans were jamming in the filler as they moved off the rotating machine and onto the conveyor for sealing. As was common at this plant, operators would often reach into the area of the machine where filled cans discharged in order to remove jams or product that had spilled from the filler and accumulated in this area. This process was typically done while the machine was running. To accomplish this task, the operator must reach over a nylon gear approximately 1 foot in diameter that acts as a guide for the cans as they discharge from the machine. Judy had reached across the gear to remove a can that had jammed in the machine. She was wearing a long-sleeved uniform shirt with the cuffs unbuttoned. As she reached into the machine, her shirt caught on the nylon wheel, pulling her arm and eventually her upper torso into the machine. As a result of her accident, Judy's left arm was amputated and she suffered permanent brain damage as a result of oxygen deprivation that occurred when her chest was compressed by the machine, rendering her unable to breathe for several minutes.

Judy's employer paid her the workers' compensation benefits allowable under the state statutes where she lived. The settlement for permanent total disability, however, was barely enough to cover all the expenses a single mother of two children incurs, especially since Judy requires assistance in managing many everyday tasks such as bank accounts, bills, and normal expenses that are more difficult for her to understand today.

Judy's attorney has decided to pursue a product liability suit against your employer. The basis of the suit revolves around your company's design of the equipment and the machine guards. Her attorney alleges that your employer built a machine with known hazards and failed to follow generally accepted industry practices in designing machine guards for the machine. Additionally, the attorney alleges that, since your company employed professional engineers for design purposes, your company's engineers violated their professional duty by failing to exercise due care relative to the design of the machine.

The suit asks for damages in the amount of $5 million. It names both your employer and individual officers as defendants. You have been requested to aid in the investigation and in preparation of a defense. You will be working with your company's legal counsel and the company's insurance carrier in the investigation.

While this scenario is a compilation of different events known to the authors, it reflects the very real exposure all companies face in developing and selling products or services. As a safety professional you have an obligation to your employer and your profession to participate in the assessment of your company's products and services. In today's evolving workplace, where downsizing is commonplace, the safety professional must assume a broader role in all phases of risk assessment. Where should the safety professional begin?

Product safety management is viewed differently from company to company. The level of importance that management gives to product safety is dependent on the type of products or services provided and their related hazards. Therefore, while some companies pay relatively little attention to product safety, others implement a comprehensive product safety management (PSM) program.

Customer misuse of products is difficult to eliminate completely. The manufacturer can, however, take certain steps to minimize the likelihood of liability suits. The manufacturer's best opportunity to prevent claims is to produce and sell products (including instructions when necessary) that are reasonably safe. The product safety management concept can help minimize a company's exposure to product liability because safety is incorporated into the design, manufacturing, and marketing phases of the product.

A typical product safety management program consists of people from a variety of departments, including design engineering, manufacturing, marketing, and distribution. Other team members represent quality assurance, purchasing, legal, insurance, human resources, safety, and the service organization. Each department is important to the overall success of the program and, therefore, to the overall success of the company.

It is important for management to demonstrate support for the program through commitments made in strategic or tactical plans, policy statements, and other objectives of the company that support the company's mission statement. Management must set the goals and objectives related to product safety and liability and designate overall responsibility for the program to an individual. The safety professional is often delegated the responsibility in small- to medium-size companies. In larger companies, and when multiple product lines are involved, the assignment may be given to someone in engineering. In some cases the company hires a product safety director. Whatever the organizational structure, the safety professional will play an important role in the successful program. The safety professional's broad base of knowledge and experience will allow him/her to offer valuable insight across departmental lines. The safety professional's understanding of applicable regulatory requirements will help shape the control measures implemented by the company to minimize product liability. The safety professional can and should be involved in each phase of a product's life.

## PRODUCT DEVELOPMENT

From the moment an idea is brought forth within your company, the safety professional plays a crucial role in the development of the new product or service. Usually, the process begins at the conceptual stage where discussions of products or services under consideration take place. The safety professional in any organization should take an active role in these discussions, offering informed input into the product or service being developed. Regulations that could offer barriers to the development of the product or service should be fully explored. Too often companies overlook the significance of safety and environmental regulations that may affect their goods or services. These regulations should be considered both from the impact they will likely have on your company and from the impact they could have on your customers. A product is not worth developing if its use exposes your customers to increased risk of regulatory intervention. This is especially true in the professional services industries where certification or licensing requirements may exist. Often these requirements are not uniform between states and could create problems for your company or your customer.

For example, in many states businesses confronted with regulations affecting underground storage of petroleum products may use licensed geologists to conduct assessments of possible leaks. While this is appropriate in most states, some states, such as North Carolina, require that certain assessment documents be signed and stamped by a professional engineer, a discipline that may not be employed by many consulting firms. A company that offers consultation services to a customer in one state may find their efforts and their customer's resources wasted in another. While this may not result in a product liability lawsuit, it does threaten your company's standing within the industry, where reputation is as important as the service provided.

## MATERIALS AND MANUFACTURING

Once an idea has passed from the drawing board to the prototype stage it becomes important for the safety professional to be involved in the process. Are the materials selected for use in manufacturing the product suitable? Has the engineering group considered possible uses of the product other than that intended by your company? Have assessments of possible failure modes and their effects been performed?

Materials selected for use in your product are critical to your company. While the engineering department is more suited to materials selection, the safety professional can offer valuable input. Regulations such as hazard communication, process safety, confined space entry, etc., should all be considered in the process of selecting the materials for construction. One large manufacturing company recently embarked on a multi-million-dollar project to install pollution control technology only to discover that a material used in the pollution control process, ammonium hydroxide, would require the company to implement process safety management techniques and install further pollution control technologies. This critical error went undiscovered by the professional engineers hired to design the process. The safety professionals employed by the company quickly recognized this problem early enough in the design process to allow for redesign, thereby avoiding a potentially costly error.

Materials selection alone is not the only important aspect of reducing risk for your company's product. Suitability for use is another. Here the safety professional can assist his/her company by employing proven techniques for evaluating the process or the product. Fault tree analysis, failure mode and effect analysis, and "what if" analysis are examples of hazard identification/analysis techniques available to the safety professional. While it is beyond the scope of this chapter to explore these techniques in detail, the safety professional can find ample information on these methods and their application in texts and professional journals.

## PRODUCTION

The manufacturing department is responsible for turning a reasonably safe and reliable design into a finished product. The safety professional can provide assistance in manufacturing by performing preliminary and specific hazard analysis, including the analysis of the process and materials used to complete fabrication of the product. The need for additional operating permits, i.e., air emissions permits, etc., must be evaluated well in advance of the production process. Similarly, employee training and educational needs should be assessed and implemented prior to the initial trial run of the product.

Working closely with the quality assurance staff, the safety professional can assist in the development of quality techniques and measurement tools that will help track the manufacturing process. While product rejects may be indications of failures or restraints in the manufacturing process, they may also help identify product design deficiencies and other product-related deficiencies.

## PRODUCT PACKAGING

While often overlooked in the process, product packaging is an important part of risk assessment for the safety professional. Errors in product packaging can lead to accidents and injuries on the user's end and can trigger product liability claims.

The safety professional can help determine if product packaging is acceptable. Will the packaging prevent damage to the product during shipment? Will the packaging prevent product contamination, corrosion, and deterioration during shipment and while on a retail outlet's shelf? The adequacy of the packaging should be assessed for suitability through the entire distribution channel. This assessment should consider conditions during manufacturing, transit, warehousing, and display.

The manner in which a product is presented through labeling and graphics can be interpreted by the end user as representation for use or suitability for a particular purpose not intended by your employer. Although labeling and graphics decisions are made by the marketing and product development groups and most often undergo legal scrutiny, the safety professional should also review the material and provide insights about the product packaging. It helps avoid embarrassment later if the safety professional discovers that the graphics on the product contain items that are clear violations of safety and health regulations.

A producer of industrial training films recently offered the authors an opportunity to preview a new safety video training program that was in the final stages of development. The production company was quite proud of the new product and its marketing materials as well as their ability to produce it ahead of their competitors, a clear marketing advantage. They were somewhat embarrassed to learn that after a great deal of time, effort, and editing a critical segment of the video showed actors engaged in activities that were correct and in compliance with the regulation the topic addressed, but clearly in violation of other long-standing safety regulations. While this example relates to a product-related problem, it also illustrates the need for the safety professional to evaluate product packaging. The use of clip art as a graphic in packaging promotion, for example, could communicate a message not intended by the producer of the product.

## OPERATING MANUALS AND OTHER MATERIALS

If your company produces a product that includes or requires operating instructions, maintenance manuals, or related training materials, you as the safety professional should be engaged in the review and development of these documents. Much like the packaging and labeling issues, manuals and other materials developed by your company should be reviewed to identify instructions or directions that fail to warn of possible hazards or that may mislead the user of your product to engage in activities that put him/her at risk.

Instruction or operating manuals related to the product should repeat hazard warnings and address how those hazards can be reduced or avoided by the user. Instructions for product inspection, assembly, and maintenance should be addressed. Any potential hazards associated with troubleshooting should be indicated, and preventive maintenance guidelines should include a discussion of pertinent safety issues. Although the safety professional is usually not responsible for drafting these documents, a concise review could prevent unnecessary and avoidable problems after the product is sold.

Many courts have held that a manufacturer has a duty to warn customers about any reasonably foreseeable use of a product beyond the purposes for which it was designed. An adequate warning must advise the user of the hazards involved in the product's use (and in some cases, possible misuse), avoidance of these hazards, and the potential consequences of failure to abide by those warnings.

## MARKETING AND SALES

While addressed during the early stages of a product or service's life cycle, understanding the market for your company's products or services is necessary for success in risk assessment. Knowing who the target market is, what channels of distribution will be used to get your product to the consumer, and whether the product will be pushed to the consumer or the marketing department will pull the consumer to the product will help the safety professional develop a more comprehensive assessment.

It is also important to understand your company's sales strategy. How will the sales force sell the product? What type of training does the sales force receive? Is there a clear understanding of how the product is intended to be used? Do sales personnel understand the product's limitations? What type of sales support material is provided? Will your company offer free trials? If your product will be sold to the retail consumer through a distribution channel of retail outlets, how have the retailers trained their sales force? Do the retailers understand the proper uses and limitations of your company's products? If your company is selling services, does the consumer or customer understand the scope of services your firm is capable of providing and what limitations on those services exist?

Product safety management includes a review of sales brochures and advertising materials. These materials should be evaluated by the product safety management team to ensure that the product's features and capabilities are accurately depicted and illustrate only safe operating and maintenance procedures. The company's legal counsel should be included in this review cycle.

## PRODUCT RECALL

Despite a company's best efforts, circumstances may arise that require a recall of all or a portion of the products it has sold. Other situations may require field modifications to products already in the hands of the end consumer. Here again, the safety professional may play a critical role. Does your company have a contingency plan to recall products? Is the chain of command during these crisis situations clear? Who will address the press during this period? What will be your company's approach in dealing with the public? A successful product safety management program will address these questions.

Companies initiate product recalls or field-related modification programs when a substantial performance or safety defect exists in one or more of its shipped products. Such programs may also be considered when a company either is made aware of or has developed a substantial product improvement or change.

Product recalls and modification programs can prove costly to a company — both in dollars and in public perception. Either effort requires a substantial commitment of resources. It should be recognized that the basis for implementing either product recall or modification can often be traced to inadequate product safety management efforts.

It is easy to understand why your company should develop a contingency plan for rapid implementation of product recalls or field modifications. Quick action can often minimize the costs and improve a company's public perception during times of adversity. McNiel-PPC, Inc., the maker of Tylenol®, is an excellent example of a manufacturer with well-planned and well-implemented strategy for product recall. However, the reader should recognize that a successful product safety management program strives to identify and eliminate the product hazards and misuses and, therefore, the need for most product recalls and field modification programs.

## CONCLUSION

Regardless of size, any company producing products or supplying services should assess its exposure to potential product liability issues and take proactive steps to ensure that a product safety management program of appropriate size and effort is in place to reduce or eliminate associated risks.

The interaction of representatives from many departments is essential to the success of a product safety management program; the safety professional plays a significant role through involvement in many program areas. The safety professional must utilize his/her expertise in the field of safety and health and participate assertively in the program.

As with a conventional safety and health program, senior management must actively demonstrate support for the product safety management program. Senior management must undertake the effort to ensure that the program is set up properly and is sufficiently funded. Finally, senior management must either establish or approve the goals and objectives associated with the program and actively communicate these goals and objectives to all employees.

## REFERENCES

Allison, W. *Profitable Risk Control*. Chicago, IL: American Society of Safety Engineers (ASSE), 1986.

Anderson, R., Fox, I., and Twomey, D. *Business Law, Principles, Cases, Legal Environment*. Cincinnati, OH: South-Western Publishing, 603–604, 1987.

Coughlin, G. *Your Handbook of Everyday Law*. New York: Harper Collins Publishers, 341–350, 406–407, 1993.

Jablonski, J. *Prosper through Environmental Leadership: Succeeding in Tough Times*. Albuquerque, NM: Technical Management Consortium, 55–62, 1994.

National Safety Council. *Accident Prevention Manual for Business and Industry: Administration and Programs,* 10th ed. Itasca, IL: National Safety Council, 1992.

Porter, M. *Competitive Strategy: Techniques for Analyzing Industries and Competitors*. New York: The Free Press, 3–52, 1980.

Schonberger, R. *World Class Manufacturing: The Lessons of Simplicity Applied*. New York: The Free Press, 3, 4, 7, 124–133, 201–203, 1986.

## FURTHER READING

### Books

Hammer, W. *Product Safety Management and Engineering*. Chicago, IL: American Society of Safety Engineers (ASSE), 1994.

National Safety Council. *Product Safety Management Guidelines*. Itasca, IL: NSC, 1989.

Petersen, D. *Techniques of Safety Management*. Goshen, NY: Aloray, 1989.

Slote, L. *Handbook of Occupational Safety and Health*. New York: Wiley-Interscience, 1987.

### Other Publications

Colling, D. Materials and product safety. *Prof. Saf.* 36(4): 17–19, 1991.

Kitzes, W. Safety management and the Consumer Product Safety Commission. *Prof. Saf.* 36(8): 25–30, 1991.

Nassif, G. Products liability and the Economic Community. *Prof. Saf.* 36(1): 21–23, 1991.

# Section VI
# Risk Management Aspects

# OVERVIEW OF RISK MANAGEMENT AND ITS APPLICATION TO SAFETY AND HEALTH

## 20

**Gerard C. Coletta**

## TABLE OF CONTENTS

## INTRODUCTION

Risk management is a term used to describe various operational and financial activities aimed at minimizing the impact of accidental or unplanned loss. As a process, risk management focuses on

- Identifying and monitoring an organization's exposures to possible loss; such exposures range from natural perils and manufacturing hazards to kidnapping/ransom and contractual omissions
- Assessing the possibility (the risk) that each identified exposure will result in losses and then quantifying the likely financial and operational impacts
- Controlling or reducing these risks through administrative, procedural, and engineering methods

1-56670-054-X/96/$0.00+$.50
© 1996 by CRC Press, Inc.

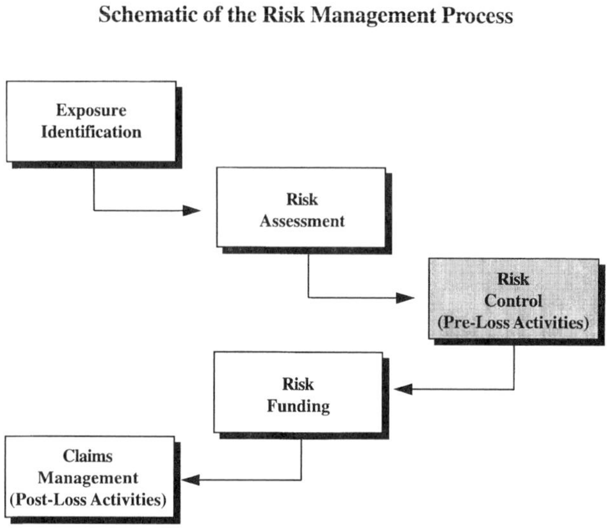

**Schematic of the Risk Management Process**

**Figure 20-1**  Schematic of the risk management process.

- Funding the losses that actually do occur by use of internal resources like self-insurance, by risk transfer through a commercial insurance program, or by other formal fianacing options
- Managing claims aggressively to contain and minimize the overall cost of incurred losses

Thes five steps are illustrated as a flowchart in Figure 20-1.

The risk management process is obviously much broader than the purchase of insurance and, subsequently, the processing of claims. Many businesses are focusing more and more on the third segment, risk control (i.e., employee and third-party safety, property protection, emergency preparedness, contract stipulations, etc.) as the principal method for reducing both the frequency and severity of losses over the long term.

Such a broad approach to risk management is not a new concept. It has, however, gained increasing attention and sophistication over the past few years. Restrictions on the availability and changeable prices of commercial insurance, a move toward higher deductibles, and heightened responsibility for risk control imposed upon senior managers by employees, by governmental regulators, and by society at large have all provided impetus for this change.

Many organizations, regardless of whether or not they fund losses through commercial insurance or private insurance programs like self-insurance or "captive" insurance companies, are finding the costs of many risks to be strategically unacceptable because of the potential for unexpected, substantive disruptions in operations.

The five technical segments of the risk management process should be developed equally as part of an integrated system. Each of the segments is discussed more fully in the following sections and, where appropriate, illustrative action items are presented. This discussion concludes with brief comments on overall program management. A focus on planning and communications is the best approach to ensuring that the program is integrated into usual day-to-day operations of a business.

## EXPOSURE IDENTIFICATION

Most organizations should be identifying exposures to possible loss on a regular, in-depth basis. This includes looking at single-incident exposures as well as multiple-incident exposures that could lead to accumulation of significant aggregate losses.

Key exposure categories are

- Property, equipment, and facilities
- Use of hazardous/toxic materials
- Restricted use of assets and resources
- Adherence to contractual/legal obligations
- Health, safety, and availability of employees
- Synergy or integration of individual operations
- Geographical and business competitiveness
- Errors and omissions

Each of these categories should be cross-matched against various perils to identify whether an exposure actually exists for a particular business. This cross-matching, illustrated by Figure 20-2, is subjective and is not intended to quantify the risks. Rather, it is intended to "flag" categories that should be explored more fully.

The process is best accomplished in multidiscipline discussions. For example, in reviewing exposures to employees, representatives from product groups, maintenance, security, and risk management should all be involved. Initial discussions should be on as broad a basis as possible.

Although the exposure identification process can be conducted formally on an annual basis, it also can be structured to easily accommodate changes throughout the year. The risk manager would be an active participant in the addition of new business segments, major changes in operating procedures, planning a safety and health program, monitoring security practices, and so on.

## RISK ASSESSMENT

This step should actually be viewed as a continuation of exposure identification and represents the quantification of the identified exposures. It is intended to estimate the value of losses under various conditions and events deemed probable by management. Both financial and legal consequences

| EXPOSURE CATEGORY* | KEY PERILS | | | | | | | | |
|---|---|---|---|---|---|---|---|---|---|
| | A | B | C | D | E | F | G | H | I |
| Property, Equipment, and Facilities | X | X | | X | | | | | |
| Use of Assets and Resources | X | X | | X | | | | X | X |
| Adherence to Contractual/Legal Obligations | X | | X | | X | | | X | X |
| Health, Safety, and Availability of Employees | X | X | | | X | | X | | |
| Work with Hazardous/Toxic Materials | | | | | X | | | | |
| Synergy or Integration of Operations | X | X | | X | | | X | | |
| Geographical and Business Competitiveness | X | | | | | X | X | X | |
| Errors and Omissions | | | | | | | X | | X |

\* Categories can be subdivided into individual exposures for a more thorough analysis.

**Legend**

A=Natural Disaster (flood, earthquake, windstorm)
B=Fire and Explosion
C=Injury and Illness
D=Theft, Dishonesty, Vandalism
E=Negligence

F=Economic/Business Conditions
G=Project Design Deficiencies
H=Contractual Oversight
I=Litigation

**Figure 20-2** Exposure identification matrix.

should be considered. This dual approach is especially important when assessing compliance with government regulations like those promulgated by the Occupational Health and Safety Administration (OSHA) and the Environmental Protection Agnecy (EPA), with societal expectations like disaster planning, and with good business practices like reducing overall loss costs and process disruptions.

Figure 20-3 presents a second matrix that illustrates one method for assessing the order of magnitude of losses and, thereby, prioritizing risks whose impact should be considered in much more detail. The risk assessment process should be taken to enough depth to produce an accurate view of potential losses. This, then, provides the roadmap for both risk control activities and, later, risk funding options.

## RISK CONTROL

Risk control is in many ways the most important part of the risk management process. If done well, it permits a business to effectively anticipate and prevent many, if not most, substantive losses. Many organizations do not include risk control as an element in strategic planning. Such an exclusion represents a failure to recognize continuing growth of loss costs as well as tightening regulatory liabilities.

Risk control should focus on a number of issues:

- Employee safety and health
- Hazardous materials management
  - Environmental affairs
  - Storage and use practices
  - Industrial hygiene
  - Waste generation
- Property protection
  - Buildings and facilities
  - Equipment (including computers)
  - Product and supplies
  - Cargo
- Security systems
  - Buildings and facilities
  - Employees
  - Information and data
- Disaster planning
  - Emergency preparedness
  - Crisis management
  - Business recovery
- Contractual liability
  - Subcontractor oversight
  - Appropriateness of commitments

## Estimate of Possible Losses

| EXPOSURE CATEGORY** | FINANCIAL LOSS* | | | | | | NON-FINANCIAL LOSS* | | | |
|---|---|---|---|---|---|---|---|---|---|---|
| | to $1M | to $10M | to $25M | to $50M | to $100M | above $100M | Loss of Key People | Damaged Reputation | Legal Action | Loss of Market Share |
| Property, Equipment, and Facilities | | | | | X | | | | | X |
| Use of Assets and Resources | | | | | | X | | | | X |
| Adherence to Contractual/Legal Obligations | | | X | | | | X | X | X | |
| Health, Safety, and Availability of Employees | | | X | | | | X | X | X | |
| Work with Hazardous/Toxic Materials | | | X | | | | | X | X | X |
| Synergy or Integration of Operations | | | | X | | | X | | | X |
| Geographical and Business Competitiveness | | | | | | X | X | | | X |
| Errors and Omissions | | | | | X | | X | X | X | X |

\* Both single incidents and multiple incidents should be considered.

\*\* Categories can be subdivided into individual exposures for a more thorough analysis.

**Figure 20-3** Estimate of possible losses.

Employee safety and health provides an excellent model for describing benefits that can be derived from aggressive risk control. A safety and health program should be viewed as a function that addresses the interests of employees, officers, directors, and shareholders in preserving this segment of a company's assets. Such a program can be an important element in productivity and quality improvement efforts. But most of all, effective safety and health should be looked upon as a way to prevent bottom-line dollars from slipping away. For example, stress and repetitive motion claims are becoming a major drain on many U.S. businesses and now represent a significant portion of many workers' compensation bills.

A strong safety and health program should have four primary goals.

- To prevent accidents that result in injury or illness or in related damage to property and equipment
- To manage regulatory exposures and potential liabilities created by an organization's duty to exercise due diligence in worker safety and health
- To integrate safety and health into day-to-day activities and management responsibilities
- To cast "safe performance" as a strategy for efficient and productive operations

Many aspects of safety and health fit well as part of quality improvement efforts like those included within the ISO 9000 protocols and under Baldridge Award guidelines.

## RISK FUNDING

Risk funding naturally follows exposure identification, risk assessment, and risk control, all of which are preincident, proactive sets of activities. Risk funding represents a reactive activity; it sets up financing for losses after they have already occurred.

The most effective risk funding activities are based on a number of factors, not the least of which is an organization's posture in setting deductibles and self-assuming various levels of risk. The process includes use of commercial insurance, self-insurance, captive insurance companies and other (in some cases, creative) mechanisms to find those losses that do occur.

## CLAIMS MANAGEMENT

Claims management becomes important in those loss areas with high frequency and/or high severity rates. Again, employee safety and health via workers' compensation provides an illustrative model for this process. It is important to note that the model can be adjusted to fit most property and casualty claims — the model is not limited only to workers' compensation or only to insured claims.

Workers' compensation claim costs can be brought under control by pursuing five key activities:

1. **Diligent case management.** For those accidents or situations that do occur and become claims, effective case management requires thorough investigation, medical utilization review, realistic case reserving, and aggressive negotiation for proper settlements. These actions should be directed in a planned and consistent manner by qualified, experienced claims professionals.
2. **Monitoring employee lost time.** Successful control of indemnity costs requires a team approach to managing an injured or ill employee's lost time. The team approach, backed by specific target dates for return to work, is used to ensure that individual activities are coordinated. This means close cooperation and open communications between the injured employee, the treating physician, other medical care providers, and the claim technician.
3. **Reviewing medical treatment.** Workers' compensation should pay only for related and necessary medical treatment at reasonable fees. Every medical charge should be reviewed to confirm its relationship to the compensable injury, its reasonableness, and its necessity.
4. **Overseeing legal services.** Costs in workers' compensation are not limited to idemnity and medical issues. As workers' compensation claims have become more complex and their values have risen, fees for outside defense attorneys have also grown substantially. As with the other aspects of case management, case litigation requires methods and procedures aimed at controlling associated costs.
5. **Regular use of light duty.** The final element in a workers' compensation cost containment package is light duty. Regular use of a well-planned light duty program can generate meaningful savings by returning an employee to a productive position. This is the alternative to paying for stay-at-home, do-nothing time. Care must be taken to ensure that light duty is clearly defined and meaningful.

Consistency in managing claims can be highly effective. Reductions of 25% or more have been demonstrated by following four actions:

- Assigning clear responsibility for various aspects of the claims process, especially within a human resources function
- Fostering controls to avoid "double dipping", such as can occur with workers' compensation benefits and health benefits
- Supporting a comprehensive accident investigation process that is tied to the safety and health function
- Providing risk management monitoring for all activities relating to claims

## PROGRAM MANAGEMENT

Integration of all segments of the risk management process is important in building an effective, productive program with long-term stability. Centralized coordination and oversight are appropriate in many organizations.

# A Template for Risk Management Growth

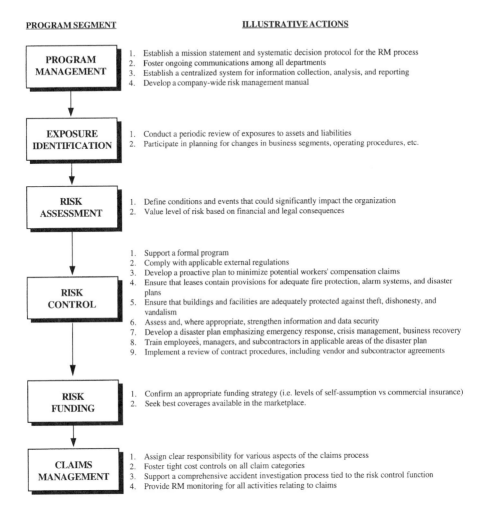

**PROGRAM SEGMENT**

**ILLUSTRATIVE ACTIONS**

**PROGRAM MANAGEMENT**

1. Establish a mission statement and systematic decision protocol for the RM process
2. Foster ongoing communications among all departments
3. Establish a centralized system for information collection, analysis, and reporting
4. Develop a company-wide risk management manual

**EXPOSURE IDENTIFICATION**

1. Conduct a periodic review of exposures to assets and liabilities
2. Participate in planning for changes in business segments, operating procedures, etc.

**RISK ASSESSMENT**

1. Define conditions and events that could significantly impact the organization
2. Value level of risk based on financial and legal consequences

**RISK CONTROL**

1. Support a formal program
2. Comply with applicable external regulations
3. Develop a proactive plan to minimize potential workers' compensation claims
4. Ensure that leases contain provisions for adequate fire protection, alarm systems, and disaster plans
5. Ensure that buildings and facilities are adequately protected against theft, dishonesty, and vandalism
6. Assess and, where appropriate, strengthen information and data security
7. Develop a disaster plan emphasizing emergency response, crisis management, business recovery
8. Train employees, managers, and subcontractors in applicable areas of the disaster plan
9. Implement a review of contract procedures, including vendor and subcontractor agreements

**RISK FUNDING**

1. Confirm an appropriate funding strategy (i.e. levels of self-assumption vs commercial insurance)
2. Seek best coverages available in the marketplace.

**CLAIMS MANAGEMENT**

1. Assign clear responsibility for various aspects of the claims process
2. Foster tight cost controls on all claim categories
3. Support a comprehensive accident investigation process tied to the risk control function
4. Provide RM monitoring for all activities relating to claims

**Figure 20-4**  A template for risk management (RM) growth.

For risk management to be effective, it must be structured to quickly address changes in both internal and external business environments. Most organizations benefit from risk management activities that in one way or another penetrate to every part of the business and motivate staff to actively manage the businesses' risks of doing business.

To stimulate a suitable level of involvement, several steps are necessary:

1. Convene a senior-level steering committee to oversee all aspects of the risk management process.

2. Establish a fucused mission statement supported by a systematic decision protocol for all strategic aspects of the process.
3. Foster ongoing communications among all departments that benefit from the risk management process.
4. Establish a centralized system for information collection, analysis, and reporting (especially to senior management).
5. Develop a company-wide risk management manual with specific chapters on the five segments:
   - Exposure identification
   - Risk assessment
   - Risk control
   - Risk funding
   - Claims management

These steps provide a solid foundation for an effective risk management program that successfully minimizes an organization's cost of risk. An overall template for the process is provided in Figure 20-4, along with a reiteration of the action items suggested throughout this chapter.

## FURTHER READINGS

### Books

Head, G. (Ed.) *Essentials of the Risk Management Process.* Malvern, PA, Insurance Institute of America, 1989.

Head, G. (Ed.) *Essentials of Risk Control,* Vol. 1 and 2. Malvern, PA, Insurance Institute of America, 1989.

### Other Publications

*The Risk Report,* a monthly publication of the International Risk Management Institute, Dallas, TX.

*Risk Management,* a monthly publication of the Risk and Insurance Management Society, New York.

# Section VII
# Management Aspects

# 21

# MODERN MANAGEMENT PRINCIPLES AND THEIR APPLICATION TO SAFETY AND HEALTH

**Richard W. Lack**

## TABLE OF CONTENTS

1-56670-054-X/96/$0.00+$.50
© 1996 by CRC Press, Inc.

## INTRODUCTION

It is beyond the scope of this chapter to present a complete in-depth review of the art and science of management. For this reason, the objective will be to provide the reader with more of an overview of the management process and its application in the safety and health setting.

The chapter will first deal with definitions and the basic functions of professional management. The following sections will address key points for special emphasis, problem solving, and setting safety goals, standards, and measurements before closing with some new and emerging trends.

The importance of "management" is emphasized by the fact that, of the 8 special interest groups among the 30,000-plus membership of the American Society of Safety Engineers (ASSE), the largest is the Management Division, with over 15,000 members. The next largest division is the Risk Management and Insurance Division, which is of course a management-related function.

Regulatory agencies are also increasingly emphasizing the need for effective management of the safety and health process. The federal Occupational Safety and Health Administration (OSHA) published their "Safety and Health Program Management Guidelines" in 1989. In 1991, the Health and Safety Executive in the United Kingdom published their excellent booklet, "Successful Health and Safety Management". This was followed a year later by their "Management of Health and Safety at Work Regulations 1992 Approved Code of Practice".

## DEFINITIONS

The following are some common definitions of terms used in the profession of management. The source of each is indicated in parentheses.

*Management*
A social process of getting things done effectively through people, organizing the elements of productive enterprise (money, materials, equipment, and people) in the interest of economic ends. The functions of management are a series of activities often described as planning, organizing, staffing, directing, coordinating, and evaluating. (American Society of Safety Engineers [ASSE])

*Planning Function*
The work a manager performs to identify and group the work to be done so it can be accomplished effectively by people. (Louis Allen)

*Leading Function*
The work a manager performs to cause people to take effective action. (Louis Allen)

*Controlling Function*
The work a manager does to assess and regulate work in progress and to assess results secured. (Louis Allen)

### Accountability Management

The obligation to perform responsibly and exercise authority in terms of established performance standards. (Louis Allen)

### Authority

The powers and rights of a person or a position. The more authority people can be given to make decisions concerning the work assigned to them, the more completely they can do that work without going to a higher level and the more interest and initiative they will tend to show in the work. (Louis Allen)

### Responsibility

Work which is a continuing obligation of a position. Responsibility is the physical and mental effort or work a person is obligated to perform in order to fulfill the requirements of the position. (Louis Allen)

### Line Relationship

The command relationship between those persons and positions directly accountable for accomplishing key objectives and therefore vested with the authority necessary to achieve those objectives. (Louis Allen)

### Staff Relationship

The relationship of those positions and components that provide advice and service to others which are accountable for direct accomplishment of key objectives. (Louis Allen)

### Line Function

Management activity directly related to the production operations of an organization. Persons performing line functions have the responsibility and authority for accomplishing the mission or objectives of the organization and are held accountable for its effective operation. (ASSE)

### Staff Function

An auxiliary service which operates in support of production or line management. Rather than performing auxiliary functions themselves, line management personnel depend upon the advice, guidance, or service of specialized staff departments while retaining responsibility, authority, and ultimate accountability for production success. (ASSE)

## THE WORK OF MANAGEMENT

According to Peter Drucker in his classic book on management, *The Practice of Management,* there are five basic operations in the work of the manager.

Step 1: The manager *sets objectives* and decides what has to be done to reach those objectives.

Step 2: The manager *organizes.* This is accomplished through analyzing the activities, decisions, and relations needed. These activities are then divided into manageable jobs which are grouped into a unit organization structure. People are then selected for the management of these units and for the jobs to be done.

Step 3: The manager *motivates and communicates.* Through constant communication and good leadership practices the manager builds a team that will enthusiastically and effectively accomplish their objectives.

Step 4: The manager performs *job measurement.* To accomplish this, the manager establishes performance standards which are agreed upon and accepted by the team members. Once standards have been established, the manager can analyze the performance, appraise it, interpret it, and communicate the findings to the team and individual members as well as to higher levels of management.

Step 5: In perhaps the most crucial step, the manager *develops people.* The effective manager is constantly seeking ways to help the team members develop their professional knowledge and skills by means of training, both on the job and through seminars, conferences, study courses, etc.

## THE FUNCTIONS OF PROFESSIONAL MANAGEMENT

A checklist of management functions based on the Louis Allen system is as follows:*

1. Management planning
   * Forecasting
   * Establishing objectives
   * Programming (resources and people)
   * Scheduling
   * Budgeting
   * Developing policies
   * Establishing procedures
2. Management organizing
   * Establishing organizational structure
   * Delegating
   * Establishing relationships (between units)
3. Management leading
   * Decision-making
   * Communicating
   * Motivating
   * Selecting people
   * Developing people
4. Management controlling
   * Establishing performance standards
   * Measuring/evaluating performance standards
   * Correcting performance deviations

* The management duties outlined above should be considered as a general guideline. In organizations which are managed with greater emphasis on employee participation in the management process, some of these duties may shift or receive greater emphasis.

For example, the duties involved with *planning* may consist more of telling what and less of explaining how. *Organizing* duties may concentrate more on providing the resources. *Leading* may be focussed more on developing people, while *controlling* may be more concerned with checking results. Those readers not actually practicing these functions are strongly encouraged to increase your knowledge by reading (see Further Reading at the end of this chapter), attending seminars, or enrolling in courses offered by local colleges and universities.

## KEY POINTS FOR SPECIAL EMPHASIS

In other chapters, and in earlier sections of this chapter, we have discussed safety and health organization, the functions of the safety and health professional, line and staff responsibilities, and the various elements of effective safety and health systems. The purpose of this section, therefore, is to highlight the distinctions between some important terms and to provide a central focus on priorities, since the theme of this entire book is the *essentials* of safety and health management.

I am indebted to Dr. John Grimaldi for drawing my attention to the distinction between *administration* and *management* of the safety and health process. Readers are referred to Dr. Grimaldi's book *Safety Management* (5th edition) for further information.

The issue does appear somewhat divided. For example, Louis Allen defines administration as the "total activity of a manager". The ASSE *Dictionary of Terms Used in the Safety Profession* defines management and administration similarly. Readers may reach their own conclusion, and for what it is worth, I share Dr. Grimaldi's opinion. In my view, *administration* of the safety and health process refers to the development of safety and health policy, procedures, and systems, their *assessment,* and their documentation.

Further to this discussion, I am also indebted to Dr. Michal Settles for her research on this subject, which is outlined in Chapter 24 of this book. Dr. Settles points out that an administrator is more focussed on his/her specialty area than a manager, who usually has a more "across the board" application.

Readers are also referred to Dr. Settle's chapter for further information on leadership skills. Stephen Covey, Warren Bennis, and others have conducted considerable research on the traits and competencies of managers and leaders. Managers tend to be more analytical problem solvers and leaders more visionary, inspiring, and risk taking. A final comment on "leadership" comes from Anita Roddick's book entitled *Body and Soul,* from which I quote in part her comments on this subject, which she calls "leading from behind". Ms. Roddick states,

> Leadership is not commandership — it is about managing the future. The principal forces which motivate a leader are an incredibly high need for personal achievement and a different vision of the world — one who marches to a different drum-beat, who does not see himself as part of the mainstream, is essentially an outsider. You don't necessarily have to be charismatic, you just have to believe in what you are doing so strongly that it becomes a reality....

Ending on those stirring words seems to be a fitting point to close this discussion and allow the reader to ponder these critical concepts.

*Management* of the safety and health systems refers to their *implementation* by management *action* utilizing basic professional management techniques (planning, leading, organizing, and controlling). In short, management is the control of resources with the objective of maximizing their value and contribution for the good of the enterprise.

The practice of effective safety management thus presents some unique challenges to its practitioners. In addition to their technical knowledge, they not only must develop administrative skills in organizing and assessing the system. They also must practice broad professional management skills so they are able to communicate, sell, and persuade others to accept and implement the process.

Another aspect that requires special emphasis is the matter of *risk*. The ASSE *Dictionary of Terms Used in the Safety Profession* defines risk as

> A measure of both the probability and the consequence of all hazards of an activity or condition. A subjective evaluation of relative failure potential. In insurance, a person or thing insured.

In Grimaldi and Simond's *Safety Management,* risk is defined as

> The assumed effect of an uncontrolled hazard, appraised in terms of probability it will happen, the maximum severity of any injuries or damages, and the public's sensitivity to the occurrence.

Fred Manuele has discussed this issue in several articles and in his book *On the Practice of Safety.*

The important point for safety and health professionals to keep in mind is that it is vitally important to help their management stay focussed on the identification and control of severe risks or, more accurately, high-risk hazards. This approach is incorporated in the process safety management approach which is now mandated by OSHA in the U.S. for chemical process and related industries. Nevertheless, other industries and organizations should adapt this concept in order to minimize their exposure to disastrous or very severe events.

A simplified risk evaluation and control process is as follows:

1. *Investigation* to identify risks. What risks do we have? An aggressive inquiry by a top management committee is conducted to determine high-risk hazards and their loss possibilities. This stage is not left until every likelihood of something going seriously wrong is investigated.
2. *Elimination* of very severe risks, if feasible
3. *Assurance* that all severe risks are controlled, preferably through engineering or, if not feasible, through appropriate management hazard control systems
4. *Correction* of uncontrolled severe risks
5. *Follow-up* periodically to assure maintenance of controls and to ascertain that no new severe risks have been introduced

Safety and health professionals should systematically work with their management to be sure that recognized severe risks are under control. For example, if your employer is in the retail or lodging business, do you have adequate security measures and systems to protect employees and the public from acts of criminal violence? How about the dangers of heavy mobile equipment, aircraft jet blast, hazardous atmospheres in confined spaces, excavations in unstable soils, natural gas- or propane-fired systems located where leaking gas might accumulate, crane operations ... the list could go on. Suffice it to say be sure you are aware of the most severe recognized risks in your workplaces and that systems are in place for their control.

## PROBLEM SOLVING AS IT RELATES TO SAFETY AND HEALTH MANAGEMENT

### Introduction

The word "problem" seems to be falling into disfavor in modern professional management parlance because of its negative connotations. Some readers may, therefore, choose to substitute another, more positive or opportunistic word such as "situation", "issue", "question", or "challenge". For the purpose of this discussion, however, and to avoid any confusion, we will continue to use the term "problem".

Surveys conducted among persons in a management position, particularly those at the first line or team leader level, indicate that identifying and solving problems is a major part of their daily activities. Indeed "problem solving" is the essence of "prevention". Obviously, it is better to try and find the problems in a proactive sense rather than wait for the problems to be actualized and then have to react to them. Among all the qualities of effective professional management, problem solving and problem prevention are some of the most essential skills.

The following summary of the principal elements in this process is based on the teachings of Homer K. Lambie.

## Definition of a Problem

For the purpose of the safety and health aspects of professional management, a problem may be defined in the following manner.

*A problem is any action or condition, actual or potential, that may produce adversity in terms of loss, injury, or poor human relationship.*

## Types of Problems

All problems can be classified into one of the following categories:

- Engineering/environmental — physical/chemical/biotechnical
- Human element — behavioral
- Human element — mental/emotional

## Methods of Identifying/Recognizing Problems

Problems may be all around you, but you may be unaware of them until you deliberately and systematically take the time to search and discover them. The key is to continuously practice and develop your skills in problem recognition. Here are areas in which to begin the process:

1. *Deviations from Standards.* Please refer to the section in this chapter on safety standards and measurements. Standards may be related to safety engineering, safety systems, or safety performance standards. Using an audit approach will lead you to deviations in all these categories.
2. *Abnormal Situations (Human Behavior).* Finding this type of problem taxes your powers of observation. For example, if you observe a number of improper or incorrect actions by people following your unit's lock and tag procedure, you should anticipate that if this problem is not addressed the potential exists for a serious event to occur.
3. *Loss/Injury Experience — In-House and Others.* Another excellent way to identify problems is to review your organization's past experience. Scan your workplace injury/illness records (OSHA log in the U.S.). Also look at your accident investigation reports for accidents involving damage and/or a significant incident (near miss) for more clues. Another very helpful area to find problems is by looking at the experience of others. Regulatory agencies publish injury experience statistics and reports. National and trade associations publish reports. The National Safety Council (U.S.) also publishes statistics and reports. Finally, Your own professional "network" of fellow professionals is a great resource for accident experience information.

4. *Communication/Dialog with Employees.* How many times has it happened *after* an accident that someone comes forward and says, "We knew something like this would happen; we have reported this problem many times and nobody paid any attention."? To avoid "surprises" like this, it is vitally important for managers to cultivate a climate that encourages employees to communicate problems. You can do this a number of ways, but always remember that *perception* is half the battle. You must demonstrate by your actions that you want to hear about problems. Here are some things to do to build up positive communications:

- **Ask** people about problems.
- **Tell** them frequently that you want to hear about problems.
- **Act promptly** on problems when your people bring them to you.
- **Follow through** systematically on the resolution of all pending problems.
- **Provide recognition** for those who identify significant problems.

5. *Looking/Searching Intensively.* The old saying goes, "You can look and not find, but you will never find if you don't look!" This point has been discussed already; however, it is so important that it deserves a special category to ensure proper emphasis. Many of us become so wrapped up with our daily tasks that soon we are in danger of developing "blind spots" to problems — problems of a magnitude and so glaring that we almost stumble over them, and yet we still do not recognize them for what they really are. It takes discipline to systematically set aside time to walk around and concentrate on looking with your full mind to see if you can uncover problems. To aid your search, it is best to program your surveys, using the "ten point approach" as a guide. These ten points are

1. Work area
2. Portable tools
3. Machines
4. Equipment
5. Materials — health
6. Materials — traumatic
7. Electrical
8. Vehicle
9. Multicategory
   - Lockout-tagout
   - Confined space entry
   - Excavations
   - Ergonomics
10. Related
    - Medical health
    - Security
    - Fire protection
    - Emergency procedures

Refer also to the sections on hazard control and the safety manual in this chapter plus the chapters on health for more information on hazard sources and causes.

## Evaluation of Problems

As a problem becomes evident to you, the most important first step is to write it down. There is an old Chinese saying, "The finest memory is worse than faded ink." Cultivate the habit of carrying a notebook with you at all times so you can jot down problems as they flash into your mind. Many a serious accident has occurred where the potential problem had been previously recognized, but the manager concerned forgot to write it down. Homer Lambie always says, "The beginning of the solution of any problem is when it is defined in writing." He also cautions that a general problem is not solvable until it is dissected down to the specific problems that collectively produce the general situation.

The hardest step is Step 1 in the problem-solving process, which is defining the problem in simple and specific terms. To help with this process, go to Step 2, which is to define the source and/or cause of the problem.

- **Source** always relates to a physical *condition.*
- **Cause** relates to the human *action* or lack of action that produces the problem.

Once the source and/or cause are identified, go to Step 3, which is to take any immediate temporary control (ITC) measures that may be appropriate. Step 4 is to measure the problem by the severity of its ultimate potential to produce adversity. In hazard control, this is known as risk rating or risk assessment. There are numerous techniques for risk assessment, some of which are discussed in the section on hazard control in this chapter. A more technical system is also discussed in Chapter 9, "Process Safety Management".

A simple approach is to consider the criteria, frequency, severity, and probability. If any of these factors is high, some measure of control action is necessary.

## Controlling Problems

The first consideration in problem solution, which was already mentioned in the evaluation stage, is to always consider what ITC measures must be taken to avoid loss and/or injury. After this you must decide on a plan of action by identifying the number of steps and methods of controlling the problem. Considerations will include effort, money, and risk. The priority for control will be determined by the risk rating after temporary control measures are in place.

Processing any problem involves three options, known for short as CS or P.

- **Correct** the problem.
- **Schedule** the problem for correction.
- **Pass** the problem on to the person having the authority and/or the ability to solve it.

### Summary and Final Hints on Problem Defining and Solving

One of the key qualities of effective leaders is their skill at problem solving and problem prevention. There is nothing wrong with coming up with a general problem. In fact, it is often a good place to start. However, it is just the *start;* after this you must go on to dissect or expand the problem into its specific components and then solve them one by one.

In the realm of problems involving the human element, be sure you have the facts and concentrate on what the person is doing wrong or is likely to do wrong. This is necessary to be sure you are dealing with a realistic situation versus an inaccurate one. In other words, separate the facts from fantasy.

Convert problems of a mental or emotional nature to the underlying source(s) or cause(s). Look for what is really triggering the emotional response. Then convert the "I don't like...", "I feel...", etc., to the specific hazard(s) and solve them through engineering or training.

Another very important principle is to always give priority to high-risk issues. Remember the old saying, "Don't worry about ants when tigers are in your path."

To recap, then, here are the steps:

- Problem — must be specific
- Source/cause
- Risk value
- Priority
- ITC
- Control — CS or P
- Follow-up

## SETTING SAFETY GOALS, STANDARDS, AND MEASUREMENTS

The material in this section is based on research conducted by the author and contained in a presentation on this topic at the American Society of Safety Engineers Annual Professional Development Conference, Las Vegas, Nevada, June 1994.

With all the trends that have been discussed earlier in this chapter, it will be vital for safety and health professionals to continuously develop their skills in the long-range strategic planning process and, in particular, goal setting, standards, and measurements.

First, here are some definitions for the terms used in this process and examples of their application.

### Definitions

*Ethics* — A code of moral principles that guide people in the organization. Test questions:

- Is it legal?
- Is it balanced? Is it fair?
- Does it promote win-win relationships?
- How would it make us feel if it were published in the newspapers or known in the community?

*Goal/Objective* — The end result toward which effort is directed: a desired outcome.

*Mission* — A clear, definite, and motivational point of focus — an achievable goal; a finish line to work toward. Mission is purpose-based.

*Mission Statement* — States what business the organization is in and may include ranking. It provides the overall direction and scope of the organization. Key words are accomplishments, contributions, values, and principles.

*Purpose* — The broad ongoing aim or intention of an organization. This may have two components:

- External — what the people in the organization want to do
- Internal — what the people in the organization want it to be

*Strategic Plan* — The process that must be completed to achieve a goal. This includes assignment of responsibility and completion dates.

*Values* — The core qualities that people in the organization consider most important and essential for the success of their mission. Values form the basis for beliefs.

*Vision* — Seeing a desired state as though it exists today; seeing the potential in or necessity of opportunities right in front of you. Visualization is a way of seeing the "what might be done". Vision is values-based.

## Definition Examples

*Purpose — External* — *Giro Sport Design:* "Design of great products that change the face of the industry, improve safety, and make people happy."

*Purpose — Internal* — *Circuit Technology Group, Hewlett-Packard:* "... To continually work towards building a work environment that helps employees attain a happy meaningful existence — to attain self-actualization."

*Mission Statement* — The mission of the American Society of Safety Engineers is to foster the technical, scientific, managerial, and ethical knowledge, skills, and competency of safety, health, and environmental professionals for the protection of people, property, and the environment and to enhance the status and promote the advancement of the safety profession.

The American Society for Industrial Security will establish, develop, and promote excellence in the security profession by assuring

- High-quality educational programs
- Responsiveness to members' needs
- Standards for professional and ethical conduct
- A forum for the debate and exchange of ideas
- Promotion of the organization and the profession
- Strategic alliances with related organizations

*Values* — Examples are family, health, career, religion, honesty, integrity, credibility, justice, liberty, freedom, and respect for the rule of law.

*Vision* — The American Society for Industrial Security will be the foremost organization advancing the security profession worldwide.

### SAFETY SYSTEM STANDARD — Example
### Title: Lockout-Tagout

Key elements
- Written procedure
- Standardized devices
- Step-by-step procedure
- Group lockout-tagout
- Contractors

Training requirements
- All affected employees trained
- Authorized employees receive specific training
- Entry supervisors trained

Management controls/documentation
- Documentation of energy sources and isolation points for each piece of equipment
- Training records
- Annual audits

### SAFETY PERFORMANCE STANDARD — Example
### Title: Safety Inspections

Key elements
- Written procedure
- Inspection responsibility areas assigned

Minimum performance standard
- Each supervisor inspects assigned areas each month
- Reports findings and action

Management controls/documentation
- Immediate supervisor evaluates and rates
- Immediate supervisor trains as needed
- Upper management conducts periodic audits

## Goal Setting

There are many different techniques for defining goals. A simple system that works well is the so-called "SMART" technique. The criteria for an effective goal using this system are that each goal must be

| | |
|---|---|
| Specific — | Generalities mean zero! Dissect a general goal into its *specific* elements which must be the *exact outcomes* to be a accomplished. |
| Measurable — | The goal must be measurable by quantity and by time (how much and by when). |
| Achievable — | The goal must be believable and conceivable. (Is the goal within your group's reach?) |

Realistic — The goal must be results oriented, practical, and controllable. (Can it be accomplished in the "real world?")

Time Bounded — There must be a target or deadline for accomplishment of the goal.

The following are some other tips on goal setting based on the teachings of Brian Tracy:

- There must be a strong degree of *desire* to accomplish the goal. The opposite of desire is fear, which is a powerful demotivator. Dwelling on desire will help override the natural inertia which prevents people from getting started.
- The people must *believe* it is possible to achieve the goal.
- It is said that a wish is a goal without any energy in it. *Writing* the goal intensifies the desire and strengthens resolve to achieve the goal.
- List all the *benefits* of achieving the goal. The more "reasons why" that are identified, the more people will be motivated to achieve the goal.
- Be sure to define the *starting point* because this gives a baseline from which to measure progress.
- It is said there are no unrealistic goals, merely unrealistic deadlines. Set *deadlines* on all tangible (measurable) goals. Do not worry if the deadline is missed; set another one.
- *Planning backward* is a powerful exercise; start from the goal and work backward, identifying the steps involved.
- Identify *obstacles* that stand in the way of the goal. Focus attention on removing the biggest major obstacle first.
- Identify *limiting steps*. Is there some additional knowledge or information needed to achieve the goal? Can the training be done in-house or must it be outsourced?
- Make a list of all the people needed to *help* achieve the goal.
- Having done all of the above, next make a *plan* and write it out in detail. List the actions by time and priority.
- As Stephen Covey says, "Begin with the end in mind."
- The next step is to organize the list of all the things that will have to be done to accomplish the goal — Brian Tracy calls this project management. Arrange these things to be done either in *sequential* order (dependent on each other being completed) or as *parallel* activities that can be worked on at the same time.
- Next, *identify the team* needed to accomplish the goal and *delegate* to them the various parts of the project.
- Continually monitor and measure to ensure that every part of the project is on track.
- Encourage all involved to *visualize* the goal so their subconscious minds go to work to program them toward the goal — remember the saying, "What you see is what you get!"

- Final words of wisdom: *never, never give up!* Encourage the group to discipline themselves to systematically keep doing something that moves them toward the goal. Nothing succeeds like success, and this helps to keep everyone positive and motivated.

## Safety Standards

Safety standards may be divided into three broad categories. First are *safety engineering standards.* These standards relate to the physical environment of the people. Standards of this type include those established by the various regulatory agencies. Many others are established by national organizations such as the American National Standards Institute, engineering societies, trade associations, etc. Other standards are found in industrial hygiene standards for employee exposure, environmental standards, fire, security, and building codes, and the list goes on and on.

The second category includes *system safety standards.* System safety standards are those safety systems established by organizations as necessary for the implementation of an effective safety plan. Examples might be safety inspections, hazard/risk assessment and control, job hazard analysis, or accident investigation. Each of these standards must outline the procedures for implementation, training requirements, documentation, and management controls.

The third category of safety standards is the *safety performance standard.* A performance standard tells the performers what must happen for them to accomplish the goal successfully.

Taking safety inspections as an example to illustrate the practical application of this process, the goal might be to establish a process that will ensure that all workplaces are systematically inspected and maintained to established safety standards and procedures. The applicable engineering standards such as OSHA regulations would be identified and may be encapsulated into simple, easy-to-use checklists.

Next a safety system standard must be developed that will provide a guideline for how this system will be implemented. Elements that this procedure must address will be assignment of responsibility areas, a self-survey system, and a follow-up control system to ensure that the inspection findings are corrected in a timely fashion.

Finally, an integral part of this system will be safety performance standards. These standards must describe who does what, when. Figure 21-1 illustrates typical performance standards for each level in the basic safety action systems.

In the example we are discussing, the standard will state who inspects and at what frequency. Also, how do the inspectors report and follow through on their findings?

**SAFETY PERFORMANCE STANDARDS**

| MANAGEMENT LEVEL | RESPONSIBILITIES | | METHODS BY WHICH ACCOMPLISHED | |
|---|---|---|---|---|
| DEPARTMENT MANAGER | 1. | Provide policy procedures and ensure they are reviewed and accepted by all supervisors. | 1. | Delegate to dept. safety coord. the responsibility to assist with administration of dept. safety program (Reports-problems-objectives-audits-controls). |
| | 2. | Provide necessary dept. procedures defining who does what to ensure unit policy is implemented in the dept. | 2. | Conduct dept. meeting to discuss problems and review progress. |
| | 3. | Set up management controls to ensure the programs are effective. | 3. | Random inspections to audit effectiveness |
| | 4. | Sample the 2nd line's work and provide advanced training and guidance as needed. | | |
| | 5. | Identify dept.-wide problems and set up objectives for continued improvement of safety results. | | |
| SECOND LINE SUPERVISOR | 1. | Provide training in the basic safety action programs. | 1. | Review HC logs weekly. |
| | 2. | Evaluate safety activity and provide necessary training and guidance. | 2. | Evaluate all supervisors safety activity reports and measure progress. |
| | 3. | Assist supervisors in solving problems beyond their ability or authority to control. | 3. | Review all accident reports and investigate all those considered significant. |
| | 4. | Set up management controls to ensure the supervisors safety action programs are effective. | 4. | Conduct spot-audits of self-surveys and HC notebooks. |
| | 5. | Identify problems with the safety action programs and determine methods for their solution. | 5. | Conduct spot-audits of "critical" procedures (lock and tag, eye protection). |
| | | | 6. | Inspect area for major housekeeping and hazard problems. |
| | | | 7. | Random audit of supervisor safety meetings. |
| | | | 8. | Random audit JHA/JSA compliance. |
| | | | 9. | Plan and schedule area JHA/JSA manuals program. |
| FIRST LINE SUPERVISOR OR TEAM LEADER | 1. | Systematically look for problems (hazards), identify,risk rate and process their control or correction. | 1. | Use of HC notebook and HC log. |
| | 2. | Guide and assist the employee through encouraging joint discussion to solve mutual problems. | 2. | Monthly safety self-survey. |
| | | | 3. | Monthly JHA/JSA meeting. |
| | 3. | Provide repetition and emphasis on recognized high-risk hazards. | 4. | Monthly crew safety meeting. |
| | | | 5. | Accident investigation. |

EMPLOYEE                          Recognize, understand, and control job
                                   HAZARDS (Unsafe acts and conditions)

**Figure 21-1**   This diagram illustrates the typical safety performance standards in the basic safety action systems for various levels in the organization.

The training requirements must be identified for each level involved in that system. Finally, the standard must define what management does to monitor the activity and results.

## Measurements

The final phase of the goal-setting process is measurement. The Deming total quality management (TQM) philosophy points out that measurement is a vital step in that it provides data that can serve as a basis for system improvement. Without measurement, how can we tell if the goal has been achieved?

Tom Peters says, "What gets measured gets done!" This is a succinct implication that without measurement our systems will be "hit or miss" at best.

Using the safety inspections example mentioned above, if the standard requires every first-line supervisor or work team to do a self-inspection of their workplace each month, what are the chances that this will consistently happen in all areas unless there is measurement?

There are many techniques for measurement of safety systems. The principal ones involve measurement of system results such as injury experience, accident costs, and trends.

Measuring results is most commonly used, but those in the profession have long since recognized that this is a very crude measurement technique because it does not address the performance failures that produced these results.

Measurement of safety system activity can be on a quantitative or a qualitative basis. Using the safety inspection report example (Figure 21-2), the quantitative measurement would be the receipt of the report each month as evidence of system implementation. Hence, quantitative measurement is satisfied as 100% compliance with the standard. The quality of the report in terms of hazards identified and controlled, however, may be 100 or it may be zero! A qualitative measurement as shown on the form enables the inspector and/or a reviewer to better assess the effectiveness of their inspection.

Other measurement techniques involve going "beyond the paper" and evaluating what is actually happening in the workplace. This type of measurement "compares the paper to reality" in the workplace and is, of course, the *audit.*

In Figure 21-3 there is illustrated an example audit format for the self-audit of a typical safety management system. On page 3 of this audit, Paragraph B, Safety System Activity Questions 1 through 5 are aimed at assessing the effectiveness of the safety inspection system.

So far in measurement we have looked at the quantity, quality,and actual results of the system. How about the perception of the people implementing the system? Also, how about underlying issues, such as breakdown in human communications or such indirect factors as lack of knowledge and/or understanding due to inadequate training?

To assess these "root cause" aspects, a number of techniques have been tried. Two that are commonly utilized are attitude and perception surveys and interviews. These surveys can often provide valuable clues to weak links that may be lurking in your safety systems. I would caution, however, that these surveys can have a negative impact if they are not professionally administered and interpreted. These aspects of program measurement will be discussed in other chapters of this book.

## Safety System Performance Assessment

Figure 21-4 illustrates a comparison between the more traditional management control systems and a performance management approach as described by Daniels and Rosen in their book of the same title. This type of management system

**SAFETY INSPECTION**

| DEPARTMENT | INSPECTION RESPONSIBILITY AREA /EQUIPMENT ASSIGNED | INSPECTED BY | DATE |
| --- | --- | --- | --- |

| PROBLEM | | | CONTROL ACTION | | FOLLOW UP | |
| --- | --- | --- | --- | --- | --- | --- |
| Describe WHAT is wrong and where it is located | Give reasons WHY it exists (source/cause) | Priority Code | Describe actions taken or planned to control the problem and its source/cause | | Code | Name/Date |

| PROBLEM | Is the problem clearly described and located? Are the reasons why the problem exists identified? | 10 30 |
| --- | --- | --- |
| PRIORITY | Do the priorities reflect an understanding of the potential effects of the problems listed? | 10 |
| CONTROL | Did the action control the problems and the reasons for their existence? | 40 |
| FOLLOW-UP | Does the follow -up specify who will do what, when, to ensure action completed? | 10 |
| | TOTAL | 100 |

**REVIEWER'S COMMENTS**

Signature:                    Date:

**PROBLEM PRIORITY CODE**
**E** Emergency
**A** 1 week
**B** 1 Month
**C** 3 Months
**D** 6 Months
**FOLLOW-UP CODE**
**C** Corrected (Date)
**S** Scheduled (Date)
**P** Passed to (Name)

**DISTRIBUTION:** SAFETY   ORIGINATOR   DEPARTMENT

**Figure 21-2** Format for a safety inspection report with a qualitative evaluation.

# SELF AUDIT

SAFETY MANAGEMENT SYSTEM

Organization _____ Date _____

Branch, Division or Subsidiary _____

| QUESTION | ANSWER | | REMARKS |
|---|---|---|---|
| | YES | NO | |
| **Basic Safety Management System Audit** | | | |
| **A. Administration** | | | |
| 1. Are the unit safety objectives defined in writing? | | | |
| 2. Has one person been designated as safety and health coordinator? | | | |
| 3. Is there a unit executive safety committee or group, and do they meet on a regular basis? | | | |
| 4. Does the executive safety committee review all new or revised policies? | | | |
| 5. Is there a unit safety manual? | | | |
| 6. Is the manual available in each department? | | | |
| 7. Are supervisors familiar with the manual? | | | |
| 8. Is there a formal safety orientation program for new employees? | | | |
| 9. Have supervisors received formal training in the unit safety program responsibilities? | | | |

# SELF AUDIT

SAFETY MANAGEMENT SYSTEM

Organization _____ Date _____

Branch, Division or Subsidiary _____

| QUESTION | ANSWER | | REMARKS |
|---|---|---|---|
| | YES | NO | |
| 10. Is there a formal system which includes safety performance in salary reviews? | | | |
| ● Is the program in writing? | | | |
| ● Does each supervisor have objectives? | | | |
| ● Does the program ask for completion of: | | | |
| ➡ safety inspections? | | | |
| ➡ hazard control action? | | | |
| ➡ accident & incident investigation reports? | | | |
| ➡ job safe practices? | | | |
| ➡ safety meetings? | | | |
| ● Does each program have an established minimum program activity standard? | | | |

**Figure 21-3** Example format for self-audit of safety management systems.

# SELF AUDIT

### SAFETY MANAGEMENT SYSTEM

Organization _____  Date _____

Branch, Division or Subsidiary _____

| QUESTION | ANSWER | | REMARKS |
|---|---|---|---|
| | YES | NO | |
| **B. Safety System Activity** | | | |
| 1. Is there a written safety inspection program? | | | |
| 2. Is the unit divided into departmental safety inspection responsibility areas and these in turn divided into specific areas for each supervisor? | | | |
| 3. Are there regular safety inspections? | | | |
| 4. Are safety inspection reports written? | | | |
| 5. Is there a follow-up and control system for identified safety problems? | | | |
| 6. Is there a written hazard control system with a standard for supervisor action? | | | |
| 7. Are supervisory hazard surveys documented? | | | |

# SELF AUDIT

### SAFETY MANAGEMENT SYSTEM

Organization _____  Date _____

Branch, Division or Subsidiary _____

| QUESTION | ANSWER | | REMARKS |
|---|---|---|---|
| | YES | NO | |
| 8. Is there a system with priorities assigned for the follow-up and control of all identified hazards? | | | |
| 9. Is there a procedure for handling employee suggestions or complaints regarding safety and health? | | | |
| 10. Do members of top management conduct regular safety inspections? | | | |
| 11. Are there written procedures, such as: Job safe practices, job safety analysis or similar to cover all jobs involving high risk potential hazards? | | | |
| 12. Are there procedures available for systematic re-training and review with new employees? | | | |
| 13. Do all employees attend a safety meeting a minimum of once per month for field personnel or quarterly for office personnel? | | | |
| 14. Are minutes kept of all safety meetings? | | | |

**Figure 21-3 (continued)**

# SELF AUDIT

## SAFETY MANAGEMENT SYSTEM

Organization _____ Date _____

Branch, Division or Subsidiary _____

| QUESTION | ANSWER | | REMARKS |
|---|---|---|---|
| | YES | NO | |
| 15. Is there a written procedure to follow when an accident occurs? | | | |
| 16. Are no-injury (damage and/or significant incident) accidents investigated? | | | |
| 17. Are all accident reports reviewed and signed by at least one level above the originator? | | | |
| 18. Do top management review at least all lost workday injury accident reports? | | | |
| C. Critical Safety Procedures | | | |
| 1. Are inspections, tests and preventive maintenance conducted on the following: Are employees authorized and trained? | | | |
| ➡ Scaffolds | | | |
| ➡ Man-lifts/powered platforms | | | |
| ➡ Forklifts | | | |
| ➡ Cranes, hoists and slings | | | |

# SELF AUDIT

## SAFETY MANAGEMENT SYSTEM

Organization _____ Date _____

Branch, Division or Subsidiary _____

| QUESTION | ANSWER | | REMARKS |
|---|---|---|---|
| | YES | NO | |
| ➡ Wire rope | | | |
| ➡ Compressors and air receivers | | | |
| ➡ Elevators | | | |
| 2. Are the following procedures established, reviewed and audited for compliance. Are employees authorized and trained? | | | |
| ➡ Lockout/Tagout | | | |
| ➡ Permit Required Confined Space Entry | | | |
| ➡ Hot Work Permit | | | |
| ➡ High Voltage Electrical Procedures | | | |
| ➡ Pipe Blinding and Blanking | | | |
| ➡ Excavations | | | |

Figure 21-3 (continued)

is commonly used in production operations, but its adaptation to safety performance management has been slow. This will be a trend to watch in the future.

## Conclusion

Safety goals, standards, and measurements are literally the "key to the future" of safety management. It must be the goal of all safety professionals to continuously develop their skills in this aspect of professional management.

If safety and health professionals wish to remain "on the team" in their respective organizations, it is absolutely vital to serve by helping management to set up and maintain comprehensive and effective safety systems which will ensure that the enterprise is successful in an increasingly competitive world.

## IMPLEMENTATION OF SAFETY MANAGEMENT SYSTEMS

One of the frustrating difficulties most organizations have to deal with is that, having adopted a particular type of safety system, they still have to get the system implemented and producing effective results. To help ensure that this process takes place with a minimum of delays and misunderstandings, a step-by-step method follows:

| Step | Description |
| --- | --- |
| Procedure | The system procedure must be reviewed with the various management levels and accepted by them. |
| Basic training | All levels must be trained in the basics of the system, including their responsibilities and the minimum action required. The training will also cover paper flow of reports or other documentation. |
| Advanced training | Advanced training must be provided for management levels above the first line level to explain their responsibilities and the minimum action required. This will include evaluation of reports and feedback to their first-line supervision. |
| Management controls | The procedure must include systems for auditing the effectiveness of the system. |

Figure 21-5 illustrates a typical format for implementing safety systems within individual departments. Refer also to Figure 21-1 for a sample illustrating the specific actions in the basic safety systems by various management levels.

## NEW AND EMERGING TRENDS IN PROFESSIONAL MANAGEMENT

Management is a dynamic evolving process, and in today's highly competitive business climate there has been an almost constant stream of new techniques and systems, each espoused by the latest "guru".

## ACCIDENT PREVENTION PERFORMANCE ASSESSMENT

| Employee Performance Action Step | Antecedents | Employee Perceived Consequences | Management Control Systems | | |
|---|---|---|---|---|---|
| | | | Data System | Feedback System | Review System |
| 1. Not looking for hazards. | • Job description<br>• Job instruction<br>• Tools and materials<br>• Work environment | • More time to do the job<br>• Avoids "paperwork"<br>• "Makes sense to them"<br>• Risk of supervisory reprimand<br>• Accidents may occur<br>• Unfavorable reaction when report another person's unsafe actions | | | |
| 2. Looking for hazards | • Supervisor observing and providing guidance<br>• Safety Manual<br>• Procedures<br>• Training | • "Fault finding" after accidents occur<br>• More "hassle" follow up, etc.<br>• Jobs take longer<br>• Promotion/pay increase may be affected if do not comply<br>• "More paperwork"<br>• May be less accidents | • Safety Inspections Reports<br>• Accident Reports<br>• JHA JSA/JSP<br>• Safety Meeting Reports<br>• Hazard Control System | • Periodic discussion | • Quarterly MBO<br>• Annual Performance Appraisal |
| PERFORMANCE REINFORCEMENT PLAN | • Set goal<br>• Meet w/employees, explain measures, goal and baseline<br>• Ask employees for suggestions | • Supervisor personally says something positive to each employee whose performance is above baseline*<br>• Set up graph<br>• Set up incentives<br>• Set up awards<br>• During the week surprise those groups with coffee and donuts who have a higher performance than previous week*<br><br>* Reinforcements which produce Positive Immediate Certain Respose | • Each employee keeps daily tally of hazards recognized<br>• Manager and supervisor randomly check reports and audit for accuracy | • Group maintains graphs of safe hours worked and hazards recognized<br>• Individual graphs for each supervisor<br>• Manager maintains graphs and writes reinforcing statements on them<br>• Manager explains baseline and graph to all when posting<br>• Quarterly performance appraisals | • Manager conducts review meetings with each supervisor each month<br>• Supervisor holds monthly meeting with crew, asks them about reinforcers and ideas |

**Figure 21-4** Illustration of traditional safety performance controls versus over performance management techniques which increase employee involvement.

**SAFETY MANAGEMENT SYSTEM ACTION PLAN**

Dept. _____ Program _____     Plan  originated _____
                                                            reviewed _____
Dept. Manager approved _____     revised _____

| BASIC STEP | NO. | STEP DESCRIPTION | RESPONSIBILITY | COMPLETION DATE | |
| | | | | SCHED. | ACTUAL |
|---|---|---|---|---|---|
| PROCEDURE | 1 | Procedure reviewed and accepted by all team leaders first line supervisors . | | | |
| | 2 | Dept manual accessible to all team leaders/supervisors. | | | |
| IMPLE-MENTATION TRAINING | 3 | Basic training program developed in writing stating what the team leader/first line supervisor must know. | | | |
| | 4 | All second line supervisors trained in basics. | | | |
| | 5 | All team leaders/supervisors trained in basics. | | | |
| MINIMUM ACTION REQUIRED | 6 | Dept. procedure issued to state the specific minimum action required by each teamleader/first line supervisor. | | | |
| FOLLOW-UP | 7 | Dept. procedure issued to state who does what by when with the paper. | | | |
| ADVANCED TRAINING FOR REVIEWER | 8 | Advanced training program developed in writing stating what the reviewer does. | | | |
| REVIEW OF THE SYSTEM ACTION | 9 | Dept. procedure issued to state what the second line supervisors do with the paper and by when. | | | |
| SAMPLING OF THE REVIEWER'S WORK | 10 | Dept. procedure to state how second line supervisors set up management controls to ensure that the system is effective. | | | |
| ADDITIONAL MANAGEMENT CONTROLS | 11 | Additional system steps, e.g. index and review schedule and line employee training schedule. | | | |

**Figure 21-5**  Sample format for the implementation of safety systems by the various levels of department management.

The 1980s saw an increasing interest in the total quality management (TQM) movement based on the teachings of Deming, Juran, and others. Many companies and organizations went through a process of reorganizing or restructuring in order to reduce costs and improve profitability. These efforts usually involved reducing layers of management and/or combining job functions.

The 1990s have brought some refinements to the scene through the work and teachings of such proponents as Stephen Covey *(Seven Habits of Highly Effective People),* Tom Peters *(In Search of Excellence),* Peter Senge *(The Fifth Discipline),* and Peter Block *(Stewardship and the Empowered Manager),* to name but a few. Nevertheless, to quote John Parkington of the Wyatt Company, "Downsizing is still the diet of corporate choice."

The latest technique is termed "reengineering", as championed by authors Michael Hammer and James Champy in their best-selling book entitled *Reengineering the Corporation.*

The most positive outcome of all this turmoil in the management scene is that organizations are now more focussed on employee involvement and empowerment, and the old bureaucratic "top-down" management culture is being replaced by a much more open collaborative teamwork approach.

## SOME CONCLUSIONS FOR THE SAFETY AND HEALTH PROFESSIONAL OF THE FUTURE

"Flexibility" and "continuous learning" will have to be the watchwords of tomorrow's safety and health professional. It is highly likely that you will find yourself with new partners, both internal and external. Furthermore, you will need to be constantly looking at ways you can outsource and/or offload some of your responsibilities in order to concentrate on the functions that will provide you with maximum leverage and results.

To be successful, the safety and health professional should concentrate on being recognized as a policy setter, auditor, and internal adviser. Communication skills will be vital as you work with your management to define the organization's needs and then go out and sell your solutions in terms of improvements to their management systems.

The goal should be to become recognized as a valued resource. Do this by looking for ways in which you can contribute to the bottom line. Always keep in mind that your management's agenda is your agenda. They are your customer, so be sure to blend your agenda to theirs.

As we said, be flexible and creative, try different ways, be prepared to compromise, and take it step by step. To quote Peter Drucker, "It is more important to do the right things than to do things right." If you want to change results, you may need to change your strategy.

Here are some final hints based on the teachings of security management consultant Dennis Dalton:

- Identify your agenda; make sure it compliments rather than competes with management's agenda.
- Understand your business.
- Speak in business managers' terms.
- Meet regularly with managers.
- Walk around and observe — is Safety and Health helping?
- Solicit customer feedback.
- Listen to what your employees are saying.
- Be aware of new approaches and ideas — find out what others are doing.

- Show you care — the human touch.
- Really *listen.*
- Focus on the *positive.*
- Maintain *focus on what you do best.*

Finally, in terms of safety and health systems, be looking regularly at how these systems can be improved by using either state-of-the-art computer technology or new management tools such as flow diagrams, fishbone diagrams, or new computerized training and simulation systems.

## FURTHER READING

### Books

Adams, J. D. et al. *Transforming Leadership from Vision to Results.* Alexandria, VA: Miles River Press, 1986.

Ailes, R. *You Are the Message.* New York: Doubleday, 1988.

Albrecht, K. *Successful Management by Objectives.* Englewood Cliffs, NJ: Prentice-Hall, 1978.

Allen, L. A. *Common Vocabulary of Professional Management.* Palo Alto, CA: Lois A. Allen Associates, 1969.

Allen, L. A. *The Management Profession.* New York: McGraw-Hill, 1964.

Allesandra, A. J. and Hulnsaker, P. L. *Communicating at Work.* New York: Simon & Schuster, 1993.

American Society of Safety Engineers. *The Dictionary of Terms Used in the Safety Profession.* Des Plaines, IL: American Society of Safety Engineers, 1988.

Bellman, G. M. *Getting Things Done When You Are Not in Charge.* San Francisco, CA: Berrett-Koehler, 1992.

Bennis, W. *On Becoming a Leader.* Reading, MA: Addison-Wesley, 1989.

Bennis, W. and Nanus, B. *Leaders — The Strategies for Taking Charge.* New York: Harper & Row, 1985.

Bird, F. E. and Germain, G. L. *Practical Loss Control Leadership.* Loganville, GA: International Loss Control Institute, 1990.

Block, P. *Stewardship.* San Francisco, CA: Berrett-Koehler, 1993.

Block, P. *The Empowered Manager.* San Francisco, CA: Jossey-Bass, 1987.

Blanchard, K. and Johnson, S. *The One Minute Manager.* New York: Berkley Publishing Group, 1984.

Blanchard, K. and Peale, N. V. *The Power of Ethical Management.* New York: Fawcett Crest, 1988.

Byham, W. C. and Cox, J. *Zapp! The Lightning of Empowerment,* New York: Ballantine, 1988.

Cohen, A. R. and Bradford, D. L. *Influence without Authority.* New York: John Wiley & Sons, 1990.

Connellan, T. K. and Zemke, R. *Sustaining Knock Your Socks Off Service,* New York: American Management Association, 1993.

Covey, S. R. *The Seven Habits of Highly Effective People.* New York: Simon & Schuster, 1989.

Covey, S. R., Merrill, R. A., and Merrill, R. R. *First Things First.* New York: Simon & Schuster, 1994.

Dalton, D. R. *Security Management: Business Strategies for Success.* Woburn, MA: Butterworth-Heinemann, 1994.

Daniels, A. C. and Rosen, T. A. *Performance Management — Improving Quality and Productivity through Positive Reinforcement.* Tucker, GA: Performance Management Publications, 1984.

Dawson, R. *Secrets of Power Persuasion.* Englewood Cliffs, NJ: Prentice-Hall, 1992.

Decker, B. *The Art of Communicating.* Los Altos, CA: Crisp Publications, 1988.

Drucker, P. F. *The Practice of Management.* London: Heinemann, 1963.

Drucker, P. F. *The Frontiers of Management.* New York: Harper & Row, 1986.

Drucker, P. F. *The Effective Executive.* London: Heinemann, 1967.

Findlay, J. V. and Kuhlman, R. L. *Leadership in Safety.* Loganville, GA: Institute Press, 1980.

Fulton, R. V. *Common Sense Supervision.* Berkeley, CA: Ten Speed Press, 1988.

Gellerman, S. W. *Motivation in the Real World.* New York: Penguin Books, 1993.

Gordon, H. L. et al. *A Management Approach to Hazard Control.* Bethesda, MD: Board of Certified Hazard Control Management, 1994.

Grimaldi, J. V. and Simonds, R. H. *Safety Management.* Boston, MA: Irwin, 1989.

Grose, V. L. *Managing Risk — Systematic Loss Prevention for Executives.* Englewood Cliffs, NJ: Prentice-Hall, 1987.

Hammer, M. S. and Champy, J. *Re-engineering the Corporation.* New York: Harper Collins, 1993.

Health and Safety Executive. *Successful Health and Safety Management.* London: Her Majesty's Stationery Office, 1992.

Hopkins, T. *How to Master the Art of Selling.* Scottsdale, AZ: Warner, 1982.

Hughes, C. L. *Goal Setting — Key to Individual and Organizational Effectiveness.* New York: American Management Association, 1965.

Jaffe, D. T., Scott, C. D. and Tobe, G. R. *Rekindling Commitment.* San Francisco, CA: Jossey-Bass, 1994.

Janov, J. *The Inventive Organization.* San Francisco, CA: Jossey-Bass, 1994.

Kazmier, L. J. *Principles of Management.* New York: McGraw-Hill, 1974.

Knowdell, R. L., Branstead, E., and Moravec, M. *From Downsizing to Recovery: Strategic Transition Options for Organizations and Individuals.* Palo Alto, CA: CPP Books, 1994.

Kouzes, J. M. and Posner, B. Z. *The Leadership Challenge: How to Get Extraordinary Things Done in Organizations.* San Francisco, CA: Jossey-Bass, 1987.

Krause, J. R., Hidley, J. H., and Hodson, S. J. *The Behavior Based Safety Process.* New York: Van Nostrand Reinhold, 1990.

Ladou, J. et al. *Occupational Safety and Health.* Itasca, IL: National Safety Council, 1994.

Lund, D. R. and Finch, L. W. *Lessons in Leadership.* Staples, MN: Nordell Graphic Communications, 1987.

Manuele, F. A. *On the Practice of Safety.* New York: Van Nostrand Reinhold, 1993.

Nirenberg, J. *The Living Organization — Transforming Teams into Workplace Communities.* New York: Irwin, 1993.

Pater, R. *Making Successful Safety Presentations.* Portland, OR: Fallsafe, 1987.

Pater, R. *The Black-Belt Manager.* Rochester, VT: Park Street Press, 1988.

Peters, T. *Thriving on Chaos.* New York: Alfred A. Knopf, 1987.

Peters, T. and Austin, N. *A Passion for Excellence.* New York: Random House, 1985.

Petersen, D. *Safety By Objectives.* Goshen, NY: Aloray, 1978.

Petersen, D. *Safety Management.* Goshen, NY: Aloray, 1988.

Petersen, D. *Techniques of Safety Management.* Goshen, NY: Aloray, 1989.

Pinchot, G. and Pinchot, E. *The End of Bureaucracy and the Rise of the Intelligent Organization.* San Francisco, CA: Berrett-Koehler, 1994.

Pope, W. C. *Managing for Performance Perfection: The Changing Emphasis.* Weaverville, NC: Bonnie Brae Publications, 1990.

Ray, M., Rinzler, A. et al. *The New Paradigm in Business.* New York: Tarcher/Perigee, 1993.

Rose, C. *Accelerated Learning.* New York: Dell Publishing, 1985.

Roddick, A. *Body and Soul.* New York: Crown Publishers, 1991.

Rosenbluth, H. F. and McFerrin, Peters, D. *The Customer Comes Second.* New York: William Morris & Co., Inc., 1992.

Sashkin, M. and Kiser, K. *Putting Total Quality Management to Work.* San Francisco, CA: Berrett-Koehler, 1993.

Senge, P. M. *The Fifth Discipline.* New York: Doubleday, 1990.

Senge, P. M. et al. *The Fifth Discipline Fieldbook — Strategies and Tools for Building a Learning Organization.* New York: Doubleday, 1994.

Taylor, H. L. *Delegate — The Key to Successful Management.* New York: Warner Books, 1989.

Thomas, H. G. *Safety, Work, and Life — An International View.* Des Plaines, IL: American Society of Safety Engineers, 161–185, 1991.

Tracey, W. R., Ed. *Human Resources Management and Development Handbook.* New York: Amacom, 1994.

Walsh, T. J. et al. *Protection of Assets Manual.* Santa Monica, CA: The Merritt Company, latest printing.

Walton, D. *Are You Communicating?* New York: McGraw-Hill, 1989.

Walton, M. *The Deming Management Method.* New York: Putnam Publishing, 1986.

Wheatley, M. J. *Leadership and the New Science.* San Francisco, CA: Berrett-Koehler, 1992.

## Other Publications

Barenklau, K. E. Developing standards for safety work activities. *Occup. Hazards Natl. Saf. Manage. Soc. Focus,* 145–148, October 1989.

Barenklau, K. E. Effectively measuring safety involves consequences and cause. *Occup. Saf. Health,* 41–49, March 1986.

Barenklau, K. E. Measurement: key to professionalism. *Occup. Hazards Natl. Saf. Manage. Soc. Focus,* December 1973.

Carder, B. Quality theory and the measurement of safety systems. *Prof. Saf.,* 39(2), 23–28, February 1994.

Carr, C. Ingredients of good performance. *Training,* 30(8), 51–54, August 1993.

Cooper, D. Goal setting for safety. *Saf. Health Pract. (U.K.),* 11(11), 32–37, November 1993.

Creswell, T. J. Safety and the management function. *Occup. Hazards,* 31–34, December 1988.

Crutchfield, N. and Waite, M. A vision and mission for safety. *Occup. Hazards, Natl. Saf. Manage. Soc. Focus* 65–68, September 1990.

Deacon, A. The role of safety in total quality management. *Saf. Health Pract (U.K.),* 12(1), 18–21, January 1994.

Earnest, R. E. What counts in safety? A non-injury based measurement system. *Natl. Saf. Manage. Soc. Insights Manage.,* 2–6, 2nd quarter 1994.

Esposito, P. Applying statistical process control to safety. *Prof. Saf.,* 38(12), 18–23, December 1993.

Kouzes, J. M. and Posner, B. Z. *The Leadership Challenge: How To Get Extraordinary Things Done in Organizations.* San Francisco, CA: Jossey-Bass, 1987.

Krause, T. R., Hidley, J. H., and Hodson, S. J. Measuring safety performance: the process approach. *Occup. Hazards Natl. Saf. Manage. Soc. Focus,* 49–52, June 1991.

Lack, R. W. Safety management — accountability or sideshow? *Saf. Pract. (U.K.),* 3(8), 4–7, August 1985.

Le Clerg, R. E. Why the "safety" function? *Occup. Hazards Natl. Saf. Manage. Soc. Focus,* May 1989.

Minter, S. G. Safety audits — a measured approach to progress. *Occup. Hazards,* 51–33, June 1988.

Moravec, M. From re-engineering to revitalization. *Executive Excellence,* 12(2), 18–19, February 1995.

Northage, E. I. Auditing: a closed control loop. *Occup. Hazards Natl. Saf. Manage. Soc. Focus,* 87–90, May 1992.

Petersen, D. Integrating safety into total quality management. *Prof. Saf.,* 39(6), 28–30, June 1994.

Ragan, P. and Carder, B. Systems theory and safety. *Prof. Saf.* 39(6), 22–27, June 1994.

Swartz, G. Consider a safety audit program. *Prof. Saf.,* 28(4), 47–49, April 1983.

Top, W. N. What makes safety management flourish? *Occup. Hazards Natl. Saf. Manage. Soc. Focus,* 67–70, June 1992.

Tracy, B. *Action Strategies for Personal Achievement.* Series of Tapes. Niles, IL: Nightingale Conant, 1993.

Tracy, B. *The Effective Manager Seminar.* Series of Tapes. Niles, IL: Nightingale Conant, 1994.

# 22

# HOW TO BE A SUCCESSFUL SAFETY ADVISOR

**Robert A. Lapidus**

## TABLE OF CONTENTS

1-56670-054-X/96/$0.00+$.50
© 1996 by CRC Press, Inc.

## INTRODUCTION

*Technical expertise does not guarantee success.*

You can know all there is to know about a given subject, but if you do not know how to build long-lasting, credible relationships, no one will listen to you. Your expertise will be of no use to anyone. This lesson is one that many safety people fail to learn.

We fail to learn this lesson because we believe that our technical safety knowledge is the reason we have been hired. In many cases, what we believe is the truth. We *have* been hired for our technical expertise.

The typical impetus for instituting new safety programs and putting someone in charge of the safety function is management's desire to reduce losses, prevent serious injuries, and comply with governmental mandates. To implement such programs requires people who know the standards, regulations, techniques, and methods so that programs can be designed for specific needs.

What the new employer fails to tell us, and probably does not consciously recognize, is that they also expect us to *work with others* in the organization to achieve safety success. Consequently, when we begin to espouse our knowledge and lay down our expectations, the workforce, those people who have to apply our remedies to their daily work lives, starts to feel put upon. We may find doors closed to us if we push too hard to achieve our goals.

Accordingly, to do our safety advisory jobs in a professional manner, we must have current technical knowledge and skills, but we cannot assume that such expertise will in itself get safety incorporated into the work environment. That takes other knowledge and skills — that which is associated with how we relate as human beings and how we seek to meet expectations. This chapter discusses 18 steps that can help you become a more successful safety advisor.

## STEPS TO SUCCESS

*Advisory success comes from giving yourself to others.*

### Step 1: Develop a Trusting Relationship from the Beginning

Establish a base for management to learn to trust you. Discuss your role with management. Listen to what they want to accomplish, and massage your goals to fit their goals.

In safety, there are specific actions we need to take when we start to create a safety program. These actions are usually nonthreatening to anyone in the organization because they are simply information retrieval, and they include the identification of the

- Causes of why losses are occurring
- Exposures to loss that could result in additional losses, including compliance requirements

# ADVISORY SUCCESS STEPS

1.  Develop a trusting relationship from the beginning

2.  Ask Questions

3.  Get Input Before Taking Action

4.  Keep Communication Open

5.  Tailor Your Advice

6.  Be Prepared To Explain Why

7.  Put Management's Priorities First

8.  Suggest Alternatives

9.  Avoid the Enforcement Role

10.  Stay Behind the Scenes

11.  Be Timely With All Activities

12.  Be Responsive To All Requests

13.  Be Positive In Your Daily Communications With Others

14.  Apologize

15.  Learn To Foresee Problems Before They Happen

16.  Avoid "I Told You So"

17.  Recognize Others' Successes

18.  Show You Are Proud To Be On The Management Team

**Figure 22-1** Advisory success steps.

- Current activities being done to target the organization's losses and exposures
- Organization's style of management and culture so that future programs will fit

Obtaining this information can be an excellent opportunity to start to build relationships within the organization. Begin by having dialogue with others by sitting down with top, middle, and first-line managers, employees, and union representatives and talking with them about the current safety program, the future as they see it, and what they believe would work in the organization.

Developing a trusting relationship means that the advisor needs to listen. In the beginning, refrain from giving your opinion on every subject that is discussed. Build your knowledge base by gathering the knowledge, opinions, and feelings of others.

Step 1 should be initiated at the commencement of a new opportunity in safety, but can also be initiated after an advisor has been in the job for some time. The latter situation requires the advisor to set aside ego and make a concerted effort to reinitiate the advisor/management relationship. To accomplish this mission, we need Step 2.

## Step 2: Ask Questions

Both Steps 1 and 2 are part of the building of personal credibility. Safety advisors need to be seen by others as people who are easy to talk to, are open to new ideas, and can be trusted to hold confidences.

No single person has all the answers. Therefore, probably one of the most effective means to obtain buy-in on the part of all managers, supervisors, and employees is to gain their involvement in the creation and implementation of the safety program. Instead of doing an "information dump" (one-way communication) on everyone else, ask questions to discover what is going on in the organization and how others see the safety program being designed. Upon starting a new job, ask others: "What are your expectations of me and my function?" If you are in a current job and have never asked this question, ask it now.

After 3 or 4 months of working on those expectations, return and ask, "Do you feel that I am meeting your expectations?" If they do not feel that you are achieving their expectations, find out what you can do to change that. Keep pursuing this line of questioning.

Those to whom you give advice are your customers. Treat them with respect. Let them know that their opinions, comments, and ideas are important to you. They need to know that you will recommend the implementation of safety programs and activities that are based upon the input you have received from those who will be affected.

## Step 3: Get Input Before Taking Action

Others may be privy to information crucial to your job. Getting input about what you plan to do before you actually do it will strengthen your relationship through a mutual sharing of information.

Get input on all programs that are going to affect other people. Such input should come from top and middle management, first-line supervision and employees, and special interest groups that are represented in the organization.

## Step 4: Keep Communication Open

Continually apprise management of what your objectives are and how you are going to attain them. Use the communication tools that management prefers. Communication techniques may include face-to-face contacts, short, simple communiqués, formal periodic reports, or simply telephone contacts anytime something needs to be communicated. In the latter case, management may then decide what kind of follow-up documentation they need. The key is to use the mode of communication that your recipients want, not what you feel more comfortable using.

Keep a dialogue going with as many parties as possible. No one wants to be blindsided with surprises. Get input and provide input.

## Step 5: Tailor Your Advice

Avoid canned answers or traditional approaches that worked for other organizations, but that do not fit the reality of your own. Be creative. Whatever advice you provide, be sure to tailor it to the climate of the organization, to the nature of the way things are. You should also seek creative new ideas from others. Those who actually do the main line of work have a pretty good idea of what will and will not work.

## Step 6: Be Prepared to Explain Why

"Because" and "the law requires it" are not the answers that most of us appreciate receiving to the question of why something has to be done. Whenever you provide a suggestion, have as many facts as possible at your fingertips. Describe how your suggestion will affect the nature of the activity that is being done and how the suggestion will improve the workings of the organization. Provide actual examples of situations where the failure to have your recommended action in place resulted in a loss, such as an occupational injury or a property damage incident. Obtain backup data and real-life anecdotes from internal and external sources that will support your propositions. Know how your recommendations might provide a positive financial reward.

## Step 7: Put Management's Priorities First

Conceptually, certain organizations may put safety first as the most important goal, but in reality safety is not first. Safety should be an integral part of how the work gets done, but a perfectly safe environment rarely exists. Everyone has to establish for themselves their own level of safety, what they consider to be an acceptable level of risk.

In this regard, sometimes management does not recognize certain elements of the safety advisor's work as being necessary. Management may not see your activities fitting within their priorities.

If your objectives appear to be important for the welfare of the organization, educate management on the importance of the objectives or activities that need to be done. Plant seeds (suggestions) in a manner that, in time, management may see the importance of the concept or need.

If you are not getting the response from management that you are seeking, you may wish to back off. Tell management, "I guess I should rethink my ideas". Then return to your office and rethink what you are trying to do and how you are trying to do it.

You may be on the right track, but not explaining it well. Then again, this time may not be right to pursue this particular activity. It may be necessary for you to let go of some portion of your job, objectives or some recommendation for change if management will not allow you to do it or will not support you in doing it.

One of the most difficult concepts for safety advisors to accept is that sometimes it is better not to force an issue for the time being, but to wait until the time is right to reinstate the activity or recommendation.

## Step 8: Suggest Alternatives

Refrain from coming across as if you are the only one with *the* answer. Seek to be a true advisor (helper) by avoiding playing a game of "gotcha."

Working in a technical profession, safety advisors have to balance the need to comply with regulatory requirements to the need of maintaining a working environment that permits employees to have input into their daily activities. The inspector mentality tends to cite the regulatory standards and require compliance regardless of the impact on the work. The helping or servant mentality tends to cite the regulatory standards, but then discusses with others how best to comply while maintaining the ability to get the job done. Be straightforward in your comments from a helping standpoint. Ask people for ideas about how problems might be solved.

## Step 9: Avoid the Enforcement Role

If at all possible, avoid the role of enforcer. Instead of you playing the role of police officer, train line people to play that part. Work with them to develop the technical information they need to comply with external and internal mandates; then let line managers, supervisors, and employees enforce their own rules and regulations. Let them establish their own parameters and enforce proper compliance. They would then be doing their own self-monitoring.

Try to avoid the enforcement role by doing positive things instead:

- Seek to observe people doing their work in a safe manner, both in terms of safe practices and maintaining a safe environment. Let people know that you appreciate their positive efforts.
- Through line management, publicly acknowledge successful activities and results.

## Step 10: Stay Behind the Scenes

Advisors should avoid putting themselves in the limelight, either as the receiver of praise or the giver of recognition. Allow management the privilege of taking the public credit and giving appropriate commendation to their people. You will get "advisor gratification" by seeing your recommended programs implemented by operational management. When your advice is taken and others are successful, you will also reap the fruits of success.

## Step 11: Be Timely With All Activities

Agree to mutually acceptable timetables for getting things done. If you do not do this, the parties who are expecting you to respond to them will normally have a shorter deadline in their minds than you will have in your mind. You will then be in trouble. If you think you will not be able to meet the deadline or target date that has been mutually agreed upon, inform the expecting party of the problem prior to the established date. Again, no one likes surprises.

## Step 12: Be Responsive to All Requests

Keep those who have asked you for something informed as to the status of their request and your response. Let them know you have received their request. Work out a mutually acceptable timetable, and be responsive. Do not let a request drop into a "black hole." Your credibility will be greatly hurt.

## Step 13: Be Positive In Your Daily Communications With Others

The advisor needs to be seen as a helper, someone whom others feel comfortable coming to and obtaining support from. To achieve this positive status in the organization, avoid sarcastic or abrasive comments. Be careful about joking around; someone could easily take offense. Guard your temper. Do not lose your self-control, no matter how right you think you are.

## Step 14: Apologize

There may be situations in your work life where you will do or say something that hurts another's feelings or that hurts the final outcome of a project. Be prepared at all times to repair relationships by apologizing for what you said or did. You may think you were correct in what you did, but if the relationship is suffering, be the first to apologize and seek to make things right.

## Step 15: Learn to Foresee Problems Before They Happen

Advisors, especially safety advisors, are supposed to protect their organizations from potential problems. Keeping up-to-date on the latest new regulations is critical for knowing what type of compliance will be required. Keeping posted on accidents and incidents that have occurred in similar types of organizations will help you prevent those things from happening in your organization.

## Step 16: Avoid Saying "I Told You So"

When management has a problem, and you had previously recommended a way to have avoided it, be there to help out and rectify the problem. Avoid saying "I told you so" in any way whatsoever. This includes side comments to others, facial expressions, body language, and direct statements that show that those in charge did not listen to you.

## Step 17: Recognize Others' Successes

When other people succeed, recognize their success. Be on the lookout for anyone who has done something special. Keep congratulatory and thank you cards in your desk, and send them out in a timely manner. Those who receive your notes will appreciate you and your thoughtfulness even more.

## Step 18: Show You Are Proud to Be on the Management Team

Let management know you are proud to be part of their team, working together to achieve mutual goals. Tell them. Write thank you letters. Be there to participate in celebrations and special events. If you are not proud to be on the team, if you awake each workday with a headache or stomach ache, and dread going to work, move on, get out of that situation. You are not doing yourself or management any favors by staying around.

## SUMMARY

*Seek to be a credible resource.*

We need to have the knowledge necessary to help others solve safety-related problems. We need to be current on all subjects that relate to our work,

# SPECIAL ADVISORY TECHNIQUES

- Establish a base for management to learn to trust you.

- Ask: What are your expectations of me and my function?

- Facilitate getting input from others prior to taking action.

- Use the communication tools that management prefers.

- Avoid canned answers or traditional approaches.

- Be creative within the culture of the organization.

- Have backup data to support your recommendations.

- Do not force an issue; be prepared to let an idea ride.

- Refrain from coming across as if you are the only one with *the* answer.

- Try to avoid the enforcement role by doing positive things.

**Figure 22-2** Special advisory techniques.

but such knowledge does not guarantee success. In support of this knowledge we need to have a reputation for being dependable, honorable, reliable, and trustworthy; that is, we need to be credible.

The best way to achieve this credibility is to follow the Golden Rule: *Do unto others as you would have them do unto you.*

How do you want to be treated? Most of us have similar answers to this question. We want to be treated respectfully, honestly, and as if our ideas are worthy. We want to be treated fairly and be recognized for doing a good job. We want to have input into how our jobs and work activities will be done.

This latter desire is probably the most important element of safety advisory success, as has been described throughout this chapter. If we are to be successful advisors, we need to facilitate the involvement of others in the development and maintenance of the safety program itself. Everyone will then have an opportunity to become part of the safety effort, buy into the safety program, be treated respectfully, and look to the safety advisor as a credible resource.

## FURTHER READING

### Books

Barker, J. A., *Discovering the future — The Business of Paradigms.* St. Paul, MN: ILI Press, 1985.

Blanchard, K. and Peale, N. V., *The Power of Ethical Management.* New York: William Morrow & Company, 1988.

Kouzes, J. M. and Posner, B. Z., *Credibility.* San Francisco, CA: Jossey-Bass, 1993.

Petersen, D., *Safety Management — A Human Approach.* Goshen, NY: Aloray, 1988.

### Other Publications

Greenleaf, R. K., *The Servant as Leader.* Indianapolis, IN: The Robert K. Greenleaf Center, 1991.

# DEVELOPING EFFECTIVE LEADERSHIP SKILLS — HOW TO BECOME A SUCCESSFUL CHANGE AGENT IN YOUR ORGANIZATION

## 23A

**Paula R. Taylor**

## TABLE OF CONTENTS

1-56670-054-X/96/$0.00+$.50

## INTRODUCTION

For millenniums scholars and ordinary folks have been trying to define leadership. The result is a multitude of definitions and no consensus. One reason may be that leadership is demonstrated in many ways and in many places. There is personal leadership, political leadership, artistic leadership, scientific leadership, world leadership, and organizational leadership. A cry often heard today is that there is a lack of leadership in all areas. Perhaps the tremendous amount of change that is occurring requires more or stronger leadership than ever before. Our specific concern is the leadership role of the health and safety manager and how that manager can be a positive force in times of change and transition.

As organizations and the nature of work change, leaders are needed more than ever. Leaders demonstrate specific characteristics. Fortunately, each of these characteristics has associated skills and these skills can be learned. As each characteristic is discussed you will have the opportunity to assess your current strength in it, identify where you need to develop your skill, and learn specific steps you can take to increase your ability to lead and to be perceived as a leader. No matter where you are in your organization, or your life, you can make a difference.

Use the **Assessment Checklist** below to measure your current strength in each characteristic. If you want additional feedback other than your self-perception, ask members of your team, peers, or employees to complete an Assessment Checklist on you. Use the Assessment Checklist again in 6 months and at various intervals to check your progress.

## ASSESSMENT CHECKLIST — LEADERSHIP CHARACTERISTICS

As you read through each leadership characteristic, assess your current level of strength or that of the person for whom you are completing the assessment. Use a 1 to 10 scale. A "1" indicates a deficiency, no strength at all, and a "10" means fantastic, could not be better, no need for improvement.

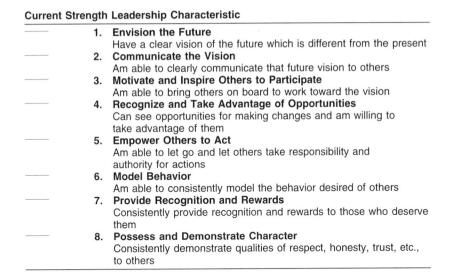

**Current Strength Leadership Characteristic**

| | | |
|---|---|---|
| ——— | 1. | **Envision the Future**<br>Have a clear vision of the future which is different from the present |
| ——— | 2. | **Communicate the Vision**<br>Am able to clearly communicate that future vision to others |
| ——— | 3. | **Motivate and Inspire Others to Participate**<br>Am able to bring others on board to work toward the vision |
| ——— | 4. | **Recognize and Take Advantage of Opportunities**<br>Can see opportunities for making changes and am willing to take advantage of them |
| ——— | 5. | **Empower Others to Act**<br>Am able to let go and let others take responsibility and authority for actions |
| ——— | 6. | **Model Behavior**<br>Am able to consistently model the behavior desired of others |
| ——— | 7. | **Provide Recognition and Rewards**<br>Consistently provide recognition and rewards to those who deserve them |
| ——— | 8. | **Possess and Demonstrate Character**<br>Consistently demonstrate qualities of respect, honesty, trust, etc., to others |

The first four of the eight characteristics are *creation characteristics.* These generate action and ideas. Characteristics 5, 6, and 7 are *implementation characteristics.* They make things actually happen. The eighth characteristic, the ability to demonstrate character, is in a category of its own. It is typical to be naturally stronger in either the *creation characteristics* or the *implementation characteristics.* However, both categories of skill are required.

Before leadership characteristics and skills are discussed in more depth, it is important to understand the environment in which today's manager operates and what is expected of him/her in that environment. The environment is one of change, and the manager is expected be a leader and a change agent. Managers are expected to move their organization through change, which is likely to be continuous and constant, and to create a creative, productive, effective group that will look different in structure, output, and style than anything they have known or imagined in the past.

Organizational change is, and will continue to be, the norm, not the exception. Some will find it exciting; many will find it unsettling and frightening. What should a leader know about change?

## UNDERSTANDING CHANGE

Change, in all aspects of our lives, whether perceived as positive or negative, causes stress and anxiety. Change means leaving the past behind and going into an unknown future. Regardless of how much the past was disliked, it was familiar, and familiar is comfortable. Change feels uncomfortable and unfamiliar. Since organizational change is usually imposed upon us, it often feels as though our past is being taken away. This feels like a loss. Most people go through five stages as they make the transition in a changing environment.

## Five Stages to Move to the Future

1. **Denial** — Regardless of what is said or done, people believe nothing will change or, if it does, it will have nothing to do with them. "They don't mean it." "It's just a rumor; we've heard this before." "Even if it happens it won't affect me (or this group)."
2. **Resistance** — At this stage people are very emotional. They are angry, depressed, resentful, self-pitying, and blaming. "Why me?" "After all I did for the company, I can't believe I'm being treated this way." "No way. I'm not changin' nothin'!"
3. **Bargaining** — People start making deals with themselves or God. "Okay, just let me stay until my retirement in 2 years." "I'll even take a pay cut; just don't move me."
4. **Depression** — People start to feel sad and anxious. Productivity declines.
5. **Acceptance** — The final stage is acceptance of what is happening and the beginning of a future orientation. "Hey, this is really happening. What do I need to do now?"

## Actions for the Leader

1. **Recognize and accept your own movement through the five stages.** Leaders are not immune from passing through the five stages. If you anticipate and expect that you will experience the same feelings as everyone else, you will move through the stages quicker, will understand the process, and will be more empathetic to others.
2. **Open and expand the communication channels.** Keep communicating. Provide every possible means of getting out information — from group meetings to newsletter to hotlines — even if the information is constantly changing. The void of no information will be filled with rumors and speculation.
3. **Be the first to give out the news.** Do not allow your employees to hear about what is happening to them from the *Wall Street Journal,* the local newspaper, or another department.
4. **Communicate with honesty and clarity.** Make every effort to be clear and honest. This is not the time for ambiguity and "corporate-speak". If you do not know, say you do not know.
5. **Encourage others to speak.** Allow and encourage others to speak their minds, express feelings, and ask questions. Accept feelings, answer questions, and do not get defensive.

## LEADER FOR CHANGE

In the midst of a major organizational change, strong leadership is more important than ever. Consider the eight characteristics of leaders and what you can do to strengthen your ability as a leader.

## Characteristic 1: Envision the Future

All leaders without exception possess this characteristic. They are able to envision the future. They can see possibilities; they can imagine what could be. Leaders often see the future as they would create it or improve it. Much has been written about our ability as human beings to think in future terms. It is this vision, this hope of a future — a better future — that keeps us alive and keeps us going. Once we have passed through the acceptance stage, we ask, "What can I do? What will it be like?"

How strong is your ability to envision the future? If you want to develop skills in this characteristic, here are some specific steps you can take.

### *Steps to Increase Your Ability to Envision the Future*

1. **Spend quiet time alone.** Finding time is often the hardest thing to do these days. It is important to set aside 20 minutes a day, an hour a week, and half a day or more once or twice a year.

   **INSTANT EXERCISE**

   **STOP**! Stop right now and make that commitment. Take out your calendar. Give yourself 20 minutes every day for the next month. The best time is usually in the morning, right after you wake up, or right before you go to sleep. If that does not work for you, take it out of your lunch hour. (Yes, take time for lunch.) What are you going to do in that time with yourself? You are going to sit alone in a quiet place, get relaxed and calm, and go inside your head and find out what is going on in there. Then you move to the next step.

2. **Ask yourself what you would like to be different in your life, your work, your organization.** Spend time just thinking about what you would like to be different. This is personal brainstorming. The rule is no censorship. After you have considered for a while, write your thoughts in a notebook. This will be a valuable reference book for you. You are now ready for the final step.

3. **Ask yourself what this vision would look like if it were in place.** Close your eyes, stay relaxed, and imagine the future. The more vividly you can imagine the outcome the more clear your vision will be. Again, do not censor your vision. Make it big. Imagine you in your role; imagine how you would feel, how others would feel. Imagine the accomplishments of the organization, the impact on people's lives, the additional dollars made or saved by the organization. Include as much detail as possible. Engage all your senses. Finally, write the details in your notebook. Imagine first; write second.

Every dream and vision, every change starts in the mind of one person. This vision comes from a leader. That brings us to the second characteristic.

## Characteristic 2: Communicate the Vision

Leaders can communicate their vision. Having a good idea and not being able to communicate it is as useless as having no idea or vision at all. Being able to articulate your vision is critical. Nothing produces more frustration in a group, particularly during times of change, than a leader without a vision or one who cannot communicate it. The result is the same. Think back on your own experience. Don't you find this to be true?

The vision is communicated verbally, in writing, or by example. Great leaders use all three. The skill required here is one of communication.

### Steps to Increase Your Ability to Communicate the Vision

1. **Write out your vision or mission statement.** The very act of taking pen and paper and writing out what you see and really want for yourself and your organization will have the effect of breathing life into your vision. It will become more real and more possible to you, and ultimately to others.
2. **Say it out loud.** Speak into a tape recorder. Tell your mirror. Articulate your vision before a video camera. Communicate your vision out loud in a meeting. People want to know how you envision the future. Remember how frustrating it is not to know.
3. **Walk your talk.** If you imagine an organization where people speak out, try out new ideas and support each other, behave as you speak.
4. **Assess your communication.** As you listen to the tape recorder, watch yourself in the mirror or on the videotape, or read what you have written, ask: "Is my vision concise, clear, enthusiastic, and congruent?" If it is not, people will not understand you or will not believe you. If possible, get feedback from a good friend or a colleague. Does he/she understand what you are trying to say?

#### INSTANT EXERCISE

Take a few minutes right now. Take out a piece of paper or an index card. Write out your vision for your organization or group. Make your first draft as long as you need it to be. Then edit it down to no more than three sentences. Use this first write-out as you follow the steps listed above. The more you articulate and assess your vision, the clearer it will become to you and to others.

Having a vision and being able to communicate it are still not enough. Leaders possess the next characteristic.

## Characteristic 3: Motivate and Inspire Others to Participate

Leaders have followers. How well do you move and motivate other people? Are you able to get people on board, to buy into your ideas? The special communication skill of influencing plays an important role with leaders who are strong in this characteristic.

Each person is motivated and inspired by something different, but great leaders are able to appeal to a wide range of people. The ability to influence others goes back to how well, how clearly and concisely you communicated your vision initially. The more dramatically you present your vision, the more passion and belief you demonstrate in your ideas, the more people you will bring on board with you. People respond strongly to leaders who show confidence, passion, and commitment.

Many potential leaders are unable to motive and inspire others because they are unwilling to share their ideas completely. If you give only partial information, if you hold back your enthusiasm, if you think you have an idea that is too good to share, you may be left with an idea and nothing else. Your desire to have other people join you must be genuine.

### Steps to Increase Your Ability to Motivate and Inspire Others

1. **Own your vision, but be willing to share it.** Once people are inspired, they share the vision. They feel it is theirs. This is when things happen, when people are working on a common goal. A sense of shared vision is critical.
2. **Communicate the benefit of being involved.** Since each person is motivated by something different, the benefit to pursuing the shared vision will be different for everyone. Understand and be able to communicate the benefits of being involved with you and your organization.

#### INSTANT EXERCISE

Reread your vision statement. Take out a sheet of paper and draw a vertical line down the center. Label the top "Benefits." Above the left column write "For Me." Above the right column write "For Them." List the benefits to you of accomplishing the vision. List all the benefits you can think of for others involved in working on the vision. Benefits can be both tangible and intangible. Keep these benefits in mind and communicate them as you are discussing your vision.

Moving the vision and talk into action brings us to the last of the *creation characteristics* and the fourth leadership characteristic.

## Characteristic 4: Recognize and Take Advantage of Opportunities

Leaders notice what is going on. They are willing to act and willing to take a risk. Leaders are not careless and they are not reckless. They are action oriented. Strong leaders are up-to-date on their company, their industry, and the external environment. They also frequently have interests in seemingly unrelated events and organizations. Often opportunities and ideas come from an atypical and unexpected source.

Leadership strength in this characteristic is built by communication, participation, education, self-assessment, and risk. If this is an area in which you would like to be stronger, here are some things you can do.

### Steps to Increase Your Ability to Recognize and Take Advantage of Opportunities

1. **Communicate and participate.** In order to identify opportunities you have to be out there listening and talking to people. The more people you know and who know what your vision is, the more possibilities will come your way.
2. **Practice listening attentively to what other people are saying.** As you hear what other people need and want, you will recognize where your plans and ideas fit. You will hear benefits for you and them.
3. **Be prepare to act.** Know your risk threshold. Think about just what you would be willing to do if the opportunity presented itself. Would you be willing to change departments, jobs, or companies or move to another city to see your vision in place? If you have a sense of what you are willing to do when the opportunity presents itself, you can take action or you can create your own opportunity. We all have stocks we should have bought, things we should have said, actions we should have taken; leaders minimize the "if onlys" and act. Dr. Maxwell Maltz expressed it well: "Often the difference between a successful man and a failure is not one's better abilities or ideas but the courage that one has to bet on his ideas, to take a calculated risk — and to act."

Take a moment to review the first four characteristics of leaders, the *creation characteristics*. These leadership skills sow the seeds necessary to create the future. They begin the process of change, the process of creating the future.

#### INSTANT EXERCISE

Consider your own strength in the creation characteristics. Identify the characteristic in which you are the weakest. What has not happened as a result of your not being as strong as you might be in that characteristic? Make a list of the specific things you will do differently that will make you more effective in that characteristic.

It is important to remember that although these characteristics are necessary in all leaders, each individual has a different style and a different approach. Some leaders will be noisy and flamboyant; others will be quiet and steady. One style is not inherently better than another. Each person must operate within his or her own personality while consciously doing what is necessary.

Let us look now at the *implementation characteristics* of leaders. If you can envision the future but can do nothing else, you may be known as a visionary. If you can communicate your vision, inspire others to join you, and seize every appropriate opportunity, you may be known as a group of visionaries. Visionaries are needed during times of change, but in order to make the changes work, a manager must be able to keep employees on board and moving in the right direction. If you cannot do that, the vision and the desired changes for the future will never happen.

The manager and the leader as a loner is not the model for today. The age of the leader as a loner is gone. A manager who cannot create a team, be part of a team, and keep that team together will not succeed in today's changing world. That brings us to leadership characteristic number five, the first of the *implementation characteristics.*

## Characteristic 5: Empower Others to Act

A real leader today empowers other people. The leader shares the vision and shares the planning. As things are changing around them, people are comforted by having a hand in the solution. The leader not only asks the question, "How are *we* going to make this happen?" but listens to the answers.

Empowering others means sharing responsibility and authority. People must be given the authority to decide what is to be done and then must face the resulting consequences of their decision. The manager provides the team the tools with which to do the job, with training in skills such as problem solving, listening, and presenting, with the latitude to accomplish something, and with support and assistance. Although time often seems like a commodity in short supply in transitional periods, it is the wise leader who finds the time and money to develop the team. It pays off in the end.

A leader also knows his/her team and knows the strengths and weaknesses of every team member. A leader who is capable of empowering others to act utilizes and encourages the strengths of each person and ignores or works around their weaknesses. Look for what people do well: use it, praise it, benefit from it. Allow and encourage people to learn new skills and improve weak areas, but go for the strengths.

### INSTANT EXERCISE

Create a strength profile of your group. Use a separate sheet of paper for each member of your team. Write their name at the top. On one side list their strengths, on the other their weaknesses. Draw an "X" through the weaknesses. List how the strengths can benefit the group at this time. Praise the strengths, utilize them, and make assignments as appropriate.

*Variation*
A variation on this exercise, and one that would be very beneficial to the team, is to share with the group what you want to do. In the whole group (if the level of trust is high) or with each individual, ask them to list their strengths and weaknesses. You and/or the other team members can add to each person's list.

Again, put an "X" through the weaknesses — this is about strengths — and identify how each person can best benefit the group at this time. It may be that people decide to work together on tasks they once did separately, or they may shift duties altogether.

### Steps to Increase Your Ability to Empower Others to Act

Communication is the key here; that means talking, listening, providing feedback, developing trust.

1. **Create goals of "how we get there" as a group.** Although it is the responsibility of the leader to present the vision, it is the responsibility of the group to figure out how to accomplish it. Being part of the solution helps people feel more involved and more committed to the changes that will take place.
2. **Create clear expectations of each person's responsibility.** Clarity is critical. Change produces enough ambiguity. The job of the leader is to make things as clear as possible.
3. **Trust others to represent the best interests of the team.** If the vision, goals, and roles are clear, people can represent each other. Everyone does not have to attend every meeting.
4. **Utilize and encourage the strengths of the group members.** When in doubt, refer to your list.
5. **Allow freedom to make mistakes.** Mistakes will be made because they always are. Changing times require new approaches, and this often means trial and error. Do not look for whom to punish, but to what can be learned. If people are allowed to make mistakes then they will try new things. That is what is needed.
6. **Help people stretch to the next level.** The natural tendency during change is to become more cautious and to withdraw when what is needed is to stretch and expand. Encourage others (and yourself) to go to the next level in ideas and action.

Goethe said, "Treat people as if they were what they ought to be and you help them to become what they are capable of becoming."

## Characteristic 6: Model Behavior

A manager today must be not only the leader we are describing, but also a member of the team. In many ways that means modeling the behavior that you desire in others. The leader is not the exception to the rule but the rule itself. How a manager communicates, solves problems, spends money, manages time, and treats other people is observed by the group. People see what you do and take their lead from that regardless of what you say.

## Steps to Increase Your Ability to Model Behavior

1. **Do what you want others to do.** It is rather simple: do not do anything you would not want a member of your group to do. You all represent each other.
2. **Keep the end in mind.** It is the responsibility of the leader to keep the vision alive. Remember the desired outcome and keep reminding the group so you all stay on track.
3. **Stay positive.** Things may get tough, but if you — the manager, the leader — do not stay positive, who will?

### INSTANT EXERCISE

By yourself, or better yet with your group, create a list of "norms of behavior". Ask the following series of questions, answer them, and post the answers as the norms of the group.

1. How do we communicate with each other?
2. How do we approach problem solving?
3. How do we manage our time?
4. How do we spend our dollars?
5. How do we treat other people?
6. How do we run our meetings?
7. What do we do when things go wrong?

## Characteristic 7: Provide Recognition and Rewards

Everyone needs recognition for what he/she accomplishes. Leaders never forget this. Creating ceremonies and celebrations, particularly during times of change, helps remind people that they are important and valued. These rewards can be both public and private, monetary and nonmonetary. They can be for individuals and for the entire team.

## Steps to Increase Your Ability to Provide Recognition and Rewards

1. **Reward the behavior you want.** Reward the behavior and accomplishments you want or movement toward it. This encourages people to keep going.
2. **Share the wealth.** If your group is recognized for an accomplishment, share the glory. Praise the team and all those involved in the success. If you do not you may not get many other opportunities.
3. **Shoulder the blame.** If there is a mistake or a problem, the leader takes responsibility. Do not point the finger or shift the blame. In private, with the team, review the problem and learn from the mistake.

4. **Give the recognition and give it yourself.** Good intentions will buy you nothing. Do it. Do not just plan to do it. Give the recognition. Also, do it personally. As the leader your feedback and praise mean a great deal to people. Make it personal.

### INSTANT EXERCISE

Make a list of the people who deserve recognition for a job well done, a favor, a great idea, etc. Next to their names write what you are going to do (call, write a note, send a gift, etc.) and the date by which you will do it. Then follow through with the action.

## Characteristic 8: Possess and Demonstrate Character

Leaders have character. Although each person may have his/her own definition of character, it is the essence of an individual. It has to do with the respect they give to people and the respect they receive. Character is demonstrated by ethics, honesty, and sincerity. Character is the bedrock on which all the other characteristics of a leader rest.

Two traits that may best demonstrate leaders of character are that they trust people and that they possess empathy. Strong leaders have a positive view of humankind. They trust that people, given the opportunity, will do a good job. Their view is not "everyone is stupid except me," but rather "I know they can do it." Strong leaders also have empathy. They remember what it was like when they were in a similar position, or they have the capacity to imagine how someone in a given situation may be feeling. It is also their ability to communicate trust and empathy to others that results in the respect and loyalty required for a leader to lead a team through change, toward its vision.

### INSTANT EXERCISE

Think back among the many people in your life to someone you really admire. This may be a friend, a relative, a business acquaintance, or someone you met for just a moment. Think about how that person makes you feel when you are with him/her, what he/she does or says that impresses you, what he/she has done that you admire. List the specific characteristics he/she possesses that makes him/her special to you. Put a check next to those that you would like to possess. Decide what you need to do differently and take action.

## SUMMARY

In order to be a change agent in your organization, your role as a manager requires you to be a strong leader capable of making the difference in your organization. Strong leaders (1) envision the future, (2) communicate the vision, (3) motivate and inspire others to participate, (4) recognize and take advantage of opportunities, (5) empower others to act, (6) model behavior, (7) provide rewards and recognition, and (8) possess and demonstrate character. Each of these characteristics has associated skills that you as a manager can learn and utilize.

## FURTHER READING

Bass, M. *Leadership and Performance Beyond Expectations.* New York: The Free Press, 1985.

Batten, D. *Tough-Minded Leadership.* New York: AMACOM, 1989.

Belasco, A. and Stayer, C. *Flight of the Buffalo.* New York: Warner Books, 1993.

Bennis, G. and Nanus, B. *Leaders.* New York: Harper & Row, 1985.

Bennis, G. and Nanus, B. *Leaders: The Strategies for Taking Charge.* New York: Harper Perrenial, 1985.

Byham, C. and Cox, J. *Zapp! The Lightning of Empowerment.* New York: Faucet Columbine, 1988.

Cleary, T. *The Book of Leadership and Strategy: Lessons of the Chinese Masters.* Boston, MA: Shammbala, 1992.

Covey, S. *The Seven Habits of Highly Effective People.* New York: Fireside, 1990.

Covey, S. *Principle-Centered Leadership.* New York: Summit Books, 1990.

Cox, A. *Straight Talk for Monday Morning.* New York: John Wiley & Sons, 1990.

Cox, D. and Hoover, J. *Leadership When the Heat's On.* New York: McGraw-Hill, 1992.

Cribbin, J. *Leadership.* New York: AMACOM, 1981.

DePree, M. *Leadership is an Art.* New York: Dell Publishing, 1989.

DePree, M. *Leadership Jazz.* New York: Doubleday, 1992.

Dilenschneider, L. *A Briefing for Leaders.* New York: Harper Collins, 1992.

Gardner, N. *On Leadership.* New York: The Free Press, 1990.

Kotler, P. *A Force for Change.* New York: The Free Press, 1990.

Kouzes, M. and Posner, B. Z. *Credibility.* San Francisco, CA: Jossey-Bass, 1993.

Kouzes, M. and Posner, B. Z. *The Leadership Challenge.* San Francisco, CA: Jossey-Bass, 1987.

Leavitt, J. *Corporate Pathfinders.* Homewood, IL: Dow Jones-Irwin, 1986.

Loden, M. *Feminine Leadership, or How to Succeed in Business Without Being One of the Boys.* New York: Times Books.

McLean, J. W. and Weitzel, W. *Leadership: Magic, Myth, or Method,* New York: AMACOM, 1991.

Nanus, B. *Visionary Leadership.* San Francisco, CA: Jossey-Bass, 1992.

Peters, T. and Austin, N. *A Passion for Excellence.* New York: Random House, 1985.

Phillips, T. *Lincoln on Leadership.* New York: Warner Books, 1992.

Richardson, J. and Thayer, S. *The Charisma Factor.* Englewood Cliffs, NJ: Prentice-Hall, 1993.

Tichy, N. M. and De Vanna, M. *The Transformational Leadership.* New York: John Wiley & Sons, 1986.

Yeomans, W. N. *1000 Things You Never Learned in Business School.* New York: Signet, 1985.

# 23B INTERPERSONAL COMMUNICATIONS

**Richard W. Lack**

This section serves as a short supplement to Paula Taylor's discussion on the safety and health practitioner's function as a leader and change agent (Chapter 23A).

So much of a safety and health practitioner's role involves communication that some thoughts on interpersonal communications seem appropriate.

"Professional behavior" may not be a requirement of your employer, but, as I am sure readers will agree, it is a vitally important factor in one's success as a communicator in the work setting.

What are some of the characteristics of "professional behavior"? I recently attended a training class on the subject of interpersonal communications conducted by management consultant, Carole Ellison. Here are some of the characteristics our group identified:

- Polite and courteous
- Focuses on business issues versus personalities
- Competent
- Broad knowledge of the subject
- Self-confident — builds and inspires others
- Good communicator
- Has respect for others
- Is a team player
- Appearance and manner create a professional attitude
- Listens well

Much of your success as a communicator in one-on-one situations will depend on your ability to listen. Here are some tips to improve your active listening:

- Focus on the other person with your whole mind. Make eye contact and try to read their body language as well as the spoken message.
- Repeat what you have heard them say to verify your understanding.
- Ask open-ended questions for clarification. Ask "What?", "How?".
    - What are your alternatives?
    - What are you going to do now?
    - How are you going to solve it?

Remember always that You *Are* The Message. People "tune in" much more to your visual message than your verbal one.

To help make your interpersonal communications more effective, besides being a good listener you need to tailor your message according to the receiver's communication style. The listener may be visual, auditory, or kinesthetic.

| | |
|---|---|
| Visual (See) | — These people need maps, diagrams, written instructions, manuals, photographs, and other visual aids to ensure imprinting. |
| Auditory (Hear) | — These people have a mind like a tape recorder. You will need to tell them and repeat as necessary until it is imprinted. |
| Kinesthetic (Do/Feel) | — These people need to feel what it is like first. They need to do it for themselves before it is imprinted. |

Remember also that people tend to view things more from an emotional than from a factual standpoint. As the saying goes, the "chemistry" has to be right for any relationship to really succeed. Be sure then that you keep your "emotional bank account" on the positive side because this is what will define your relationships. For example:

**Emotional Bank Account**

| Positive (Deposits) | Negative (Withdrawals) |
|---|---|
| Giving sincere compliments | Withholding information |
| Sharing information | Embarrassing others |
| Seeking feedback | Treating others unfairly |
| Courteous and respectful | Sarcastic remarks |
| Offering constructive suggestions | Ignoring employees |
| Living up to commitments | Interrupting others |
| Listening attentively | Aggressive/abrupt responses |

As a good general approach, practice the "Golden Rule" and treat people with respect at all times.

Finally, professional image is a vital factor for your success as a communicator. Besides the one-on-one situation, you can reinforce your image and, therefore, your message in many other little ways such as by sending thank-you notes, remembering occasions such as birthdays and anniversaries, giving small gifts, or entertaining for lunch, dinner, etc.

The overall process can be summarized in one word: RESPECT.


## FURTHER READING

Ailes, R. *You Are the Message.* New York: Doubleday, 1988.

Allesandra, T. and Hunsaker, P. *Communicating at Work.* New York: Simon & Schuster, Fireside Books, 1993.

Decker, B. *The Art of Communicating.* Menlo Park, CA: Crisp Publications, 1988.

Elsea, J. G. *First Impression, Best Impression.* New York: Simon & Schuster, 1984.

Frank, M. O. *How to Get Your Point Across in 30 Seconds or Less.* New York: Simon & Schuster, 1986.

Hamlin, S. *How to Talk so People Will Listen.* New York: Harper & Row, 1988.

Nothstine, W. L. *Influencing Others.* Menlo Park, CA: Crisp Publications, 1989.

Swets, P. W. *The Art of Talking so that People Will Listen.* New York: Prentice-Hall, 1983.

Walton, D. *Are You Communicating?* New York: McGraw-Hill, 1989.

# 24 THE ROLE OF A SAFETY MANAGER AND SAFETY LEADER

**Michal F. Settles**

## TABLE OF CONTENTS

> "Managers do things right ... leaders do the right thing."
>
> *Warren Bennis*

## OVERVIEW

The safety industry has faced continuous challenges to meet existing and newly passed legislation in recent years. What is needed by the industry to address these demands? Some say more management is needed. I ask you to consider the roles of leadership as the safety profession undertakes existing and newly developing challenges. Safety professionals face catastrophes, technology changes, injury control, environmental issues, and ergonomic influences, to name a few. The role of management with any of these issues is, of course, of grave concern.

## ROLE OF MANAGERS

Management, as defined in Mondy's *Supervision,* is the process of getting things done through the efforts of other people. Managers are vital to any organization's success. Without managers, an organization would be faced with a void in service, execution, and delivery. Successful managers are results oriented and focus on goal attainment. A safety manager ensures that the details are not missed or overlooked. Many organizations have moved from the title of "manager" to a "team leader" or "coach".

## ROLE OF LEADERS

Leadership can be defined as getting people to follow your example because they want to. So how would you describe a safety leader? When I think of leadership, some concepts that come to mind are

- Visionary — Leaders see what the potential possibilities are ... what can be.
- Risk takers — Leaders are willing to act on their visions (turn them into realities).
- See the big picture — Leaders view a world without limitations and see the relationship to other things in the environment.
- Taps available resources — Leaders maximize all existing potential resources (human and capital).
- Creative — Leaders are constantly challenging existing paradigms.

People usually respond to leaders because they elect to do so. Successful leaders usually have a clear purpose and are associated with improvements.

## MANAGER OR LEADER

What is the difference between leadership and management? At a recent seminar, participants brainstormed this question. The quotations which surfaced were classic:

"Managers are selected for their positions while leaders rise to the occasion."
"People respond to leaders because they want to. They respond to managers because they have to."
"Managers have a function while leaders have a purpose."

The contributing works of Dr. Stephen Covey (*The Seven Habits of Highly Effective People*) and Dr. Warren Bennis (*Leaders: The Strategies for Taking Charge*) are major works for the study of organizational leadership. Dr. Covey's book highlights effective leadership factors. The seven identified habits are

1. Be proactive
2. Begin with the end in mind
3. Put first things first
4. Think win-win
5. Seek first to understand, then to be understood

6. Summarize
7. Sharpen the saw

Dr. Bennis' book is considered a classic. Bennis' work focuses on four leadership strategies:

1. Attention through vision
2. Meaning through communication
3. Trust through positioning
4. Deployment of self

One of the most important questions answered by Bennis is, "Where do leaders get their ideas, visions, and dreams?" Both Covey's and Bennis' books are insightful and describe in detail the power of leadership. Anyone interested in the topic of leadership should avail themselves of these resources.

The American Society for Training and Development (ASTD) document "Fundamentals of Leadership" distinguishes a manager from a leader:

| Manager | Leader |
| --- | --- |
| Knows how to follow directions from above | Inspires people to cooperate |
| Takes modest risks | Teaming with others, provides tools for |
| Sees people as employees, not partners | success, then gets out of the way |
| Creates and realizes a vision | People follow out of choice |
| Uses power cautiously | Takes risks |
| Committed to the organization | |
| Proficient at planning, organizing, and controlling | |

Nine leadership competencies were also identified:

1. Technical and tactical proficiency (know your job and the jobs of others)
2. Communication (speaking, writing, and listening)
3. Professional ethics (do the right thing)
4. Planning (forecasting)
5. Use of available systems (technology and people systems)
6. Decision-making (make tough choices and involve others)
7. Teaching/counseling (train your successors)
8. Supervision (to plan, direct, evaluate, coordinate, and even control [if necessary])
9. Team development (share the knowledge, involvement, and responsibility freely)

## SUMMARY

It is my contention that a successful organization needs both managers and leaders. Can an individual be both a leader and a manager? Absolutely! Can most people accomplish both with ease and success? For most people, achieving both can be very difficult and complex at best. The continuous visionary abilities of leaders allow an organization to move from a short-term day-to-day focus to a long-term focus. Managers, on the other hand, are the key to making the vision a reality. Organizations need both (leaders and managers) to succeed. An organizational recipe without one or the other will not be a healthy organization for long.

Safety professionals have an opportunity to shape their organizations in both capacities. Which role will you play?

## FURTHER READING

Bennis, W. *Leaders, the Strategies of Taking Charge,* Perennial Library, New York, 1985.

Mondy, W. *Supervision,* Random House, New York, 1983

Covey, S. R. *The Seven Habits of Highly Effective People,* Simon & Schuster, Fireside Books, New York, 1990.

# 25 ESSENTIALS OF EFFECTIVE INFLUENCE

**John D. Adams**

## TABLE OF CONTENTS

## INTRODUCTION

We are living today in a world of frequent, unpredictable changes. New technologies are springing up daily which pose new challenges and new opportunities for safety and health professionals. As organizations restructure and as the marketplace changes, we must always be ready to update the safety and health policies in our workplaces.

Safety and health professionals seldom have the formal authority to implement changes in the programs and policies they oversee. Rather, we must use whatever personal and political skills we have to get our ideas across to others. This chapter provides a checklist and a road map for improving our effectiveness in establishing "win-win" influential communications with others.

Most of us already know much of what we should do to be effective in situations calling for influence, but we frequently fail to put our knowledge into

1-56670-054-X/96/$0.00+$.50
© 1996 by CRC Press, Inc.

practice. For example, have we not already learned that we should listen actively, set clear and mutually agreed upon goals, establish clear agendas, and so on?

Much has been written about effective influence and negotiation techniques, but most people do not employ these techniques with any regularity. Instead, we assume that if we would just "try harder" with the techniques that are not working, we would eventually break through the resistance and be successful.

People's learned styles of operating and ways of processing information have a lot of inertia. The more often a thought or behavior is repeated, the more difficult it becomes to replace it with a new thought or behavior that would be more effective. It takes a lot of commitment and discipline to "reprogram" our autopilots!

## SIX ESSENTIALS

I have been collecting information on the ingredients of successful, win-win influence engagements for over 15 years. In the course of this work, I have found that the essentials for effective influence can be nicely sorted into six categories: **mindset, preparation, awareness, skills, models,** and **stage management.** You can use the following paragraphs as a self-assessment and a checklist for preparing yourself for significant influence engagements.

### Mindset

- **Versatility of thinking and behavior.** Versatility of thinking means being able to hold multiple perspectives simultaneously (e.g., immediate and long-term focuses). Versatility of behavior means having a repertoire of styles and skills from which to draw.
- **Relationship.** It is becoming more and more apparent that things happen in organizations through effective relationships much more rapidly than via the hierarchy. Trust takes a lot of effort to establish and is easily lost forever.
- **Optimistic outlook.** The self-fulfilling prophecy never takes a break! Get a clear sense of how you would be at the end of a perfect influence and adopt that "posture" from the outset.
- **Continuous learning.** Plan to learn something new about your influencing style every chance you get.
- **Be yourself.** Never put on an act or play a role. The other person can always spot it.
- **Stay centered and relaxed.** Be sure to relax yourself and keep breathing normally throughout influence engagements.

### Preparation

- **Attend to all details.**
- **Clearly envision successful, positive outcomes.**

- **Update your knowledge of the subject matter.**
- **Style.** Reflect on the style of the person(s) you plan to influence. Always be ready to align with their focus and their energy. Do not force the other person to adopt your focus at the outset.
- **Skill practice.** Always identify a skill that you will practice as a part of your preparation. This is an extension of the continuous learning point above.

## Awareness

- **Attend to emotions and procedure as well as the task.** Emotional and procedural issues will always block work on the task.
- **Emotions are sometimes more important than the task.**
- **Reflect on your own influence "assets and liabilities".** What are you good at, and where do you often "shoot yourself in the foot" when you are engaged in influencing someone?
- **Fear.** Remember that the person you want to influence probably has fears that are the same or similar to your own.

## Skills

- **Active listening.** This especially includes paraphrasing, summarizing, and asking open-ended questions.
- **Flexibility.** This means being open to feedback, willing to be influenced, and able to let go of no-win situations.
- **Language.** Less effective influencers use more qualifiers (i.e., "sort of …"), self-discounts ("I'm only a …."), intensifiers ("Wow, super!!!"), and "gotcha" questions ("Don't you think…?").
- **Assertiveness.** This should be coupled with patience and tolerance. It is *not* the same as aggressiveness.
- **Style differences.** It is normal for each of us to want others to set their priorities and see the world in the same way that we do. You will have a lot more success in your influence activities if you first gain an awareness of the priorities, focus, and world views of the person you want to influence and then "match" or support those positions regardless of how different they are from your own. For example, if your priority is urgency and getting on with things, you are going to have trouble establishing a positive influence relationship with someone whose priority is having all the data and being deliberate, unless you are willing to shift from the viewpoint of "hurry up" to one of "systematic analysis".

## Models

- **Systematic.** Having a systematic model of the influence process guides the development of the interaction and ensures that you will not overlook anything essential to your success.

In the section that follows, a systematic model for "stage managing" your influence interactions is developed. With practice, this model becomes an

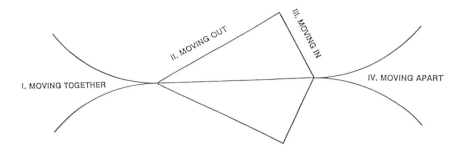

**Figure 25-1** Stages for exerting influence. (Adapted by John D. Adams from a model developed by Dennis Kinlaw.)

automatic guide that you do not have to think about to make use of. **The best way to make it automatic is to practice the various elements in the model, one step at a time.** After some period of practicing the bits and pieces, you will find that you are automatically using the whole thing.

## Stage Management

The diagram shown in Figure 25-1 provides a graphic portrayal of a systematic way to stage manage your influence interactions. There are four aspects to the model, which should not be taken as linear or sequential. In other words, your goal is to eventually learn to monitor **all four** of the aspects **all** of the time.

### *Stage I. Moving Together*

This aspect is about **contacting** and **contracting,** and it has to do with building the necessary relationship with the person you want to influence. Neglecting to establish and maintain rapport and not to establish a clear contract for working together are the most frequently overlooked aspects of the influence process. When these things are overlooked, a successful influence interaction is unlikely.

Whenever there is tension in an interaction it is easy to fall out of rapport. It is essential that you pick the right time and place, ensure that there is enough time to talk, eliminate interruptions, and do whatever is necessary to ensure that both you and the other person are paying attention to each other.

An easy way to establish a working contract with someone is to remember the acronym TEA, which stands for "time, expectations, and agenda." A single sentence can establish a solid working agreement. "I understand that you have 10 minutes available right now (time) and we both understand how important the elimination of the toxic fumes is (expectations), so let's focus right now on identifying the source of the fumes and clarifying our roles relative to getting them stopped (agenda)."

## Stage IV. Moving Apart

This aspect is about **finishing** and **following up,** and it has to do with sustaining the relationship you have built with the person you are influencing. Neglecting to summarize agreements and repair "damages" to the relationship and establish specific follow-up accountabilities are the second most frequently overlooked aspects of the influence process. When these are overlooked, a successful influence interaction is unlikely.

As the influence process progresses, many agreements are likely to be reached. These are often lost if they are not summarized from time to time, especially at the close of the interaction.

One reason that agreements do not get carried out is that they often require the person to act in nonhabitual ways. When this is the case, it is exceedingly easy to forget or to gradually regress. Following up is essential to keep this from happening.

## Stage II. Moving Out

This aspect is about **diverging** and **diagnosing,** and it has to do with exploring the content of the influence. Most of us move to taking action too quickly. We also expect the other person to get what we are after and to buy in far more quickly than is usually possible.

It is essential that you maintain an exploratory, divergent stance for longer than you think necessary. Always ask, "Is there anything else we need to consider?" at least once more than you think is necessary.

There are five elements that are useful to know about in this aspect if you are expecting there to be any resistance to your influence. When these five elements are developed fully, most nonpolitical resistance will dissipate. You may find them to be useful guidelines for doing your homework prior to the influence meeting.

- **Belief** that what you are asking of them is both desirable and possible
- **Sufficient dissatisfaction** with the status quo or present situation
- **Goals** for change that are understood and agreed upon
- **First steps** that are guaranteed to be successful
- **Other people** — Who needs to be informed, included, excluded, negotiated with, etc.?

## Stage III. Moving In

This is the stage of deciding among the alternatives and taking action. Little needs to be said about this aspect if the above three are attended to sufficiently. In our culture we are all too often in a hurry to get to this aspect, at the expense of the other three — and at the expense of success!

In preparation for a crucial influence interaction, look over the above sections and select one or two points to concentrate on. Use these ideas to build your own personal plan for the event. While working at improving your general ability to influence others, do not try to simultaneously adopt every point on the list. Pick one or two key ideas to work on in a conscious way in every influence situation into which you enter. When those items become a natural part of your approach, select a few more. After a few months, you will have dramatically improved your ability to influence others in a positive, win-win fashion.

## FURTHER READING

Bellman, G. M. *Getting Things Done When You Are Not in Charge.* San Francisco, CA: Berrett-Koehler Publishers, 1992.

Deep, S. and Sussman, L. *What to Say to Get What You Want.* Reading, MA: Addison-Wesley, 1992.

Laborde, G. Z. *Influencing with Integrity.* Palo Alto, CA: Syntony Publishing, 1987.

McCann, R. *How to Influence Others at Work.* Oxford, U.K.: Heinemann, 1988.

Northstine, W. L. *Influencing Others.* Los Altos, CA: Crisp Publications, 1989.

# 26

# DEVELOPING EFFECTIVE PRESENTATION SKILLS

**Robert S. Fish and Ken Braly**

## TABLE OF CONTENTS

## INTRODUCTION

Giving a terrific presentation is not easy, but it need not be a cause of great anxiety, worry, and dread. One reason it is torturous for so many of us is that we received little training in how to speak to groups at any point in our education. It is bad enough to feel inept, but to have to stand up in front of others and parade our ineptness goes beyond the bounds of decency.

What can one do? First, understand that it is not a genetic defect; you really can become an accomplished and confident speaker. Second, understand that the route to that end is to be motivated, persistent, and willing to make mistakes. Remember when you learned a sport? You had to accept that for a time you would fall down, or miss the ball completely, or just look silly. It is part of the process; accept it and get on with it.

Make an earnest attempt to apply as many of the techniques suggested in this chapter as you can over time. As Winston Churchill said in what may be the shortest commencement address ever given: "Never, never, never, never give up."

## ENTHUSIASM — THE MAGIC WEAPON

Probably no characteristic of a speaker can do more to keep the attention and interest of an audience than enthusiasm. Yet enthusiasm is often misunderstood.

Some points to consider include the following:

- *Enthusiasm is not volume.* We have all heard advertising on radio and television where the announcer shouts at the top of his lungs about what a great deal is being offered and how you had better not miss it. Many presenters, especially technical presenters, fear that showing enthusiasm means coming across like this.
- *Enthusiasm is not hyperactivity.* Your body language is always important when giving a presentation, but showing enthusiasm does not mean racing around and flailing your arms.
- *Let them know your topic is important.* Enthusiasm will show through naturally when you feel that your topic is interesting and important and that it is important for your audience to hear about.
- *What if you just are not enthusiastic?* What if you do not care about the subject? Or what if you cannot see how to be enthusiastic the tenth time you give the same presentation on why rectabular extrusions will not transmogrify? Well, try. Try to find something about the subject you can be interested in. Challenge yourself to find ways to bring a dull, boring subject to life for your audience. If the subject does not turn you on, be creative and get turned on with giving a terrific presentation.

## EXAMPLES AND STORIES

People remember stories, often long after the facts are forgotten. They can add a lot to keeping an audience's interest. They are the best way to add meaning to technical facts and figures and to add humanity and life to dry material.

- *Stories are not mysterious.* You do not have to be a Garrison Keillor or Bill Cosby. Just add some life to the facts by telling about what happened in real life.
- *Keep them brief.* A good story or anecdote can be just 30 to 60 seconds long.

- *Make them relevant.* A story may illustrate a point, or it may provide context for the facts, or it may simply add color, letting the audience know that behind what is being reported are real people, with real concerns and real foibles.
- *Add details.* A good story is rich with details about the people, places, and things that are described. Rather than saying "John didn't think we'd get done on time," add details about how it really happened: "John had been sitting at his computer for hours, trying to get the last tests run. When I walked by, he glanced up with a worried look and said, 'I don't think we're going to make the deadline.'"
- *Practice and rehearse a story.* Even if what you tell about happened to you, do not assume that your words will flow smoothly and succinctly without practice.

## ASPECTS OF DELIVERY THAT CONNECT WITH AN AUDIENCE

What do good speakers do during a presentation to demonstrate confidence and to foster a sense of trust and authenticity?

1. They have strong eye communication with the audience.
2. They show enthusiasm for the topic with natural gestures and movement.
3. They make their voices expressive, i.e., use fluctuations in volume, pitch, and tone rather than speaking in a monotone.
4. They smile appropriately.
5. They feel comfortable in moving about the room, rather than being "tied" to one spot.
6. They use facial expressions.
7. They give the impression that they are enjoying speaking and being with the audience.

## DEFINING THE PURPOSE AND MESSAGE

The **purpose** describes what you want to accomplish with the talk. It is vital for you to know specifically what outcome you want. A statement of your purpose usually begins with the word "to".

Examples of purpose statements include the following:

1. To inform management of the progress made on the back injury control program during the last month
2. To persuade the supervisors that their current inspection system is outdated
3. To demonstrate the functionality of the new record-keeping software

Sometimes it is helpful to distinguish between *your* purpose and the purpose you are going to tell the *audience.* For example, you might define your purpose as "to get the management review committee to fund my project". Yet you probably would not use those words when opening your talk. Instead, you might say, "My purpose today is to explain this project and illustrate why it is a good investment for our company."

The **message** — sometimes also called the *point* — is the single statement that sums up your whole presentation. If the audience remembers nothing else, you want them to remember this.

Using the examples from above, the following message statements might be prepared:

1. The back injury control program is nearly complete, but we will need additional resources to finish on schedule.
2. You will need to improve the inspection system within a year.
3. The new version is substantially more powerful than the previous version.

When you prepare a new presentation, write out the purpose and the message and keep them in front of you as you work. Make every decision with an eye to achieving your purpose and communicating your message.

## UNDERSTANDING YOUR AUDIENCE

It is your responsibility to make sure that the audience hears the message you want them to hear. You must analyze your audience to understand their needs and wants.

Questions to ask include:

1. What is their level of knowledge about your subject? Are they technical? Nontechnical? A mixed group?
2. What are their expectations? What sort of presentation, and what sort of information, are they looking for?
3. What is their attitude? Are they already in favor of what you are going to present, or are you trying to persuade some who may be skeptical or even hostile?
4. What are their biggest concerns ("hot buttons")? For example, are they most concerned about money? Schedule? Your ability to achieve what you promise?
5. What turns them off? Is there anything you should be sure you do *not* say or do?

## OPENING A PRESENTATION

"You never get a second chance to make a first impression." That cliché is especially true when it comes to giving a presentation. In your first few minutes, you can change your audience's thinking from "Oh brother, another safety talk (yawn)" to "Hey, this is going to be interesting."

Here is what the opening of your presentation should do:

- *Capture attention.* The minds of those in the audience may be wandering. You want them focused on you.
- *Identify the purpose and message.* Get the bottom line out there.
- *Clarify the agenda.* Tell them what you are going to tell them.

Here are some ideas for achieving this:

- Instead of opening with "Good morning, my name is X, and I'd like to tell you about Y," try something more intriguing — a bold statement, a question, dramatic statistics, or a brief story or example. Begin with confidence.
- If the audience does not know you, tell them a little about your background to help establish your credibility.
- Your presentation has (or should have) a message — the single most important idea you want your audience to remember. Tell them the message up front, in the form of a benefit to them.
- Explain the agenda of your presentation — what you are going to cover. The audience likes to know where you are going. Show them you are interested in them by letting them know you will address their concerns.
- Humor is okay if you know they will think it is funny and if it is relevant to the topic. Do not start with a joke just because you once heard that you should.
- Do not apologize or ever admit you are not as ready as you should be.

## BUILDING A STRONG CONCLUSION

The conclusion is arguably the most important part of your presentation. It is often the part that the audience remembers best. So it is a good idea to work to make the ending as strong as possible.

Here are some guidelines for a strong conclusion:

- Here is where you summarize your presentation, reiterating the key points. The conclusion should not contain any new material.
- This is your last chance to state (passionately, perhaps) your message.
- This is the place where you "ask for the order", in sales terms. What do you want from the audience? Tell them.
- Consider this sequence:
  1. Summarize the main ideas of the body.
  2. Conduct a question and answer period (we will cover this later).
  3. State the message once again.
  4. Ask the audience for approval, concurrence, etc.
- Try not to let the talk end with the answer to a question. Have the final say — a restatement of the message, perhaps, or of the main benefit.
- Prepare in advance the last few sentences of the presentation, so you know how you will end before you even begin.

## QUESTIONS AND ANSWERS

Some speakers fear the question and answer ("Q&A") aspect of a presentation: it is unpredictable, it introduces new threats, and it calls for thinking on your feet. Q&A can be the most valuable part of your talk, however — the time when you really sell your ideas best, when you can really connect with your audience over areas they care most about.

Keep these suggestions in mind:

- "Seek first to understand, then to be understood" — Stephen Covey
- Listen for what is not said: what are they *really* asking?
- Think. Ask for clarification if you need it.
- Get to the point quickly, and back it up with examples or data.
- Have eye contact with both the questioner and everyone else.
- Do not get caught up in technical complexity. If it is a long answer, or extremely detailed, ask to put it off for one-on-one discussion later.
- If audience members interrupt you, let them.
- If you are asked a number of questions at once, answer them one at a time.
- Avoid negative words, e.g., can't, won't. Be careful with "but" — it cancels everything you had said before it.
- Prepare and practice your answers ahead of time. One of life's little pleasures is being asked a "killer" question you are ready for.
- When you absolutely do not know, admit it with confidence, and if you offer to follow up, do it!
- Conclude your talk with a summary and restatement of your main points, not just an answer to a question.

## REHEARSALS

No professional actor would dream of going on stage to perform in front of the public without extensive rehearsal. Why do we resist practicing our presentations beforehand? Well, rehearsing can be time-consuming and boring. However, rehearsals can work *magic*.

Consider the following:

- Rehearsals give you control over your material, so you are smooth instead of stumbling.
- Rehearsals add to your confidence. You present a stronger image, and you handle the unexpected better.
- Rehearsals give you the air of professionalism.
- Rehearsals reduce stage fright and anxiety.

The big drawback is that they take time — and there are always so many other, more important things to take care of. It is hard to find time to do it.

However, you can rehearse your introduction and conclusion while commuting or taking a shower. Use any spare time. Carry your notes and hard copy of visual aids around with you. Look them over while you are standing in line at the post office.

It comes down to priorities. Those who see good speaking skills as career-enhancing will *make* time to rehearse.

How should one rehearse?

- Say the words *out loud.* We all have a tendency to be brilliant in our minds, so let the words come out to learn where the rough spots are.
- Practice with your visual aids. Again, do not just run through it mentally. Actually flip the pages, or turn the projector on and off.
- If it is an especially important presentation, practice before some colleagues whose feedback you trust.

## FURTHER READING

Clancy, J. *The Invisible Powers: The Language of Business.* Lexington, MA: Lexington Books, 1989.

Leech, T. *How to Prepare, Stage, and Deliver Winning Presentations.* New York: Amacom, 1982.

LeRoux, P. *Selling to a Group: Presentation Strategies.* New York: Barnes & Noble, 1984.

Meuse, L. *Succeeding at Business and Technical Presentations.* New York: John Wiley & Sons, 1988.

Perret, G. *Using Humor for Effective Business Speaking.* New York: Sterling Publishing, 1989.

Timm, P. *Functional Business Presentations: Getting Across.* Englewood Cliffs, NJ: Prentice-Hall, 1981.

# 27 CREATIVE PROBLEM-SOLVING TECHNIQUES

**Anne Durrum Robinson and Holland A. Young**

## TABLE OF CONTENTS

## MENTAL ENERGY: MIRED IN THE PROBLEM OR MOVED TOWARD THE SOLUTIONS?

In Chapter 21, Richard Lack addresses problem-solving techniques for safety and health professionals. This particular chapter narrows the focus to an encouraging look at the globally practiced approach offered by the Creative Education Foundation through its often-repeated Creative Problem-Solving Institute (CPSI) in Buffalo, New York. The institutes bring participants from every continent and many countries for a creative field-day (actually almost a field-week) of basic instruction in the CPSI process and hundreds of enriching adjunct sessions on many phases of creative thinking. As one adherent so aptly expressed the general feeling, "It's like being a kid in a creativity candy store!"

Because of the chapter confines we will address the process rather directly, covering roles, rules, advantages, and process steps. We will illustrate the approach to the process with concrete but limited examples. Many of its enriching capabilities must, of necessity, be left to readers to seek for themselves from the various books listed in Further Reading section at the end of the chapter.

To make the chapter more pertinent for safety and health professionals we are taking the dual view that, often, specific safety and health measures for individuals stand or fall on the basic health of an entire organization. So we are combining Anne Robinson's knowledge of the creative problem-solving process with Holland Young's expertise as planning and environmental manager for the Austin (Texas) airport.

Executives at the Austin facility face rough-rider challenges that can be compared to those of circus performers who attempt to ride two steeds at once, with one foot on the back of one horse and the other foot on the back of a different animal. After a long, involved (and often bitter) struggle among civic factions, a decision was made to move the present airport from its in-town position to the rather recently closed Bergstrom Air Force Base. Bergstrom is located on the edge of Austin proper. Complete relocation from the present site will be a very complex process which will take many years. The Austin Department of Aviation's New Airport Project Team has relocated to Bergstrom to begin the process of building new airport facilities. Questions continually arise about what to keep "as is", what to transform or renovate, what to demolish, etc. Obviously there are multiple distractions which consume energy and affect the safety as well as the mental and physical health of employees at both locations. A department which heretofore has needed only to operate one airport now must meet the challenge of building a new airport while continuing efficient and safe operations at the existing airport.

The first caveat to any problem solver is to caution him/her/them to erase the word "problem" and, instead, substitute

- Challenge
- Opportunity
- Learning event

or any other word which connotes a positive outlook, a belief that "… there are solutions and all we have to do is find them!" Such an approach conveys a much more encouraging message to the subconscious mind, which can then work more effectively with the conscious mind to uncover productive directions of effort.

Research has shown that any tried-and-true problem-analysis or problem-solving process offers a much greater possibility of successful outcomes than repeated gnawing on the problems themselves. As our chapter heading indicates, with such a process we move the energy from the problems mire toward the higher (and safer) ground of beneficial solutions.

There are any number of proven processes, including Synectics (Cambridge-based but world renowned) and Kepner-Tregoe (also globally taught and recognized). Challenge-meeters will find that some processes work best in some instances, other processes or techniques in others. In this chapter, we are dwelling only on the CPSI process and, as indicated, primarily on its structure.

The CPSI process involves three principal roles:

- The CLIENT (or decision-maker) — who has the challenge, knows the facts, and can have *some* impact on the outcome.
- The LEADER (or facilitator) — who knows the process and is adept at leading client and participants through it, but who has the least possible stake in the process outcome. (This neutral attitude is essential. There must be no feeling that the leader is manipulating the process or the people involved.)
- The PARTICIPANTS — those who have gathered to help the client (or clients) meet the challenge.

The process can be effectively used by an individual who conscientiously follows the steps, by partners (one of whom takes the role of leader and participant while the other takes the role of client and participant), or by a reasonable-sized group (of perhaps eight people). Anne Robinson *has* given CPSI-process demonstrations to much larger audiences (even one where part of the participants were blind and part were visually impaired). She also mentors a process-learning group from a large state organization which often approaches 15 or more attendants. However, she finds a group of six to eight most effective. A group of six allows for a client, a leader, and four helpers; a group of eight for a client, a leader, and six helpers. Such a group is recommended during the initial learning of the process. Once one is comfortable using and leading the process, one can profitably move to individual or partnership creative problem-solving efforts.

Necessary supplies include a flip-chart easel and paper (perhaps the 3M® chart paper which adheres to a wall or paper similar to Wall-Right), black and colored markers, possibly Post-It® notes and/or colored sticky dots, paper, and pencils or pens for participants.

Participant materials should include a description and expectations of the roles, a listing of the rules, a listing and explanation of the process steps, and a sample of process-step pages (so learners can duplicate those for their own future use and guidance).

Where a team or group is involved on a regular basis, individuals should rotate roles as much as possible. Thus each will learn to lead, to participate, and to act as an effective client.

Experienced leaders can, when necessary, handle more than one client. However, the process works more cleanly and moves more quickly when only one decision-maker is involved at a time.

The usual brainstorming rules are

- All contributions are honored and recorded exactly the way the donor wants them.
- Suggestions may be embellished or built on.
- No criticism or editing (direct or implied) is allowed.
- Attention is focused on the leader and the flip chart. No side conversations are permitted. Experience has shown that these take the focus away from the front, funnel energy off into wasteful "tributaries".
- Participants may contribute or pass if they wish.
- Steps in the process are taken in order — no leaping to ideas or conclusions before the proper time.
- Complete records of the process are kept on the flip chart up front.
- These records are given to the client at the close of the session.
- Participants are encouraged to have fun with the exercise as this fosters a more relaxed and creative atmosphere.

Steps in the process include:

- The FUZZY MESS or POSSIBLE PROBLEM — This is a one-sentence statement of the worry or wonder or concern or challenge or opportunity, as it now seems.
- FACT FINDING — The client states the concern or challenge as he/she sees it, gives all the pertinent facts, says why it is a group or personal problem, and states all the things he/she has tried or thought of it to resolve. In this step participants may ask questions for clarification. After this step participants are discouraged from asking further questions. Too often a question is, in itself, a "safety measure". The participant wants to be "safe" in making a suggestion. *Creative problem-solving encourages the making of suggestions without fear of being ridiculous or wrong.*
- PROBLEM FINDING — In this step participants (and the client) look at various aspects of the problem and state them in several ways:
  - How to (H/2)....
  - In what ways might (IWWM)....
  - I wish (I/w)....

Participants may have been thinking of these problem analysis statements as facts are being given. The leader may encourage random suggestion of such statements or follow some order of his/her choosing. A leader may also involve a scribe to help with fast posting so momentum is maintained. Posting can be done with Post-It® Notes (large ones).

A leader continues to press for more problem statements because these give more range for later ideas. Often a group's creativity is enhanced toward the end of a process step.

When the group has exhausted its problem finding (or the time it can allot to that), the client is asked to select the problem analysis statements to which he/she wants ideas addressed. He/she may combine statements and may select as many as appeal. This is the client's decision alone; he/she need not explain a choice as criticism. Sometimes a client knows why such a suggestion or problem statement is not feasible.

- IDEA FINDING — Now participants suggest specific, behavioral ideas which may help solve or improve the problem statements selected by the client. Anyone (other participants or the client) may build on or add to the suggested idea. Participants are not there to get credit or shoot down other people's ideas. They are there to help the client move toward solutions of the problem aspects.

  Again, the leader presses for the greatest quantity. Quality can be determined later. Any type of judgment (verbal or implied) slows the idea flow.

  A group (or a person) will tend first to offer more practical, obvious, standard ideas. As the flow increases and the group warms to its collective creative task, ideas tend to become more imaginative, more free. Here is where the leader may inject some of the ideation techniques listed on a later page. Such techniques encourage the thinking to move across the corpus callosum into the right brain's more playful, risk-taking, visionary, and intuitive realm. When group trust is high, these ideas come more freely.

  When the idea finding has apparently run its course (or when the client signals that he/she has a sufficient number) the group stops. The client then again selects (without explanation or excuse) the ideas which he/she wishes to incorporate in his/her action plan.

- SOLUTION FINDING — This step involves making a decision table with a list of a number of favored options (identified by key words) which the client will rate against a chosen list of criteria. Options are listed vertically on the left. Criteria (musts and wants) are listed across the top. Each option is judged against the same criterion in sequence. Against the MUST criteria the options are checked only to see if they pass that *necessary* hurdle. Against the other criteria each option is rated on a scale of 1 to 10, with 10 as the best. Again, each option is measured against the same criterion in sequence. After all the options have been tested and rated against all the criteria, the leader totals the ratings. If the ratings tend to cluster too closely together, the client can weight the criteria by assigning a relative numerical importance, multiply ratings by weights, and get new weighted totals.

  The decision table is not a "table of stone". It simply indicates how the client subjectively feels about the various options and which he/she might want to pursue first.

- ACTION FINDING — Once the client has defined preferred options, he/she should record a definite action plan, stating *which option* should be done *by whom at what time* or *by when.*

- ACCEPTANCE FINDING — This step allows the client to look at his/her action plan and determine who might help with it, who might hinder, and how to follow up with either help or hindrance. The client might take the action plan through a who, what, where, when, how, and why catechism and alter or improve the plan accordingly.

We have taken this process outline and — in the confines of a chapter — attempted to show how it might be employed to remedy some of the dilemmas faced by the Austin Aviation Department as previously discussed. Please bear in mind that actual use of the process would go much further in the problem analysis and idea-finding steps. It would also be greatly enriched by the addition of such ideation stimulators as those listed in the Methods of Ideation section.

## SIMULATION

### Fuzzy Mess or Task

The original problem statement offered by Holland was the following: We have a big problem with communication.

### Fact Finding

- Aviation Department employees have problems with intradepartment communication.
- They also have a problem with airlines and other businesses that operate in the airport (interdepartment communication).
- We are not all informed as to what others are doing.
- We often are not aware of where other people are focusing their efforts — what they are trying to accomplish.
- The lack of communication cuts into the ability to run an efficient and cost-effective operation.
- Personal friction evolves when different parties do not understand what others are doing and why.
- We have done some things with the tenants:
  - Have held regular meetings
  - Instituted a series of mail boxes for tenant communication
- Because we are building a new airport we have part of our staff in two different physical locations.
- People tend to establish a territory and do not know that they might be stepping over the boundaries of others' territory. This is related to organization as well as to communication.
- We do have regular staff meetings, but they are not well organized.
- We have been through the total quality management (TQM) process and are very deeply involved in TQM implementation.
- A management consultant is now reviewing the organization.
- We are in the process of building a new airport, but our organization was set up to run our existing airport.
- Because of the change involved in taking on a project like the new airport, the lack of communication leads to some fear among staff members that their individual positions might not be needed. This really is a problem for us. It is siphoning off energy and saps people's initiative and their desire to do a good job.
- Some people see the group that is building the new airport as the "elite" group, one which gets favored treatment.
- The very fast change and fast-paced nature of the business demands that we all know what all are doing for us to work well.

- Generally speaking, we have a very high level of individual excellence in the department.
- There has never been any sort of formal or informal communication process other than departmental newsletters.
- Holland is doing a new airport newsletter, *NEW AIRPORT NEWS,* which goes out to the public. People need to be informed and we need to work toward consensus.
- We have established a computer network between the two locations, with everyone linked by E-mail. It is a slick system — easy to use, can print, can attach documents. It has already eased the situation somewhat.

## Problem Finding

- I wish (I/w) that everyone had more clearly defined areas of responsibility.
- I/w there were only two people in the department so it would be easier to talk. (Note: This is an example of a problem statement where the person making it knows it is an impossible wish but hopes something might come out of this perception.)
- I/w a solution would fly in on a plane. (Note: This is another absurdity which might open some thinking avenues.)
- How to (H/2) give clear-communication training to all involved staff.
- I/w people would not first jump to conclusions without getting the facts.
- H/2 make sure all the people get the right facts at the right time.
- I/w people would take a little more time to communicate with one another.
- I/w people would be a little more sensitive to how their information affects others.
- I/w each airport could define its own vision.
- I/w the two airports could define a shared vision.
- H/2 keep people aware of the existing vision.
- H/2 translate the existing vision into concrete action.
- I/w we could define the best ways for staff to communicate: media? verbal? written? electronic?
- I/w the entire staff could be taught a good problem-solving system.
- H/2 get teams to work more rapidly and/or efficiently.
- H/2 teach teams idea-getting methods and see that they use them.
- I/w all levels of employees would be involved in the problem-solving experience, especially for those problems that affect them.
- H/2 conserve personal energy for the direct demands of each person's current job.
- I/w people would be less negative and focus more on the good things we have and how to improve them.
- I/w the airport would buy some helpful books and audio/video tapes for their staff to use on their own time or in groups.
- I/w that people would realize the high quality of others among us.
- I/w the most productive relationship between the two airports could be envisioned and built on.

- I/w people had a more clear understanding of their career paths in the department.
- I/w everybody involved had a clear picture of the perfect outcome.

Holland was then asked to check over the problem-finding (problem analysis) statements to determine to which ones he wanted ideas addressed. He was reminded that ideas should be specific and behavioral. We both acknowledged that, obviously, problem finding could have gone on much longer. However, in the interest of brevity, we decided that we should begin the judgment/focusing part of the process. Holland then decided that, inasmuch as many of the ideas might be grouped, and inasmuch as there were two of us instead of a larger group of participants, he would like ideas addressed to the entire list of problem statements; hence the following:

## Idea Finding*

- Redefine the organization.
  - Have a new chart end product.
  - Look at how the department functions, where logical work units are and how they are related.
  - Try to involve staff in developing a new organization so people have some control of their own destiny and see what is happening and why.
  - Have a contest among staff for new organizational suggestions.
  - Give awards for helpful suggestions.
  - Have a contest for best overall organizational chart.
  - Appoint a committee to judge the contest.
- Encourage wider use of E-mail.
- Add training on use of E-mail.
- Increase E-mail implementation. (You can copy other people on E-mail.)
- Encourage people to communicate more quickly and more often when they have something that is important.
- Encourage creative ideas for improvements in any area.
- Have a regular schedule for publication of good suggestions.
- Be more accurate in information that circulates.
- Determine the level of communication skills the staff possesses.
- Develop a program to enhance speaking/writing/listening skills.
- Get staff trained in meeting skills.
- Get staff trained in facilitating skills.
- Check on communication consultants.
- Make sure people get facts at the proper time by
  - Formal communication procedures such as newsletters
  - Update papers
  - Hiring someone for communication control
- Have a routine process for looking at each employee and celebrating his/her best skills. (Get employee suggestions as to how this can be done.)

* Both Holland and Anne submitted ideas.

- Get feature stories in the local paper. (Contact Lifestyle editor.)
- Define human assets available at each airport.
  - Get someone from outside.
  - Tap a staff member.
  - Investigate a way to get a communication specialist on staff.
  - Assign this to the communication division.
- Implement regular and structured management/staff communications.
- Redesign employee-supervisor performance.
- Develop a clear outline of the department career paths.
- Develop "imagineering" abilities in all employees in a series of workshops.
- Reiterate the visions of both airports.
- Simplify and integrate those into culture, perhaps with
  - Ball caps
  - Letterhead
  - Bottom of the newsletter
  - T-shirts
  - Everywhere
- At fun gatherings wear "rumperstickers" with pertinent slogans.
- Integrate staff into the development of a shared vision for the two airports.
- Stress that visions of both airports are of equal importance.
- Translate vision into specific goals and objectives.
- Encourage each employee to devise a course of individual action leading toward the goals.
- Ask each employee to define what he/she thinks the perfect position for him/her would be.
- Study and analyze how individuals *want* to communicate. Find what they prefer and cater to that.
- Have monthly lunch meetings for project managers
  - at Bergstrom
  - at the existing airport
- Analyze what teams need to do to improve.
- Keep teams focused.
- Improve facilitators' skills.
- Train more facilitators.
- Make teams aware of "excursions" for getting ideas.
- Allow latitude to be on the team of one's choice.
- See that team suggestions are given sincere consideration.
- Define what "empowerment" means at the airports.
- Teach employees how to "work smart".
- Advertise the positive attributes and the successes. Find fun and clever ways to do that.
- Establish a commendation process for "positive" leaders and practice it.
- Encourage informal approbation of positive performance and outlook.
- Determine the best print/audio/video materials from available source and encourage employees to use them.

As with problem finding, idea finding could go on much longer, with Anne and Holland experimenting with the rich possibilities of the idea-getting "excursions" on the following page. Groups are often led through many of these escaping-from-the-thinking-rut techniques with extremely fruitful results. Such seemingly absurd approaches can give much wider glimpses of possibilities and encourage participants to seek answers from *all* parts of their limitless minds.

## Methods of Ideation — CPSI Method

**Pocket articles** — Each participant takes an article from a purse or pocket and suggests an idea based on that article. Suggestions can be practical, wild, funny, etc.

**Imagery** — Participants close their eyes and relax to music. Each uses the first image which comes to his/her mind to serve as the basis for an idea.

**Personal analogy** — Each participant writes a short paragraph about what he/she would think or say if he/she were some element in the problem: a person, an inanimate object, a concept, a situation … whatever. The group then attempts to derive ideas based on the analogies.

**Sensory check** — Each participant searches his/her five senses for responses which might trigger ideas.

**Attribute listing** — The group works together to list attributes of the problem or elements in the problem to see what comes to mind.

**Role play** — Volunteers from the participant group role play aspects of the problem to see if ideas arise from this dramatic approach.

**Other worlds** — The leader or the participants suggest(s) some other world (such as the world of insects or the world of space) and try to get ideas from comparisons of the problem world to the selected "other world".

**Random related words** — One participant suggests a word related to the problem. The next participant immediately says a word which that first word suggests. In the proper order participants suggest random related words. Then each picks a word or words to stimulate ideas.

**Back to the sun** — The group attempts to take some part of the problem back to its origins to find ideas.

**Morphological charting** — The group lists various types of attributes of a problem. It then assigns these lists to various sides of an imaginary stack of cubes. The leader then draws one of the imaginary cubes and asks participants to force-fit the sides (the words on the various sides) to see what ideas are produced.

**Forced relationships** — Participants call out random words for a vertical, list, then for a horizontal list. They then force-fit #1 vertical to #1 horizontal, etc. to try for ideas.

**Alphabet: what else can I do with this?** — The participants go through the alphabet, asking, for example, "What else can I do with this which begins with A?" "With B?" "With C?"

**Scamper** — Participants take for first letters the word S-C-A-M-P-E-R. (Substitute, Combine, Alter or Adapt, Modify or Magnify or Minify, Put to other uses, Elevate or Eliminate, Reverse or Rearrange). They attempt to see if these words give them any ideas.

**Purge sheet** — The leader or the participants keep(s) a purge sheet on which is put thoughts or ideas that occur at improper times.

**In-out sheet** — Participants keep a personal in-out sheet, showing what thoughts they may have when their attention wanders. This helps access the right hemisphere.

## Solution Finding

In determining the options for judging, Holland chose the options which he thought represented the best possibilities for improving departmental communications. It was apparent that several were related and would perhaps be implemented together; thus they were judged as one.

Options chosen were

- Redefinition of the organization
- E-Mail enhancements (includes encouraging wider use, training, and implementation)
- Training (includes communication/meeting skills/facilitator training)
- Vision enhancements (includes vision development and integration)
- Employee/supervisor performance/career tracks
- Training library

Criteria given were

- Must be able to be implemented
- Want it to be relatively inexpensive
- Want potential for fast results
- Want solution(s) to be easy for people to do

**Solution-Finding Matrix**

| Options | Must be implementable | Want it to be inexpensive | Want fast results | Want it to be easy | TOTAL |
|---|---|---|---|---|---|
| Redefine the organization | ✔ | 5 | 5 | 5 | **15** |
| Enhance E-mail | ✔ | 8 | 8 | 10 | **26** |
| Employee training | ✔ | 3 | 3 | 3 | **9** |
| Vision enhancements | ✔ | 10 | 8 | 10 | **28** |
| Performance evaluations | ✔ | 8 | 8 | 8 | **24** |
| Training library | ✔ | 7 | 5 | 5 | **17** |

In reviewing the totals, three options were obviously the highest ranked: vision enhancements, E-mail enhancements, and performance evaluations. Holland felt that this made sense given that all three were strongly related to communication. He believed that all three areas should receive immediate attention.

As can be seen from the example, the CPSI process has tremendous potential in the development of solutions for very complex problems.

## SUMMARY

In employing the CPSI creative problem-analysis, problem-solving method, remember that

- It can be used by individuals, partners, or groups (with the most effective groups numbering six to eight).
  - *Leader/facilitator* (one who has least stake in the outcome but is experienced in the process)
  - *Client (or decision-maker)* who is familiar with the problem and can in some way influence the outcome
  - *Participants* who are there to assist client in solving the problem
  - (Scribe, who helps the leader with recording all of the process on the flip chart)
- It is comprised of seven steps which should be followed in order:
  - FUZZY MESS/TASK — one-sentence statement of apparent challenge.
  - FACT FINDING — statement of pertinent facts by the client, including what he/she has tried or thought of. Only in this stage may participants ask questions.
  - PROBLEM FINDING — multiple statements by all concerned of alternate statements of the challenge or parts of the challenge. Everyone (including the leader) may take part. Problem or challenge statements begin with I/W, IWWM, or H/2.
  - IDEA FINDING — specific, behavioral ideas offered after client has selected the problem statements to which he/she wants ideas directed. Ideas can be added to or built on. No criticism (stated or implied, verbal or visual) is permitted. No judgment is allowed at this stage.
  - SOLUTION FINDING — key ideas selected by client are listed on a decision table, then rated and weighted by client to discover the most promising ones. In neither this nor the selection of problem statements does the client explain choices.
  - ACTION FINDING — client offers an action list, indicating who will do what, by when in carrying out highest-rated ideas.
  - ACCEPTANCE FINDING — client checks possibilities of acceptance or hindrance in projected actions.

The CPSI process offers opportunities for focused and random awareness, for quantity of suggestions, and for timely judgment about quality. It centers attention and offers equal opportunity for participation to everyone involved. It also stimulates all parts of the fertile brain/mind.

## FURTHER READING

Ackoff, R. L. *The Art of Problem Solving.* New York: John Wiley & Sons. 1978.

Adams, J. L. *The Care and Feeding of Ideas: A Guide to Encouraging Creativity.* Reading, MA: Addison-Wesley. 1986.

Albrecht, K. *Brain Building: Easy Games to Develop Your Problem-Solving Skills.* Englewood Cliffs, NJ: Prentice-Hall.

Barrett, F. D. *10 Techniques for Creative Thinking: Bionics, Synectics, Morphology, Brain-Storming, Transcendental Meditation, Forced Association, Work Improvement, Attribute Listing, Value Engineering, Scenarios.* Canada: Management Concepts Limited, 1972.

De Bono, E. *Serious Creativity: Using the Power of Lateral Thinking to Create New Ideas.* Harper Collins. 1992.

Firestien, R. L. *Unleashing The Power of Creativity: The Key to Teamwork, Empowerment and Continuous Improvement.* 41-minute video, copyright by Firestien, with workbook.

Grossman, S. R., Rodgers, B. E. and Moore, B. R. *Innovation, Inc.: Unlocking Creativity in the Workplace.* Plano, TX: Wordware Publishing. 1988.

Kepner, C. H. and Tregoe, B. B. *The Rational Manager: A Systematic Approach to Problem-Solving and Decision Making.* Princeton, NJ: Kepner-Tregoe. 1965.

Koberg, D. and Bagnall, J. *The Universal Traveler: A Soft-Systems Guide to Creativity, Problem-Solving and the Process of Reaching Goals.* Los Altos, CA: William Kaufmann. 1976.

Osborn, A. F. *Applied Imagination: Principles and Procedures of Creative Problem-Solving,* 3rd revised ed. New York: Charles Scribner's Sons. 1963.

Parnes, S. J. *A Facilitating Style of Leadership.* Bearly Limited in association with The Creative Education Foundation. 1985.

Parnes, S. J. *Visionizing: State-of-the-Art Processes for Encouraging Innovative Excellence.* East Aurora, NY: D.O.K. Publishers. 1988.

Parnes, S. J. *Source Book for Creative Problem-Solving.* Buffalo, NY: Creative Education Foundation Press. 1992.

Prince, G. M. *The Practice of Creativity: A Manual for Dynamic Group Problem-Solving.* New York: Harper & Row. 1970.

VanGundy, A. B. *Idea Power: Techniques and Resources to Unleash the Creativity in Your Organization.* New York: AMACOM. 1992.

Van Gundy, A. B. *Techniques of Structured Problem-Solving.* New York: Van Nostrand Reinhold. 1981.

Wenger, W. and Wenger, S. *Your Limitless Inventing Machine.* Gaithersburg, MD: Psychegenics. 1979.

# 28

# EFFECTIVE WRITING: GUIDELINES FOR CLEAR COMMUNICATION

**Yvonne F. Alexander**

## TABLE OF CONTENTS

## INTRODUCTION

This chapter is about writing effectively. It is a chapter that might surprise you — it is not filled with grammatical rules and parts of speech, fancy labels,

1-56670-054-X/96/$0.00+$.50

or laborious technicalities. It is about saying what you mean and meaning what you say, because if you don't, the health and safety of your readers could be at stake.

For example, imagine that you are the pilot of a large jet plane. You have just completed what you think is a successful take-off. Suddenly you hear the air traffic controller saying, "Flight 1486, you've lost an engine. Make a left turn back to runway 32." As you begin to turn, you hear, "1486, a left turn, a left turn! You have lost an engine! Any runway, 1486, any runway!"

How will the air traffic controller's instructions determine what you do next? What do you think has happened to the engine? Is the plane gushing fuel because *an engine has fallen off*, or is the plane flying too slowly because it has *lost the power in an engine?*


## CLARITY, BREVITY, SIMPLICITY

Clarity, brevity, and simplicity are the three goals of effective communication. Think of them as the larger context in which you will find the other guidelines and techniques of good writing. Let us begin by defining these three concepts; then I will explain how you can use them to make your writing more effective. At the end of the chapter you will find a list of writing guidelines.

Over 2000 years ago, Quintilian, a Roman rhetorician, defined clear writing as "writing that is *incapable* of being misunderstood." Nowhere is this guideline more important than in health and safety management. *Clarity* means there can be no ambiguity in your writing. When writing is clear, there is no room for "lost" engines.

The famous American writer Mark Twain wrote to a friend: "Sorry for the long letter. I didn't have time to write a short one." *Brevity* does not mean writing the perfect first draft; it means editing to cut repetitious and unnecessary information and "ten-dollar" words that might alienate your reader. Brevity means using as few words as possible to convey your meaning. If you can delete a word without changing the meaning, do so. For example, does "very" add meaning to "very hungry"? Of course, because you can be mildly hungry or extremely hungry. On the other hand, you cannot be very famished.

*Simplicity* means using easy words and short sentences to convey your meaning. So when you can, try to use one- and two-syllable words. Many times, when you use words with three syllables or more, you force your reader to work too hard. It is far better to write, "Grandmother, what big eyes you have" than to tell granny that her ocular implements are of an extraordinary order of magnitude. And do not worry — you will not sound like a simpleton. The fact is that simple writing is a sign of clear thinking and hard work. Writing that is wordy and rambling is a sign of a writer who does not care or does not know any better.

Using simple words does not mean you should limit your vocabulary. On the contrary, with a large vocabulary you can express yourself clearly and precisely — in as few words as possible. A good vocabulary lets you use one word to convey your meaning rather than several words to define what you are saying. With a strong vocabulary, you can say *tolerate* instead of *put up with* and *imminent* rather than *likely to occur at any moment.*

## FOGGY WRITING

This chapter might have been called "Effective Written Communication." However, isn't *writing* (one word, two syllables) the same thing as *written communication* (two words, seven syllables)? Pretentious, overblown writing is also called the "official style," "weasel words," and "fog." Writing becomes foggy when we use too many words or use big words inappropriately: medication for medicine, utilize for use, purchase for buy.

Foggy writing is imprecise and ambiguous, and it can be costly or dangerous if your reader misunderstands your meaning. Compare these two examples. Which is easier to understand? What might happen if your reader does not follow the directions because they are unclear?

1. These special procedures for manual door operation must be continued and performed prior to each flight until electrical restoration and operation are resumed and reinspection of the lock sectors has been accomplished.
2. Until the electrical locking system is repaired, the doors must be manually locked before each flight.

Why do we write in this complicated, "official" style? Often it is because we think we are supposed to. The truth is that readers have a difficult time understanding writing that is unnecessarily complex. If your reader has to work too hard, your writing is ineffective, and the results will be poor communication and wasted time on everyone's part. It is your job as the writer to make reading easy. Do not try to impress your reader with fancy words; write to express, not impress. *Remember that the more technical or complicated something is to explain, the more simple the writing needs to be.*

## READABILITY FORMULA

"Readability" is the term used to describe how difficult something is to read. It might surprise you that an appropriate level for most business writing is the eighth to tenth grade. For scientific or highly technical writing, you might write at the 12th grade level. Anything beyond grade 13 will be too difficult for most readers. Most major metropolitan newspapers, for example, are written at the sixth grade level.

How can you reduce your readability level? Simplify, simplify, simplify. Avoid too many big words (three syllables or more), and keep your sentences short (an average of 15 words per sentence; a maximum of 25).

To measure your readability level, use the grammar checker on your computer or do the math yourself. Using a passage of at least 100 words, add the average number of words per sentence to the percentage of words with three or more syllables. (Treat the percent as a whole number.) Multiply the total by 0.4.

For example:

$$12 \text{ average number of words per sentence}$$
$$+10 \text{ percentage of big words}$$
$$\underline{22} \text{ total} \times 0.4 = \text{9th grade}$$

How can you practice reducing fog? Just imagine that you are talking on the phone rather than writing a memo. On the phone, we usually speak in a clear, simple style: "Bring your umbrella because it might rain." Often when we write, we obscure the message with fog: "It is highly recommended that you take into consideration the utilization of appropriate foul weather gear due to the fact that precipitation is anticipated."

Now that we have covered how to simplify and clarify your message, let us look at the other aspects of good writing that you will need to understand.

## Appearance

First impressions are crucial because your writing is judged by its appearance before the message is even read. Carefully select letterhead, logo type, ink, and paper. Use an appropriate size and style typeface. Always use the computer spell checker if you have one; proofread slowly and carefully.

## Seize Your Reader's Attention

STOP! DANGER! KEEP AWAY! One of the most critical aspects of health and safety writing can be getting your reader's attention quickly and conveying a sense of urgency. You can do that by using words that appeal to your reader's self-interest and are filled with emotion. So you might consider using no, don't, always, never, caution, stop, danger, or beware.

## Focus

Unless you have bad news, get to the point immediately to get your reader's attention. If you ramble, you risk losing your reader's interest.

## IDENTIFYING YOUR PURPOSE

Effective writing means that you accomplish your goals. To do so, before you begin writing, it is important that you identify your purpose, your reason for writing. Unfortunately, identifying your purpose is not always as easy as it sounds. Are you trying to explain, persuade, promote goodwill, analyze, warn, justify? Knowing your purpose will help you organize your ideas before you begin to write.

Your writing will seldom have only one purpose. Since you will usually have a combination of purposes, you will need to clarify and separate them first. For example, perhaps you want to promote goodwill by thanking your customer for the new order for the fire extinguishers and expressing your appreciation for the business. You might also want to remind your customer about the importance of following the safety rules for using the extinguishers. In addition, maybe you want to persuade him or her to take advantage of your volume discount. After you have separated your purposes, you will need to prioritize them; doing so will help you organize your thoughts before you begin writing.

In a letter or memo, it is a good idea to start by telling your reader what your purpose is: "This letter is to let you know that your order for the fire extinguishers will be delivered on Friday, May 2." Do not make the mistake of thinking you are merely conveying information. Always ask yourself what purpose the information serves. Also, do not avoid identifying the purpose altogether; chances are if you do not know where you are going, you will end up somewhere else.

## TARGETING YOUR AUDIENCE

"Audience" means the reader. As with the purpose, we often have more than one audience, which makes writing even more difficult. Keep in mind your principal audience and target your message to him or her. To identify your audience, you will need to answer two questions:

1. What does she know? Is she familiar with the policies, procedures, ratio-nale, methods, industry jargon, technical concepts? What is her back-ground? Is she a generalist or a specialist? Will she need just the big picture or the details?
2. How does he feel? Is he open-minded, resistant, skeptical, enthusiastic, hesitant, reasonable? How much of your writing will have to be persuasive? Will you need to be especially diplomatic or tactful, or can you be blunt?

### Point of View

Write from your reader's point of view, not your own. You know why you are writing this, but why should she read it? What is in it for her? How will she benefit? How will her health and safety (or that of her employees or customers) be improved or assured?

## Tone

Tone conveys an underlying feeling to your reader. It is often the "writing between the lines". Like facial expressions or gestures in speech, tone often sends a message that the actual words might not. Match the appropriate tone with the circumstances. Do not sound friendly when you want to reprimand. Do not sound official when you want to be cordial. Be sure to sound professional; remember that anger and disrespect are inappropriate in business writing. If you question the appropriateness of your tone, ask a colleague for his or her opinion.

You will notice that my tone is warmer and more personal than you might expect from the average textbook or desk reference. To get this tone, I use a lot of personal pronouns: I, you, we. I chose to write this way to get your attention and encourage you to read about a topic that otherwise might not hold your interest.

Tone is particularly important in health and safety management. You will want to identify your desired tone before you begin writing. Perhaps you want to use a tone that conveys a sense of urgency, immediacy, caution, or alarm. Notice how the meaning changes when tone differs:

- Use caution on left turns
- Left turns not advised
- Turn left at your own risk
- No left turn
- Never turn left

First decide what meaning you want to convey to your reader; then choose the words that match the tone with your meaning. Remember that you cannot please all the people all the time with any particular tone. What one reader considers straightforward and matter-of-fact might be seen by another reader as blunt or even abrasive.

## Jargon

Jargon is a specialized vocabulary known to those in a particular profession. For instance, in medical journals, cardiologists sometimes refer to heart attacks as "events." Jargon is good shorthand and can save time as long as your reader understands it. For example, "Training is handled through the 550 process at the OPM."

The problem is that most of us use far too much jargon (and too many abbreviations and acronyms). So use jargon only if you are sure it will clarify rather than obscure your message.

## Cultural Diversity

Be sensitive to communication styles in other cultures. Americans tend to be direct, to-the-point, and informal — qualities that can be offensive in some cultures. You might need to modify your style to make it appropriate for your reader. For example, "I think you made a mistake" might be too harsh; instead you can write "It appears there has been an error." Personal pronouns (I, you, we, they) can create closeness, so avoid them if you want distance, and use them cautiously when you have bad news. In some situations you will want to avoid using personal pronouns; they can seem excessively personal, informal, harsh, or inappropriate in many cultures.

## WRITING CLEAR DIRECTIONS

Writing directions can be one of the most important aspects of health and safety management, yet directions are some of the most difficult things to write well. They should be written in simple language, in a clear style, and in chronological order. Be sure that the logic of what you are writing is clear to your reader. Does your reader want to know only *what* to do? Or does he or she need to know *how* to do it or *why* it should be done this way? Be sure the directions are complete; do not assume your reader will know your meaning if you leave something out. When you say "push the button," do you mean push and let go or push and hold? Sometimes things that are simple or obvious to you may be confusing to your reader.

## STYLE

Style means putting words together in a way that is unique to you. When you like a certain writer's work, you usually mean that his or her style appeals to you. Style is not quite the same thing as tone, which you can change, depending on your message. It is hard to change your style, which is much like your signature. Think of your style as your "voice."

We all have our own style, and we are entitled to sound like ourselves, as long as the style is appropriate for the topic and audience. There is no ideal style; some people are terse, others verbose. Although some uniformity in a company or department is desirable, people should not be expected to sound alike.

## EDITING

We are all sensitive about our writing, which is an extension of ourselves. So if you edit someone else's work, do not change his or her style unless it is inappropriate. Also, be sure not to change the meaning. Sometimes changing a word or inserting or deleting punctuation can change the writer's intent. It is important to put your ego aside when you edit another's work and not change something just because it is *different;* the change must be *better.* Try to make as few changes as possible when you edit someone else's work.

## A LIFELONG JOURNEY

Keep in mind that writing is a process, not a goal. By using these guidelines you can make your writing clearer and stronger every time you write. Think of writing not as a destination but as a *journey*. And have a pleasant trip.

## WRITING GUIDELINES

- Always be clear.
- Be brief.
- Use simple words.
- Avoid fog.
- Keep the readability level below grade 13.
- Limit the number of big words (three or more syllables).
- Write short sentences (average 15 words per sentence; maximum 25).
- Make appearances count.
- Seize your reader's attention.
- Get to the point immediately.
- Identify your purpose.
- Target your audience.
- Write from your reader's point of view.
- Use the proper tone.
- Avoid jargon.
- Be sensitive to other cultures.
- Write directions in a clear, simple style.
- Respect other people's writing styles.
- Edit another's writing with care.
- Think of writing as a process, not a goal.

## FURTHER READING

### Books

Bernstein, T. *The Careful Writer,* New York: Atheneum, 1965. A 500-page guide to usage that clarifies words such as "assume" and "presume."

Gordon, K. E. *The New Well-Tempered Sentence,* New York: Tichnor & Fields, 1993. A grammar and punctuation handbook with a refreshing, whimsical approach.

Rico, G. *Writing the Natural Way,* Los Angeles, CA: J. P. Tarcher, 1983. Using right-brain techniques to release your expressive powers.

Strunk, W. Jr. and White, E. B. *The Elements of Style,* New York: Macmillan, 1979. An absolute must! Fewer than 100 pages. Read it once a year.

*The Chicago Manual of Style,* Chicago, IL: University of Chicago Press, 1993. Primarily for publishers and editors. Answers esoteric questions about writing. Nice to have, but not a necessity.

Venolia, J. *Write Right!,* Woodland Hills, CA: Periwinkle Press, 1988. A small book. Easy to use. Punctuation, grammar, and style.

Zinsser, W. *On Writing Well,* New York: HarperPerennial, 1994. An informal guide to writing nonfiction.

*Writing to Learn,* New York: Harper & Row, 1988. How to write and think clearly about any subject.

## Other Publications

### *Dictionary*

Dictionaries vary considerably. The word "Webster" is a generic term like aspirin. It tells you nothing about the quality because anyone can use the word. Be sure to get a dictionary that has the derivations of words so you can build your vocabulary by learning roots. Here are my favorites:

A. *Webster's New World Dictionary,* 2nd college ed.
B. *American Heritage Dictionary*
C. *Random House Dictionary*

### *Thesaurus*

I like Roget's hardback desk version.

### *Word Guide/Speller*

An electronic spell checker is ten times faster than using a dictionary for checking the spelling and syllabication of words.

# 29

# MANAGING DIVERSITY IN THE WORKPLACE*

**Sondra Thiederman**

## TABLE OF CONTENTS

## INTRODUCTION

"It drives me crazy," one frustrated manager said. "No matter how often I tell my staff to ask for help if something is too heavy to lift, many will still go ahead on their own. I've lost count of the number of injuries I've seen because of this sort of thing over the last few years." A safety officer at a large company in southern California had another complaint. His main concern was the fact that a high percentage of his employees refused to wear the new safety masks that the plant purchased some months before. He would get comments like, "We did fine before without them. Why do we have to start wearing them now?"

* The material in this article is based on that found in Sondra Thiederman's books, *Bridging Cultural Barriers for Corporate Success: How to Manage the Multicultural Work Force* (San Francisco: Lexington Books, 1990) and *Profiting in America's Multicultural Marketplace: How to Do Business Across Cultural Lines* (San Francisco: Lexington Books, 1991).

1-56670-054-X/96/$0.00+$.50

Because this article is about how to manage employees who have come to the United States from other countries, you might be wondering why I have begun with these particular examples. Why, for example, didn't I start out with something to do with language differences or perhaps differences in values or etiquette? The reason I chose these more subtle examples, both of which involve immigrant employees, was to illustrate the fact that cultural factors can come into play in the most unexpected ways. In these instances, for example, it was later discovered that the first complaint grew out of a value shared by the manager's Mexican employees. That value was the culturally rooted desire to appear strong and masculine in front of others. In the second case, employees — most of whom were from Vietnam — explained their reluctance to change with comments such as, "In my country we believe that when something works well, there is no need to change it. We did fine without the masks before — my friends and I just don't see the point."

When working with a culturally diverse workforce, good cross–cultural management skills become synonymous with good safety management skills. As most of you have discovered, when morale is up, safety claims are down. When people feel respected and valued, cooperation, satisfaction, and concentration increase and, in turn, safety improves. The purpose of this piece is to examine two key values that managers need to understand in order to create this type of atmosphere. The values, varying attitudes toward authority and the importance of saving face, are central to the confusion that has arisen in workplaces all over America.

## ISSUE I: ATTITUDES TOWARD AUTHORITY

### The Challenge

American managers have been complaining for years that employees do not have enough respect for them. "It's not like the old days," many say. "It would sure be nice to be looked up to once in a while like my father was." The paradox of this complaint is that these same managers are now confused by the amount of respect that they are seeing in many of their immigrant employees. What many fail to realize is that, with almost one million immigrants entering the country each year, workplace values are changing. One of these values is respect for authority. Most cultures throughout the world are hierarchical in structure. Showing respect for one's boss is considered a virtue. In the United States, on the other hand, such behavior can be perceived as "kissing up" or, at least, as evidence of a neurotic lack of self-esteem.

On the surface, this kind of respect may sound like good news to many of you. The problem is that the current climate of participative management can make this degree of respect impractical and frustrating. Employees, for example, from authoritarian cultures tend not to complain to superiors about faulty equipment, inefficient procedures, or confusion over safety regulations. This reluctance stems not, as we might first assume, from passivity, stupidity, or lack of interest in the job, but more likely from the belief that the boss has, as a Thai proverb clearly states, "been bathed in hot water". In short, the manager knows what is right and to disagree with him or her is to communicate disrespect by implying the the employee is more knowledgeable than the superior.

## The Solution

One of the fascinating things about cultural differences is the way that seemingly complex problems often have simple solutions. In the case of what seems for native-born managers to be excessive respect for authority, the answer lies in the very hierarchical structure that creates the problem in the first place. Because of the respect for hierarchy which so many immigrants have, there is apt to be, somewhere among the workers, an individual who has been informally designated the leader of the group. This individual functions as an informal leader, a role model, and, in the best of workplaces, as a liaison between manager and staff.

Most of you have informal leaders in your workplaces. They are the persons to whom others turn for advice, to whom they go if they do not understand an instruction, and who provide general guidance to those who are less experienced in the American workplace. The following tips will help you identify these leaders:

- They tend to be bilingual. This is important because language facility means that the individual can communication directly — without the aid of an interpreter — to both you and foreign–born staff.
- In a group of mixed gender, the leader will generally, although not always, be male.
- Informal leaders are usually the eldest of the group. Seniority and respect for age are dominant values among many immigrant groups and are manifested most clearly in their choice of leadership.
- Frequently, informal leaders held a prestigious position in the homeland. A leader, for example, might have been a military officer in Vietnam, a government official in South America, or a physician in India.

It is a simple matter to approach informal leaders with the goal of gaining their cooperation in achieving compliance with safety regulations. These steps will help you overcome any initial hesitation you may feel.

1. Acknowledge your respect for their status. This means to clearly state that you understand the importance of the individual's leadership position.
2. Explain clearly what you want. If you need employees to wear their goggles, to follow safety procedures more carefully, or to let you know if a piece of equipment is faulty, spell out the details so that there is no risk of misunderstanding.

3. Explain why your request is important. One of the ways in which cross–cultural management can fail is by asking for behaviors which seem nonsensical to employees. This is not to say that the request is nonsensical, but simply that it involves new concepts or approaches that require explanation. This applies, I believe, more to safety than almost any other area. When people have worked, either in their homeland or in the United States, for years without having to worry about safety procedures or special equipment, it takes some convincing to persuade them that what you are asking is more than a bureaucratic whim.

4. Ask the leader to pass your explanation on to the group. Do not be surprised if his or her message is far more persuasive than yours. Do not take this personnally. Sharing the culture and the language of the employees as well as being in a position of respect is a magical combination which can be very effective in getting the message across.

5. Ask the leader to model the behavior which you desire. Words are one thing, but action is far more powerful. When employees see their leader complying with the rules that used to seem so silly, you will be amazed how rapidly they begin to cooperate.

## ISSUE II: THE IMPORTANCE OF SAVING FACE

### The Challenge

Saving face is another value that can interfere with cross–cultural safety management. Although traditionally associated with the Far East, this value is also found in Hispanic, Middle Eastern, and Southeast Asian cultures. Saving face may seem like a complex philosophical concept, but it is really very simple. The doctrine of saving face places priority on seeing to it that no one in a relationship is being embarrassed or humiliated. Here is a brief list of the ways in which saving face can be manifested in the workplace.

1. *Reluctance to disagree with the boss.* Although sometimes employees fail to disagree simply because they are afraid or do not care, more often, especially in the multicultural workplace, this hesitancy has to do with saving face for the boss. The concern is that if the employee disagrees, the boss will be embarrassed and suffer loss of face, especially if that disagreement takes place in front of others.

2. *Hesitance to admit lack of understanding.* We have all felt this one. How often have you pretended to understand something said, for example, by your tax accountant or auto mechanic, just because you did not want to look stupid? It is not that you actually *were* stupid, nor that you did not care about learning — more likely you just wanted to save face by pretending you knew something that you did not.

3. *Reluctance to ask questions.* In America we are taught that "there is no such thing as a stupid question". You might be surprised to learn that this view is not shared by much of the rest of the world. In schools and workplaces

all over the globe, asking questions is more apt to be considered a sign of ignorance than of intellectual curiosity. Saving face also affects the immigrant's willingness to ask questions. For many, to ask a question of a superior not only makes the asker appear stupid; it also risks loss of face for the boss if he or she is unable to answer the question.

4. *Reluctance to admit mistakes out of fear of appearing inadequate.* This may apply to instances like losing a pair of protective glasses, doing a procedure wrong, or carelessly lifting something that is too heavy.

5. *Sensitivity to being reprimanded.* Discomfort when being reprimanded or coached is certainly worse when conducted in front of others and can result in extreme loss of face and emotional withdrawal.

We have all felt these emotions at one time or other. For those, however, from cultures in which saving face is a dominant value the discomfort is apt to be far worse. The Chinese proverb, "If there is no face, there is no life," points out the depth of importance that this value holds for many immigrant groups. It is not to be taken lightly and, as I am sure many of you would agree, is probably the dominant value that distinguishes many immigrants from native-born Americans.

## The Solution

As in the case of respect for authority, the challenges resulting from the value of saving face can best be solved by using the value itself. Saving face is probably the most deeply rooted value you are apt to encounter — it affects the ability to sustain good communication in the face of language barriers; it impacts your ability to coach and counsel effectively; and, most tragically, it can lead to the loss of good employees because of hurt feelings and humiliation.

It is, however, because of how deeply rooted this value is that you can use it to your advantage. When you feel that an employee is not giving you what you need — information, admissions of errors, etc. — because of fear of losing face, the following steps will help.

1. Express an understanding of their point of view. One of the big mistakes which managers of multicultural employees make is to believe that it is wrong to acknowledge that a cultural value exists. As long as it is done with respect, mentioning cultural differences can be an effective management strategy. In the case of losing face, it is perfectly acceptable to say that you understand how embarrassing it can be to disagree or admit lack of under-standing and that you have often felt the same way and know how difficult it can be.

2. Tell employees that you are concerned about losing face for yourself. Point out that you cannot do your job effectively if they do not give you the information you need, admit lack of understanding, or let you know if a

piece of equipment is faulty. In other words, you will be embarrassed in front of your superiors and colleagues — you, too, will lose face — if the employee does not help you out. What you are doing here is stating a simple truth, but using terms — the terms of the employee's central value — to which he or she can readily relate.

3. Reinforce the behavior once it happens. This does not mean that you further embarrass employees by overdoing it, but simply thank them when they bring you a problem, admit a mistake, or make a suggestion. I mention this rather simplistic step because it is all too easy to forget to reinforce behaviors which we regard as easy. For most Americans, for example, asking questions is fairly straightforward — we would never expect anyone to reward us for what seems to us to be a basic tenet of proper employee behavior. The challenge comes in recognizing that what is easy for one person can be difficult for another, and we need to stay alert to what behaviors require, and deserve, our reinforcement.

## CONCLUSION: TAKING IT TO THE WORKPLACE

Managing cultural diversity can be a daunting task. Language barriers, differences in values, varying ideas about proper employee behavior — all of these can render even the most experienced manager frustrated and confused. It will help if you remember that beneath the cloak of culture all human beings are after the same things: human dignity, physical comfort, and social support. By building on this commonality we can work to find the compromises that will make even our most multicultural companies productive and safe places to work.

## FURTHER READING*

### Presses Specializing in Diversity Publications

Brigham Young University, David M. Kennedy Center for International Studies, 280 HRCB, Provo, UT 84602, (801) 378-6528.
Gulf Publishing Company, P.O. Box 2608, Houston, TX 77001, (713) 529-4301.
Intercultural Press, P.O. Box 700, Yarmouth, ME 04096, (207) 846-5168.
Sage Publications, Inc., P.O. Box 5084, Newbury Park, CA 91359, (805) 499-0721.

### General Works on Managing Workplace Diversity

Fernandez, J. P. *Managing a Diverse Work Force: Regaining the Competitive Edge* (New York: Lexington Books, 1991).

*Note: The emphasis in this bibliography is on those items which address the issue of cultural and ethnic diversity and how this diversity affects the way we manage and do business. Although several of the books deal with international business, much of the material contained within them is applicable to those ethnic and immigrant cultures found within the borders of the United States.

Jamieson, D. and J. O'Mara. *Managing Workforce 2000: Gaining the Diversity Advantage* (San Francisco: Jossey-Bass, 1991).

Loden, M. and J. B. Rosener. *Workforce America!: Managing Employee Diversity as a Vital Resource* (Homewood, IL: Business One Irwin, 1991).

Thiederman, S. *Bridging Cultural Barriers for Corporate Success: How to Manage the Multicultural Work Force* (San Francisco: Lexington Books, 1991).

Thomas, R. R. *Beyond Race and Gender: Unleashing the Power of Your Total Work Force by Managing Diversity* (New York: Amacom, 1991).

## General Works on Cross-Cultural Business

Axtell, R. E. *Do's and Taboos Around the World* (New York: John Wiley & Sons, Inc., 1985).

Chesanow, N. *The World-Class Executive: How to Do Business Like a Pro Around the World* (New York: Bantam Books, 1986).

Fisher, G. *International Negotiation: A Cross-Cultural Perspective* (Yarmouth, ME: Intercultural Press, 1980).

Foster, D. A. *Bargaining Across Borders: How to Negotiate Business Successfully Anywhere in the World* (New York: McGraw-Hill, 1992).

Kennedy, G. *Doing Business Abroad* (New York: Simon & Schuster, 1985).

Thiederman, S. *Profiting in America's Multicultural Marketplace: How to Do Business Across Cultural Lines* (San Francisco: Lexington Books/Macmillan, 1991).

## Books Focusing on Specific Cultures and/or Countries*

### American Culture

Stewart, E. *American Cultural Patterns: A Cross-Cultural Perspective* (Yarmouth, ME: Intercultural Press, 1972).

### Arab Culture

Nydell, M. K. *Understanding Arabs: A Guide for Westerners* (Yarmouth, ME: Intercultural Press, 1987).

### Black Culture

Kochman, T. *Black and White Styles in Conflict* (Chicago: University of Chicago Press, 1981).

### Chinese Culture

Seligman, S. D. *Dealing with the Chinese: A Practical Guide to Business Etiquette in the People's Republic Today* (New York: Warner Books, 1989).

*Note*: The volumes listed here are a sampling of the dozens that are available on specific countries and cultures. Many of the books mentioned elsewhere in this Bibliography also include extensive material on specific groups.

### European Cultures

Hall, E. T. and M. R. Hall. *Understanding Cultural Differences: Germans, French, and Americans* (Yarmouth, ME: Intercultural Press, 1990).

### Filipino Culture

Gochenour, T. *Considering Filipinos* (Yarmouth, ME: Intercultural Press, 1990).

### Hispanic Culture

Condon, J. C. *Good Neighbors: Communicating with Mexicans* (Yarmouth, ME: Intercultural Press, 1985).
Knouse, S. et al. *Hispanics in the Workplace* (Newbury Park: Sage Publications, 1992).

### Japanese Culture

Moran, R. T. *Getting Your Yen's Worth* (Houston: Gulf Publishing, 1985).
Rowland, D. *Japanese Business Etiquette: A Practical Guide to Success with the Japanese* (New York: Warner Books, 1993).

### Korean Culture

Leppert, P. *Doing Business with the Koreans: A Handbook for Executives* (Chula Vista, CA: Patton Pacific Press, 1987).

# 30

# MANAGING THE HUMAN ELEMENT OF OCCUPATIONAL HEALTH AND SAFETY

**E. Scott Geller**

## TABLE OF CONTENTS

1-56670-054-X/96/$0.00+$.50
© 1996 by CRC Press, Inc.

## INTRODUCTION

Along with the design of safer industrial operations and environments has come an increased realization that the human element in a safety achievement system needs further attention. Safety professionals are becoming increasingly aware that the next successive approximation to achieving zero work injuries requires concerted efforts in the realm of psychology. However, the theory, research, and tools in psychology are so vast and often so complex that it can be an overwhelming task to choose a particular approach or strategy to apply. Indeed, there is an apparently endless market of self-help books, audiotapes, and videotapes addressing concepts seemingly relevant to the human element in industrial health and safety. How can one make an informed decision about which psychological perspective to take without earning an advanced degree in psychology? Can one trust the opinions of a psychologist-consultant who is selling a training or intervention program that represents his or her own perspective?

This chapter offers some guidelines and recommendations for selecting a particular psychological paradigm for industrial safety. Obviously, there is no quick-fix answer to managing the human elements of occupational health and safety. My particular training, research, and scholarship in psychology over the past 35 years has influenced my perspective, and this chapter reflects my professional experience and biases. I have attempted, however, to role-play "professor" rather than "psychologist-consultant" in preparing these recommendations. First, let us consider two divergent approaches to understanding and managing the human element of health and safety.

## BEHAVIOR-BASED VS. PERSON-BASED APPROACHES

Most of the myriad opinions and recommendations given to address the human element of occupational health and safety can be classified into one of two basic approaches to producing beneficial change in people — a person-based approach and a behavior-based approach. Indeed, most of the numerous psychotherapies available to treat developmental disabilities and psychological disorders (from neuroticism to psychosis) can be classified as essentially person-based or behavior-based. That is, most psychotherapies focus on changing people either from the inside out (e.g., "thinking people into acting differently") or from the outside in (e.g., "acting people into thinking differently"). In other words, person-based approaches attack individual attitudes or thinking processes directly (e.g., by teaching clients new thinking strategies or giving them insight into the origin of their abnormal or unhealthy thoughts, attitudes, or feelings). In contrast, behavior-based approaches attack the clients' behaviors directly (e.g., by changing relationships between behaviors and their consequences).

Many clinical psychologists use both person-based and behavior-based approaches with their clients, depending upon the nature of the problem. Sometimes the same client is treated with both person-based and behavior-based intervention strategies. I am convinced that both of these approaches are relevant in certain ways for improving health and safety. This chapter provides a rationale for using both approaches to manage safety and health in the workplace and offers a framework and guidelines for doing so.

### The Person-Based Approach

Imagine observing two employees pushing each other in a parking lot as a crowd gathers around to watch. Is this aggressive behavior, horseplay, or mutual instruction for self-defense? Are the employees physically attacking each other to inflict harm, or does their physical contact indicate a special friendship and mutual understanding of the line between aggression and play? Perhaps longer observation of this interaction, with attention to verbal behavior, will help determine whether this physical contact between individuals is aggression, horseplay, or a teaching/learning demonstration. However, a truly accurate account of this event might require an assessment of each individual's intentions or feelings. It is possible, in fact, that one person was aggressing while the other was having fun or that the interaction started as horseplay and progressed to aggression (from the perspective of personal feelings, attitudes, or intentions of the two individuals).

This scenario illustrates a basic premise of the person-based paradigm — namely, that focusing only on observable behavior does not explain enough about the situation. People are much more than their behaviors. Concepts like intention, creativity, intrinsic motivation, subjective interpretation, self-esteem, and mental attitude are essential to understanding and appreciating the human element of a situation. Thus, a person-based approach applies surveys, personal interviews, and focus-group discussions to find out how individuals feel about certain situations, conditions, behaviors, or personal interactions.

A wide range of therapeutic approaches fall within the general framework of person-based therapy, from the psychoanalytic techniques of Sigmund Freud, Alfred Adler, and Carl Jung to the client-centered humanism developed and practiced by Carl Rogers, Abraham Maslow, and Viktor Frankl (cf. Wandersman, Poppen, and Ricks, 1976). Humanism is the most popular person-based approach today, as evidenced by the current market of "pop psychology" videotapes, audiotapes, and self-help books, although some popular industrial psychology tools (e.g., the Myers-Briggs Type Indicator and other trait measures of personality, motivation, or risk-taking propensity) stem from psychoanalytic theory and practice.

The key perspectives of humanism included in most "pop psychology" approaches to increasing personal achievement are (1) everyone is unique in numerous ways, and the special characteristics of individuals cannot be understood or appreciated by applying general principles or concepts (e.g., the behavior-based principles of performance management or the permanent personality trait perspective of psychoanalysis); (2) individuals have far more potential to achieve than they typically realize and should not feel hampered by past experiences or present liabilities; (3) the present state of an individual in terms of feeling, thinking, and believing is a critical determinant of personal success; (4) one's self-concept influences mental and physical health, as well as personal effectiveness and achievement; (5) ineffectiveness and abnormal thinking and behavior result from large discrepancies between one's real self ("who I am") and ideal self ("who I would like to be"); and (6) individual motives vary widely and come from within a person. Readers familiar with the writings of W. Edwards Deming (1982, 1993) and Stephen R. Covey (1989, 1990) will recognize that these well-known industrial consultants would be classified as humanists (or person-based).

## The Behavior-Based Approach

The behavior-based approach to applied psychology is founded on behavioral science as conceptualized and researched by B.F. Skinner (e.g., 1938, 1971). In his experimental analysis of behavior Skinner rejected *for scientific study* unobservable inferred constructs such as self-esteem, cognition strategies, intentions, and attitudes. He researched only overt behavior and its environmental, social, and physiological determinants. Therefore, the behavior-based approach starts with an identification of overt behaviors to change and environmental conditions or contingencies (i.e., relationships between designated target behaviors and their supportive consequences) that can be manipulated to influence the target behavior(s) in desired directions.

This approach has been used effectively to solve environmental, safety, and health problems in organizations and throughout entire communities by first defining the problem in terms of relevant observable behavior and then designing and implementing intervention programs to decrease behaviors causing the problem and/or increase behaviors that can alleviate the problem (e.g., Elder, Geller, Hovell, and Mayer, 1994; Geller, Winett, and Everett, 1982; Goldstein and Krasner, 1987; Greene, Winett, Van Houten, Geller, and Iwata, 1987). The behavior-based approach to occupational safety is reflected in the research and scholarship of several safety consultants (e.g., Geller, Lehman, and Kalsher, 1989; Krause, Hidley, and Hodson, 1990; Peterson, 1989) and is becoming increasingly popular for industrial applications.

## Considering Cost-Effectiveness

When people act in certain ways they usually adjust their mental attitude and self-talk to be consistent with their actions, and when people change their attitudes, values, or thinking strategies, certain behaviors change as a result. Thus, person-based and behavior-based approaches to changing people can influence both attitudes and behaviors, either directly or indirectly. Furthermore, most parents, teachers, first-line supervisors, and safety captains have used both of these approaches in their attempts to change other persons' knowledge, skills, attitudes, or behaviors. Thus, when we lecture, counsel, or coach others in a one-on-one or group situation, we are essentially using a person-based approach, and when we recognize, correct, or discipline others for what they have done, we are operating from a behavior-based perspective. Unfortunately, we are not always effective with our person-based or behavior-based change techniques, and often we do not know whether our intervention techniques have worked as intended.

In order to apply person-based approaches to psychotherapy, clinical psychologists receive specialized therapy or counseling training for 4 years or more, followed by an internship of at least 1 year. Such intensive experiential training is necessary because tapping into an individual's perceptions, attitudes, and thinking styles is a demanding and complex process. Also, these inside dimensions of people are extremely difficult to measure reliably, making it cumbersome to assess therapeutic progress and obtain straightforward feedback regarding one's therapy skills. Consequently, the person-based therapy process can be very time-consuming, involving numerous one-on-one sessions between professional therapist and client.

In contrast, the behavior-based approach to psychotherapy was designed for administration by individuals with minimal professional training. From the start, the idea behind the behavior-based approach was to reach people in the settings where their problems occur (e.g., the home, school, rehabilitation institute, workplace) and teach the managers or leaders in these settings (e.g., parents, teachers, supervisors, friends, or co-workers) the behavior-change techniques most likely to work under the circumstances (cf. Ullman and Krasner, 1965). Thirty years of research has shown convincingly that this on-site approach is cost-effective, primarily because behavior-change techniques are straightforward and relatively easy to administer and because intervention progress can be readily monitored by ongoing observation of target behaviors. Thus, intervention agents can obtain objective feedback regarding the impact of their intervention techniques and accordingly refine or alter components of a behavior-based process.

## A Need for Integration

A common perspective, even among psychologists, is that humanists (as in person-based) and behaviorists (as in behavior-based) represent opposite poles of an intervention continuum (Wandersman et al., 1976). Behaviorists are considered cold, objective, and mechanistic, operating with minimal concern for people's feelings, whereas humanists are thought of as warm, subjective, and caring, with limited concern for *directly* changing another person's behaviors or attitudes. In fact, the basic humanistic approach to therapeutic intervention is termed "nondirective" or "client-centered", referring to the paradigm that therapists, counselors, or coaches do not directly change their clients but rather provide empathy and a caring and supportive environment for enabling clients to change themselves (i.e., from the inside out).

Given the conceptual foundations of humanism and behaviorism, it is easy to build barriers between person-based and behavior-based perspectives and assume one must follow either one or the other approach when designing an intervention process. In fact, many consultants in the safety management field market themselves as using one or the other approach, but not both. An integration of these approaches is not only possible but necessary for optimal safety and health management. In other words, all factors contributing to the safety of an organizational culture can be classified as *environmental* (e.g., equipment, tools, machines, engineering, housekeeping), *personal* (e.g., attitudes, motives, thinking strategies, personality), or *behavioral* (e.g., working safely or unsafely, acting on behalf of the safety of others), and aspects of both person-based and behavior-based psychology are relevant for addressing the two human dimensions of this "safety triad" (Geller et al., 1989).

## Building Bridges Between Person-Based and Behavior-Based Approaches

The founder of contemporary behaviorism, B.F. Skinner, is probably among the most frequently misunderstood researchers and scholars of this century. Most people, from psychologists to the general public, have presumed that Skinner's behavior-based approach to psychological research and societal application has no regard for feelings, attitudes, or thinking styles (i.e., human factors inside the person), but this is just not the case. Skinner's paradigm was that only observable behavior should be the research domain of psychology, but he certainly did not deny inside reactions or interpretations to outside (observable) events. In fact, Skinner (1971) claimed that perceptions of freedom (today we might substitute "empowerment" for "freedom") are determined by the nature of the response-consequence contingencies controlling one's behavior. When we are performing to achieve rewarding consequences (i.e., controlled by a positive reinforcement contingency) we feel "free" (or empowered); but when we are working to avoid unpleasant or punishing consequences (technically referred to as control by negative reinforcement), we feel "controlled" (i.e., our empowerment is sapped). This is the main reason proponents of the behavior-based approach advocate the use of positive consequences to motivate behavior change.

The behavior-based approach also advocates that reinforcing consequences be soon and certain. However, behaviorists also recognize that delayed consequences can influence behavior if their occurrence is certain. For example, employees work for weekly wages; students study for end-of-the-term grades; professors write research reports for delayed professional recognition, tenure, and promotion; and consultants initiate work on a contract for eventual reimbursement for expenses and fees. To explain the control of behavior with delayed consequences, behaviorists again consider person or inside factors. More specifically, it is presumed that we use rules (e.g., by talking to ourselves) to connect behaviors with delayed consequences, and then through self-talk (to remind ourselves of the rules) we can maintain our personal motivation for earning eventual rewards (cf. Malott, 1992).

Probably the most important characteristic of consequences is "certainty", a prime reason why the safety professional's job is so challenging. If a response-consequence is certain, we are motivated to perform even if the other consequence characteristics are negative and delayed. Consider, for example, the great amount of societal control achieved with the threat of delayed penalties for various undesirable behaviors (e.g., from drunk driving to tax evasion). Such threat approaches are effective as long as we believe the delayed and negative consequences have a high probability of occurring. Of course, when we are controlled by *delayed* and *certain* negative consequences, we *feel* controlled, and as a result we might procrastinate, accomplish inferior work, or attempt to beat the system.

Negative consequences for unsafe behaviors are quite rare. In fact, individuals can work unsafely for years and never receive an injury. As a result most people internalize the rule or response-consequence contingency as, "This unsafe behavior gives me convenience, comfort, or a faster job and never gives me a negative consequence." Thus, it is natural to develop the attitude, "It will never happen to me."

If people look beyond their individual contingencies to the plant population as a whole, the undesirable consequences (i.e., injuries) for unsafe behavior become much more certain. In other words, although the probability of an
injury to a certain individual is minuscule, the probability that someone at the
industrial site will get hurt in the next 6 months as a result of unsafe behavior
is probably near certainty at most large plants. When employees truly care
about this group contingency and believe their own actions can reduce the
probability that a co-worker will get injured on the job, new levels of health and
safety excellence are achievable.

A caring attitude for others and a belief that one can make a difference
relate directly to certain person factors from humanistic psychology, including
self-esteem ("I am valuable"), self-efficacy ("I can do it"), optimism ("I expect
the best"), and a sense of personal control and belongingness. Research psychologists have shown that these person characteristics (or belief states) can be
increased or decreased by environmental and interpersonal conditions, and
promoting these states increases the probability that one individual will help
another. Consequently, "actively caring" for the safety of others can be enhanced by establishing conditions, situations, or contingencies to cultivate key
person factors of humanistic psychology. The highest level on Maslow's need
hierarchy, for example, is self-transcendence, a need state quite analogous to
the "actively caring" concept (Maslow, 1971). Thus, optimal safety and health
management can be achieved by integrating person-based and behavior-based
approaches in special ways. Indeed, B.F. Skinner himself said, "Behaviorism
makes it possible to achieve the goals of humanism more effectively" (Skinner,
1978, pp. ix–x). Let us get more specific with regard to processes from
behavior-based and person-based psychology that need to be developed in
corporate cultures to improve safety and health.

## THE DO IT PROCESS

The objective of a behavior-based and person-based approach to safety
management is to change behavior directly in such a way that the resultant
attitude is positive; to achieve this objective, a continuous process is followed,
as typified by the acronym "DO IT" which my associates and I use to teach this
process. The behavior-change process involves the following four steps: D =
Define the target behavior to increase or decrease; O = Observe the target
behavior during a preintervention baseline period to set behavior-change goals,
and perhaps to understand natural environmental or social determinants of the
target behavior; I = Intervene to change the target behavior in desired directions; T = Test the impact of the intervention procedure by continuing to
observe and record the target behavior during the intervention program; and
then evaluate the cost-effectiveness of the intervention and decide whether to

continue the program, implement another intervention strategy, or define another behavior for the DO IT process.

The DO IT process for safety and health achievement is easier said than done. It begins with the clear and concise definition of a behavior (e.g., using equipment safely, lifting correctly, locking out power appropriately, looking out for the safety of others) or the outcome of behaviors (e.g., wearing personal protective equipment, working in a clean and organized environment, using a safety belt). If two or more people obtain the same frequency recordings when observing the defined behavior or behavioral outcome during the same time period, the definition is sufficient for an effective DO IT process. Baseline observations of the target behavior should be made and recorded before implementing an intervention program.

## Designing Interventions

When designing interventions to change behavior, behavior managers follow a simple ABC model (A for Activator, B for Behavior, and C for Consequence). Activators direct behavior (as when the ringing of a telephone or doorbell signals certain behaviors from residents), and consequences motivate behavior (as when residents answer or do not answer the telephone or door depending on current motives or expectations developed from prior experience at telephone or door answering). As discussed above, the most motivating consequences are soon, sizable, and certain. Safe behaviors are not usually reinforced by soon, sizable, and certain consequences. In fact, safe behaviors are often punished by soon and certain *negative* consequences (e.g., inconvenience, discomfort, slower goal attainment). The consequences that motivate safety professionals to promote safe work practices (i.e., the reduction of injuries and associated costs) are delayed, negative, and uncertain (actually improbable) from an individual perspective. Moreover, unsafe behaviors avoid the discomfort and inconvenience of most safe behaviors and usually allow people to achieve their production and quality goals faster and easier. Indeed, sometimes production supervisors inadvertently activate and reward unsafe behaviors in their attempts to achieve more production.

Because activators and consequences are naturally available throughout our everyday existence to support unsafe behaviors in lieu of safe behaviors, safety management can be considered a continuous fight with human nature. Hence, the development and maintenance of safe work practices often requires the implementation of intervention strategies to keep people safe. These intervention strategies serve as activators or consequences, or they combine both approaches, and they focus on increasing safe behaviors and/or decreasing unsafe behaviors. Let us consider some actual intervention techniques which were quite effective at increasing safe work practices. The interventions which included consequences were more effective (although usually more costly), and those which involved employees in designing, implementing,and refining the behavior-change procedures resulted in larger and longer-term change.

## Worker-Designed Safety Slogans

Signs and slogans are activators with only limited behavioral influence unless they signal the occurrence and availability of motivating consequences. Furthermore, the same signs (e.g., "Caution: Hazardous Area", "Please Hold Handrail", "Buckle Up") eventually "blend into the woodwork" and are not noticed. However, changing signs periodically can increase attention and awareness, and if employees design the messages or slogans (perhaps as a result of a company contest), awareness and attention can be increased before and after sign display. In 1985, employees and visitors driving into the main parking lot for Ford World Headquarters in Dearborn, Michigan, passed a series of four signs arranged with sequential messages. The messages were changed periodically as selected from a pool of 55 employee entries in a limerick contest for safety belt promotion. Two creative examples were

Sign 1: Mary, Mary Quite Contrary
Sign 2: Just Wouldn't Buckle, You Know
Sign 3: She Had a Fuss with a Greyhound Bus
Sign 4: Now She's Planted All in a Row.

Sign 1: Think of Those Who Died in Cars
Sign 2: And How Their Family Felt
Sign 3: How Much Trouble Can It Be
Sign 4: To Tug and Click a Belt.

## Near Miss and Corrective Action Reporting

A proactive stance to occupational safety requires discussion and problem solving regarding "near misses" (i.e., workplace incidents that could have resulted in an injury had the timing or body positioning been slightly different). Motivating near-miss reporting and discussion is not easy, especially if the reporting process is inconvenient and the instigators or victims of an injury or near miss are viewed as "careless", "dumb", "accident prone", "thoughtless", etc. In the mid-1980s, Air Products and Chemicals in Allentown, Pennsylvania attempted to encourage near-miss reporting by recognizing near-miss reporters in the company newspaper, and the entire Body and Assembly Division of Ford Motor Company promoted attention to near-miss reporting by having reporters reenact their near misses on videotape for later analysis and corrective action. It is noteworthy that these near-miss reports considered factors in each dimension of the "safety triad" (i.e., environment, person, and behavior) when searching for root causes of the incidents *and* when deriving recommendations for corrective action.

## Group Safety Share

At the start of group meetings for the Hoechst Celanese Plant in Rock Hill, South Carolina, facilitators ask participants to report something they have done for safety since the last meeting. Because this "safety share" is used to open all kinds of meetings, safety is given a priority status and integrated into the various other functions of the organization (from marketing to quality control). Employees expect to be asked about their safety achievements and thus they prepare accordingly, sometimes going out of their way for safety in order to report an impressive safety share. This awareness booster focuses on achieving safety (i.e., "What have you done for safety?") and contributes to teaching the culture that safety is not only loss control (i.e., an attempt to avoid failure) but belongs in the same achievement system as productivity, quality, and profits.

## Private and Public Commitment

Intentions are verbal statements to ourselves (i.e., private) or to others (i.e., public) that we plan on doing something. Whether private or public, such statements activate personal commitment to emit particular behaviors. However, when intentions are given publicly, the commitment is strengthened through peer influence. Public intentions can activate peer support as reminders (e.g., "Didn't you say you were going to use your ear protectors?") and as consequences (e.g., "I see you're reaching your personal safety goal; congratulations!").

Years ago, when the nationwide use of safety belts was below 15% (i.e., before state belt-use laws), many companies more than doubled the use of safety belts in company and private vehicles by distributing "buckle-up promise cards" and encouraging their employees to sign them (e.g., General Motors, Ford Motor Company, Corning Glass in Blacksburg, Virginia, Burroughs Welcome Company in Greenville, North Carolina, and the Reeves Brothers Curon Plant in Cornelius, North Carolina). Logan Aluminum in Russellville, Kentucky instituted a "Public Safety Declaration" whereby employees signed a poster at the plant entrance which specified a safe-behavior commitment for the day (e.g., "We wear safety glasses in all designated areas.").

## Actively Caring Thank-You Cards

Kal Kan of Columbus, Ohio and two Hoechst Celanese plants (Celco in Narrows, Virginia and Celriver in Rock Hill, South Carolina) have provided employees with "thank you" cards for distribution to co-workers when observing them actively caring (AC) or going beyond the call of duty for another person's safety. For some variations of this process, the observers described the AC behaviors they saw on the thank-you cards and thereby increased their operational understanding and checklists of AC behavior. In one program, the AC thank-you cards could be exchanged for a beverage in the company cafeteria; for another program, the AC thank-you cards contained a removable recognition sticker that could be attached to a variety of objects, the most popular being a hard hat.

Two of these intervention programs offered a unique consequence that extended the AC concept beyond the workplace. Specifically, when deposited in a special collection container, each thank-you card was worth 25¢ toward corporate contributions to a local charity or to needy families in the community. The employees who designed one of these programs called their AC rewards START cards for **S**afety **T**hrough **A**wareness, **R**ecognition, **T**eamwork. This intervention process was an incentive/reward program, since the employees knew they could receive a "thank you" card for emitting certain behaviors (an incentive), and they actually received this reward when meeting an AC criterion. However, this intervention was very different from the typical incentive/reward program used in industry to manage safety and health. Let us turn now to a discussion of guidelines for designing an effective incentive/ reward intervention to manage safety and health.

## INCENTIVES AND REWARDS

Referring to the three-term contingency of behavior-based psychology (i.e., **A**ctivator-**B**ehavior-**C**onsequence), an *incentive* is an activator which announces the availability of a particular pleasant consequence (i.e., a reward) following the occurrence of a certain desired (e.g., safe) behavior. Activators which announce the availability of unpleasant consequences (i.e., a penalty) following the occurrence of certain undesired (e.g., unsafe) behaviors are termed *disincentives.* Note that the power or motivational influence of incentives and disincentives is determined by consequences. Rules or policies (i.e., disincentives) which are not consistently enforced (e.g., with penalties for noncompliance) are often disregarded, and if promises of rewards (i.e., incentives) are not fulfilled when the correct behavior is emitted, subsequent incentives might be ignored.

### Doing it Wrong

The definitions of incentives and rewards imply that specific target behavior(s) must be specified to complete the three-term contingency of activator-behavior-consequence. However, most incentive programs for occupational safety do not specify behavior. That is, the most common application of incentives for safety management specifies a certain reward which employees can receive by avoiding a work injury (or by achieving a certain number of "safe work days"). Many of these nonbehavioral, outcome-based incentive programs implicate substantial peer pressure because they use a group-based contingency (i.e., if anyone in the company or work group is injured, everyone loses their reward).

So what behavior is motivated by such an outcome-based, group-contingency incentive/reward program? Obviously, if workers link certain safe behaviors directly with a high probability of avoiding an injury, then an outcome-based incentive program can have a beneficial impact. However, as discussed above, most safe and unsafe behaviors are rarely followed by supportive consequences (i.e., injury avoidance following safe behaviors and injury following unsafe behaviors). Thus, the most likely behavior to be influenced by an outcome-based incentive/reward program is *injury reporting*.

If an injury results in the loss of one's reward (or worse, the reward for an entire work group), there is pressure to avoid reporting an injury if possible. For example, I have observed co-workers cover for an injured employee in order to keep accumulating "safe days" and not lose their reward possibility. Hence, these incentive programs do decrease the numbers (i.e., the injury rate), at least over the short term, but corporate safety is obviously not improved. Indeed, such outcome-based incentive programs often lead to a detrimental attitude of apathy or helplessness regarding safety achievement. In other words, employees develop the perspective that they cannot really control their injury record, but must cheat or beat the system to celebrate the "achievement" of an injury reduction goal.

## Doing it Right

This discussion leads logically to seven basic guidelines for establishing an effective incentive/reward program for managing the human element of industrial health and safety.

- The behaviors required to achieve a safety reward should be specified and perceived as achievable by the participants.
- Everyone who meets the behavioral criterion should be rewarded.
- Groups should not be penalized (e.g., lose their rewards) for failure by an individual *unless* the group can control the individual's performance.
- It is better for many participants to receive small rewards than for one person to receive a big reward.
- Contests should not reward one group at the expense of another.
- Progress toward achieving a safety reward should be systematically monitored and publicly posted for all participants.
- The rewards should be displayable and represent safety achievement (e.g., coffee mugs, hats, shirts, sweaters, blankets, or jackets with a safety message).

## An Exemplary Incentive/Reward Program

In 1992, I consulted with the Safety Incentives Committee of a Hoechst Celanese company of about 2000 employees in the development of an exemplary plant-wide incentive program that followed each of the guidelines given above. The Safety Incentives Committee, including four hourly and four salaried employees, met several times to identify specific behavior-conse-quence contingencies (i.e., what behaviors would earn what rewards). The resultant program plan was essentially a "credit economy" whereby certain desirable safe behaviors achievable by all employees earned certain numbers of "credits", and at the end of the year the participants could exchange their "credits" for their choice among a number of different prizes (all containing a special safety logo). The variety of behaviors earning credits included atten-dance of monthly safety meetings; special participation in a safety meeting; leading a safety meeting; writing, reviewing, and revising job safety analyses; and conducting periodic audits of environmental and equipment conditions and certain work practices. For a work group to receive its credits, the results of environmental and personal protective equipment (PPE) audits had to be posted in the relevant work area. Only one behavior was penalized by a loss of credits — the late reporting of an injury.

At the start of the new year, each participant received a "safety credit card" on which individual credit earnings could be tallied per month. Some indi-vidual behaviors earned credits for the person's entire work group, thus pro-moting teamwork and group cohesion. It is noteworthy that this kind of incentive/reward program exemplifies a basic behavior-based principle for health and safety management — observation and feedback. In other words, for this intervention employees were systematically observed with regard to their performance of certain desirable behaviors, and they received quick, certain, and positive feedback (i.e., a reward) after emitting a target behavior. An incentive/reward program is only one of several methods to increase safe work practices with observation and feedback. Let us turn now to a more generic discussion of behavioral observation and feedback as a key principle for managing the human element of occupational health and safety.

## BEHAVIORAL OBSERVATION AND FEEDBACK

Ask any safety manager, industrial consultant, or applied psychologist whether he/she has heard of the "Hawthorne effect" and it is likely the answer will be "yes". He/she might not be able to describe any details of the studies which occurred between 1927 and 1932 at the Western Electric plant in the Hawthorne community (near Chicago) that led to the classic "Hawthorne effect". It is likely, however, that he/she will be able to paraphrase the infamous finding from these studies that the hourly output rates of the employees studied increased whenever an obvious environmental change occurred in the work setting. The explanation of the Hawthorne results is also well known and is recited as a potential confounding factor in numerous field studies of human behavior. Specifically, it is commonly believed that the Hawthorne studies showed that people will change their behavior in desired directions when they know their behavior is being observed. The primary Hawthorne sources (Mayo, 1933; Roethlisberger and Dickson, 1939; Whitehead, 1938) leave us with this impression, and in fact this interpretation seems intuitive. The fact is, however, this interpretation of the Hawthorne studies is not accurate — it is nothing but a widely disseminated myth.

## Hawthorne Workers Got Feedback

H. McIlvaine Parsons conducted a careful reexamination of the Hawthorne data and interviewed eyewitness observers, including one of the five female relay assemblers who were the primary targets of the Hawthorne studies. Parsons' findings were published in a seminal article entitled "What Happened at Hawthorne?". What happened was the five women observed systematically in the Relay Assembly Test Room received regular feedback about the number of relays each had assembled. "They were told daily about their output, and they found out during the working day how they were doing simply by getting up and walking a few steps to where a record of each output was being accumulated" (Parsons, 1980, p. 58).

From his scientific detective work, Parsons concluded that performance feedback was the principal extraneous, confounding variable that accounted for the Hawthorne effect. The performance feedback was important to the workers (i.e., they were apt to attend to it) because their salaries were influenced by an individual piecework schedule — the more relays each employee assembled, the more money each earned.

## Feedback for Safety

For anyone who has studied the behavior-based approach to performance management, the only surprise in Parsons' research is that the critical role of performance feedback had not been documented in the original reports of the Hawthorne studies. To these folks the finding that feedback was the critical change variable is common sense. For example, numerous research studies have shown that posting the results of behavioral observations related to safety, production, or quality yields beneficial change in the targeted work behaviors. If desired work behaviors are targeted, they increase in frequency of occurrence; when undesired behaviors are targeted for observation and feedback, they decrease in probability (e.g., Geller, Eason, Phillips, and Pierson, 1980; Kim and Hamner, 1982; Komaki, Heinzmann, and Lawson, 1980; Krause et al., 1990; Sulzer-Azaroff, 1982).

It is noteworthy that, when asked about their preference for working in the test room rather than the regular department, the five Hawthorne employees did not mention the feedback intervention, but gave a variety of other reasons, including "smaller groups", "no bosses", "less supervision", "freedom", and "the way we are treated" (Roethlisberger and Dickson, 1939, pp. 66–67). This suggests that it is inadvisable to rely on only verbal reports to discover the factors influencing work performance. Sometimes people are not aware of the basic contingencies controlling their behavior. Through systematic and objective behavioral observation these factors can be uncovered and corrective feedback given.

Most of the published research that showed significant benefits of observing work practices and then posting feedback charts used outside observers (i.e., research assistants) to do the behavioral observing and data posting. Hiring extra staff for such tasks is certainly not practical for most organizations and in fact is not an optimal approach to safety management. In actuality, using outsiders as observers instead of peers can make the process seem threatening ("Will the data be used against me?") and can decrease group cohesion or teamwork. Therefore, the employees themselves should (1) decide what behaviors to observe, (2) conduct regular behavioral observations, (3) deliver one-on-one feedback sessions with fellow employees, (4) calculate daily percentages of safe behaviors in circumscribed work areas, (5) post the group data in conspicuous locations, (6) design and implement strategies (e.g., training sessions, goal setting, incentives, recognition celebrations) to motivate participation in the observation feedback process, and (7) refine components of the process with a continuous improvement perspective.

## Employee Involvement Is the Key

The idea of watching fellow employees to observe their safe and unsafe work practices (and being watched by fellow employees for the same behaviors) is a novel concept to many individuals and can appear threatening at first. Accomplishing such a paradigm shift can be quite challenging in some work cultures and can only be accomplished by establishing conditions and contingencies that build and nourish interpersonal trust throughout the work culture. The process should not include any punishment contingencies (e.g., no names recorded for unsafe behaviors), and participation should be strictly voluntary (although reward or recognition contingencies can be implemented to motivate involvement). In addition, the employees (e.g., a representative design team) should have as much choice as possible in the operational details of *their* observation/feedback process. Providing opportunities for choice is motivating and helps in the development of interpersonal trust.

## The Process Is Critical

Obviously, there is no "best" protocol for an observation/feedback process for safety management. The process must fit the culture, and this can only happen if the employees themselves decide on the procedural details consistent with certain principles and guidelines. It is possible, for example, that a full-blown observation/feedback process will not be feasible at first, but successive approximations to a complete, plant-wide process are possible. Thus, a plant might start small (e.g., with brief and infrequent observation/feedback of a select number of behaviors in one work area) and then expand the process as confidence and interpersonal trust develop through experience. It is easy to get caught up in the particulars of an observation/feedback process (e.g., duration, frequency, and sampling of observations; timing, location, and length of feedback sessions), but it is more important to get the process going. The true value of the process is not in the behavioral data themselves (which are no doubt biased by numerous confounding variables), but in the behavior-focused interactions between employees.

I have seen an employee safety observation/feedback process work well at several companies, and the implementation procedures have varied widely. The employees at an Exxon Chemical plant, for example, designed an innovative observation/feedback process whereby each month every employee schedules three to five observation/feedback sessions with any two other employees. Thus, on days and at times selected by the observee, two observers show up at the observee's work site and use a standard behavioral checklist to conduct a systematic 30-minute observation session, followed by interactive feedback. Those concerned with obtaining a representative and objective sample of work practices will readily criticize this procedure of allowing employees to choose their own observers and observation times. However, the critical ingredient here is the process of people willingly observing and coaching others in order to increase safe work practices. Although those observed are "on their best behavior" during assigned observation sessions, unsafe acts are observed, and often an observee's demonstration of safe behavior becomes a valuable learning experience for the observers. Also, when people choose to be observed for constructive feedback they are apt to accept suggestions for making their work performance safer.

A very different observation/feedback process was developed by employees at a Hoechst Celanese plant. Termed the "Planned 60-Second Actively Caring Review", all employees attempt to complete one feedback card and brief coaching session every day. The feedback card is completed after a 1-minute observation of an employee's safe and unsafe work practices in five general categories: body position, personal apparel, housekeeping, tools/equipment, and operating procedures. The feedback cards are collected each day and the results are tallied and posted. The number of feedback cards collected per department is exhibited in a large display case at the plant entrance, along with weekly departmental goals.

These 1-minute observation sessions are more brief than I had advised, but the employees felt that plant-wide acceptance of longer sessions would be unlikely at the time. Again, the process of employee involvement in behavior-focused observation and feedback is the most important feature of this intervention; perhaps longer sessions will come later. The members of the design team who developed their process reminded me that 1 minute was sufficient time for goal setting, praising, and reprimanding (Blanchard & Johnson, 1982), so why not use 1-minute safety observations?

After 1 minute of behavior-based observing, employees are urged to review the feedback card with the observee. This one-on-one interaction is termed "coaching" and is based on principles from both behavior-based and person-based psychology. The next section reviews these principles and presents guidelines for effective safety coaching. When large numbers of employees practice effective safety coaching, interpersonal trust and teamwork increase naturally.

## COACHING FOR SAFETY AND HEALTH MANAGEMENT

The coaches of winning athletic teams practice the basic observation and feedback processes necessary for effective safety management and follow most of the guidelines reviewed here. In other words, the best team coaches observe the behaviors of their players, record their observations in a systematic fashion (e.g., on a team roster, behavioral checklist, or on videotape), and then offer specific behavioral feedback (both correcting and supporting) to the team members. Often this behavioral feedback is given both individually (in one-on-one communication) and in group sessions (e.g., by reviewing videotapes of the team competition), and the most effective coaches give this feedback so the team members learn from the exchange *and* are motivated to continuously improve. The five letters of the word COACH can be used as a mnemonic to remember the basic characteristics of the most effective coaches, whether coaching for a winning athletic team or for a work group seeking improved safety performance.

### "C" for Care

Safety coaches truly care about the health and safety of their co-workers, and they act on such caring. In other words, they "actively care". In the next section, ways to increase actively caring are discussed; for now, just consider the value of initiating a one-on-one coaching process with an actively caring attitude. When people realize by observing other persons' words and body language that they care, then they will listen to advice. In other words, when people know you care, they will care what you know.

Covey (1989) explained the value of caring interdependence (exemplified by appropriate safety coaching) with the metaphor of an "emotional bank account". The idea is that individuals establish an emotional bank account with others through personal interaction. Deposits are made when the holder of the account perceives a particular interaction to be positive (e.g., they feel recognized, appreciated, or listened to). Withdrawals from an individual's emotional bank account occur whenever that individual feels criticized, humiliated, or less appreciated (e.g., from personal interaction). It is sometimes necessary, of course, to offer constructive criticism to others or even to state extreme displeasure with another person's behavior; however, if such negative discourse occurs on an "overdrawn or bankrupt account", corrective feedback will have limited impact. In fact, continued withdrawals from an overdrawn account can result in defensive or countercontrol reactions (from simply ignoring the communication, to emitting overt behavior to discredit the source or undermine the process or system implicated in the communication). Thus, safety coaches need to demonstrate a caring attitude through their personal interactions with others, thereby maintaining healthy emotional bank accounts (i.e., operating in the "black").

### "O" for Observe

Safety coaches observe the behavior of others objectively and systematically with an eye for supporting safe behavior and correcting unsafe behavior. Behavior that illustrates "going beyond the call of duty" for the safety of another person should be especially noted for supportive coaching. This is the sort of behavior that contributes significantly to safety achievement and improvement and can be increased through rewarding feedback.

Observing behavior for supportive or corrective feedback is easy if the coach knows exactly what behavior is desired and what is undesired (an obvious requirement for athletic coaching) and takes the time to observe occurrences of these behaviors in the work setting. It is often advantageous (usually essential) to develop a checklist of safe and unsafe behaviors in the various work settings of an industrial complex and rank these in terms of risk. As discussed above, the workers themselves should develop their behavioral checklist. Risk rankings can be derived from careful study of injury records (from first-aid cases to lost-time injuries) and reports of "near misses". Promoting the reporting of "near misses" to learn ways of preventing injury is another important challenge for the effective safety coach.

### "A" for Analyze

Safety coaches appreciate the ABC model (for **A**ctivator, **B**ehavior, **C**onsequence) in interpreting their observations. In other words, they understand that there are usually observable reasons for behaviors. As discussed earlier, activators (e.g., signs, memos, instructions, policies, mission statements) direct behavior, whereas consequences (e.g., praise, feedback, reprimands, recognition ceremonies) motivate behavior. Coaches realize, for example, that certain unsafe behaviors occur because they are directed by such activators as work demands, unsafe example-setting by peers, and inconsistent or mixed messages from management, and unsafe behaviors are often motivated by one or more consequences (e.g., comfort, convenience, work breaks, and management or peer approval). Of course, safety coaches also realize that their behavioral observation, analysis, and communication can activate the occurrence of safe behaviors (e.g., through instructions, reminders, individual and group discussions) and motivate their reoccurrence (e.g., through verbal feedback, recognition, and group celebrations). This brings us to the next letter of COACH.

## "C" for Communicate

Effective coaching requires basic communication skills, including strategies for active listening and persuasive speaking. Indeed, communication training sessions for employees which incorporate role-play exercises can be invaluable in developing individuals' confidence and competence in sending and receiving behavioral feedback. Such training should illustrate the need to separate behavior (i.e., actions) from person factors (i.e., attitudes and feelings), thus enabling corrective feedback to occur at the behavioral level without "stepping on" feelings. For example, people need to understand that anyone can act unsafely without even realizing it (as in "unconscious incompetence"), and performance can only improve with behavior-specific feedback. Thus, even corrective feedback for unsafe behavior is appreciated when it is given appropriately, regardless of the work status of the feedback sender.

Whether supportive or corrective, feedback should be specific (with regard to a particular behavior) and timely (occurring soon after the target behavior), and it should be given in a private, one-on-one situation to avoid potential interference or embarrassment from others. In addition, corrective feedback is most effective if the alternative safe behavior is specified and potential solutions for eliminating the at-risk behaviors are discussed. Whether giving supportive or corrective feedback, the communicator of feedback must actively listen with appropriate body language and verbal responses when the feedback recipient reacts to support or correction. Through active (or empathic) listening the safety coach shows sincere concern for the feelings and commitment of the feedback recipient. The best listeners give empathic attention with facial cues and posture, paraphrase to check understanding, prompt for more details (e.g., "Tell me more"), accept feelings as stated without interpretation, and avoid autobiographical statements that can stifle self-disclosure (e.g., "When I worked in your department, I always worked safely.").

## "H" for Help

The word "help" summarizes the essential mission of a safety coach. In other words, the goal of safety coaching is to help an individual prevent an injury by supporting safe work practices and correcting unsafe work practices. It is critical, of course, that a coach's helping communication is accepted by the feedback recipient. The four letters of HELP offer another mnemonic for remembering four words which suggest ways to obtain acceptance of a coach's advice, directions, or feedback.

**"H" for humor** — Safety is certainly a serious matter, but sometimes a little humor can be inserted in our safety communications as a way of increasing interest and acceptance. In fact, some research has indicated that laughter can reduce stress and even benefit an individual's immune system (*National Safety News,* 1985), and we know from personal experience that humor can enhance the constructive gains from a communication by increasing our attention and decreasing our resistance to an appeal for change.

**"E" for esteem** — Esteem, or more specifically "self-esteem", is the personal perception of self-worth ("I am valuable") and is critical to increasing safe work practices. People who feel inadequate, unappreciated, or unimportant in a particular work setting are not as likely to go beyond the call of duty to benefit the safety of themselves or others as are people who feel capable and valuable. Thus, the most effective coaches choose their words carefully (i.e., emphasizing the positive over the negative) in an attempt to build or avoid lessening another person's self-esteem. In other words, effective coaches make many more deposits than withdrawals in people's emotional bank accounts (Covey, 1989).

**"L" for listen** — One of the most powerful and convenient ways to build self-esteem is to listen attentively to another person. This sends the signal that you care about the person and his or her situation; the person can in turn interpret this active listening as a reflection of self-worth (e.g., "I must be valuable to the organization because my opinion is considered and appreciated."). Also, after a coach actively listens, his or her message is more likely to be heard and accepted.

**"P" for praise** — Praising another person for specific accomplishments is another powerful way to build self-esteem, and if the praise targets a particular behavior, the probability of the behavior occurring again increases. This reflects the basic behavioral science principle of positive reinforcement and motivates people to continue their safe work behaviors and to look out for the safety of their associates. In other words, behavior-focused praising is a powerful rewarding consequence which not only increases the behavior it follows, but also increases a person state (i.e., self-esteem) which in turn increases the individual's willingness to act for the safety of others.

In summary, safety coaching is a key process for managing industrial safety and health. Obviously, training and practice are essential to the development of safety coaching skills, but also critical is a work culture which enables and supports interpersonal coaching for safety improvement. For some cultures, this requires some dramatic paradigm shifting (e.g., from fault finding to fact finding, from a focus on discipline for safety violations to a focus on recognition and support for safety achievements, and from a fatalistic perspective to a "make-a-difference" attitude). Thus, a large-scale safety coaching process may take significant time and resources to achieve, but the long-range outcome will be well worth the effort — an organization of people who coach each other consistently and effectively to increase safe work practices. This coaching process increases people's sense of self-esteem, empowerment, and belongingness. The next section discusses these three person states and explains the value of building these states among employees at all levels of an organization.

## AN ACTIVELY CARING MODEL

In order to facilitate the design of the most cost-effective intervention process for a particular target behavior, setting, and audience, Geller et al. (1990) proposed a systematic scheme for classifying and evaluating behavior-change techniques. This system classifies intervention programs (which usually combine several behavior-change techniques) into multiple tiers or levels, each tier defined by its overall cost-effectiveness and personal intrusiveness. At the top of the "multiple intervention hierarchy" (i.e., Level 1) the interventions are the least expensive per person. At this level, intervention techniques (e.g., activation through signs, billboards, and public service announcements) are designed to have broad-based appeal with minimal personal contact between target individuals and intervention agents. Geller et al. hypothesized that normally people uninfluenced by initial exposure to these types of interventions (i.e., Level 1) will be uninfluenced by repeated exposures to interventions at the same level. These people require a more intrusive and costly (i.e., higher level) intervention program.

A key proposition in the multiple intervention level model proposed by Geller et al. (1990) and refined by Geller (1992) is that people influenced by an intervention program (at a particular level of cost-effectiveness and intrusiveness) should not be targeted for further intervention, but instead they should be enrolled as intervention agents for the next (i.e., higher) level of behavior-change intervention. In other words, "preaching to the choir" is not as beneficial as enlisting the "choir" to preach to others. This section explores ways of identifying those individuals most likely to become intervention agents for industrial safety and health, as well as ways to increase participation as an intervention agent or intervention target.

### Actively Caring

Intervention agents are individuals who care enough about a particular problem or about other people to implement an intervention strategy that could make a beneficial difference. In other words, intervention agents are individuals who actively care. Actively caring (AC) behaviors can be classified into three categories, depending upon the target of the intervention — environment, person, or behavior (Geller et al., 1994). Thus, when people intervene to reorganize or redistribute resources in an attempt to benefit the safety of others (e.g., clean up another's work area, report an environmental hazard, practice appropriate energy control and power lockout procedures), they are AC from an *environment* focus.

Actively caring from a *person* focus is behaving in an attempt to make another person feel better (e.g., intervening in a crisis situation, actively listening in one-to-one communication, verbalizing unconditional positive regard for someone, sending a "get well" or "thank you" card). Finally, *behavior*-focused AC is doing something to influence another individual's behavior in desired directions (e.g., giving rewarding or correcting feedback, demonstrating or teaching desirable behavior, developing or implementing a behavior-change intervention program).

## Actively Caring States

With nonhumans as experimental subjects, behaviorists have influenced marked changes in performance when altering certain physiological states of their subjects (e.g., through food, sleep, or activity deprivation). Similarly, behaviorists have demonstrated significant behavior change in both normal and developmentally disabled children as a function of simple manipulations of the social context (e.g., Gewirtz and Baer, 1985a,b) or the temporal proximity of lunch and response-consequence contingencies (Vollmer and Iwata, 1991). Although behaviorists typically refer to these manipulations of physiological (e.g., food deprivation) or psychological (e.g., social deprivation) conditions as "establishing operations" (Michael, 1982), these independent variables are certainly analogous to the person-based concepts of expectancies, personality states, and intrinsic motivation. In other words, certain operations or environmental conditions (past or present) can influence (or establish) physiological or psychological states within individuals, which in turn can affect their behavior. From the behavior-based approach discussed above, the basic mechanism of this impact is through enhancing the quality and quantity of positive consequences achievable by designated target behaviors.

I have proposed that certain psychological states or expectancies affect the propensity for individuals to actively care for the safety or health of others and, furthermore, that certain conditions or establishing operations (including activators and consequences) can influence these psychological states (Geller, 1991). These states are illustrated in Figure 30-1, a model my associates and I have used numerous times to stimulate discussions among industry employees of specific situations, operations, or incidents that influence their willingness to actively care (e.g., to participate actively in safety achievement efforts).

Factors consistently listed as determinants of self-esteem include communication strategies, reinforcement and punishment contingencies, and leadership styles. Participants have suggested a number of ways to build self-esteem, including (1) providing opportunities for personal learning and peer mentoring, (2) increasing recognition for desirable behaviors and personal accomplishments, and (3) soliciting and following up on a person's suggestions. Common proposals for increasing an atmosphere of belongingness among employees have included decreasing the frequency of top-down directives and "quick-fix" programs and increasing team-building discussions, group goal setting and feedback, group celebrations for both process and outcome achievements, and the use of self-managed (or self-directed) work teams.

In the management literature, empowerment typically refers to delegating authority or responsibility, or sharing decision making (Conger and Kanungo, 1988). In contrast, the person-based perspective of empowerment focuses on

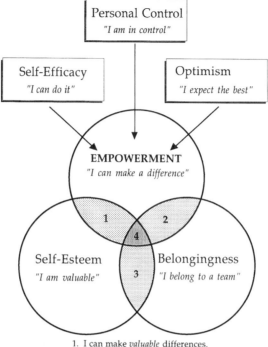

1. I can make *valuable* differences.
2. We can make a difference.
3. I am a *valuable team* member.
4. We can make *valuable differences.*

**Figure 30-1**    The actively caring model for occupational safety. (Adapted from Geller, E. S., Roberts, D. S., Gilmore, M. R., and Pettinger, C. B., Jr. *Achieving a Total Safety Culture through Employee Involvement,* 3rd ed. Newport, VA: Make-A-Difference, 1994. With permission.)

the reaction of the recipient to increased power or responsibility. In other words, this view of empowerment requires the personal belief "I can make a difference," and this belief is strengthened with perceptions of personal control (Rotter, 1966), self-efficacy (Bandura, 1977), and optimism (Scheier and Carver, 1985; Seligman, 1991). Such an empowerment state is presumed to increase motivation (or effort) to "make a difference" (e.g., to go beyond the call of duty), and there is empirical support for this intuitive hypothesis (e.g., Bandura, 1986; Barling and Beattie, 1983; Ozer and Bandura, 1990; Phares, 1976).

Employees at my AC training sessions have listed a number of ways to increase empowerment, including (1) setting short-term goals and tracking achievements; (2) offering frequent rewarding and correcting feedback for process activities (e.g., behavior that reflects AC for health or safety) rather than only for outcomes (e.g., total recordable injury rate); (3) providing opportunities to set personal goals, teach peers, and chart "small wins" (Weick,

1984); (4) teaching employees basic behavior-change intervention strategies (e.g., response feedback and recognition procedures) and providing them with time and resources to implement and evaluate intervention programs; (5) showing employees how to graph daily records of baseline, intervention, and follow-up data; and (6) posting response feedback graphs of group performance.

## Research Support for the Actively Caring Model

There are actually a number of empirical studies, mostly in the social psychology literature, that support the individual components of the AC model depicted in Figure 30-1. The bystander intervention paradigm (Darley and Latane, 1968) has been the most common (and rigorous) laboratory technique used to study variables related to AC behaviors. With this approach, one or more of the person states presumed to affect AC behavior (i.e., self-esteem, empowerment, and belongingness) were measured or manipulated among subjects, and subsequently these individuals were placed in a situation where they had an opportunity to help another individual who presumably encountered a personal crisis (e.g., falling off a ladder, dropping personal belongings, or feigning an illness or heart attack). The latency in attempting to help the other person was the primary dependent variable, studied as a function of a subject's social situation or personality state. All AC behaviors studied in these experiments were person-focused (i.e., helping a person feel better) or environment-focused (i.e., redistributing resources to benefit someone else), but the AC behaviors were never behavior-focused (i.e., attempting to change another individual's behavior in beneficial directions).

### Self-Esteem

Michelini, Wilson, and Messe (1975) and Wilson (1976) measured subjects' self-esteem with a sentence completion test and then observed whether they helped another individual in a bystander intervention situation. Subjects with high self-esteem were significantly more likely than subjects with low self-esteem to help another individual pick up dropped books (Michelini et al., 1975) and to exit an experimental room to help someone in another room who screamed that he had broken his foot following a mock "explosion" (Wilson, 1976). Analogously, subjects with higher self-esteem scores were more likely to help a stranger by taking his place in an experiment that would presumably give them electric shocks (Batson, Bolen, Cross, and Newinger-Benfiel, 1986).

### Belongingness

By systematically manipulating a person's sense of belongingness in groups of two and four, Rutkowski, Gruder, and Romer (1983) found group cohesion to reverse the usual bystander intervention effect (i.e., the finding of more individual intervention in smaller groups). Cohesiveness was manipulated by having groups discuss topics and feelings related to college life. Both frequency and speed of helping a "victim" (confederate) who had ostensibly fallen off a ladder were greater for the cohesive groups. Indeed, the most AC behavior was found among subjects in the high-cohesion/four-person group condition.

In a retrospective real-world study, Blake (1978) studied real-world relationships between group cohesion and the ultimate in AC behavior — altruistic suicide. He collected data from official records of Medal of Honor awards given during World War II and in Vietnam. The independent variable was the cohesiveness of combat units (estimated by group training and size), and the dependent variable was percentage of "grenade acts" — voluntarily using one's body to shield others from exploding devices. The smaller, more elite, especially trained combat units (e.g., the Marine Corps and Army airborne units) accounted for a significantly larger percentage of "grenade acts" than larger, less specialized units (e.g., Army non-airborne units).

### Personal Control

The personal control factor of the AC model represents one of the most extensively researched individual difference variables in psychology and refers to a general expectancy regarding the location of forces controlling an individual's life (i.e., internal vs. external factors). Persons with an *internal* locus of control believe they normally have personal control over important life events as a result of knowledge, skills, and abilities. In contrast, individuals with an *external* locus of control believe factors like luck, chance, or fate have significant influence in their lives (Rotter, 1966; Rushton, 1980). From a behavioral perspective, externals generally expect to have less personal control over the pleasant and unpleasant consequences in their lives than do internals.

Those high-esteem subjects who showed more AC behavior than low-esteem subjects in Wilson's (1976) bystander intervention study (discussed above) were also characterized as *internals,* in contrast to the lower self-esteem *externals,* who were less apt to display AC behavior. Similarly, Midlarsky (1971) observed more internals than externals willing to help another person perform a motor coordination task that included receiving electric shocks. In addition, those who helped at an accident scene scored significantly higher on personal control (i.e., internals) and self-esteem than those who only stopped and watched (Bierhoff, Klein, and Kramp, 1991).

Sherrod and Downs (1974) asked subjects to perform a task while hearing loud, distracting noise. They manipulated subjects' perception of personal control by telling half of the subjects that they could terminate the noise (if necessary) by notifying them through an intercom. The subjects who could have terminated the noise (but did not) were significantly more likely to comply with a later request by another individual to take some extra time (with no extrinsic benefits) to solve math problems.

### Self-Efficacy

Self-efficacy refers to people's beliefs that they have the personal skills and resources to complete a task successfully (Bandura, 1977). Other individual difference factors relate significantly to this construct, including self-esteem (Rosenberg, 1965), locus of control (Rotter, 1966), and learned hopefulness (Zimmerman, 1990). Thus, research that showed more AC behavior from internals with high self-esteem (e.g., Bierhoff et al., 1991; Midlarsky, 1971) indirectly supported this factor as a potential determinant of AC behavior. Zimmerman (1990) defined "empowering experiences" as experiences providing opportunities to learn skills and develop a sense of personal control. He proposed empowerment to be a product of learned hopefulness. In other words, people become empowered as they gain control and mastery over their lives and learn to use their skills to affect life events.

### Optimism

Optimism is the learned expectation that life events, including personal actions, will turn out well (Scheier and Carver, 1985; Seligman, 1991). Researchers have manipulated optimistic states (or moods) among individuals by giving them unexpected rewards or positive feedback and then observing the frequency of AC behaviors. Isen and Levin (1972), for example, showed that individuals who found a dime in the coin return slot of a public phone (placed there by researchers) were more likely to help a stranger who dropped a folder of papers than were individuals who did not find a dime. Similarly, students given a cookie while studying at the university library were more likely than those not given a cookie to agree to help another by participating in a psychology experiment. Carlson, Charlin, and Miller (1988) reviewed these and other studies that showed direct relationships between an optimistic mood state and AC behavior.

## Direct Tests of the Actively Caring Model

My students and I have been conducting a series of studies to test the AC model in field settings, and so far the results have been quite promising. We have developed a "safety culture survey" (SCS) for industrial application which includes measures of each person factor hypothesized to influence AC (see Figure 30-1). The SCS also assesses the respondent's willingness to AC in various ways (i.e., from a person, environment, and behavior focus). To date we have administered this survey at nine industrial sites and have shown rather consistent support for the AC model. The stepwise regression analyses from these assessments have resulted in high regression coefficients, e.g., 0.54 (n = 262), 0.57 (n = 307), and 0.71 (n = 207) at the three plants studied by Geller and Roberts (1993) and 0.52 (n = 328) and 0.68 (n = 202) at the two plants studied by Geller, Roberts, and Gilmore (in press). The personal control factor was usually the most influential in predicting willingness to AC, with group belongingness predicting significant independent variance in AC propensity at all but one of the plants. Self-esteem, optimism, and self-efficacy have always correlated highly with each other and with willingness to AC, but usually only one of these factors predicted *independent* variance in AC (i.e., above that predicted by personal control and belongingness).

In one test of our AC model (Roberts and Geller, 1995), we studied relationships between workers' on-the-job AC behaviors and prior measures of their self-esteem, optimism, and group cohesion. More specifically, employees (n = 65) were instructed to give their co-workers special "actively caring thank-you cards" (redeemable for a beverage in the cafeteria) whenever they observed a co-worker going beyond the call of duty (i.e., actively caring) for another person's safety. Those employees who gave or received an AC thank-you card scored significantly higher on measures of self-esteem and group cohesion than those who neither gave nor received an AC thank-you card.

Most recently, five of my students asked individuals (n = 156) who had just donated blood at a campus location to complete a 60-item survey which measured each of the five person factors in Figure 30-1. The high return rate of 92% was consistent with an AC profile, but most remarkable was that this group scored significantly higher ($p < 0.01$) on each of the five subscales than did a group of students (n = 292) from the same university population (Buermeyer, Rasmussen, Roberts, Martin, and Gershenoff, 1994).

## SUMMARY

This chapter has attempted to illustrate the value of integrating behavior-based and person-based psychology to manage the human element of occupational health and safety. The purpose of this integration is to develop a culture in which everyone shares the responsibility of organizational safety by looking out for conditions and behaviors that are risky and then intervening to make them safer. Techniques for increasing the occurrences of safe behavior and decreasing the frequency of at-risk behaviors have been developed and tested by behavior-focused psychologists. However, these techniques are only useful if people use them on a regular basis, and this requires consideration of issues and concepts in person-based psychology. For example, certain person states or individual conditions increase one's willingness to apply behavior-based techniques (e.g., observation, feedback, and rewards) for the benefit of another person's safety. In other words, these person states influence "actively caring".

Understanding these person states (i.e., self-esteem, belongingness, self-efficacy, personal control, and optimism) can lead to the development of action plans to increase these states among members of a workforce. Incorporating behavior-based strategies in these action plans can make them more effective.

Interpersonal coaching is a key strategy for increasing both safe behaviors and the person states that facilitate actively caring (or other-directed behaviors). In other words, when people interact effectively with others to support safe behaviors and correct unsafe behaviors, they not only activate beneficial behavior change; they also send a message to others that they are valuable team members (thereby building self-esteem and belongingness) and that people can make beneficial differences in the safety of an organization (thereby building self-effectiveness, personal control, and optimism). The five letters of "coach" represent a useful mnemonic for remembering an invaluable sequence of events that integrate behavior-based and person-based psychology to manage effectively the human element of occupational health and safety (i.e., C = Care, O = Observe, A = Analyze, C = Communicate, and H = Help).

The first word is *care* because of the person-based principle that people will care what you know after they know you care. When people care enough to give supportive or corrective feedback on behalf of another person's safety, and such caring is recognized among individuals in a work group or throughout an entire plant, the next step *(observation)* is accepted and appreciated. Careful and systematic behavioral observations are followed by an ABC *analysis.* In other words, the directive activators and motivational consequences of the situation are considered in order to understand causes of the observed behaviors and to develop interventions for increasing safe behaviors and decreasing at-risk behaviors.

After observation and analysis, the useful information is *communicated* effectively and objectively to the person observed. If the communication is constructive and accomplished in an atmosphere of win/win collaboration and mutual trust, it will be accepted and appreciated. Such actively caring communication will truly *help* the individual and the management of occupational health and safety.

## ACKNOWLEDGMENTS

The author is grateful for numerous opportunities to learn occupational safety issues and challenges from the professionals at many Fortune 500 companies, most recently at Exxon Chemical Company, Hoechst Celanese, James River Corporation, Sara Lee Knit Products, and Westinghouse. The support of dedicated and creative colleagues, especially Harry Glaser, Steve Roberts, and Michael Gilmore, facilitated the development and presentation of the concepts introduced in this chapter.

## FURTHER READING

Covey, S. R. *The Seven Habits of Highly Effective People.* New York: Simon & Schuster, 1989.

Deming, W. E. *The New Economics.* Boston, MA: MIT Press, 1993.

Elder, J. P., Geller, E. S., Hovell, M. F., and Mayer, J. A. *Motivating Health Behavior.* New York: Delmar Publishers, 1994.

Geller, E. S., Lehman, G. R., and Kalsher, M. R. *Behavior Analysis Training for Occupational Safety.* Newport, VA: Make-A-Difference, 1989.

Goldstein, A. P. and Krasner, L. *Modern Applied Psychology.* New York: Pergamon Press, 1987.

Seligman, M. E. P. *Learned Optimism.* New York: Alfred A. Knopf, 1991.

## REFERENCES

Bandura, A. *Social Foundations of Thought and Action.* Englewood Cliffs, NJ: Prentice-Hall, 1986.

Bandura, A. Self-efficacy: toward a unifying theory of behavioral change. *Psychol. Rev.,* 84, 191–215, 1977.

Barling, J. and Beattie, R. Self-efficacy beliefs and sales performance. *J. Org. Behav. Manage.,* 5, 41–51, 1983.

Batson, C. D., Bolen, M. H., Cross, J. A., and Neuringer-Benefiel, H. E. Where is altruism in the altruistic personality? *J. Pers. Soc. Psychol.,* 1, 212–220, 1986.

Bierhoff, H. W., Klein, R., and Kramp, P. Evidence for altruistic personality from data on accident research. *J. Pers.,* 59(2), 263–279, 1991.

Blake, J. A. Death by hand grenade: altruistic suicide in combat. *Suicide Life-Threatening Behav.,* 8, 46–59, 1978.

Blanchard, K. H. and Johnson, S. *The One Minute Manager.* West Caldwell, NJ: William Morrow & Co., 1982.

Buermeyer, C. M., Rasmussen, D., Roberts, D. S., Martin, C., and Gershenoff, A. B. Red Cross Blood Donors vs. a Sample of Students: An Assessment of Differences Between Groups on "Actively Caring" Person Factors. Paper presented at the Virginia Academy of Science, Harrisonburg, VA, May 1994.

Carlson, M., Charlin, V., and Miller, N. Positive mood and helping behavior: a test of six hypotheses. *J. Pers. Soc. Psychol.,* 55, 211–229, 1988.

Conger, J. A. and Kanungo, R. N. The empowerment process: integrating theory and practice. *Acad. Manage. Rev.,* 13, 471–482, 1988.

Covey, S. R. *The Seven Habits of Highly Effective People.* New York: Simon & Schuster, 1989.

Covey, S. R. *Principle-Centered Leadership.* New York: Simon & Schuster, 1990.

Darley, J. M. and Latane, B. Bystander intervention in emergencies: diffusion of responsibility. *J. Pers. Soc. Psychol.,* 8, 377–383, 1968.

Deming, W. E. *Out of the Crisis.* Cambridge, MA: Massachusetts Institute of Technology, Center for Advanced Engineering Study, 1982.

Deming, W. E. *The New Economics.* Boston, MA: MIT Press, 1993.

Elder, J. P., Geller, E. S., Hovell, M. F., and Mayer, J. A. *Motivating Health Behavior.* New York: Delmar Publishers, 1994.

Geller, E. S. Actively caring for occupational safety: extending the performance management paradigm. In C. M. Johnson, W. K. Redmon, and T. C. Mawhinney (Eds.), *Organizational Performance: Behavior Analysis and Management.* New York: Springer-Verlag, in press.

Geller, E. S. Applications of Behavior Analysis to Prevent Injury from Vehicle Crashes. Monograph Series: Progress in Behavioral Studies, Monograph #2, Cambridge Center for Behavioral Studies, Cambridge, MA, 1992.

Geller, E. S. *Corporate Safety Belt Programs.* Blacksburg, VA: Virginia Polytechnic Institute and State University, 1985.

Geller, E. S. If only more would actively care. *J. Appl. Behav. Anal.,* 24, 607–612, 1991.

Geller, E. S. Prevention of environmental problems. In L. Michelson and B. Edelstein (Eds.), *Handbook of Prevention* (pp. 361–383). New York: Plenum Press, 1986.

Geller, E. S., Berry, T. D., Ludwig, T. D., Evans, R. E., Gilmore, M. R., and Clarke, S. W. A conceptual framework for developing and evaluating behavior change interventions for injury control. *Health Educ. Res. Theory Pract.,* 5, 125–137, 1990.

Geller, E. S., Eason, S. L., Phillips, J. A., and Pierson, M. D. Interventions to improve sanitation during food preparation. *J. Org. Behav. Manage.,* 2, 229–240, 1980.

Geller, E. S. and Lehman, G. R. The "buckle-up promise card": a versatile intervention for large-scale behavior change. *J. Appl. Behav. Anal.,* 24, 91–94, 1991.

Geller, E. S., Lehman, G. R., and Kalsher, M. R. *Behavior Analysis Training for Occupational Safety.* Newport, VA: Make-A-Difference, 1989.

Geller, E. S. and Roberts, D. S. Beyond Behavior Modification for Continuous Improvement in Occupational Safety. Paper presented at the FABA/OBM Network Conference, St. Petersburg, FL, January 1993.

Geller, E. S., Roberts, D. S., and Gilmore, M. R. Predicting Propensity to Actively Care for Occupational Safety. *J. Saf. Res.,* in press.

Geller, E. S., Roberts, D. S., Gilmore, M. R., and Pettinger, C. B., Jr. *Achieving a Total Safety Culture through Employee Involvement,* 3rd ed. Newport, VA: Make-A-Difference, 1994.

Geller, E. S., Winett, R. A., and Everett, P. B. *Preserving the Environment: New Strategies for Behavior Change.* Elmsford, NY: Pergamon Press, 1982.

Gewirtz, J. L. and Baer, D. M. Deprivation and satiation of social reinforces as drive conditions. *J. Abnorm. Soc. Psychol.,* 57, 165–172, 1985a.

Gewirtz, J. L. and Baer, D. M. The effects of brief social deprivation on behaviors for a social reinforcer. *J. Abnorm. Soc. Psychol.,* 56, 49–56, 1985b.

Goldstein, A. P. and Krasner, L. *Modern Applied Psychology.* New York: Pergamon Press, 1987.

Greene, B. F., Winett, R. A., Van Houten, R., Geller, E. S., and Iwata, B. A. *Behavior Analysis in the Community: Readings from the Journal of Applied Behavior Analysis.* Lawrence, KS: University of Kansas, 1987.

Isen, A. M. and Levin, P. F. Effect of feeling good on helping: cookies and kindness. *J. Pers. Soc. Psychol.,* 21, 384–388, 1972.

Kim, J. and Hamner, C. Effect of performance feedback and goal setting on productivity and satisfaction in an organizational setting. *J. Appl. Psychol.,* 61, 48–57, 1982.

Komaki, J. A., Heinzmann, T., and Lawson, L. Effect of training and feedback: component analysis of a behavioral safety program. *J. Appl. Psychol.,* 67, 334–340, 1980.

Krause, T. R., Hidley, J. H., and Hodson, S. J. *The Behavior-Based Safety Process.* New York: Van Nostrand Reinhold, 1990.

Malott, R. W. A theory of rule-governed behavior and organizational behavior management. *J. Org. Behav. Manage.,* 12(2), 45–65, 1992.

Maslow, A. H. *Motivation and Personality.* New York: Harper & Row, 1970.

Maslow, A. *The Farther Reaches of Human Nature.* New York: Viking Press, 1971.

Mayo, E. *The Human Problems of an Industrialized Civilization.* Boston: Harvard University Graduate School of Business Administration, 1933.

Michael, J. Distinguishing between discriminative and motivational functions of stimuli. *J. Exp. Anal. Behav.,* 37, 149–155, 1982.

Michelini, R. L., Wilson, J. P., and Messe, L. A. The influence of psychological needs on helping behavior. *J. Psychol.,* 91, 253–258, 1975.

Midlarsky, E. Aiding under stress: the effects of competence, dependency, visibility, and fatalism. *J. Pers.,* 39, 132–149, 1971.

*Nat. Saf. News.* Laughter could really be the best medicine, p. 15, 1985.

Ozer, E. M. and Bandura, A. Mechanisms governing empowerment effects: a self-efficacy analysis. *J. Pers. Soc. Psychol.,* 58, 472–486, 1990.

Parsons, H. M. Lessons for productivity from the Hawthorne Studies. In Proceedings of Human Factors Symposium sponsored by the Metropolitan Chapter of the Human Factors Society. New York: Columbia University, pp. 57–67, 1974.

Parsons, H. M. What happened at Hawthorne? *Science,* 183, 922–932, 1980.

Peterson, D. *Safe Behavior Reinforcement.* Goshen, NY: Aloray, 1989.

Phares, E. J. *Locus of Control in Personality.* Morristown, NJ: General Learning Press, 1976.

Roberts, D. S. and Geller, E. S. An actively caring model for occupational safety: a field test. *Appl. Prev. Psychol.,* 4, 53–59, 1995.

Rosenberg, M. *Society and the Adolescent Self-Image.* Princeton, NJ: Princeton University Press, 1965.

Rotter, J. B. Generalized expectancies for internal versus external control of reinforcement. *Psychol. Monogr.,* 80 (1), 1966.

Roethlisberger, F. J. and Dickson, W. J. *Management and the Worker.* Cambridge, MA: Harvard University Press, 1939.

Rushton, J. P. *Altruism, Socialization, and Society.* Englewood Cliffs, NJ: Prentice-Hall, 1980.

Rutkowski, G. K., Gruder, C. L., and Romer, D. Group cohesiveness, social norms, and bystander intervention. *J. Pers. Soc. Psychol.,* 44, 545–552, 1983.

Scheier, M. F. and Carver, C. S. Optimism, coping, and health: assessment and implications of generalized outcome expectancies. *Health Psychol.,* 4, 219–247, 1985.

Seligman, M. E. P. *Learned Optimism.* New York: Alfred A. Knopf, 1991.

Sherrod, D. R. and Downs, R. Environmental determinants of altruism: the effects of stimulus overload and perceived control on helping. *J. Exp. Soc. Psychol.,* 10, 468–479, 1974.

Skinner, B. F. *The Behavior of Organisms.* Acton, MA: Copley Publishing Group, 1938.

Skinner, B. F. *Beyond Freedom and Dignity.* New York: Alfred A. Knoff, 1971.

Skinner, B. F. *Reflections on Behaviorism and Society.* Englewood, NJ: Prentice-Hall, 1978.

Sulzer-Azaroff, B. Behavioral approaches to occupational health and safety. In L. W. Frederiksen (Ed.), *Handbook of Organizational Behavior Management* (pp. 505–538). New York: John Wiley & Sons, 1982.

Ullman, L. P. and Krasner, L. *Case Studies in Behavior Modification.* New York: Holt, Rinehart & Winston, 1965.

Vollmer, T. R. and Iwata, B. A. Establishing operations and reinforcement effects. *J. Appl. Behav. Anal.,* 24, 279–291, 1991.

Wandersman, A., Popper, P., and Ricks, D. *Humanism and Behaviorism: Dialogue and Growth.* New York: Pergamon Press, 1976.

Weick, K. E. Small wins: redefining the scale of social problems. *Am. Psychol.,* 39, 40–49, 1984.

Whitehead, T. N. *The Industrial Worker.* Cambridge, MA: Harvard University Press, 1938.

Wilson, J. P. Motivation, modeling, and altruism: a person X situation analysis. *J. Pers. Soc. Psychol.,* 34, 1078–1086, 1976.

Zimmerman, M. A. Toward a theory of learned hopefulness: a structural model analysis of participation and empowerment. *J. Res. Pers.,* 24, 71–86, 1990.

# 31

# PSYCHOLOGICAL RISK FACTORS AND THEIR EFFECT ON SAFE WORK PERFORMANCE

**Barbara Newman**

## TABLE OF CONTENTS

## INTRODUCTION

In 1931, researcher H. W. Heinrich noted that there are two basic factors leading to accidents: "unsafe mechanical or physical conditions" and "unsafe acts of persons", or the human factor. Many safety programs focus on the physical protection of workers, ignoring the human factors that may lead people to disregard or neglect the procedures that could prevent accidents from happening, yet studies show that employee mistakes account for 50% of all accidents at work.

Psychological risk factors in accidents are the potentially troublesome thoughts, feelings, and behaviors that may occur as a result of stressful events.

1-56670-054-X/96/$0.00+$.50

PERSONAL

- Marital problems/Divorce/End of relationship
- Issues affecting significant other (illness, loss of job, etc.)
- Illness
- Death in the family
- Accident
- Moving
- Alcohol/drug use/abuse
- Financial problems
- Marriage
- Birth of child
- School/graduation

OCCUPATIONAL

- New job/job transfer/promotion
- New responsibilities/expectations
- Cutbacks/downsizing/co-workers leaving
- Problem with co-worker or supervisor (harassment; cultural differences)

SOCIETAL

- Natural disaster:  Fire, flood, earthquake, etc.
- Man-made disaster:  Burglary, robbery/mugging, riot, kidnapping
- Hard times:  Recession, depression

"CONTAGION":  Reaction to an event unrelated to self

- Reading about disaster/television coverage
- Disaster befalling a friend or relative
- Witnessing an accident or disaster
- Responding to/reporting a disaster

**Figure 31-1**  The four areas of stress-related events.

These stressors occur in four areas of our lives: personal, occupational, societal, and "contagion" (Figure 31-1). The events listed under these four categories are examples of many of the stressors we experience. When stress levels rise and the effects of stressors persist, coping skills may become depleted, contributing to a loss of concentration and affecting the ability to perform tasks safely.

## LEVEL OF STRESS

In 1967, the Life Change Index was formulated by T. H. Holmes and R. H. Rahe. Given a list of 43 life events, individuals are asked to check the number of occurrences of each event over the past year. Every occurrence is ranked according to its numerical stress-value impact, from a low of 11 to a high of 100. Adding up the score provides the total amount of stress experienced. The Index contains many of the stressful events that can become strong contributors to the "human factor" in work-related accidents. However, the primary determinant in evaluating the impact of stressors on behavior will be the reaction of each individual to the stressful situations that occur.

## DURATION OF STRESS

In estimating how long a stressor may affect an individual, consideration must be given to the event, its after-effects, and the individual's response to the circumstances. An earthquake or flood may not last very long as an actual stressful incident, but the emotional consequences may last much longer and be more disruptive. On the other hand, an individual tending someone with a chronic illness may experience more consistent stress, but the caregiver may adjust well by learning healthy coping mechanisms and calling on resources for assistance. The post-traumatic stress resulting from the emotional reaction to the earthquake or flood consequently would be more significant in terms of becoming a psychological risk factor in accidents.

As individuals with distinctive personality styles, each of us will respond to a given stressor in an idiosyncratic way. In general, however, short-term problems will not have as great an impact on performance levels as longer-term concerns. Being involved in a minor car accident may create a temporary distraction of attention. Ongoing, more severe issues, such as drug/alcohol addiction, may create a more critical loss of concentration and have more profound effects as a risk to safe work performance.

## ABILITY TO COPE

Everyone differs in how well he or she handles stressful situations and the accompanying internal stress messages (Figure 31-2). The most notable factor that interferes with successfully managing a stressful event is the sense of loss of control. The more an individual feels a loss of control, the greater the level of stress. This out-of-control feeling may lead to confusion and preoccupation with the stressor, resulting in increased negative self-talk and anxiety. A rising sense of frustration and an accompanying loss of concentration may lead either to taking unnecessary risks or to overcompensation for the lack of attentiveness by becoming overly cautious and hesitant. Indecision, uncertainty, and anxiety levels may then escalate further, precipitating an unhealthy stress cycle and contributing to increased accident risk.

Because many people feel uncomfortable discussing the problems that may be plaguing them, a method of assessment is needed to appraise the factors involved in an employee becoming "at risk" for an accident. The most effective assessment comes from observation of behavior or performance level.

When someone is distracted, performance and productivity go down. Stressful events trigger behavioral changes. Some of these changes may be

MESSAGES

- Generalized fear and anxiety
- Confusion
- Preoccupation with problem, loss of concentration (loss of productivity)
- Hypervigilance (overcompensation for distractedness)
- Grief process
- Negative self-talk
- Alienation
- Isolation
- Fear of reprisal
- Fear of showing weakness

REACTIONS

- Post-traumatic stress disorder, including
    - fatigue, low energy
    - irritability, agitation
    - anxiety
    - apathy, emotional numbness
    - headache, nausea, dizziness
    - minor illness
    - difficulty concentrating, distractedness
    - confusion, forgetfulness
    - insomnia
    - nightmares
    - increased drug and alcohol use
    - jumpiness
    - recklessness
- Fearfulness
- Anger
- Feeling overwhelmed
- Nervousness
- Sleeplessness

**Figure 31-2**  Internal stress messages and emotional/physical reactions.

overt and obvious, others less so. It takes a willingness to get to know employees and to learn their habits in order to notice changes in functioning, many of which are listed in Figure 31-3. The longer these changes in conduct continue, the more "at risk" the individual becomes.

Once you have identified an employee as being an accident risk, the issue must then be addressed directly. There are three levels of confrontation — personal, performance, and policy (Figure 31-4).

## LEVELS OF CONFRONTATION

### Personal

If you are willing to leave the realm of your professional capacity, try approaching the individual as a friend who is concerned and who cares. Ask if there is a problem and what you can do to help. Attempt to become an active

- Change in affect (demeanor)
    - Mood changes, mood swings
    - Teariness
    - Withdrawal
    - Forced cheerfulness
    - Bursts of anger; overreaction to situations
    - Agitation
    - Sluggishness
- Distractedness, daydreaming
- Lateness; no excuse
- Excessive use of sick days or personal days
- Long lunches; returning from lunch with smell of liquor on breath
- Personal telephone calls; conducting personal business
- Workaholism
    - Working late or overtime, with no previous pattern
    - Adding to already taxed work load
    - Unwillingness to leave work
- Frequent breaks and bathroom time
- Forgetfulness, passiveness
- Performance drop--more mistakes than usual; minor mistakes
- Erratic performance
- Secretiveness
- Denial of problem when confronted

**Figure 31-3** Behaviors that contribute to accident susceptibility.


PERSONAL

- Awareness of gender differences and employees' behavior patterns
- Approachability ("open door" policy)
- Willingness to take a risk and intervene
- Concern for safety (must be sincere)
- Privacy must be ensured
- Increased cooperation, decreased defensiveness
    - Confidentiality
    - Openness, willingness to listen
    - Availability

PERFORMANCE

- Performance reviews
- Professional, impartial evaluation of conduct
- Behavioral modification alternatives
- Negotiation, mediation skills
- Cross-training of employees

POLICY

- Written policy
- Training and wellness programs, classes, information
- Guidelines for safe behavior

**Figure 31-4** Levels of intervention.

support system for the individual if he or she desires it. The effort must be sincere, of course. The personal level of contact may be very beneficial for people who can appreciate your sincerity, know that confidentiality will be respected, and trust you to listen, empathize, and help. The risk you run is that the problem may be denied, and you may be rebuffed as being intrusive. Extreme tact is necessary if you decide to intervene on a personal level. You may prefer to remain in the professional arena.

## Performance

From a professional stance, your tactic will be to discuss performance level and behavior changes. Note the problem areas and address them specifically. Try to find the reason behind the decline in productivity or change in conduct. If you encounter denial that a problem exists, keep the conversation focused on specific behavior, informing the employee how he or she can improve performance and how that improvement will be documented. Express your concerns regarding safety issues. Cross-training of employees will ensure that job duties are covered if it should become necessary to remove someone from a potentially hazardous situation.

You may notice gender differences occurring at the personal and performance levels. Women, in general, are more likely to be willing to discuss personal problems and accept solutions and offers of help. They are inclined to respond positively to compassion and an honest desire to listen. Men, often raised in an atmosphere which emphasizes problem solving and privacy, may believe that acknowledgment of difficulty or stress is an admission of weakness. They usually will respond more positively to an intervention at the performance level, with clear expectations and guidelines for improvement.

## Policy

Policy and performance go hand in hand. Written expectations of safe behaviors, with posted policy guidelines, are a legal requirement. A specific policy, combined with a comprehensive training program, prepares a company to take action when conduct crosses the line and becomes unsafe. Along with safety regulations, policy can address other behavioral issues such as lateness, personal and sick days, personal telephone calls, and lunch and break times. A good written policy fosters safe behavior.

Naturally, the best course of action is prevention. The best case scenario is for accidents never to occur. If an accident should take place, immediate debriefing is necessary to prevent the contagion that can create an atmosphere conducive to further accidents (Figure 31-5). Open, honest discussions of the event are essential; secrecy breeds anxiety. Although the need for counseling is not always immediately apparent, sharing of feelings helps to contain the sense of helplessness. Additional safety training with respect to the accident

- Debrief to prevent contagion; provide counseling if necessary
- Promote open, honest discussion and education
- Share knowledge and information--secrecy breeds anxiety
- Ensure confidentiality

*Note: A written program should be in place--the need for debriefing is not always apparent*

**Figure 31-5** When an accident occurs.

will help other employees feel more aware and in control. There is no such thing as too much information.

Successful intervention in controlling accidents depends on the nature of the corporate culture. Dissemination of information, thorough training, and wellness programs provide opportunities for the corporate culture to encourage the values that promote safe performance. Emphasis on communication, accountability, involvement, flexibility, priorities that put employees first, and a positive perception of the workplace all contribute to preventing emotional issues from adversely affecting performance. When coupled with keen observation of behavior and timely intervention, healthy corporate values create a safe and responsible environment.

## REFERENCES

### Other Publications

Holmes, T. H. and Rahe, R. H. Life change index. *J. Psychosomatic Res.* 11: 213–218, 1967.

Jones, J. W. Breaking the vicious stress cycle. *Best's Rev.* 88(3): 74, 1988.

Scherer, R., Brodzinski, J. D., and Crable, E. A. The human factor (human failings as a main cause of workplace accidents). *HR Mag.* 38(4): 92, 1993.

## FURTHER READING

### Books

Blanchard, M. and Tager, M. *Working Well: Managing for Health and High Performance.* New York: Simon & Schuster, 1985.

O'Donnell, M. P. and Ainsworth, T., Eds. *Health Promotion in the Workplace.* New York: John Wiley & Sons, 1986.

### Other Publications

Anon. Sleeping on the job. *Small Bus. Rep.* 16(1): 27, 1991.

Borofsky, G. Pre-employment psychological screening. *Risk Manage.* 40(5): 47, 1993.

Halgrow, A. A trauma response program meets the needs of troubled employees. *Personnel J.* 66(2): 18, 1987.

Krause, T. R., Hidley, J. H., and Hodson, S. J. Measuring safety performance: the process approach. *Occup. Hazards* 53(4): 49, 1991.

Reher, R., Wallin, J. A., and Dubon, D. I. Preventing occupational injuries through performance management. *Public Personnel Manage.* 22(11): 301, 1993.
Van Wagoner, S. I. Stress' many disguises. *Across Board* 24(3): 51, 1987.
Verespej, M. A. Wired for disaster: drug and alcohol abuse can wreak havoc on a safety program. *Occup. Hazards* 52(4): 107, 1990.

# 32 IMPROVING SAFETY PERFORMANCE THROUGH CULTURAL INTERVENTIONS

**Rosa Antonia Simon and Steven I. Simon**

## TABLE OF CONTENTS

## INTRODUCTION

Safety and health managers face increasing pressures from government and from consumer and community groups to meet tougher environment, safety, and health demands. This, coupled with downsizing, streamlining, and trends toward employee empowerment, have created the need for safety professionals to develop new frameworks. They must redefine their roles in flatter organizations and find new models to diagnose complex problems, design solutions, and do more with less.

Traditional reliance on safety engineering and identification of human error is no longer sufficient to meet these demands. In response, tremendous cultural changes are taking place right now in our organizations. Some companies are responding by moving tasks that were formerly performed by safety staff over to line managers. Others are moving responsibilities like accident investigations, audits, and communication to safety teams. In both these cases, the safety professional serves in a new role of facilitator, advisor, or team member.

The cultural interventions described in this chapter have been used by companies throughout the United States to reduce accidents. This approach focuses on culture for two reasons: first, because reengineering and implementing a self-directed workforce require cultural change in most organizations, and second, because more and more research indicates that there is a correlation between safety performance and how well an organization manages its social aspects like communication and cooperation. This makes it imperative for the safety manager to understand how culture affects organizations and to understand its impact on safety performance in an organization.

## RESISTANCE TO CHANGE

One of the most difficult issues to manage in a change effort is the resistance to change. One obstacle is the rule-bound tradition inherited from a time when tight control over policies and procedures through enforcement and training was thought to be the ultimate answer to accident reduction. This tradition keeps organizations from making many of the changes needed to adapt to evolving conditions. Another obstacle is that many of the changes needed are viewed negatively by one or more levels of the organization. Some examples of the comments taken from interviews with employees, supervisors, plant managers, and corporate executives may make the dilemma more concrete (Simon, 1991, 1992, 1993).

Employees say that management is the problem. "Safety departments aren't providing the kind of support they used to provide. 'They' (managers) cut back on safety staff and gave safety duties to line managers, but managers and supervisors aren't adequately trained to spot safety problems." Furthermore, "they've instituted cross-functional teams which means sending employees out to do jobs they're not properly trained to do safely. This sets up accidents waiting to happen."

Supervisors say that safety concerns are out of balance. "Sure, safety is important, but we have to keep our jobs, too. We don't want anyone to get hurt, but it's impossible to eliminate all risk. Now 'they' want us to take work time out for behavior observation and positive reinforcement. Safety is its own reward. That ought to be enough."

Plant managers say one problem is corporate headquarters and another is the workforce. "Corporate sets tough production and cost-cutting goals that don't always support safety. Staff and budget cutbacks make it harder and harder to cover all bases even though the staff is working 60-hour weeks or more." Regarding the workforce, "Studies show that accidents are avoidable for the most part. It's just that people don't take personal responsibility for their own safety. With all the time we spend on safety, I don't understand why accidents still happen."

Corporate executives say the same pressures exist in all plants, but some plants have more accidents. "Since the same programs are available to all facilities, the problem must be in implementation. If we could just get all the plants to do what our best plants do, we wouldn't have a safety problem." Yet, programs that are successful in one plant too often fail when they are transplanted to another.

Strong delineation between "Us" and "Them" (i.e., union vs. management, plant management vs. corporate) is common across companies and blocks the ability of all parties to work together to find solutions to common problems. Bringing people together requires understanding how these attitudes or beliefs come into existence and how they are maintained. It is necessary to learn under what circumstances it is possible for an entire group to reexamine its attitudes and change them. Many of the conflicts described have their roots in the cultural norms and assumptions of organizational members. Thus, we need a framework to understand how culture forms and how it can change, and we need the skills to facilitate the change.

Some safety behavior consultants reject the notion that you should manage attitudes, and yet they speak about the importance of "culture". They say, "Manage behaviors, not attitudes." "You can't see attitudes, but you can observe behavior." "Get people to behave the way you want them to behave and their attitudes will change." These beliefs are based on Pavlov's and Skinner's behaviorism theories, as well as a fear of invading people's privacy.

This concern has merit, but it does not represent a complete picture of the cultural approach. It is not the intention of the authors to suggest that management involve itself in changing individual attitudes. Yet research indicates that group attitudes are formed by culture and that culture plays a crucial role in the productivity, quality, and safety performance of our organizations.

The problem with many behavioral safety programs is that they fail to take into account the influence of culture on individual behavior and the role of free will or consciousness on individual decision-making. This may be the reason that many behavior-based programs end up being perceived as "I Spy" programs where co-workers are expected to "tell on each other". (Simon, 1994)

The cultural approach, on the other hand, helps people realize that individual actions spring from group norms and traditions, as well as individual traits. Once people are aware of these influences, they can make better choices. Only then can they consciously create safer organizations without the use of manipulation or coercion.

In the following section, "Changing the Safety Culture", the cultural approach provides safety managers with a new lens to get at cultural problems and design comprehensive solutions. Next, "Changing the Organization to Support the New Culture" outlines the key areas that must be evaluated and aligned to support cultural changes.

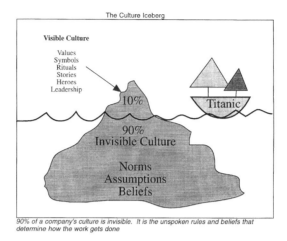

90% of a company's culture is invisible. It is the unspoken rules and beliefs that determine how the work gets done

**Figure 32-1** The Culture Iceberg.

## CHANGING THE SAFETY CULTURE

> Instead of pouring knowledge into people's heads, you need to help them grind a new set of eyeglasses so they can see the world in a new way. That involves challenging the implicit assumptions that have shaped the way people have historically looked at things.
>
> *John Seeley Brown, Vice President, Zerox Parc*

The Culture Iceberg (Figure 32-1) illustrates the components of culture and shows why changing the culture is a great challenge. The visible part (values, symbols, rituals, leaders, stories, language), which translates into policies, procedures, and accident records, is only a small percentage of the whole. The much larger invisible part, norms and assumptions, is the real driver in organization behavior. The unaware manager, like the captain of the *Titanic,* who focuses only on the visible and ignores the invisible culture, will crash.

Starting with the visible culture, **values** are guiding principles, spoken or written, of a culture. **Symbols** such as posted achievements or **rituals** such as safety meetings would alert the newcomer to the idea that safety is important here. **Stories** might be told that represent positive or negative experiences with safety. They usually involve either **heroes** and **heroines** or **villains.**

**Leadership** is highly visible and is the most powerful component of culture. People look to leaders to set the example and set the tone for trust and open communication. Depending on the size of the company, several levels of leadership (including hourly workers) are needed to create a high-performance culture.

Below the iceberg, people experience their beliefs or attitudes about safety as personal and originating from within. However, there is substantial research showing that people's beliefs about how things work arise from the norms or expectation of the groups in which they hold membership.

Research on the importance of norms (defined as any uniformity of attitude, opinion, feeling, or action shared by two or more people) in determining attitudes toward production is well documented by Robert Blake and Jane S. Mouton in *Productivity: The Human Side*. It was their opinion that the same phenomena, "convergence" and "cohesion", which shape attitudes and/or norms toward productivity shape attitudes toward safety.

Briefly, convergence is the tendency of human beings to adopt the point of view of people whom they perceive as part of their group. Studies show that even if a group member initially disagrees with the group, given enough time he or she will adopt the group's point of view (or leave the group). The second force, cohesion, is our tendency to form groups with people we like and who hold opinions, attitudes, and ideas similar to our own.

This research on norms has deep implications for understanding how safety attitudes form and what one has to do to change them. If you train individuals in new behaviors or offer individual rewards for changing, but leave the group norms unchanged, the individual will most likely revert to the old behavior. Thus, to change organizational behavior, focus on changing the group norms, not the individual.

Culture as a management concept is fully explained in Edgar Schein's *Organizational Culture and Leadership* (1991). Schein goes beyond group norms to explain the fabric of social dynamics that shape and control the behaviors of the entire organization. Schein defines culture as the assumptions that a given group has invented, discovered, or developed for survival. Cultural assumptions can be thought of as explanations of how the world works, which is why cultural change is frequently extremely difficult.

For example, it was assumed at one time that the sun rotated around the earth. This assumption made people feel secure because it provided an explanation for something that was unverifiable at that time. When Copernicus tried to challenge that assumption, he was ostracized and ignored. The reaction to Copernicus is typical of our resistance to changing many assumptions. Our assumptions stabilize the world even if they are wrong. Giving them up often means dealing with anxiety until new ones take root. This anxiety is disruptive because people have difficulty concentrating on primary tasks while key assumptions are in transition. It is natural, then, for people to resist changes that produce anxiety.

How this dynamic applies to safety can be seen in the following example. Printers worked with certain inks and solvents that for years were thought to be safe. Respiratory illnesses began to show up in retired printers and were traced back to these chemicals, so respirators became required equipment for printers when they used those chemicals. Years after the medical information surfaced, printers at a major newspaper were refusing to wear the respirators. When interviewed, they stated that the equipment was too cumbersome to wear. When pressed about the long-term effects, they said, "I know how to take care of myself. I've been doing this for 30 years and nothing has happened."

In other words, work groups form assumptions about the way to stay safe based on their experience. These "discoveries" are passed on to new members as the correct way to do things. Once an assumption becomes part of the culture, it drops into the unconscious, and rejecting it would require invalidating years of experience. This is unfortunate when it is based on faulty experience, as was the case in our printer example.

If it is difficult to change an assumption where medical evidence is available, imagine the difficulty in changing an assumption that is counter-intuitive. One such example is changing the belief that safety is the safety department's responsibility. Because safety professionals are the "experts", they are commonly viewed as the people most likely to enforce safety and implement changes. In reality, line management holds the authority and resources to enforce safety and line workers are the final implementers, but the old belief is held in place by years of experience and tradition.

Until now, we have discussed the importance of norms and assumptions, but not how to change them. It is extremely difficult to change assumptions because they are so deep within us that changing them feels like we are giving up our freedom or a part of our inner selves. So forget the notion of getting the behavior you want by establishing a written policy. Policies, like laws, do not determine behavior unless, of course, the policy is already a reflection of a cultural norm. Prohibition is a good example of a failed attempt to change behavior through laws. Seat belt usage, on the other hand, increased after laws were passed because many, many years of concerted effort had won significant support for seat belt use before the laws were enacted. If the culture is to change, a specific set of circumstances must exist.

First, there must be an urgent need to change and it must be communicated to people. Second, the resources and capabilities to change must exist. Third, an action plan to guide the transition from the old to the new culture must be developed. Leadership is responsible for detecting the need for change, communicating it to the organization, gaining consensus on solutions, and guiding implementation.

The safety manager's role in cultural change is to be a catalyst for change, a collaborator, advisor, and facilitator, but he or she cannot bring about change alone because the authority and resources for change lie in the hands of line management. Nevertheless, his or her specialized knowledge and experience play a critical role in designing and implementing the strategies to improve the safety culture. Two proven analysis and action planning models are described next — the norms-changing workshop and the organizational influences model.

## Norms-Changing Workshop

The cultural change technology that works to change norms and assumptions beneath the iceberg consists of six steps:

1. Help people identify unsafe behaviors and articulate the unstated norms and assumptions behind those behaviors, why they came into existence, and why they need to change.
2. Get agreement on desired norms and assumptions from key players.
3. Create an action plan to change the norms and assumptions.
4. Acquire commitment and involvement of all levels of the organization for a specific action plan to change the norms and assumptions.
5. Set up the systems to support the new norms and assumptions.
6. Implement and follow up.

The norms-changing worksheet (Figure 32-2) can be used for steps 1 to 4. It is best to gather a group of 8 to 10 people who work in the same area. Keep managers and employees in separate groups because each group has different norms. Use the examples in Figure 32-2 to get people started or use your own while following these instructions:

A. Write out a specific goal to improve a specific area of safety.
B. Identify unsafe behaviors/actions.
C. Identify the unstated norm(s) that drives the undesired behavior.
D. State why the undesired norm exists.
   - What are the rewards for following the undesired norm?
   - Are there negative consequences for following the norm?
   - Are there negative consequences for going against the norm?
   - Are there problems in the design of the guards, procedures, or equipment that push people toward unsafe actions?
E. Specify desired norms (which is also the desired culture).

## CHANGING THE ORGANIZATION TO SUPPORT THE NEW CULTURE

Here we wish to expand on step 5 of the norms-changing process: set up the systems to support the new norms and assumptions. For the new culture to take root, the formal organization must be examined and changed to support the "new culture". This task requires a comprehensive model or map of the various components of an organization that influence safety performance because attempting to change the safety culture without getting support from the rest of the organization is not possible. Figure 32-3 is one model that breaks out the organizational factors that must be considered when reshaping a culture. Culture would be at the heart, while the supporting elements that produce the company's product or service surround the culture.

There are five organizational components surrounding the culture: (1) structure, (2) technology, (3) rewards, (4) measurement systems, and (5) social processes. The arrows indicate that these components are a set of closely connected systems. Therefore, each component must support the others, much like the parts of a motorcycle must work together. It is not enough to have the best engine, transmission, wheels, and so on. The various components of the

| Unsafe Behaviors/Actions | Unstated Norms | Why Norms Exist | Desired Norms/Behaviors |
|---|---|---|---|
| Not locking out | I've always done it this way 25 yrs | We have actually been trained to take shortcuts | Lockout-Tagout |
| Reaching into moving equipment | I don't get done with my work as soon | Get longer breaks | Train the safe way |
| Bypassing the guards | Production is the first priority (Get the numbers) | Get more production, get more leeway | Production, quality, safety & dedication are all equal |
| Taping blades open | | Design of some equipment goes against the job | |
| Climbing on equipment | Keep the line going | Belief it's not going to happen to me | Everyone work together |
| Not wearing gloves | Must be OK if no one says anything | | Look out and speak up for each other |
| Using wrong tools | I've done it so long and nothing has happened | People pick up shortcuts | Use proper tool |
| | | Negative Peer pressure--"Don't go too fast" | Pride in job |

\* This worksheet is an unaltered document that was prepared by hourly employees

**Figure 32-2**  Norms-changing worksheet. The goal is to eliminate shortcuts and assume a safe attitude about hand safety. (This worksheet is an unaltered document that was prepared by hourly employees.)

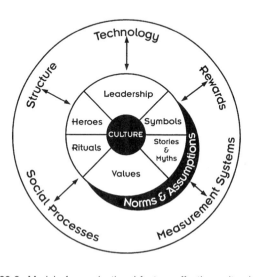

**Figure 32-3**  Model of organizational factors affecting cultural change.

motorcycle must fit with each other and complement each other. Likewise, the various parts and design elements of organizations must fit with each other to achieve the desired results.

Although all five organizational components are important, a recent research study conducted by the authors indicates an especially strong correlation between the perception of managers and employees of the leadership, social process, and technology organizational components and the actual safety performance. "Culture ratings" were obtained from nine sites of a national organization and then compared to workers' compensation costs at each site. The "perception" of employees and managers of the role of leadership, the degree of open communication, and the positive fit between work flow/work assignments and safety all correlated strongly to positive safety performance as measured by low workers' compensation costs. This research suggests that companies with high scores in these three areas tend to have excellent safety records.

A brief description of the function and content of each component follows with examples of what types of organizational changes might be required to develop safety excellence through a cultural change process.

1. **Structure.** These are the policies, procedures, and formal communication structures (such as safety departments, committees, safety meetings, and training) designed to get individuals to perform tasks in compliance with organizational objectives. An example of structural changes to support a new culture is recognizing that safety is a line responsibility or empowering employee-led safety teams to handle safety functions formerly done by safety personnel.

2. **Technology.** This represents the way the work gets done, the work flow, the type of equipment used, and how jobs are designed. An organization's safety technology focuses on safety equipment, job planning, and hazards correction. Larger companies are additionally focused on information systems. An example of technological change is using computers and videos to communicate in place of the training and administrative tasks usually performed by the safety function.

3. **Rewards.** National surveys indicate that people do not feel rewarded or recognized for safety performance. The rewards system provides a great deal of leverage for change because people perform in accordance with their expectation for rewards which may be extrinsic (bonuses, incentives) or intrinsic (job satisfaction, one's own health). For example, when rewards for production outweigh rewards for safety or quality, production will be the higher priority. Care must be taken in changing reward structures to reward desired outcomes. Improper incentive designs have been known to encourage employees to hide minor accidents and suppress information.

4. **Measurement Systems.** A recent survey at a major chemical company indicated that safety managers felt that using the injury incidence rate as the sole measure of safety performance was "antiquated" and a "morale detrac-

Do we have the tools, equipment
and knowledge that we need to
support the desired culture?

Are the current organi-
zations for safety,
including policies and
procedures, set up to
support the desired
culture?

Is line management
responsible for safety?

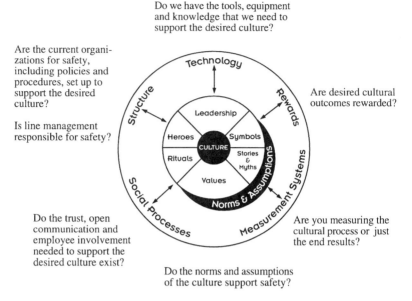

Are desired cultural
outcomes rewarded?

Do the trust, open
communication and
employee involvement
needed to support the
desired culture exist?

Are you measuring the
cultural process or just
the end results?

Do the norms and assumptions
of the culture support safety?

**Figure 32-4** Organizational factors influencing the safety culture.

tor". They felt workers' compensation costs and severity rates were good
after-the-fact measures, but did not contribute to accident prevention. Changing these after-the-fact measures to process measurements such as perception surveys and statistical process control can help prevent accidents by
pointing out weaknesses in the system. These systems can also be used to
measure progress in culture change efforts.

5. **Social Processes.** These include the informal communication systems, the
degree of employee empowerment and participation in safety, cooperation,
conflict, politics, openness, and trust. The social processes of an organization deeply affect morale, productivity, and safety performance. Any attempt to improve performance requires examination of these processes and
change as necessary. An example of changing a social process to improve
safety is empowering employees to issue work permits, a task normally
done by supervisors.

Figure 32-4 applies our model to safety performance. It lists five analytical
questions to help determine if the organizational factors are supporting the
desired safety culture. Going through the model asking these questions will tell
you which systems are out of balance. The next step is to determine how each
factor needs to be changed.

We will illustrate how to use this model to change the culture of an
organization by showing the results from a petrochemical division meeting
where safety supervisors met to discuss how to successfully implement an
accident investigation process to conduct accident root cause analysis called
"TapRoot".

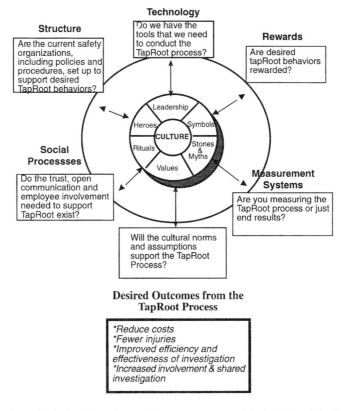

**Structure**
Are the current safety organizations, including policies and procedures, set up to support desired TapRoot behaviors?

**Technology**
Do we have the tools that we need to conduct the TapRoot process?

**Rewards**
Are desired tapRoot behaviors rewarded?

**Social Processses**
Do the trust, open communication and employee involvement needed to support TapRoot exist?

**Measurement Systems**
Are you measuring the TapRoot process or just end results?

Will the cultural norms and assumptions support the TapRoot Process?

Leadership
Heroes — Symbols
**CULTURE**
Rituals — Stories & Myths
Values

**Desired Outcomes from the TapRoot Process**

*Reduce costs
*Fewer injuries
*Improved efficiency and effectiveness of investigation
*Increased involvement & shared investigation

**Figure 32-5**   Analyzing the culture and systems to support installation of the TapRoot investigation process to achieve desired outcomes.

The TapRoot objectives are (1) discovering the root cause of accidents; (2) fact finding, not fault finding; (3) increasing near-miss reporting; and (4) employee and supervisor involvement. The safety supervisors anticipated cultural as well as organizational resistance, since the new process would require employee involvement and plant resources.

As Figure 32-5 shows, they examined each aspect of the model and asked what would have to change to support TapRoot to achieve the desired outcomes from the TapRoot process. By way of illustrating the next step, here are their answers to the Social Process question and the steps the safety managers felt would have to be taken in order to align the social process to support TapRoot:

***Do the trust, open communication, and employee involvement needed to support TapRoot exist?***

- Current state of social processes:
  1. There is a climate of fear (vs. trust).
  2. There is a lot of communication of inaccurate information.
  3. Our personnel subscribe to the "sooner or later" myth ("sooner or later there will be an accident").

    4. Our employees view their relationship to management as US vs. THEM.
    5. Our safety meetings have the tone of a "Spanish inquisition".
    6. Injured people are viewed as heroes or martyrs.
    7. Leadership needs to establish safety as a core value, not just a compliance requirement.
- Alignment of social processes needed:
    1. Break down the fear and begin to build trust, communicate history of incident occurrences, and show lack of recriminations.
    2. Get top management to enforce value of accuracy and fact-finding during incident investigation; communicate their endorsement to mid-level managers/supervisors with history of "management by fear" climate.
    3. Communicate success stories from last year's investigations to create new stories.

## MAKING THE CHANGE HAPPEN

- Involvement is the key to cultural change. Involve members in identifying current norms and assumptions and in exploring alternative solutions. Include the people affected by the change in discovering why the change is needed, what the obstacles are to changing, and how they might be overcome.
- Change leadership is in the hands of those who control the resources. A cultural change effort must include all the leaders of the existing culture. That usually means top management, union leaders, and key opinion leaders from the ranks. Without strong support from leadership, the culture will not change.
- There has to be a convincing reason for the change. Leaders must answer the question, "Why change?" The only acceptable answers will arise out of business necessity. As much as we would like to think that "doing the right thing" as a reason for change should work, it does not always seem to. This is because change is hard work; it is disruptive and it is expensive.

    The successful change efforts we have been involved in have all been based on a business need such as lowering workers' compensation costs, reducing Occupational Safety and Health Administration (OSHA) and Environmental Protection Agency (EPA) fines, fear of damage to reputation in the community, or loss of business to competitors.

    One of the key roles for a safety professional is making management aware of the need to change. Tools to accomplish this task include educating managers on the effect of their behavior on the safety culture, the cost of accidents to overall morale and productivity, and the benefits to the bottom line of accident prevention. If management will not listen, bring in an outside consultant to support you, use perception surveys to gather cultural data, and educate yourself with case histories of the successes and failures of other companies facing similar issues.
- Have a plan to get there. You need a vision of where you want to go. This is the end result. Then, collect objective data to determine where you are now. Next, make a step-by-step plan to get where you want to go. One

valuable tool for determining where you are now is a perception survey that measures norms and assumptions as well as management systems.

- No one program "fits all". There are many factors to consider, such as where the organization is starting from, before choosing an approach. If the organization has not allowed much participation, designing campaigns may be the place to start. There is a continuum that must be followed before attempting employee empowerment.
- Work on the norms, beliefs, and assumptions of the invisible culture as well as the technical and administrative aspects of the organization.

## SUMMARY

The cultural view of the organization described in this chapter presents a number of implications for improving safety performance. First, it implies a challenge to safety professionals to become agents of behavioral and cultural change. This is based on the recognition that competence in managing the social forces in organizations is equally important to technical competence in achieving safety excellence.

Second, this is a contingency view — that is, the right design for one company, division, or plant may not necessarily be the right one for another. What techniques or methods you choose depends on the technology, the people and assumptions of your organization, and challenges your company is facing.

Third, it implies that safety managers must utilize a comprehensive organizational framework to analyze safety performance and determine the root causes of problems. No longer can they rely on single-factor methods or theories to prevent accidents. Only from a comprehensive systems perspective is it possible to understand how to affect long-term safety improvements. This perspective will help managers realize why simplistic approaches such as changing a job description or changing a policy are ineffective.

Finally, this chapter provides a conceptual framework for safety leadership that extends beyond the traditional realm of management. It stretches the manager from policy maker to culture shaper. Yet, no leader stands alone. Without followers there can be no leadership. So, while it is true that a company would be unable to change its culture without a leader, a leader cannot change a culture without the support of the people. It takes everyone working together to make the change. The challenge to the leader is to involve, coach, and empower the employees to understand, accept, and act on the need to change.

## FURTHER READING

Beckhard, R. and W. Pritchard. *Changing the Essence.* San Francisco: Jossey-Bass, 1991.

Nadler, D. A., M. S. Gerstein, R. B. Shaw et al. (Eds.). *Organizational Architecture: Designs for Changing Organizations.* San Francisco: Jossey-Bass, 1992.

Perrow, C. *Normal Accidents; Living with High Risk Technologies.* New York: Basic Books, 1984.

Schein, E. H. *Organizational Culture and Leadership.* San Francisco: Jossey-Bass, 1991.

Simon, R. A. and S. I. Simon. *Grassroots Safety Leadership: A Handbook for Designing and Implementing Culture-Based Safety Improvement Strategies.* Seal Beach, CA: Culture Change Consultants, 1993.

## REFERENCES

### Books

Beckhard, R. and R. T. Harris. *Organizational Transitions,* 2nd ed., Reading, MA: Addison-Wesley, 1978.

Blake, R. and J. S. Mouton. *Productivity: The Human Side.* Self-published, 1981.

Bryson, J. M. *Strategic Planning for Public Nonprofit Organizations.* San Francisco: Jossey-Bass, 1988.

Cummings, T. G. and E. Huse. *Organization Development and Change.* St. Paul, MN: West Publishing, 1989.

Nadler, D. A., M. S. Gerstein, R. B. Shaw et al. (Eds.). *Organizational Architecture: Designs for Changing Organizations.* San Francisco: Jossey-Bass, 1992.

Nadler, D. A. and M. L. Tushman. *Strategic Organization Design.* Glenview, IL: Scott, Foresman, 1988.

Perrow, C. *Normal Accidents; Living with High Risk Technologies.* New York: Basic Books, 1984.

Schein, E. H. *Organizational Culture and Leadership.* San Francisco: Jossey-Bass, 1991.

Thompson, A. A., Jr. and A. J. Strickland III. *Strategic Management Concepts and Cases,* 5th ed., Homewood, IL: Richard D. Irwin, 1990.

### Other Publications

Asch, S. E. Studies of independence and conformity: a minority of one against a unanimous majority, *Psychol. Monogr.,* 70(9), Whole No. 416, 1956.

Sherif, M. A study of some social factors in perception. *Arch. Psychol.,* 187, 1935.

Simon, R. A. and S. I. Simon. The Relationship of Culture Perception to Worker Compensation Costs at Nine Sites. Culture Change Consultants, Seal Beach, CA, 1994.

Simon, R. A. An Open Systems Approach to Improving Safety Performance. Master's thesis, Pepperdine University, Malibu, CA, 1993.

Simon, R. A. Unpublished Safety Culture Survey Reports, 1991, 1992, 1993, 1994.

Tarrants, W. E. Emerging trends in safety and health, in *Accident Prevention Manual for Business and Industry,* 10th ed. Itasca, IL: National Safety Council, 1992.

# 33

# PRODUCING LASTING CHANGE THROUGH AN ATTITUDINAL AND BEHAVIORAL MANAGEMENT APPROACH*

Michael D. Topf

## TABLE OF CONTENTS

## INTRODUCTION

Safety professionals and business leaders continue to search for methods to reduce the impact that work-related injuries have on both the physical well-being of employees and the bottom line of companies. Managing safety performance through the use of behavior modification techniques is an idea whose time has come. It is no longer a question of whether to address safety and environmental behaviors, but of what is the best methodology to use.

---

* This chapter is based, in large part, upon the author's article in the February 1991 issue of *Occupational Health and Safety* magazine entitled, "Behavior Modification Can Heighten Safety Awareness, Curtail Accidents". Copyright 1991, Stevens Publishing.

1-56670-054-X/96/$0.00+$.50
© 1996 by CRC Press, Inc.

Implementing effective employee safety programs that address behaviors is one way to improve safety performance, and many such programs have had an impact. The emphasis in the past has been primarily on external regulations and internal policies and procedures — what to do and how to do it. In most companies today, workers generally know what to do and what not to do regarding personal health and safety; the problem is they do not always accept and comply with what is required. Awareness and attention factors also play a large role in causing injuries and incidents. Most companies are demanding improvements beyond those that current programs have achieved.

Due to the trend in industry today to place greater levels of responsibility on individual employees, along with a high degree of unsupervised, independent working situations and off-the-job accidents and injuries, companies need an innovative approach to the self-management of risk behaviors as well as the management of others.

An integrated behavior-modification approach is required for lasting improvements. This involves a seamless process which includes three types of behavior-modification strategies which interact to uncover the attitudes and beliefs that drive unsafe behavior and then increase individual responsibility for and self-management of safety behavior. The process then continues to shape a reinforcing culture in which achieved changes can be sustained.

This "awareness approach" to behavior modification includes

- *Cognitive behavior modification* — understanding how attitudes, beliefs, values, thoughts, and other aspects of people's thought processes contribute to both "automatic" and "calculated" behaviors that are unsafe.
- *Reality behavior modification* — using leadership and management strategies with coaching and counseling techniques in dealing with current behaviors and their consequences to achieve a shift to personal responsibility and self-management that leads to realistic corrective action plans.
- *Management by commitment* — once individuals realize that they are inherently committed to safety, and management shares that commitment, they are motivated by the mutual benefit to themselves and the company to act more safely in greater cooperation with company requirements.

The interplay of these three elements — thought, action, and commitment — can help individuals and organizations become aware of the thought processes and interpersonal influences that cause people to place themselves and others at risk, by acting either automatically or in a calculated manner. The process continues to shape actions consistent with this awareness and to provide an environment in which safe behavior is supported and reinforced.

## EMPLOYEE SAFEGUARDS

Enhancing employees' ability to be safe can be accomplished by providing line employees, supervisors, and managers with the necessary skills to manage themselves and others to maintain and improve safety performance. It is important to understand that most accidents, injuries, and environmental incidents do not have exotic causes. People are not scientific in their behavior. We need simple, yet effective, common sense awareness and skills to observe and manage within ourselves and others the kinds of attitudes and behaviors that cause injuries. Behavior modification can be accelerated greatly by focusing on the following:

- *Safety and awareness* — learning techniques to observe one's own safety attitudes, thinking, and behaviors will help people learn how to identify and heed early warning signs which can lead to an incident.
- *Modifying risk behavior* — heightening awareness of attitudes, beliefs, and values can lead to behavior change. Attitude change can precede behavior change.
- *Safety and the thinking process* — learning to control and manage automatic responses for appropriate behaviors.
- *Personal responsibility* — increasing the level of "ownership" or responsibility for safety, the key to effective individual actions.
- *Leadership's role* — enabling others to manage risk behaviors constructively by providing essential skills for empowerment and behavior management. Management and labor leadership need to join forces to lead, guide, and direct employees to take responsibility for safety and the environment.
- *Safety and personal commitment* — people can be helped to recognize what is at stake for themselves, their families, their co-workers, and their company and how to make appropriate choices. No one wants to get hurt; we just have attitudes and beliefs that lead us to think we can take risks and get away with it.

## THE ROLE OF RESPONSIBILITY

Optimization of employee safeguards requires awareness and training that emphasizes safety as a function of personal responsibility. It is necessary to address the common understanding of what responsibility is and people's general resistance to accepting it regarding safety. The general perception relates responsibility to blame rather than a recognition of a person's role and power in ensuring his/her own and other people's safety, health, and well-being. Individuals can learn to take responsibility for their own safety, as well as the safety of others. When people are truly empowered, they understand their ability to protect themselves, their co-workers, and their families and take initiative to take whatever actions are necessary to accomplish this. This requires an attitude change for many people which is accomplished by providing an understanding of the beliefs that create a negative perception of what responsibility is. Structures and systems must be set up that will allow employees to take constructive actions and make a difference in their safety improvement process.

## HUMAN MECHANISMS

One key to implementing successful safety programs is training people to manage themselves by observing their own behaviors. This is not as complex as it may appear. Most people are very poor observers of their "play" because they are key characters in it. An awareness approach to behavior modification addresses the basic human mechanisms or attitudes and behaviors that lead individuals to place themselves and their co-workers at risk and enables them to observe their thought processes, actions, and consequences.

The following human mechanisms can affect a person's ability to stay safe:

- *Inattention* — daydreaming, distractions, or thinking about things other than the job; not listening to instructions and losing focus while working on repetitive or familiar activities
- *Decreased awareness* — not remaining alert to danger at all times and not being on guard for unexpected dangers because of complacency and inattention from the above noted issues
- *Lack of focus* — difficulty concentrating on tasks; increases the possibility of accidents and injuries
- *Making distinctions* — unable to see, hear, and smell potential safety hazards as a result of their thinking, attitudes, and behaviors and thus being unable to take appropriate actions to remain safe
- *Taking responsibility* — not responding when safety demands immediate action, due to the inability to understand how to manage attitudes that may support or undermine the appropriate response for safety
- *Resisting change* — resisting certain safety behaviors and feedback from others due to past experiences, attitudes, and beliefs regarding safety; not changing habits as needed for safety because of a lack of awareness of what is really at stake. Belief systems can cloud a person's ability to deal with current reality and recognize the true level of risk or one's ability to protect oneself regardless of the circumstances. Macho attitudes or the "I'll do it my way" syndrome can interfere with a common sense approach to safe behavior.
- *Heeding early warning signs* — ignoring the mental alarm system — the inner voice — that can alert people to danger and necessary actions to avoid accidents, injuries, and environmental incidents
- *Keeping safety agreements* — ignoring or not communicating and discussing safety rules and regulations that protect people or not encouraging them to be more willing to participate as part of a safety team. Once again, attitudes and beliefs that cause people to resist rules or do it their own way for a variety of reasons can contribute to noncompliance.

## EMPHASIZING ATTITUDE

In the ideal world, implementing only safety tactics, such as rules, regulations, and procedures, can be successful in reducing injuries if people adhere to them consistently. Unfortunately, this ideal world does not exist.

Changing attitude as well as behavior, by emphasizing awareness, has proven successful in increasing safe behavior and productivity on an ongoing basis. Raising awareness translates into providing knowledge in a variety of forms. Knowledge can be seen as learning new information, gaining a different perspective from which to view a situation or circumstance from, having an intuition, insights or ideas which allow for new possibilities in our thinking, attitudes and beliefs, the choices we make, and ultimately, our actions and behaviors. The result is that people have a higher regard for personal health and safety. Their ability and willingness to take personal responsibility and act in accordance with what aligns in terms of common sense and/or company requirements is increased.

## MEASURING SAFETY

Safety can be measured by qualitative as well as quantitative measures. Traditional quantitative measures such as lost times, recordable rates, minor injuries, etc., are not necessarily an accurate measure of safety performance. Behavior modification programs that focus only on observing the behaviors of others must, by definition, rely as well on quantitative methods to measure safety performance. Although these methods are very important, relying on these traditional measurement techniques for after-the-fact OSHA incident reporting and downstream behavioral tracking does not provide a complete view. A low incident rate does not mean that people are behaving safely, nor does the observation of compliance to requirements mean that the person has developed the attitudes and beliefs that will stay with that person wherever he/she goes, influencing their behavior in a positive way when working independently at work or at home, nor does it translate into other domains of quality and production.

Analyzing qualitative results is an essential method of measuring the effectiveness of safety programs, especially those that focus on attitude change. Many behavior-based methods do not account for qualitative results because there are no measurable statistics for comparison. Qualitative results which indicate a change in attitude can be seen by observing the following types of behaviors:

- Increased participation at safety meetings. Voluntary discussion of on- and off-the-job hazards, behaviors, and improvements.
- Increased natural caring and self-motivation for safety through greater use of personal protective equipment and compliance with company requirements

- Increased discussions on the importance of safety to the individual and company.
- Increased acceptance of giving and receiving feedback from management, safety representatives, and other workers regarding safety procedures and behaviors.
- Increased willingness of employees at all levels to be proactive regarding safety and environmental issues.
- Greater solicited and unsolicited communication of safety issues between line employees and management and among departments during normal work routines.
- Increased solicited and unsolicited reporting of observed hazards to maintenance and engineering departments, along with potential solutions.

## ENVIRONMENT AND CULTURE

Organizations, like individuals, possess certain traits and prevailing attitudes that are a result of historical backgrounds, morals, and philosophies. Individuals also are influenced by company values, attitudes, and priorities regarding quality, safety, and production. It is essential to assess the existing culture and environment for the various perspectives of norms, values, attitudes, and beliefs regarding safety in order to determine influences on employee thinking and behavior.

For example, the culture at some organizations may be so production oriented that risk taking is considered acceptable to get the job done. Workers often cut corners to remain productive, rather than take the time to do the job safely. In these instances, unsafe behavior often is encouraged and considered acceptable.

Eliminating an environment that condones risky behavior requires an organization to focus on changing the prevailing attitudes that influence behaviors. Rather than emphasizing statistics alone, organizations, from the CEO down, must communicate that the primary concern is the human being behind the statistic.

## COMMITMENT, COMMUNICATION

In almost every organization there is genuine concern for the safety, health, and well-being of employees. Increasing productivity while emphasizing safety requires everyone in an organization to remain committed to safety by proactively refining and improving the safety process.

To encourage commitment, it is important for top-level management to clearly define and communicate the importance of safety and what is required so that individuals take responsibility for their safety and the safety of co-workers. In addition, managers and supervisors, as well as labor or line safety representatives, must be skilled in leading others for safety. Finally, all employees must be aware that they have a direct impact on the overall safety of the organization.

## LEADERSHIP FOR SAFETY

To create the kind of safety culture where people are willing to observe and to give and receive feedback from anyone, regardless of their position, requires an attitude shift that gives every man and woman the perspective that he/she is a leader when it comes to safety and that they each have a responsibility for every person's safety.

It is essential to provide leadership training and skills for line personnel, management, and labor safety leadership as part of an overall safety program. Unlike many leadership courses, an integrated behavior-modification process also focuses on the underlying human mechanisms that cause leadership to resist or avoid following through on managing safety.

Company safety leaders are encouraged to intercede immediately upon discovering an unsafe behavior or condition through an understanding of the possible consequences of waiting. They must learn to provide constructive feedback in a coaching manner so that employees will take responsibility for correcting their behaviors. If done properly in a noncritical manner, employees tend to be less defensive and learn to take responsibility for their actions and to behave in accordance with the requirements.

Once again, management must support this process to demonstrate its effectiveness and credibility. In addition, it is important to ensure that the leadership apply the same requirements for everyone, not just the line staff.

Leadership's accountability for safety can be approached from two perspectives:

- *Preventive* — properly identifying potential hazards due to unsafe conditions, attitudes, and behaviors; reinforcing existing safety requirements; resolving problem areas and modifying unsafe behavior.
- *Post-incident/injury assessment and correction* — determining the accountability and responsibility of the individuals involved in an incident and implementing strategies to prevent further unsafe acts. It is essential that this aspect be handled from an educational perspective to ensure employee cooperation and involvement.

Leadership's role in safety awareness includes encouraging participation and sharing and developing ideas and concepts with everyone who is at risk within an organization. This provides purpose and meaning to participating in the process of increasing safety. This integrated process will assist greatly in the development of the kinds of positive safety and environmental attitudes and behaviors that open the possibility for breakthroughs in performance to occur.

## FURTHER READING

William, R. M. Values. In *International Encyclopedia of the Social Sciences,* B. Sills, Ed. New York: Macmillan, 1968.

Rokeach, *The Nature of Human Values.* New York: The Free Press, 1973.

Berger, and Sikor, J. *The Change Management Handbook: A Road Map to Corporate Transformation.* New York: Irwin, 1994.

Fournier, and Plunkett, C. *Participative Management: Implementing Empowerment.* New York: John Wiley & Sons, 1991.

Fournies, *Why Employees Don't Do What They're Supposed to Do.* City: Liberty Hall Press, 1988.

# 34

# TOTAL QUALITY MANAGEMENT AND SAFETY AND HEALTH

Richard W. Lack

## TABLE OF CONTENTS

## INTRODUCTION

Being a serious student of the art and science of management, I am naturally interested in the total quality management (TQM) process. Since its inception, I have attended numerous conferences, seminars, lectures, and meetings at which this subject has been discussed. I have also read many articles, papers, and books describing the process and results achieved at various organizations and facilities. In order to gain some perspective, I invite you to join me in a short journey of reflection on the whole issue of TQM.

Throughout history, the achievement of high-quality goods and services and a need to satisfy the customer were of paramount concern. Artists and artisans alike shared this concern. One has only to visit museums or art galleries or tour the wonders of ancient civilizations to immediately realize that TQM has been with us for a very long time.

I share Dr. Wayne Dyer's view that every human being since the beginning of time is a part of an intelligent system and that each of us has an "heroic

1-56670-054-X/96/$0.00+$.50

mission", as he terms it, to accomplish while we are on this planet. The difficulty, of course, is that few are born with this knowledge and skill. Most of us have many lessons to learn on our road to peak achievement and inner peace. Translate this situation into any human organization, be it a social club, restaurant, hotel, municipality, factory, or giant corporate business, and you are dealing with a group of people each on their various paths toward getting to their lives' purposes.

No doubt the pharaohs faced this problem when building the pyramids, and every undertaking, now and in the future, will probably do so, until a system is found whereby people will more easily be able to find their personal life mission.

Undoubtedly, progress is being made, when you consider the developments of the industrial revolution. The safety and health profession itself was born from the concern for preventing injuries and illnesses in what frequently were inherently dangerous workplaces.

In more recent times, the constant need to improve production efficiency and reduce costs, especially as automation was introduced, produced people with skills to serve these needs. Sometimes called "efficiency experts" or "time and motion studiers", these were your industrial engineers of the 1950s and 1960s. This profession has become far more sophisticated with the advent of the computer.

Quality inspection and control also grew up with the mass-production era, especially in the U.S. Frequently, the quality control and industrial engineering functions were, and still are, merged into one group serving their respective organizations.

The difficulty with this relentless search for production efficiency is that it tends to "dehumanize" the organization. People feel they have lost control, that their efforts are not recognized, and the inevitable result is a buildup of stress which finds outlets in ways that are frequently unproductive.

As has been discussed in other chapters in this book, a number of approaches and systems have been developed in recent times to help address this problem. These range from reengineering the organization to the team approach and behavior-based motivation programs. On the horizon are techniques developed by such luminaries as Scott Peck, Tom Melohn, Margaret Wheatley, Peter Senge, and Peter Block.

Another approach is the technique of "intrapreneuring" developed by Gifford and Elizabeth Pinchot. This approach is designed to help organizations transform their bureaucratic management systems into systems which make more use of their employees' intelligence and help them achieve their personal missions.

I was recently intrigued to receive a brochure on a special 3-day conference organized by the Association for Quality and Participation entitled "Practical Applications of Leading Edge Theory". The three presenters at this course are Margaret Wheatley, Peter Senge, and Peter Block. Need I say more?

## THE TOTAL QUALITY MANAGEMENT PROCESS

Readers interested in studying the specific elements of TQM in depth are encouraged to review the books and other publications listed in "Further Reading" at the end of this chapter.

The techniques developed by such pioneers as W. Edwards Deming, Joseph Juran, and others were built around creating a more empowered workforce and providing them with tools so they could improve their production efficiency and maintain customer satisfaction. This movement also helped to "demystify" the statistical process, and a number of simple techniques for pinpointing potential production problems were developed. Examples are

- *Flowcharts* — these are a useful way of outlining the steps in a procedure or process. There are many applications in safety management. For example, the chart could define the steps for confined space entry or lockout-tagout. This technique can supplement or in some cases may replace the traditional job safety analysis approach.
- *Fishbone diagrams* — cause and effect charts. This technique can be a useful aid in accident investigation.
- *Pareto charts* — to help determine priorities (the 80–20 rule). This graphic technique can help with accident analysis, e.g., type of injury and the immediate causes of this injury.
- *Histograms* — to chart how frequently something occurs. This could be used to track the shift, day, or time that accidents occur.
- *Run or trend charts* — to identify trends over a period of time (example, 12-month total case incidence rate for injuries and illnesses).
- *Control charts* — establish upper and lower control limits and track the actual results. This technique could be used to track injury experience results.

Problem solving in the TQM process has a variety of approaches, especially now that computers have come to our aid. They all involve the following broad steps:

1. Define the problem.
2. Analyze the facts as to its cause and effect.
3. Generate alternatives.
4. Select feasible solutions.
5. Take action.
6. Evaluate the results.

Brainstorming to help generate ideas for improvement or to solve a situation is now commonly used in successful organizations.

The "Six Sigma" concept, as adopted by several corporations (the best known being Motorola), has created much interest. "Six Sigma" is in effect a measure of "goodness" in that it measures defects or mistakes. Each group in the organization has to identify their customers (internal and/or external) for their service and find ways to mistake-proof their service and ensure continuous improvement.

Achieving "Six Sigma", for example, equals 3.4 defects per million opportunities for error. Translating this to safety would mean no more than four incidents per million hours worked. Note that by incident I mean four mistakes or unsafe work practices that could have caused an injury.

## APPLYING TOTAL QUALITY MANAGEMENT TO SAFETY AND HEALTH

In my opinion, the TQM approach is a natural for safety and health practitioners because it is a proactive preventive method. It will help you to identify your customers and find better ways to serve them.

TQM is also concerned with measuring performance to help ensure that results will continuously improve. This concept has already been discussed in several other chapters of this book.

The analytical techniques involved with TQM, discussed earlier in this chapter, can be used to help identify trends or root causes of accidents, incidents (near miss), or hazards (either unsafe work practice or unsafe condition).

In addition to the tools and techniques already described, readers are referred to the section on accident investigation in Chapter 5. This outlines a source and cause analysis approach. The key is to reach actionable root causes by asking "Why?" (ask "Why?" five times). Then select the root cause with the greatest probable impact.

Finally, the safety and health practitioner will need to help his or her organization set up measurement systems that the employees and work teams can use to track their own performance. The old approach of the safety department preparing and distributing the reports on supervisory and team performance can then be phased out as TQM really takes hold in your organization.

## CONCLUSIONS

For the reasons stated in the introduction, there is no doubt in my mind that every organization will eventually shift its management style to one that helps its people achieve a pattern of high performance. This, in turn, will open up a unique opportunity for safety and health practitioners to help adapt their unit's safety and health goals and systems to the unit's overall production goals and systems.

If you are not already involved in this process, I recommend that you take immediate steps to prepare yourself by reading books, attending courses, and joining organizations in the TQM and professional training fields.

I see our role as helping the people in our organization find ways to achieve error-free peak performance. Prevention is essential in order to ensure continuous improvement, and this will only be possible when the people are given the necessary tools and training and are provided with a supportive community in which they can do their jobs safely.

## REFERENCES

Here is a list of sources that are referred to in the chapter which are related to but not directly concerned with the TQM process.

### Books

Block, P. *Stewardship — Choosing Service Over Self Interest.* San Francisco, CA: Berrett-Koehler Publishers. 1993.

Dyer, W. W. *Real Magic — Creating Miracles in Everyday Life.* New York: Harper Collins Publishers. 1992.

Fletcher, J. L. *Patterns of High Performance — Discovering the Ways People Work Best.* San Francisco, CA: Berrett-Koehler Publishers. 1993.

Melohn,T. *The New Partnership — Profit by Bringing Out the Best in Your People, Customers and Yourself.* Essex Junction, VT: Oliver Wight Publications. 1994.

Peck, M. S. *A World Waiting to Be Born — Civility Rediscovered.* New York: Bantam Books. 1994.

Pinchot, G. and Pinchot, E. *The End of Bureaucracy and the Rise of the Intelligent Organization.* San Francisco, CA: Berrett-Koehler Publishers. 1994.

Senge, P. *The Fifth Discipline Fieldbook — Strategies and Tools for Building a Learning Organization.* New York: Doubleday. 1994.

Wheatley, M. J. *Leadership and the New Science — Learning about Organization from an Orderly Universe.* San Francisco, CA: Berrett-Koehler Publishers. 1992.

## FURTHER READING

### Books

Adams, E. E. *Total Quality Management — An Introduction.* Des Plaines, IL: American Society of Safety Engineers. 1995.

Camp R. C. *Benchmarking.* Milwaukee, WI: American Society for Quality Control, ASQC Press. 1989.

Gitlow, H. S. and Gitlow, S. J. *The Deming Guide to Quality and Competitive Position.* Englewood Cliffs, NJ: Prentice-Hall. 1987.

Sashkin, M. and Kiser, K. J. *Putting Total Quality Management to Work.* San Francisco, CA: Berrett-Koehler Publishers. 1993.

Walton, M. *The Deming Management Method.* New York: Perigee Books–Putnam Publishing. 1986.

## Other Publications

Deacon, A. The role of safety in total quality management. *Saf. Health Pract. (U.K.)* 12(1), 18–21, January 1994.

Lack, R. W. More on the safety-quality management issue. *Insights Manage.* 3(1), February 1991.

Manuele, F. A. Make quality the watchword of your safety program. *Saf. Health* 148(4), 106–108, October 1993.

Motzko, S. Variation, system improvement, and safety management. *Prof. Saf.* 34(8), 17–20, August 1989.

Petersen, D. Integrating safety into total quality management. *Prof. Saf.* 39(6), 28–30, June 1994.

Vincoli, J. W. Total quality management and the safety and health professional. *Prof. Saf.* 36(6), 27–32, June 1991.

BEYOND TEAMWORK:
HOW THE SAFETY AND HEALTH
PRACTITIONER CAN FOSTER THE
# 35 NEW WORKPLACE COMMUNITY

**Carolyn R. Shaffer**

When the members of a home-visit-based program at a large, urban hospital went out on the job, they did not know what they might find. Their mission was to provide emotional support and developmental training to families with severely disabled children. One hour a staff professional might be in a run-down, flatlands apartment counseling a cocaine-addicted mother with a child unable to walk. The next hour she might be visiting a luxurious, 12-room home in the hills to provide support to two usually high-performing, in-control parents, reduced to sobs by the frustration of caring for their deformed and unresponsive infant.

In most front-line health-care programs like this, burnout and staff turnover run extremely high, with some staffs turning over completely in the course of a year. The staff of this child-disability program confounded their colleagues by working with an increasingly high-stress population for 11 years without a single staff member leaving. All in the program, including the secretary, loved their work. Today, staff members agree that a major ingredient in this success was the sense of community that they had developed. "We felt more like a family than a work team," says one.

While the staff of this health-care program did not analyze data on the effect that this sense of community had on their safety and health, other workplaces experimenting with community-generating styles of management have gathered impressive quantitative evidence. A bakery division of a major supermarket chain reversed its downward slide into unacceptable rates of absenteeism and accidents and became a model for safety and health practices. It did so by shifting to a management approach that bore several striking resemblances to the above hospital program. According to Robert H. Rosen's report on this success story in *The Healthy Company,* employees in the bakery division of Safeway Stores in Clackamas, Oregon had felt "powerless, discouraged, and unimportant" under the earlier "iron-fisted" management style of the bakery manager. When the manager changed his style to give employees more chances to connect with one another and more say in shaping their work environment, the negative indicators began to fall. Among the dramatic results: absenteeism decreased from 8% to 0.2%, accidents plummeted from 1740 to 2 workdays lost in a year, and turnover dropped from 100% in some jobs to less than 10%. The flood of grievances and discrimination cases also turned into a trickle, decreasing from 75 to 80 a year to one grievance and no discrimination cases in 5 years. Besides improving the health and well-being of the employees, the Safeway bakery successes contributed significantly to a healthy bottom line for the company. Safeway calculated that, in real dollars, for every dollar invested in the new programs it generated an estimated return of $15, saving between $700,000 and $800,000 a year.

By fostering community — and the sense of commitment inherent in it — in the workplace, safety and health professionals can improve their companies' bottom lines while saving the lives, limbs, and immune systems of employees. To do this effectively, safety and health professionals need to know what community is, how it benefits health, how it generates commitment in employees, and how it can be created in the workplace. This short chapter can only touch on some of these topics and will refer readers to other sources of information — although much of the literature on the kind of workplace community that goes beyond teamwork has yet to be written. What is well documented are the health benefits of positive social connectedness (see "Further Reading," especially Shaffer and Anundsen, *Creating Community Anywhere,* Chapter 2). The primary focus here is on drawing practical lessons from success stories that safety and health professionals can apply in their organizational settings.

First, a word about "community," a term as amorphous and overused as "love." In *Creating Community Anywhere,* my co-author and I chose to define the term strongly. Here is our short version: **Community is the dynamic whole that emerges when a group of people commit themselves for the long term to their own, one another's, and the group's well-being.** In essence, community is about commitment. We included a long-term time element because commitment tends to manifest over time. Our long definition spells out the practices and sense of collective identity that flow from, and are an essential part of, this commitment. When people are in community, they

- Do things together
- Depend upon one another
- Make decisions together
- Identify themselves as part of something larger than the sum of their individual relationships

The members of the bakery and the hospital communities described above engaged in these practices and developed a sense of collective identity that withstood the test of time.

The professional staff of the child disability program, which grew from 8 to 11 members over the 11 years, met weekly for lunch at a local restaurant before going into their staff meeting. They shared stories about their lives as well as the joys and frustrations of working with their patient families. In the staff meetings, the director generated an atmosphere of mutual respect and empowerment. Without giving over her power or responsibilities, she encouraged consensus around important issues. She made sure the staff felt aligned around new hires and any changes in policy before implementing them. Occasionally, she and the staff arranged for a day-long retreat. Although the work group was technically hierarchical in structure, it felt egalitarian. Members liked and respected one another even when they disagreed, and they disagreed often in passionate discussions on the way to reaching a consensus. At the same time, they supported one another in practical as well as emotional ways. They adjusted their schedules to meet each other's parenting needs and substituted for one another when necessary. When one member was diagnosed with breast cancer, the others helped her continue to work — as she desired — while she received chemotherapy treatments. The team members also encouraged each others' individual gifts to flourish. "We became wildly creative," says a former member. "We invented new forms, set up offices, and created new programs." Through all this, the program produced results. Every grant the group applied for, year after year, it received. Not only did the members deem themselves productive and a success, but the hospital and the funding agencies agreed. Their collective identity included being "like family" and engaging in important, effective work. Even today, several years after the program and the personnel began to change and the sense of community to fade, members of the original group stay in touch and several remain close friends.

At the supermarket bakery, when the manager loosened his control, he generated a sense of connection and commitment by encouraging employees to meet in problem-solving groups focused on such common concerns as sanitation, safety, and ergonomics. He also let workers know that he wanted suggestions from them. Once employees were convinced that management was listening and responding positively, they submitted hundreds of suggestions. Soon the employees began expressing their commitment and their newly unleashed creativity in more significant ways. They built a small employee fitness center. This led to a wellness program and eventually to a 5000-square-foot fitness center not just for employees but for their families as well. The center began sponsoring classes and special fitness events that brought employees into even greater positive connection with one another and with the larger community.

In both workplaces, managers increased commitment and creativity by loosening control and inviting participation in the decision-making process. Both also created vehicles for employees to come together to talk about matters of importance to them — the weekly lunches, occasional retreats, problem-solving groups, and wellness classes. Just as important was the sense of positive collective identity that emerged. For the hospital staff, the work itself created a strong bonding. "We were very dedicated to our patient families and our work," says one member, "and were convinced that we were making a difference in people's lives." The bakery work became more than a job for employees when it began involving their general wellness and their families. They developed a fitness event called "Buns on the Run" at which t-shirts were handed out to regular exercisers. Activities such as this led to the development of a sense of camaraderie and belonging, which in turn led to a positive collective identity.

Creating community requires trust and vulnerability. These are difficult to develop and sustain in any setting, but the workplace presents special challenges. The hierarchical structure and competitive climate in most places of work are hardly conducive to self-revelation, honest feedback, and interdepartmental cooperation. A person could get frozen out of the information loop or fired for engaging in such behavior. It took the employees of the Safeway bakery some time to learn that it was safe, as well as effective, to make suggestions. Nonetheless, it is possible to create a community in the workplace that is both honest and committed, as the staff of the child disability program demonstrated. These women — during that 11 years, the director and staff were all women — not only shared their deep fears and joys with one another, but they also argued with and challenged one another. They had learned that the director would not arbitrarily fire, or hire, anyone, although she had the power to do so. They also came to know that no one else in the program would use the personal information a member shared against her, nor would small groups of members form factions and engage in malicious gossip or secretive plotting. If someone was having a problem with others in the group, she brought it to the group — an agreement that any collection of people intent on becoming a community should consider making.

One of the first steps toward creating the trust that leads to community in the workplace is to be open and honest about where the power lies and who wields it. If the director of the hospital program had engaged with the staff in a process that she labeled consensus, but then made decisions on her own without regard to their input, she would have undermined any germinating sense of community and plunged the group into betrayal-induced cynicism, effectively immunizing them against attempts at community building in the future.

Another key step is to encourage the members to align with a common vision or purpose — one strong enough to help them hold together and support the larger enterprise through the inevitable differences and disagreements ahead. This is where a well-facilitated group retreat can be effective. To maintain the openness and sense of mutual support that such a retreat can generate, the organization needs to create or encourage vehicles for ongoing conversations about matters of importance to the group. It also needs to clarify what role these groups play and be ready to follow up on decisions or recommendations that come from them. However, the vehicles are not enough in themselves. Participants also need to develop group agreements and collaborative attitudes and skills. These include agreeing on how and when members meet, make decisions, and deal with differences and conflict. A healthy group will make sure everyone has a chance to contribute, appreciate the varying gifts of each member, and welcome differences as opportunities for learning and generating creative solutions to problems. Its members will be willing to learn to communicate well, including giving and receiving negative feedback, and to work effectively with conflict. However, even the best groups will eventually grow stale or get stuck unless the members regularly take time out to renew their common purpose and agreements, looking honestly at what worked and what could be done better and celebrating their accomplishments.

Fostering community in the workplace is not easy, but it is possible. Generated skillfully, with care and honesty, workplace community can contribute significantly to the health and safety of employees — primarily by reducing stress and stress-related accidents and by boosting the immune system. It can infuse a business with vitality and creativity and lead to increased productivity, lower turnover, and healthier profits.

**A final word to safety and healthy professionals who may think this sounds fine but cannot imagine it happening in the organization they work for:** Even if all these benefits cannot help you convince top management to support community-generating programs, you need not give up. You can still educate managers and employees about the health value of community and provide them with information on how they can create this outside the profit centers of the organization — in their neighborhoods; through peer support groups, labor unions, and professional associations; and even at the workplace through volunteer activities. Perhaps the best place to begin is by giving yourself the gift of community.

## FURTHER READING

### Books

Shaffer, C. R. and Anundsen, K. *Creating Community Anywhere: Finding Support and Connection in a Fragmented World.* New York: Tarcher/Putnam, 1993.

Nirenberg, J. *The Living Organization: Transforming Teams into Workplace Communities.* Burr Ridge, IL: Irwin, 1993.

Peck, M. S. *The Different Drum: Community-Making and Peace.* New York: Simon & Schuster, 1987.

Autry, J. *Love and Profit: The Art of Caring Leadership.* New York: Morrow, 1991.

Helgesen, S. *The Female Advantage: Women's Ways of Leadership.* New York: Doubleday/Currency, 1992.

Rosen, R. *The Healthy Company: Eight Strategies to Develop People, Productivity, and Profits.* New York: Tarcher/Putnam, 1991.

Senge, P. M. *The Fifth Discipline: The Art and Practice of the Learning Organization.* New York: Doubleday/Currency, 1990.

Harrington-Mackin, D. *The Team Building Tool Kit: Tips, Tactics, and Rules for Effective Workplace Teams.* New York: AMACON/American Management Association, 1994.

# 36

# NEW DIRECTIONS IN SAFETY COMMUNICATION

**Bonita B. Zahara**

"PLANE CANCELED DUE TO MECHANICAL PROBLEMS!" You can imagine the angry passengers crowded in the airport terminal complaining in unison about the heat and the inconvenience. Yet having just come from the 1994 American Society of Safety Engineers (ASSE) conference I could only smile and try to get the waiting grumblers to realize how lucky we were that the ground crew was so efficient as to have found a serious problem and prevented an accident. "Consider the alternative!"

"What," you may ask, "can possibly be *new* in safety communications?" I would like to invite you to look outside your standard box of safety tools and back at some basic communication skills. No, communication skills are not new, but what is new is the *attitude* that we must bring to the importance of the safety messages we communicate every day. The tools of safety professionals are powerful and, for the most part, very well utilized. I would like to invite you to take a new look at communication skills and how they might influence the safety message you have to deliver.

You may recall the opening scenes of a long-running television series where the police sergeant would close his evenings roll call with the empathetic charge, "Be careful out there!" Somehow the same old message, given week after week, had to come across as "new". The new messages are the same old messages when it comes to safety. Familiar phrases like "Safety Doesn't Hurt", "Safety First", and "Safety Saves Lives" are as true and important today as they ever were. The challenge to safety professionals, as well as the general populus, is: How do we keep the simple principles of safety "heard" in an information society overloaded with messages and instructions?

Shifting attitudes does not seem like a big deal, but in fact our attitudes shape almost everything in our lives. Attitudes to "safety consciousness" are critical to the success of a safety program. While "attitude awareness" might not seem a "new direction" for safety professionals, I would like to suggest that there are lessons to be learned from a better understanding of communication skills — in particular, understanding "whole-brain thinking".

There is nothing new under the sun, yet the futurists and marketing gurus all make their fortunes by forecasting the new trends. I would like to make a more humble "new claim" and state up front that much of my thesis is indeed good old-fashioned common sense. They are the kinds of things safety professionals used to do as a routine part of their work. That new/old thing is to *"communicate the importance of life"*.

When I was a child, safety was an everyday part of our lives. My father was a personnel director and safety manager for an aluminum casting plant and rolling mill. While other families had backyard barbecues, we showed safety films and learned CPR. Mind you, the CPR practice came in handy during our teenage years! We knew Dad was serious, because we had a swimming pool in the backyard and he instilled in us the responsibility that was inherent in owning a pool. We all had to learn how to save lives.

My professional training is in the area of whole-brain learning skills. I became fascinated with how the brain/mind system worked and how we could increase human productivity. Peter Drucker's simple statement, "Effectiveness must be learned," was part of my battlecry as I set about to champion the underutilized infinite capabilities of the human brain. As a regional vice president for Evelyn Wood Reading Dynamics, I became fascinated with the brain's capacity to learn and remember. As computers and ergonomics overtook people's interest in "old fashioned reading" and we moved into the age of "information accessing", a powerful and simple safety message reappeared. **If we understand how people think and how they remember, they will better remember our safety message.**

A simple understanding of the human brain is helpful for anyone interested in improving his or her communication skills. Scientist Richard Sperry won a Nobel prize for his research in "split-brain development", and since that time extensive studies have repeatedly supported the idea that the capabilities of the brain are infinite. There are some simple aspects of human dynamics that, if understood, can help us to remember and help others to remember the messages we want to get through to them.

"When do accidents happen?" They happen when we are not paying attention, or when we lose concentration, or when our minds wander. As I studied the human brain some simple exercises for stimulating the "learning abilities" seemed to be applicable to many professions. The confusion about left and right specialization, and which side of the brain does what, can be eliminated if we focus on being "whole brained" and put the emphasis on getting it all to work together.

In working with "whole-brain stimulation" to increase our comprehension of written information, I became aware of how simple and powerful it was to truly "wake up" both hemispheres of the brain. Simple visual exercises, like following one hand at a time as you draw a large infinity sign in front of you, literally stimulate the dendritic connections in the outermost reaches of the brain. The brain wakes up with this simple stimulation exercise, and by so doing memory is enhanced. In many of the workshops I conduct across the country I teach "juggling" as a form of whole-brain stimulation. Kevin Strupe, the teacher and educator who introduced me to this simple brain exercise, has had enormous success working with students with learning disorders. By stimulating and integrating both hemispheres of the brain, reading levels and test scores have jumped enormously.

If we can learn, as individuals, how to increase our brain's effectiveness, as Peter Drucker has challenged, then our effectiveness as safety professionals will also be improved. And if we can begin taking that message home to our families and sharing not only "how to wake up", but how important some of our lessons are, then we can begin to make a strong and powerful difference in the world. It is a sad truth that seven times more accidents happen at home than in the workplace. It is important that the profession of safety begins to look beyond the workplace for ways to contribute. Learn communication skills, for they will help you not only at work, but also at home. Then take the big step into the third realm, out into your community and the world at large.

I suggest the following "new directions" to safety professionals for applying whole-brain learning theory:

- First: Understand how *you* think and remember.
- Second: Take it home. Take your safety knowledge and communication skills, along with the basic knowledge of "safe practices", home with you and spread the message.
- Third: Spread the word to your communities.

In my work across the country I often ask individuals to share what makes them proud. In about 90% of the cases "family" is part of the answer that is given. Family truly is at the heart of safety. As you learn more about yourself, you can share that with family and also at work. Work is, in fact, a form of family. Bette Friedan has written an excellent book, *The Fountain of Age*. There are two major points in this book: first, that we must stop treating *age* as a disease, and second, that we are evolving as a culture which will create "families of choice". Work is a sort of "family", and we need to learn more about communication skills to share not only with those with whom we live, but with those with whom we spend 60% of our waking time, our fellow workers.

Stephen Covey's *The Seven Habits of Highly Successful People* has become a classic in a very short period of time. Covey's seventh habit is to "sharpen the saw" — to learn; to study. My new direction for safety professionals is to learn as much as you can about the human brain, about how people think and how we can aid memory, and then use that knowledge. Use it on yourself, share it with your family, and then take it out into the world. The message of safety is sorely needed in our world today. Safety professionals for too long have been the quiet, gentle spirits with big hearts and a bad tag of "safety cops". I challenge you as safety professionals to become "guardians of the future". Come out loud and strong with your message of safety and hope and humanity. While you will forever continue to make the workplace safer, I beg you, for our children's sake, to also use your skills and knowledge and make the world safer. Let us be sure that safety does not become a message of fear, but instead a message of hope. *"Make it safe out there!"* You are the guardians of the future!

## FURTHER READING

Buzan, T. *Use Both Sides of Your Brain.* New York: E.P. Dutton. 1983.

Covey, S. *The Seven Habits of Highly Effective People.* New York: Simon & Schuster. 1989.

Friedan, B. *The Fountain of Age.* New York: Simon & Schuster. 1993.

Joyce, M. *Ergonomics: Humanizing the Automated Office.* Carlsbad, CA: Southwestern Press. 1989.

Robbins, A. *Unlimited Power.* New York: Fawcett Columbine. 1986.

# Section VIII
# Training Aspects

# 37

# MANAGING WORKPLACE SAFETY AND HEALTH TRAINING

**Michael-Laurie Bishow**

## TABLE OF CONTENTS

## INTRODUCTION

Risk exposure, liability, new technologies, media exposure, regulation, and mandates challenge even the most organized safety training manager today. The future is limited for the hand sign-in, repeating refresher/prevention courses program. Hazardous waste, chemical spills, bloodborne pathogens, environmental management, and public awareness create a volatile yet exciting training opportunity. This chapter helps you to organize your thinking with practical tips and a plan which anticipates your workforce needs. *Anticipation* of changing demands and organization objectives marks the difference between a stellar training program and a pedestrian one.

| Traditional Program Planning Model | vs. | New Program Planning Model |
|---|---|---|
| Employee request | | Organizational mission/objectives |
| Feel good exit (attenders) | | Gained a skill before needed |
| Individual registration | | Team registration |
| Stock training plans | | Customized training plans |
| Individual gains | | Team gains, competitive advantage |
| Low accountability, short term | | High accountability, long term |
| Reactive training (request-demand) | | Proactive training (anticipate need) |
| Accountable to supervisor, manager | | Accountable to organization |
| Market to employees, individual | | Industry/customer/community |
| Risk drives safety training | | Develop plan for organization, market plan |
| | | Organizational development drives training |

## FASTER ORGANIZATIONS REQUIRE NETWORKING

"Top-down organization" once meant that training topics were selected by executive order. Often the safety training topics were based on existing mandates and laws which were known to be enforced by government. At other times, a work accident created a recognition of an acute need for "antidote" training — after the fact. Prevention was often ineffective due to these firefighting responses and accompanying funding constraints on the safety training program.

Today, our organizations are more "lean and mean"; our executives are held to a higher accountability with added regulatory enforcement. Safety and health training programs are now expected to respond to a mind-boggling array of factors related to employees' safety practices. Ultimately we seek to reduce accident rates, workers' compensation costs, and health care costs, as well as comply with regulations, laws, and mandates. Today, regulatory requirements are more specific and may even define training content or length (at least 200 require training at this writing). Peripheral factors may create more demands on safety programs, such as when communication breakdowns lead to accidents; thus "safety communication" may become part of your program responsibility. Similarly, low worker literacy rates may lead to "safety literacy" when warnings or procedure signage uses too elevated a vocabulary. Employee teams, increasing in our new organizations, may need critical thinking skills to accurately evaluate a risky situation. Downsizing and hiring temporary workers further burdens the safety training officer. How can we deal with these expanded responsibilities? And where can we draw the line with our limited resources? Network with managers organization wide!

## The Safety Committee

The best safety officer cannot stay on top of it alone — so the safety committee is now a survival tool (rather than a "pain in the neck"). Be careful to select committee members who approve resource allocations and also to select employee representatives. Your safety committee members are an internal source of expertise, training, marketing, and forecasting. Members can become your advocates to integrate safety training into the process of technical change.

For written reference, consult texts such as the National Safety Council's Occupational Safety and Health Series, which includes the volumes entitled *Administration and Programs, Engineering and Technologies, Fundamentals of Industrial Hygiene,* and *Environmental Management.* Stay tuned to the media to anticipate exposure due to world incidents. Your organization will immediately turn to safety when an incident such as Chernobyl occurs. Workers may suddenly realize they are at risk when an accident occurs half a globe away. Hopefully, you have anticipated this need for training and they can watch the news and say, "Yeah, we just had that safety training update last week!" That is a stellar program.

## POSITIONING SAFETY TRAINING FOR IMPACT

Our accountability is higher than ever, and we are accountable to a wider constituency. Building program credibility depends on our attention to the concerns of all our patrons: relevance to our grassroots employees, cost-effectiveness for our managers, learning for our participants, and promoting organizational mission and objectives.

## Grassroots Relevance

No matter how many mandates you meet, your "common person appeal" must be maintained. If you doubt this, ask a safety officer how he or she would be received if an accident occurred and he or she said, "We don't have that training because there isn't a state mandate." By listening and responding to the complaints and scuttlebutt of the people doing the work, you can establish your program's credibility. To obtain this informal information, you may want to get out of your office for a safety walkabout. For the grassroots organization, *anticipation* means valuing the wisdom of people who may not give you a sophisticated technical description of risks. Your best recommendation is word of mouth. Even if the training is mandated, people vote with their feet.

## Cost-Effectiveness for Managers

Resistant managers often say they cannot afford to let people go to training on work time. Your first strategy is to develop policies which take managers off the hook by executive order; that is, a percentage of employee time is required for safety training, with the safety committee setting annual training priorities. Training on paid time is also important in union–management agreements.

If executive edict is not available, try documenting the cost of a related accident and compare that to the cost of prevention. Be accurate about the cost of training; include the employee time, training materials, trainer costs, and program development. Divide by the number of people trained to get a per-head cost. For example (without employee pay):

| Bloodborne Pathogens (3.5-hour class) | Class size: 20 (two classes per day) | Booklet/certificate $4.00 duplication per participant | Contract trainer: $1000 per day |
|---|---|---|---|
| COSTS: Administration Registrar: $70 | 34 completed (85% average) | Materials: $136 | Cost per person: $35.47 |

You can sometimes reduce costs by using in-house trainers and writing your own courses. Be careful to calculate development costs; it may cost more than you anticipate and you may not meet deadlines. A safer approach is to open your training to outside attenders from other facilities or companies. This can be a win-win collaboration for small companies or departments. Some internal training functions have a department charge-back system with pre-approved annual safety training budgets. This may help to stabilize your budget. The safety committee can advocate on your behalf to the executive and other companies as long as you have internal credibility.

Usually the cost per participant is reasonable; our public sector institute provides training for about $50.00 per half-day participant. This includes some costs (development, videos, equipment) which are recovered during the first run. Repeated programs and updates cost less.

## Learning for Participants

Changes in behavior after the training session are the reason for taking the time. If your program was short and intended to be a "wake-up call" or to "raise awareness", your constituency might call it a waste. Unless it is an after-dinner speaking engagement or a stockholders' luncheon, you might want to avoid "awareness" objectives without behavioral outcomes.

Sometimes the pressure is on to schedule 1- or 2-hour sessions. If you cannot avoid this, what can you get across in that length of time? Give your managers a choice of several objectives or a few critical behaviors to practice; restrict classes to less than ten participants. Sometimes your forecast for how long it will take to cover the material with several hundred employees will motivate them to offer a larger time block. If not, cut the exposition from your course or cut the history; use one or two theoretical concepts/scientific principles with practices and demonstrations that address the dramatic consequences of ignoring just one principle. For example, oil is lighter than water (extinguish a grease fire chemically). In short, only deliver training which results in increasing the probability of the desired safety behavior. If you cannot achieve an impact, avoid doing ineffective training by citing regulations on course length, researching courses in similar companies, or just saying "no."

Participants in safety training should always complete a written evaluation. Even after a class you thought was a dud, ask for suggestions about content, delivery, materials, exercises, and practices. Be sure the trainer gets a copy and reads the evaluations every time. This is particularly important for in-house trainers who you can develop into top-notch presenters. Be sure to provided pretraining practice sessions and coaching for your in-house trainers to reduce stress and ensure training quality.

## Promoting Organizational Objectives

Politics! Well, that's real! From a practical standpoint, the safety officer should make the organization look good in a crisis. Also, safety program planning should anticipate needs generated by new organizational objectives. Let us say your company plans to change from aluminum alloy castings to extruded plastics. Your current training reduces the probability that an employee will pour a 50-pound bag of zirconium dust into an open-air vat.* This training will be obsolete in 6 months; consequently, it is time to research airborne plastic gas ignition. If the new casting process is still a secret, schedule multiple sessions of a generic training program, e.g., "Preventing Production Pitfalls", to reserve the space and the employee time. Prepare flashy marketing for your new training and schedule its release for the same day as the secret is to be announced. With luck, an executive will also be announcing your new training. Above all, do not wait until the secret is out to design the new training. Be on the cutting edge! In addition to new technologies, new plant locations, hiring surges, reengineering, or management turnover can create an opening for safety and health training topics. If possible, have the safety committee stay on top of trends and provide a "heads up" service for the executive. Of course, always bring a training solution with the "heads up" problem.

## ORGANIZING THE TRAINING PROCESS

How far in the future are your rooms and training scheduled? One month? Two months? Six months? Let us talk about the process of managing a stellar program: needs assessment, program funding and flow, designing and conducting courses, program evaluation, and documentation. CAUTION: This is a parallel process, not a linear process. Expect to do multi-tasking.

* Zirconium is a metal that explodes and burns, as it does in flashbulbs. One manufacturing facility provides training for employees in the handling of this metal. Exposure to air and friction tend to ignite the material. A night shift employee who did not follow training precautions dumped a bag of zirconium alloy into the top of an open vat. He was seriously burned and the flash was seen all over the plant. This incident led to a rush of safety re-education sessions for all shifts.

## Needs Assessment

Who do you talk to? Who has the need? How do you weight the input from various sources? The most common pitfall is to rely only on participants' input. If you do that, you may not prepare employees for new technologies, you may repeat popular programs too often, and you may use up allotted trainee hours (or manager goodwill).

Participants like to recommend entertaining programs to their friends. Of course your programs should keep people awake, but you take the hit if people use risky practices because you did not prepare them in time. I recommend that participants' input contribute about 20% for program planning, but contribute 80% to course content or delivery.

Another pitfall is to ask only managers what their people need. If you do this, managers will often think practice is too "fluffy" and will favor cognitive classes with tests to document that the participants "got something out of it." The safety officer may become a kind of enforcer, an extension of the "strong arm of the law". This does not add much to your credibility unless you tackle the education of the managers. Your service here is to publicize the reduction in downtime if participants use specific safety behaviors. Managers also relate to reducing "hassle factors" such as endless report writing and defensive meetings. I would weight manager input at about 30%. Gather manager input by hosting an annual luncheon with a training needs assessment process (high budget) or send a survey to all managers (low budget).

A final pitfall is to rely too heavily on the safety committee for forecasts about the future of your industry and the organization. Most committee members are not high enough in the organization to forecast accurately. I would weight the safety committee input at about 30%. The safety officer should meet regularly with a visionary in a power position (hopefully the CEO/CAO) and follow the media trends carefully.

Another training needs source is a "conscientious objector" in the ranks. This person notices internal problems and has high ethical expectations of your company. She or he often seeks options from people in other divisions/companies to suggest specific solutions using design and process changes. I would look for this person in research and development or sales/marketing and listen carefully. Often they do not have enough power to correct the problem and will see the safety officer as an ally seeking to improve safety practices. Vendors will often alert you to the safety hazards in the competitor's equipment, but do your own research after that kind of news flash. I would weight your information from objectors and experts at 20% (more if you talk to the CEO).

## Program Objectives: Funding and Flow

After multisource input and your own research, forecast the top priorities for the next 2 years and sort them into 6-month "reaches". The first 6-month reach lists your continuing in-place programs; the 12-month reach finalizes which programs in development will go to pilot stage/early trials; and the 18-month reach is for marketing to the executives to inform them, to gain approval, and to access resources for training. The goal is to anticipate new organizational training demands resulting from new regulations, production changes, or accidents.

The 2-year reach may be too far-reaching for some organizations, but in our training institute we may need 2 years to get funding to certify our trainers in national training programs. We then have the course ready to go for our next funding cycle (which is annual). We project a budget and plan cuts in case we lose 10% or even 20%. This is depressing, but it is good discipline to clarify our core program and our bottom lines. If possible, we devise methods to generate revenue or cut costs by sharing resources with other companies or departments.

In your annual presentation for the budget, connect outcomes with funding for each reach, such as, "350 employees will successfully complete the BBP training in the next 6 months at a per-person cost of $34.50." Document training needs by quoting your managers' planning luncheon, safety committee recommendations, employee survey results, and new government regulations. Actually, if your involvement is this high, people who decide on the budget will have been consulted already and will be well informed. Your chances of approval are much better with key people "on board" early in the planning.

If you have to reduce your budget, ask vendors if they will cut prices for volume; then form organizational consortiums to raise your numbers. Use local safety conference presenters as a resource to reduce travel expenses. Locate a college internship program (human resources, educational technologies, adult education, business); offer team teaching and course design positions in safety and health employee development. By the way, students are older and more experienced these days; the average age at some state universities is 28, and students often have full-time work experience.

## Programs Delivered

The size of your program has great bearing on how formally organized you should be. Let us discuss a program for a medium-sized organization consisting of between 1000 and 3000 employees. This is your trainee population; your total annual attendance will be about 2000 participants in 100 class sessions per year. You personally deliver two topics twice each month, in half-day segments, and handle the administration. You share two trainers with Human Resources; they deliver most of the safety and health training. A university intern helps with course research and design. Three managers on the safety committee deliver technical safety training topics relevant to their professions. Fortunately, your staff is competent; all trainers earn evaluations above 80% in their courses. Your training clerk is skilled with desktop publishing, registration software, and spreadsheet programs for budgets.

Your primary instructional unit is the 3.5-hour segment with typical courses lasting 2 half-days or one 7-hour day. Course designs include a balance of 30% cognitive input (lecture/discussion, reading), 30% individual work, and 40% practice and demonstration. Participants are expected to demonstrate competency during training practice to earn a certificate. Often pre- and post-tests are used both to assess the difficulty level of the class and to document retention of technical information. All participants complete an anonymous course evaluation at the end of the last session, and 85% receive a certificate of completion. When necessary, passing test scores are verified for licenses. All current courses are certified for continuing education units (CEUs) through the Human Resource (HR) Department's state provider number.

Course content is considered property of the safety/HR function if it is developed there. Other courses belong to the vendor/contractor. Time needed to develop courses yourself may be estimated with these rules of thumb:

- New and technical topics/ground-up design: 4 hours per class hour
- Intern prepared/desktop publishing booklets: 3 hours per class hour
- Well-known subject/updating existing course: 2 hours per class hour
- Off the shelf/contract trainer: 1 hour per class session (setup, briefing)

After designing course materials, there will also be duplication costs by the page, plus binding. The service center requires seven working days from pickup to delivery.

The in-house process of course design is usually done by an individual trainer until materials are in final draft form; ideally, all trainers meet to review the content and walk through the training design. These practice sessions are festive and comic as well as a beneficial training for trainers. The final design is approved by the safety committee and any certification review groups.

## Program Evaluation and Documentation

The same targeted groups contacted in the needs assessment process should also have a voice in the evaluation. Participants who completed course evaluations contribute to improved course delivery, especially if they add to lively class discussions about unmet training needs. Since you surveyed the managers, you can determine if they see an adequate training impact on the job. Also, managers and executives may be asked a hypothetical question to help the safety officer select annual priorities. Here is a sample format for a manager survey:

**Dear Manager: You have a $10,000.00 training allocation for next year. We offer some "hot topics" on which to spend your allocation, but add any others you think relate to our current or future safety and health training needs. Be sure to spend all your allocation! Thanks for your Safety Program planning advice. [List current and projected classes with "other" blanks.]**

Managers' feedback seems more focused when they are asked to distribute a sum, thus choosing between favorite training topics, needs, and government requirements.

## Reporting to Executives

Organization level accountability might be easiest to document by using measures of success such as

- Number of participants who completed courses within the time frame
- Percentage of planned courses which were conducted (cancellation rate)
- Cost of courses (or per participant) within projected budget
- Percentage of participants who passed competency or were licensed
- Average of participant evaluations (or percentage at excellent/satisfactory)

Automated registration is necessary at some point to avoid long delays and to track your success measures. Whatever software program you choose should generate rosters, keep ongoing class lists, produce individual transcripts, and print 5-year histories for department training. The registration program should also easily merge class lists with a confirmation letter, a certificate of completion, placard, or other personalized mailings.

If you need to justify the cost of registration software, calculate the labor/materials needed to produce reports, rosters, certificates, and budgets. True, if you purchase software, the initial data entry is tedious. However, you will probably realize that it is a blessing when you first print a list of all department employees who complete a mandated course in the last 3 years. Now you can convert your file space to archiving the history of course content, handout masters, and evaluations. Individual records are now in the automated registrar, which you back up every night. The training clerk can even run the course sign-up process paper-free, straight from the phone to the screen roster.

## SELECTING YOUR TRAINING STAFF

For some extroverts, training is a perpetual joy. If you are careful to pick people who enjoy groups, they will train for the fun of it; however, many technically qualified people are detail oriented and not very delighted to be in front of a group. Course development might be the best contribution for the high-detail trainer; even then, during new course development, high-detail trainers will grieve the loss of every fact and figure. The safety officer helps define the level of detail which participants can comprehend in the allotted time frame.

To maintain a stable training staff, your task is to provide rewards that each trainer will appreciate. Trainer gatherings to talk about training problems over free food seem to be popular. Selecting a balanced trainer team will help blend different training styles into an excellent program. Rotating or trading topics can be appreciated if trainers get bored; this may help raise sagging evaluations caused by low trainer energy. Forming a cadre of managers who like to train, but do not have much time, provides a helpful rotating relief system.

## Recruiting Internal Trainers

Recognize that there is "a trainer mindset". Look for a person with a problem-solving focus, someone who resists the temptation to blame people. Great trainers think well on their feet, especially under group pressure or resistance. The best ones are optimistic about learning and patient with human attitudes. Be sure to avoid anyone who blames or credits by reference group (i.e., "You women are all good at that.").

A spontaneous and even slapstick sense of humor helps a trainer get over the rough spots. I have observed that most trainers have low resistance to change and high tolerance for ambiguity. I saw this clearly when I attended a national workshop for trainers. The learning groups quickly changed the exercises, rules, roles, and seating. They even provided some ambiguity of their own by adding a bouncing beach ball into the mix: this added uncertainty, whimsy, and more performance pressure — for fun! Trainers often enjoy the class process so much that they almost become a self-generating energy system. These trainers do not burn out.

## Selecting External Trainers

Selecting contract trainers adds a cost factor into your decision-making. A quality trainer who has an excellent program and a willingness to design to your participants' needs provides variety and relief. Wherever you work, there is a "going rate" for contract trainers. If you are unsure about contractors or compensation, ask your local American Society for Training and Development (ASTD) chapter for a recommendation. Your trainer candidates should have a respectable résumé with professional memberships related to their specialty and to adult learning. Be sure to ask for recommendations and call previous clients on the phone for a more candid report. Also ask for course materials and evaluations (you may have to look at these during an interview due to copyright concerns).

If a candidate seems acceptable, have him/her do a 10-minute audition for the safety committee or your training cadre, using an evaluation form to give you written feedback from your staff. Write the first contract for a short term, perhaps two to four class sessions with the possibility of an extension, even if you must pay a bit more for short term. That way you can stop a disaster if you have misjudged. In any event, be there at the first class to introduce the trainer, to watch at least part of the training, and to close the session with evaluations. Stay to debrief the trainer: provide more "good news" than corrections whenever possible.

## SUMMARY

Developing the safety and health program organizes the effort to protect employees from risk. In the faster network organization, employees may be under increasing stress and decreasing job security. Your program may be the biggest investment your company makes in the employees. Thus, your job is to provide a stellar training program which communicates concern, caring, and investment in employees. You can plan well, select an excellent staff, develop internal training resource people, stay on top of the breaking needs of your organization, and enjoy the process.

In fact, take a moment now to reward yourself with a short retrospective. What have your safety and health training accomplishments been over the last 2 years? Even write them down ... save them to read on a low-energy day. Now, think ahead; plan your stellar training program for the next 2 years, and remember to set a few goals for your personal satisfaction.

In the next chapter we discuss training delivery. Enjoy!

## FURTHER READING

### Books

Buckley, R. and Caple, J. *The Theory and Practice of Training.* San Diego: University Associates, 1990.

Friedman, P. and Yarborough, E. *Training Strategies from Start to Finish.* New York: Prentice-Hall, 1985.

Goad, T. *Delivering Effective Training.* San Diego: University Associates, 1982.

Johnson, S. and Johnson, C. *The One Minute Teacher: How to Teach Others to Teach Themselves.* New York: William Morrow, 1986.

Pfieffer, J. W. *The 1995 Annual.* Volume 1, Training; Volume 2, Consulting. San Diego: Pfieffer & Company, 1995.

### Other Publications

Caffarella, R. S. A checklist for planning successful training programs, *Training Dev. J.* 81–85, March 1985.

Gilbert, S. R. Company specific safety information. *Prof. Saf.* 44–47, June 1993.

LeBarr, G. New rules for safety training. *Occup. Hazards* 27–30, December 1993.

Bennett, J. K. and O'Brien, M. J. The building blocks of the learning organization. *Training,* 41–49, June 1994.

NOTE: Current editions of the following publications are helpful: *Training and Development Journal, Training Magazine,* and *Personnel Journal.* Also see University Associates or Pfieffer & Company publications.

# 38 SAFETY AND HEALTH TRAINING PRESENTATIONS

**Richard W. Lack and Kathleen Kahler**

## TABLE OF CONTENTS

## INTRODUCTION

This short chapter will provide an overview of the process and techniques for making effective training presentations. Your authors are fortunate in having a broad and complementary perspective on this subject. Kathleen's professional career has focused on health education, and more recently she has moved into the safety and health management field. My background is more in the field of safety management.

1-56670-054-X/96/$0.00+$.50
© 1996 by CRC Press, Inc.

My earliest training in the skills of making effective presentations was provided by Homer Lambie, former Corporate Safety Director, Kaiser Aluminum and Chemical Corporation. Over a period of a dozen years, I sat in many of his training sessions with all levels of employees and marvelled at his skill in getting the message across. Besides knowing his subject and the philosophies behind it, Homer was a great storyteller, and he always had a fund of actual cases to illustrate his points.

I have heard him present the same topic to a dozen different groups, and each session would be different. He had an almost uncanny "feel" for the needs of each group. This taught me a valuable lesson: know your audience.

Perhaps I was fortunate to have this background in training because, as those of us in the profession know only too well, safety and health can be a "dry" subject. Employees are often not exactly enthusiastic at the prospect of attending a presentation by the safety department. The subject, most likely required by law, may be viewed as dull and uninteresting. The presenter may approach the topic in negative terms, i.e., what not to do in the work setting. Finally, the presentation may be in lecture form, which provides little opportunity for group participation.

No doubt readers are familiar with the following figures from studies showing how people learn and retain instructional material. Nevertheless, I believe they are worth repeating because as trainers we need to always keep this in mind.

**Retention of Instructional Material**

| | | |
|---|---|---|
| Trainees retain | | |
| Only | 10% | of what they READ |
| | 20% | of what they HEAR |
| | 30% | of what they SEE |
| But | 50% | of what they SEE and HEAR |
| | 70% | if they can SAY what they have learned by talking it over with others |
| And | 90% | if through physical and verbal means they can DEMONSTRATE what they have learned (this means PARTICIPATION) |
| | 95% | if they TEACH someone else what they have learned |

To reinforce this point, as trainers we need to keep in mind that while in days past having the documentation as proof that your employees had completed the required training may have been sufficient to satisfy regulatory agencies, now the agency officials are likely to randomly observe and talk to employees to verify retention, understanding, and compliance with the training.

## PLANNING SAFETY AND HEALTH TRAINING PRESENTATIONS

The safety and health practitioner should develop an annual plan for training. This will identify all mandatory training and the recipient groups of employees. Setting out an annual schedule in this way will minimize the stress and pressure of preparation of training material, finding speakers, scheduling attendees, etc. Naturally, other subjects can be included during the year as specific needs may be identified.

Safety and health practitioners are frequently required to develop presentations which can be delivered in many different settings. For example, department managers may want health and safety information which can be delivered as part of staff meetings. In the interests of production efficiency, employees may not be able to leave their workplace, and you may have to literally carry your training to them and meet perhaps in a lunchroom or work break area. These situations require the safety and health practitioner to be creative and to develop effective mechanisms for delivering the required information.

## MAKING ARRANGEMENTS

Here, again, planning is of the essence. As a guide and with thanks to Michael-Laurie Bishow, at the end of this chapter we have provided a useful checklist of things to consider before the training program begins. This checklist also provides a list of preliminary information, presentation development, and administrative activities which must occur prior to the presentation date.

On the subject of time, I strongly recommend that you avoid scheduling training any later than 3:00 p.m. in the afternoon, unless the attendees are on afternoon (swing) shift.

When selecting a site for the training, here are some important conditions to consider:

- Noise level and general distractions
- Access to restrooms
- Electrical outlets
- Adequate ventilation
- Room temperature
- Comfortable furniture

The seating arrangement will obviously vary according to the size of the group, configuration of the room, style of presentation, and visual aid equipment. Small groups of, say, 5 to 15 can be seated around a conference table. A horseshoe, square U, or fan arrangement of tables works better for groups of up to approximately 40. Above this number, the classroom or banquet room approach is more suitable. For groups exceeding 75 to 80, the theater approach will probably be necessary, unless you have the benefit of a highly specialized, college-style auditorium.

## SELECTING TRAINING METHODS

When developing a training presentation, consider using a variety of methods to present the information. Use visuals, including photographs or slides, and especially involve members of the group to help generate interest and reinforce the training points. Assessment of the reading level of the attendees will be critical in the preparation of written material. Role playing and work groups can be useful ways to increase group participation. If your audience is at a management level, a concise summary of the key issues with action items is desirable.

To aid in your selection of suitable training methods, the following is a summary of the principal techniques with their advantages and disadvantages:

| Methods | Advantages | Disadvantages |
| --- | --- | --- |
| Case study | Creates interest | Too many can produce overkill |
| Demonstration | Shows trainees how to perform the task | Can lose people if group is large, or they cannot see the demonstration |
| Discussion | Feedback confirms learning | Can get out of control |
| Games | Breaks the pace Relaxes the group | Can be viewed as "gimmicky" |
| Guided practice | A great technique for human relations and technical skills | Make sure everybody can get a chance to practice! |
| Lecture | Good for getting material across | No feedback Low interest |
| Panel | Can bring in outside experts | Same problems as lectures |
| Role play | Good for development of human relations skills | Can be viewed as threatening |
| Group projects | Creates interest and builds participation (best if group size is 2 to 5). | Groups may be dominated by more forceful members |

## SELECTING AND USING AUDIOVISUAL AIDS

As already discussed, visual aids definitely increase comprehension and retention of your training material. Courtesy of Dr. Robert Fish, here are some important guidelines to keep in mind whatever visual aid you select:

- **Make them visual.** Use color, and also graphics — pictures and diagrams, not just words. Appeal to the right side of the brain as well as the left. Use a variety of colored pens.
- **Make them visible.** Place only a few words on a line and only a few lines on a page. A good guideline is six lines and six words maximum.
- **Have only one main idea per visual.**
- **Use upper and lower case.** This is much easier to read than all upper-case lettering.
- **Proofread.** Double check your visuals for errors.

To make the best use of your visual aids, here are some useful guidelines:

- Practice and rehearse.
- Turn off the power or cover the visual if you want the audience's attention on you.

- Talk to the audience, not the visual. Maintain eye contact.
- Do not read your visuals word for word.
- Use a pointer sparingly.

To aid in selection, the following are some advantages and disadvantages of principal types of visual aid equipment.

**Advantages and Disadvantages of Various Visual Aids**

| Advantage | Disadvantage |
|---|---|
| **Flip Charts or Chalkboards** | |
| 1. Spontaneous | 1. Slow |
| 2. Easy to use | 2. Poor readability |
| 3. Inexpensive | 3. Turn back to audience |
| 4. Good for brainstorming | 4. Chalkboards temporary |
| **35-mm Slide or Computer Graphics Video** | |
| 1. High quality | 1. Darkened room |
| 2. Photographic detail | 2. Usually expensive |
| 3. Very portable | 3. Slides become focal point — not speaker |
| 4. Easy to operate | 4. Can appear canned |
| 5. Any size group | 5. People are put off by visuals perceived as too "slick" |
| **Overheard Projector** | |
| 1. Fast, simple preparation of transparencies (any copier) | 1. Projector can block view unless positioned carefully |
| 2. Lights on | 2. Less portable than 35 mm |
| 3. Speaker faces audience | 3. Transparency preparation so simple that people tend to use too many and ones that are too busy |
| 4. Any size group | 4. Machine cooling fan sometimes noisy |
| 5. Spontaneous or advance preparation | 5. Traps speaker at podium |
| 6. Very flexible | 6. Canned "rap" |
| 7. Optional quality | |
| 8. Can be activated by writing on paper on wall using overhead image | |

## HANDOUTS

A handout is an essential item for any training presentation. Besides providing the attendees with material for further reference, it can also include copies of your visual aids so that the trainees can make notes as they follow your presentation. The quality of your handout will be one of the factors influencing how the audience judges your presentation, so my advice is to always go first class!

If your handout is a complete text of the paper you are presenting, you may choose to hold your handouts for distribution at the end of the session in order to minimize people drifting away in mid-session. If you do hold back your handout in this fashion, you should provide an outline of your visual aids to facilitate note taking.

Be sure to number the pages of your handout and tell the audience the page you are on. This will keep people from shuffling through the contents when they want to refer to the particular subject being discussed or to take notes.

## DELIVERING THE TRAINING

1. In planning your delivery, consider your training subject and the goals of your presentation. To deliver an effective presentation, what should your role be? Should it be trainer? Instructor? Facilitator? Courtesy of Dr. Michael-Laurie Bishow, here are some practical definitions for these very different techniques:
   - **Instructing** — An instructor is primarily an educator whose focus is to impart information. The goal of a good instructor is to stimulate and motivate people to learn, to retain knowledge from the information provided.
   - **Training** — A trainer is a coach whose focus is skill building and application. The goal of a good trainer is to help people develop necessary skills/actions/behaviors, to teach them how to **apply** the knowledge provided.
   - **Facilitating** — A facilitator is part discussion leader/part coordinator whose focus is group discussion and problem solving. The goal of a good facilitator is to help promote the generation of ideas through brainstorming and group discussion. As a facilitator, you participate in the process and help get the group to their goal. This type of a meeting can be used to "tap into" the group members' creativity and also help them reach solutions to mutual problems faced by the group or their organization. A phrase to remember is, "Steer the car, not the people!"
2. As discussed above, when preparing a training class, consider the objective. Is it skill building or is it imparting information? A training class will be more effective if you build your training outline around specific observable skills. The attendees must be able to demonstrate their knowledge and skills by doing. If the group requires a knowledge base, attach a section to the training material with this information.
3. When delivering training, a useful acronym to remember is PEDPER:

   > **P**repare
   > **E**xplain
   > **D**emonstrate
   > **P**erform
   > **E**valuate
   > **R**eview

Make sure your training is multichannel so the people SEE, READ, and DO.

Some other important delivery steps are as follows:

- Whenever possible, set up the night before so you will have time to deal with snags.
- Be there ahead of your attendees so you can greet them informally. You are the host; your friendly greetings will relax the group.
- Provide coffee and tea in morning sessions and consider providing other refreshments such as candies or cookies. This helps to show you care about the people attending and builds rapport.
- In your opening remarks, discuss breaks, note restroom locations, and explain the purpose, materials, length, and general process for the session.
- Remind attendees that no one has all the answers and that you will be relying on them to contribute their knowledge and experience to the training.
- Encourage attendees to ask questions at any time.
- Dress neatly and appropriately. You are the message!
- Solicit suggestions for improvement of the presentation. Use an anonymous evaluation sheet (an example can be found at the end of the chapter). Ask attendees for brief answers such as

> I have learned....
>
> If I could improve the program, I would....
>
> Having filled in many evaluation sheets in my time, I dislike those that ask too many questions. However, some questions are essential for analysis purposes since your objective as the trainer must be the continuous improvement of the content and quality of your presentations.

- Thank attendees for their input and participation.

In the section on training tips at the end of this chapter there are more ideas to help improve your training delivery.

## HANDLING RESISTANCE IN THE GROUP

Readers desiring in-depth information on dealing with conflict disruptive behaviors, etc., should refer to the Further Reading sources at the end of this chapter.

As a trainer and instructor for many years, I have found that your meetings will not get out of hand as long as you demonstrate your respect for the people and practice the "Golden Rule" when dealing with them.

You must be honest with your group at all times. If someone asks a question and you do not know the answer, do not waffle or guess; say you do not know, ask the audience if anyone in the group knows, and/or say you will find out and get back to them.

If someone asks a tough or controversial question, turn it back to the group and let them deal with the question. Alternatively, you can say you will look into the issue after the meeting or pass it to the appropriate person for their investigation. At all events, avoid direct confrontation and be polite at all times.

Courtesy of Dr. Michael-Laurie Bishow, here are some tips to help reduce stress. First avoid use of the following "red flag" words that tend to accelerate stress:

| The Words | | What They Convey |
|-----------|------|------------------|
| Ought | Should | Blame |
| Can't | Couldn't | Victim |
| Have to | Must | Helpless |
| Never | Always | Absolute |
| It, Them | They | Projection |
| (Person) | Is/are type | Labeling/stereotyping |

Choose the following "green flag" words to decelerate stress:

| The Words | | What They Convey |
|-----------|------|------------------|
| Choose to | Decided to | Self-determined |
| Want Wish | Hope | Nonrestrictive aim |
| Some Many | Several | Conditional |
| Few | Seldom | Conditional |
| Usually | Often | Speculative |
| Seems | Notice | Value-free |
| (Observation) | Gives me the impression | Behavior observation or neutral observation |
| If (act)... | Then _____ (consequence) | Informed choice |

## EVALUATING TRAINING

On the subject of evaluation, I am always reminded of a phrase Homer Lambie frequently used. He said, "… the value of any meeting is measured in terms of what the group attending the meeting will do differently after the meeting." This point is really the bottom line in any evaluation of a training presentation. Perhaps some aspects of your presentation were not perfect, but if your message got across and the group will put the message into action, that is really what counts.

Apart from observable behavior change, other methods for validation of training effectiveness include

- Questionnaires
- Performance assessments
- Interviews
- Observation checklists

Evaluating training is rather like problem solving. If your problem is specific, it is solvable. If it is too general, it will be unsolvable. If your training objectives were specific, you will be able to measure the results. If your training objectives were general in nature, or described in vague terms, evaluating the training will be equally vague or general.

## SUMMARY

Making effective training presentations may not be something that comes naturally to many safety and health practitioners. As part of their professional development, safety and health practitioners should attend courses and seminars on giving effective training presentations, group facilitation, and being a persuasive speaker. Developing these abilities will enhance your overall communications skills and open up new opportunities for growth and promotion.

## TRAINING TIPS

These tips are drawn in part from a talk given by Steve Friedland, management consultant and trainer, at the 1994 annual conference of the Northern California Human Resource Council.

1. Schedule minimum 15-minute breaks. Allow 1 hour for lunch.
2. Pay attention to how you dress — no distractions!
3. Playing music that fits the program you are doing helps the ice-breaking process.
4. Form small groups for better participation — a maximum of five or six.
5. Have more rather than less material. Your handout reflects you. Include references and graphics.
6. Use colors.
7. Training feedback is important. Attendees need to evaluate the program and you for
   - Delivery
   - What was learned
   - What can be improved
8. Study the evaluations and then let go of them. Some people are very judgmental.
9. Ask attendees to tell you what they want to find out. List them and be sure to address them.
10. Use anecdotal stories to support your message. Keep it simple. Talk about life experiences that relate to the material.
11. Openings and closings are very important. This is the heart of your presentation.
12. Transitions between content areas are the heart of comprehension.
13. Be very careful with humor. Test out jokes.
14. Be careful of copyright issues. There are some things you should not reproduce or have as overheads.
15. Do not call upon people who don't want to participate. People will volunteer.
16. Use "round robin" small groups. Your objective is to get as many to talk as possible.
17. Do not rely too much on visuals. Remember: you are the message. You need to be out there and keep eye contact. Less is more when it comes to visuals.

18. LISTEN carefully. Paraphrase to help the class understand the participant's answer.
19. Do not confront people while in the group — do it privately.
20. Always build the audience up, never down.
21. SMILE!

**SAFETY AND HEALTH TRAINING PRESENTATION CHECKLIST**

## Administration Issues

- ☐ Notification to attendees sent out
- ☐ Travel arrangements made
- ☐ Room & equipment confirmed
- ☐ Meals/refreshments confirmed
- ☐ Room set-up confirmed
- ☐ List of attendees/sign-in sheet available
- ☐ Handouts copied and available
- ☐ Name and address of key person for assistance on-site
- ☐ Determine if security badges or special arrangements are needed for access to site
- ☐ Maps or directions

## Three or more weeks before the training program

- ☐ Select and reserve the site
- ☐ Survey the site for size, lighting, ventilation, furniture needs
- ☐ If off-site, check on transportation arrangements
- ☐ Send a memo to potential participants (map, if off-site)
- ☐ Meet with or call participants' supervisors
- ☐ Contact appropriate person about opening and closing site
- ☐ Order the necessary number of chairs and tables
- ☐ Have handout materials printed
- ☐ Locate audio-visual equipment and aids

## One week before the training program

- ☐ Confirm number of participants
- ☐ Order refreshments
- ☐ Secure ashtrays or decide on smoking policy
- ☐ Obtain direction signs indicating designated room
- ☐ Secure a reliable timepiece
- ☐ Obtain name tags
- ☐ Obtain felt-tip markers

☐   Obtain pencils or pens and notepads
☐   Recheck handouts and aids
☐   Locate restroom facilities
☐   Recheck equipment arrangements
☐   Have a "dress rehearsal"
☐   Allot time for setting up materials
☐   Test all equipment

### Day of training session (a minimum of one hour before session)

☐   Organize materials on training table
☐   Make sure room is set up properly
☐   A clock
☐   Tape electrical cords to the floor
☐   Check for spare bulbs
☐   See that refreshments are in place
☐   Place notepads, pencils, name tags, etc., on tables
☐   Test equipment again

**You may want to include other checks in your planning.**  The following list is provided simply to jog your memory.  Consider those items that apply to your own situation.

| | | |
|---|---|---|
| Film projector | Pins | Extra notepads and pens |
| Slide projector | Glue | CEU credit forms |
| Overhead projector | Chalk | Scissors |
| Films | Meals | 3 x 5 index cards |
| Screens | Thumb tacks | String |
| Video camera | Hammer | Lens cloth |
| Video recorder | Fee arrangements | Matches |
| Television monitor | Pencil sharpeners | Screwdriver |
| Videotapes | Extension cords | Slides |
| Audio recorders | Breakout rooms | Overhead transparencies |
| Audiotapes | Certificates | Parking permits |
| Flip charts | Registration materials | Pliers |
| Extra easel stands | Photocopying facilities | Stapler |
| Extra newsprint pads | Map | Tape |
| Markers for writing on newsprint | Expense forms | |
| Markers for writing on transparencies | Evaluation sheet | |
| Blank transparencies | Black slides | |
| Handouts | 3-Way adapter | |
| Computer terminal | Extra bulbs | |
| Calculator | First-aid kit | |
| Clock | | |

## COURSE EVALUATION FORM

**COURSE TITLE**: _____

| Rating Scale | Excellent 4 | Good 3 | Fair 2 | Poor 1 |
|---|---|---|---|---|
| 1. How would you rate this course overall? | 4 | 3 | 2 | 1 |
| 2. How well did this course meet the stated objectives? | 4 | 3 | 2 | 1 |
| 3. Please rate the course handouts. | 4 | 3 | 2 | 1 |
| 4. How useful was this course? | 4 | 3 | 2 | 1 |
| 5. Please rate the facilities and course mechanics. | 4 | 3 | 2 | 1 |

6. Would you recommend this course to others?     YES _____     NO _____

7. Using the same scale, please rate the individual instructors on presentation and content.

8. Comments/Suggestions:

_____

_____

_____

_____

9. What other courses/training would you like to see offered?

_____

_____

_____

_____

_____Your name (Optional)

## TRAINING SESSION-PLANNING WORKSHEET

I.      Title:_____

II.      Trainer(s): _____

III.     Sponsoring Organization: _____

IV.     Description of Participant Group: _____

          _____

          _____

V.      Date and Times: _____

VI.     Place: _____

VII.    Training Objectives: _____

          _____

          _____

          _____

VIII.   Training Plan:

| Time Period | Content | Method | Materials Needed | Trainer |
|---|---|---|---|---|
| 1. | | | | |
| 2. | | | | |
| 3. | | | | |
| 4. | | | | |
| 5. | | | | |

IX.     Evaluation Procedure: _____

          _____

          _____

          _____

## FURTHER READING

### Books

Boettinger, H. M. *Moving Mountains — The Art of Letting Others See Things Your Way.* New York: MacMillan, Collier Books, 1969.

Colin, T. *Strategies for Adult Education.* London: Oxford University Press, 1981.

Gilbert, F. *Power Speaking: How Ordinary People Can Make Extra Ordinary Presentations.* Redwood City, CA: Frederick Gilbert Associates, 1994.

Good, T. W. *Delivering Effective Training.* San Diego, CA: University Associates, 1982.

Hoff, R. *I Can See You Naked — A Fearless Guide to Making Great Presentations.* Kansas City, MO: Andrews and McMeel, 1988.

Mandel, S. *Effective Presentation Skills.* Los Altos, CA: Crisp Publications, 1987.

National Research Council. *Improving Risk Communication.* Washington, D.C.: National Academy Press, 1989.

Pater, R. *Making Successful Safety Presentations.* Portland, OR: Fall Safe, 1987.

Peoples, D. A. *Presentations Plus.* New York: John Wiley & Sons, 1992.

Pike, R. W. *Creative Training Techniques Handbook.* Minneapolis, MN: Lakewood Books, 1989.

Thayer, L. *50 Strategies for Experimental Learning, Book One.* La Jolla, CA: University Associates, 1976.

### Other Publications

Hywel, J. and Race, P. 10 training techniques for safety practitioners. *Health Saf. Pract.,* 30(6), 18–21, August 1993.

Merrifield, J. and Bell, B. Don't give us the Grand Canyon to cross. *Adult Learn.* 6(2), 23–24, November/December 1994.

Moeller, M. Video conferencing hits the desktop. *Present. Technol. Tech. Better Commun.,* 8(5), 16–20, May 1994.

Pike, R. Handouts, a little charity to your audience goes a long way. *Present. Technol. Tech. Better Commun.,* 8(5), 30–32, May 1994.

# Section IX
# International Developments

# 39A

# GLOBAL SAFETY AND HEALTH MANAGEMENT

**Kathy A. Seabrook**

## TABLE OF CONTENTS

## INTRODUCTION

The world economy and the way businesses are shaping it are changing dramatically. Today, a corporation's research and development operations may be based in Cambridge, Massachusetts or Cambridge, England, while product

1-56670-054-X/96/$0.00+$.50
© 1996 by CRC Press, Inc.

manufacturing or distribution facilities span the globe, assuring customer service and economy of production. This poses a challenge to the safety professional who must manage health and safety in this global environment and have an appreciation for global regulatory, litigation, cultural, religious, climatic, and geographical differences in order to develop an effective global health and safety management system (HSMS).

This chapter will assist you in developing a global HSMS and provide a guide for identifying and overcoming cultural differences to ensure success of the HSMS. It will also guide you through the health and safety hierarchy and legislation in the European Union (EU), providing an overview of the existing and emerging health and safety directives and available health and safety resources.

## COMPONENTS OF A GLOBAL HEALTH AND SAFETY MANAGEMENT SYSTEM

The key to an effective global HSMS lies in the results of the domestic system in the home country. If the domestic system does not produce positive results, do not attempt to export it. An ineffective HSMS will not improve abroad. So ensure that the domestic system is effective before tackling the world!

Corporate commitment, planning, organizing, implementing, monitoring, and controlling are essential management techniques, in conjunction with a corporate HSMS policy, leadership commitment, risk assessments, standard operating procedures, auditing, consistent accident reporting, training and education, and designated employer-employee responsibilities for health and safety. While many countries have minimal workplace health and safety standards, corporations should strive for a consistent level of health and safety throughout the world. One should not forget the loss of life due to the Bhopal incident or the subsequent cost of litigation and adverse publicity for Union Carbide. The financial impact on its shareholders and subsequent breakup of this global corporation were a direct result of this incident and serve to illustrate the importance of a corporate commitment to a global health and safety management system.

### Statutory Requirements

Once the domestic HSMS is in place, the first step in developing a global plan is to research the country in which the HSMS will be implemented. Figure 39A-1 provides a list of areas to consider and incorporate into the global HSMS. The corporate HSMS must be at least as strict as the local-foreign standards and regulations. If the corporate HSMS is not as strict as is required by the foreign country, it must be updated and the stricter management system

**What to Research**

- Who is the Health & Safety authority
- Legal and regulatory requirements
- Work hours
- Language barriers
- Statutory training requirements
- Traditional clothing affecting worker safety
- Environmental factors

- Statutory accident reporting requirements
- Statutory Record keeping requirements
- Security issues
- Business customs
- Religious customs/beliefs
- Male - Female protocol

**Figure 39A-1** What to research for incorporation into the global health and safety management system.

incorporated into the global HSMS. A good example of this is in Europe, where an assessment of workplace risk is required for all systems of work. This provides a proactive approach for risk identification, quantification, and control. It is also an excellent management planning tool which positively impacts quality, productivity, efficiency, and production costs, in addition to the safety of the worker. Therefore, European-mandated risk assessments should be considered for inclusion into the corporate HSMS for worldwide implementation.

## Administrative Controls

Warmer climates can affect employee work hours and administrative controls in the workplace. Consider high temperature and repetitive strain exposures in these climates. In countries such as Spain, Mexico, or Italy, a midday siesta is common and provides an automatic rest break which also serves to control these hazards. This is a good example of why it is important to assess the existing workplace, customs, processes, procedures, etc., before implementing risk control changes to a work site. This applies to not only foreign operations, but the home country as well.

## Training

Employee and management training is essential for effective local implementation of the HSMS. Language barriers are real and must be addressed to ensure a clear and accurate understanding of workplace hazards and the controls to mitigate them. This applies to both worker and management training programs, written instruction manuals, warning labels, and signs. In the Pacific

Rim, some countries encompass a multilingual workforce where training in more than two languages is essential.

Local statutory requirements for worker health and safety training should also be implemented. In Europe, for example, the health and safety directives specify employee training requirements for exposure to display screen equipment (DSE) and manual material handling risks. There are also general operator training requirements based on the results of statutory risk assessments.

## Customs

Religious or societal customs can play a part in the basic control of hazards such as use of personal protective equipment. This can be seen in India, where the Sikhs wear turbans which would impact their use of hard hats in areas with overhead exposures. Therefore, alternatives to control the overhead exposure must be considered, including employing a non-Sikh in the area of exposure.

## Security/Terrorism

Security for the local offices, plant, workers, and products is a serious issue. Terrorism exists throughout the world as a result of the activities of religious, political, and animal rights or environmental activists. This translates to kidnapping of key staff, damage to plant property, injury to staff through bombs or other incendiary devices, and product tampering causing potential injury to the public. Staff hiring and screening procedures, physical plant security, access control (including identification badges for visitors and staff), exterior security lighting and patrols, formal emergency and disaster recovery plans, and staff training and good quality assurance programs are integral to controlling the risks associated with terrorism. Local management should also work closely with local authorities and other companies in the area to fully assess and address terrorism.

## Accident Reporting/Record Keeping

Another important area is consistent global accident reporting. One corporate statistical reporting method should be implemented worldwide to monitor and measure accident trends and frequency and severity rates. This may mean a dual reporting system to ensure adherence to local statutory accident reporting requirements. Additionally, the corporate definition of both an accident and a first aid case must be clearly communicated to local staff along with corporate accident classifications for purposes of automated accident trending.

Accident investigation should be implemented for all accidents *and* first aid cases — this, too, should be simplified and clearly communicated. Keep in mind potential language barriers and use of international symbols: remembering the Heinrich triangle is a good visual tool in any language to translate why reporting and investigation of first aid cases and near misses is important.

**Where to go for Help**

- Corporate Human resource department
- American Embassy

- Local Human resources department
- Local Chamber of Commerce

- US Department of Labour/UK Department of Trade and Industry
- International Labour Organization

- Local Health and Safety Consultants
- Networking with Peers

- Professional Health and Safety Associations

- Trade associations (Oil, Gas Chemical,

  Pharmaceutical manufacturers, Iron workers, etc.)

**Figure 39A-2** Where to go for help.

## Audit Systems

Finally, a formal, consistent, global audit system must be in place with follow-up procedures to measure audit results and monitor the implementation of corrective actions. The audit teams should incorporate multinationals. This promotes consistent global program implementation as well as educates multinational staff who audit plant locations with effective health and safety management systems in place.

## Cultural Integration

Cultural or religious differences should never be underestimated. A good example of this occurred during a safety and health audit at a plant in India. During introductions at the start of the audit, the head auditor (a Christian) shook hands with a Muslim (female) quality manager. Little did the auditor know an infidel (Christian) must never touch a Muslim woman. As a result, the audit went downhill from there and had to be rescheduled with another audit team.

Figure 39A-2 provides a list of resources to assist you in researching the foreign countries in which you will be operating. Your peers are also an excellent source of information — remember, many companies are going global and are dealing with the same issues. Contact the manufacturing association (e.g., pharmaceutical, oil and gas, construction) to which your company belongs. Many are working to build databases and networks to provide information to corporations doing business in foreign countries. Companies also employ local safety consultants or full-time safety professionals to bring the local expertise needed to get through the legal and regulatory maize. Contact the local embassy for information on the health and safety authorities and professional associations in that country. Croner's *Health and Safety Direc-*

*tory,* published in the United Kingdom, is an excellent resource, providing listings of these authorities and associations throughout the world. See the References and Additional Reading section at the end of this chapter for details on this directory.

So do your homework. Hospitality and willingness to adapt to the local customs go a long way in meeting your company's business objectives. Your local embassy, the local Chamber of Commerce, the Department of Trade and Industry, or cross-cultural consultants (e.g., Windham International, New York, or Bennett Associates, Chicago) can provide basic cultural information and brochures on doing business in the country in which your company will be operating. What does this have to do with health and safety? *Everything.* Effective implementation of health and safety management systems depends on the corporation's ability to communicate with local management and employees.

Customs in many countries are very formal. You would be considered rude and disrespectful if you were given a business card by a Japanese colleague or member of management and did not spend time reading and commenting on that individual's position or credentials — again, detrimental to your HSMS objectives since local management commitment is essential to integrating health and safety into the business environment. Also know your contact's names and how to address them. In many countries outside the United States, the surname is used to address colleagues and superiors; use of an individual's first name would be considered unacceptable. Some languages use the formal form of the pronoun "you", so remember when speaking French or German to use "vous" or "Sie," respectively.

Figure 39A-3 provides areas of cultural differences which should be investigated and considered prior to visiting your foreign subsidiaries.

**Know the local customs**

- Unusual customs (hand gestures, food, gifts, etc.)

- Visa Requirements for inter-country travel (Israel-Arabic countries)

- Personal Security issues

- Religious customs

- Formality of country

- Dress expectations

- Use of a translator

**Figure 39A-3**  Know the local customs.

The key to operating in the global health and safety arena is corporate and local management commitment and communication. This means asking a lot of questions, understanding the cultural differences, listening to your local contacts, and being flexible in implementing the global HSMS. Good luck in your endeavors!

## THE EUROPEAN UNION

The European Union (EU) includes 14 member states: Denmark, Germany, Luxembourg, The Netherlands, France, Spain, Italy, Greece, Great Britain, Belgium, Ireland, Portugal, Finland, and Austria. Europe has emerged as a leading economic market; there are 318 million consumers through the removal of trade barriers. The major events leading to the formation of the EU are outlined below.

- **Treaty of Rome, 1957** — This treaty established the European Economic Community. It meant the removal of trade barriers for the trade of goods and services within the Community.
- **Single European Act of 1986** — This act formalized the harmonization process for free movement of people, goods, and services throughout the European Community. In 1992 the real work on harmonization of standards (e.g., machinery, design, electricity, safety, and packaging) became a reality.
- **Maastricht Treaty** — With this came harmonization beyond the economic premise on which the Treaty of Rome was based. This treaty harmonizes the political, social, and military fabric of Europe — creating the European Union as we know it today.

European law is in the form of European directives. These directives are developed through a lengthy process with representatives and comments by all member states. Figures 39A-4 outlines the life cycle of a directive, with the individual components of the European Union provided below:

- **European Commission** — Proposes directives through the Directories General (e.g., Directives General V oversees Employment Social Affairs and Education, which includes health and safety legislation). Proposals originate in European countries from lobbyists, activist groups, professional associations, and existing government regulatory agencies (e.g., the Health and Safety Executive in U.K. — OSHA's counterpart) as well as current member state legislation.
- **European Parliament** — Representation by member state population. Parliament influences the final draft of the directive — put forward to the European Council.
- **European Council** — Represented by each member state, the council ratifies directives by a qualified majority vote (QMV).

Commission (proposes)
↓
Parliament (influences)
↓
Council (ratifies)
↓
Member State (implements & enforces)
↓
Code of Practice (member state)
↓
Court of Justice (interprets law)

**Figure 39A-4**  Development of a European Community directive.

- **Member State** — Implements and is the first-line enforcer of directives.
- **Code of Practice** — Each member state translates a code of practice for directive implementation. The member state can only opt for a more stringent standard.
- **European Court of Justice** — Interprets the legal premise of a directive.

## The "Six Pack" Health and Safety Directives

On December 31, 1992, the "Six Pack" Directives, that is, the primary European health and safety legislation, were set into law throughout the EU. The premise of these directives is to encourage improvements in the health and safety of workers in the EU by designating responsibility for implementation and enforcement of the directives to each member state. The "Six Pack" Directives include the Framework Directive (89/391/EEC), which sets out general principles and duties by which employers operating in the EU must abide. The remainder of the directives outline specific area of health and safety: Workplace (89/654/EEC), Work Equipment (89/655/EEC), Personal Protective Equipment (89/656/EEC), Manual Material Handling (90/269/EEC), and Use of Display Screen Equipment (90/270/EEC). The following provides an overview of the requirements for each of the above directives.

**Framework Directive** — This directive is applicable to all sectors of work activity, providing general principles and duties for employers, including assessing workplace risks, implementing appropriate preventative measures, developing a risk prevention policy, and adapting ergonomic principles to the workplace. It also requires employers to designate a "competent person" to take charge of health and safety. This individual can be an employee or outside consultant. The directive also provides for first aid, emergency procedures, and employee health and safety training and requires employees to be responsible for their own health and safety and those of their fellow workers.

**Workplace Directive** — This directive applies to permanent (fixed) workplaces and requires employees to comply with minimal health and safety requirements: keeping exit access and egress clear, conducting safety equipment checks, and maintaining the workplace in an orderly manner. It also includes minimum health and safety provisions on electrical installations, ventilation, exit and exit egress, structural stability of buildings and fixtures, fire precautions, first aid facilities, lighting, temperature, windows and skylights, restrooms and rest areas, escalators/elevators, danger areas, doors, and gates.

**Work Equipment Directive** — This directive applies to all work equipment such as machines, tools, apparatus, and installations used in the workplace. The work equipment must

- Be suitable for work without risk to employee health and safety
- Be properly maintained
- Minimize risk during work in places where inherent hazards exist (e.g., electrical hazards)
- Protect against use by unauthorized operators

Further, work equipment must also comply with the following minimal health and safety requirements:

- Protect against rupture
- Formal maintenance procedures in place
- Protect against risk of overheating and fire
- Must not discharge toxic gas, dust, fumes, vapor, noise, or liquid
- Must contain explosions
- Must mitigate all electrical hazards

**Personal Protective Equipment (PPE) Directive** — This directive depicts the employer's responsibility for PPE. PPE is designed to be used by an employee to protect against workplace hazards likely to endanger health or safety. It also sets out the employee's duty to use PPE and report defects as they occur. PPE is only recommended if the health and safety risk cannot be reengineered or substituted. The employer's responsibilities include

- Assessing workplace risk
- Selecting appropriate PPE which is compatible with the work
- Designing PPE to comply with European Union specifications on design and manufacture
- Providing employee with PPE at no charge
- Maintaining the PPE
- Training employees on the use and selection of PPE for the risk

**Manual Material Handling Directive** — Manual material handling refers to any work involving transport or support of a load by one or more employees and requiring carrying, lifting, placement, pulling, pushing, or movement which could cause a risk (e.g., back injury) to the employee(s). The directive states that employers should avoid manual handling if possible. If this cannot be avoided, a manual material handling risk assessment should be conducted and appropriate control measures implemented. The risk assessment must include

- Type of load (size, weight, shape)
- Work by employee
- The environment (hot, humid, wet)
- Task requirements (duration, repetition, frequency)

Employee training requirements are also outlined in the directive.

**Display Screen Equipment (DSE) Directive** — This directive applies to all workstations with display screen equipment, keyboards, optional accessories, and peripherals including disk drives, modems, software, printers, document holders, chairs, desks, and the general work environment. The directive outlines the employer's responsibility to conduct an assessment of health and safety risk for each workstation and apply the appropriate control measures to eliminate any risks identified. Further requirements on rest breaks and changes in activity for the DSE user are also provided. Risks highlighted include operator exposure to ergonomic hazards, such as glare, lighting, muscle and eye strain, noise, humidity, ventilation, and software and task design. Eyesight testing must be provided to the employees if they experience visual difficulties associated with their work environment. If the eyesight test indicates that an ophthalmologic exam is required, this must be provided (free of charge) for the employee, along with corrective lenses to correct a visual defect attributable to work.

### Status of Other Health- and Safety-Related European Union Directives

The following section outlines new and proposed health- and safety-related directives forthcoming from the EU.

### *Adopted Directives*

- **Construction (Design and Management)** — otherwise known as the CONDAM regulations. States it is the duty of clients, planning supervisors, and designers to provide a safe workplace. Effective October 1994. Directive 91/155/EEC.
- **Safety Data Sheets** — Manufactures, importers, distributors, and suppliers are required to provide safety data sheets for dangerous chemicals used in the workplace. This directive provides for consistent information and format. Effective September 1993. Directive 91/155/EEC.
- **Dangerous Substances/Preparation** — Outlines criteria for classifying, packaging, and labeling. Forms part of the CHIP — chemical (hazard identification and packaging) — regulation in the U.K. Implementation: July 1994. Directives 93/21/EEC and 93/18 EEC.

- **Pregnant Women** — Mandates improved work environment for pregnant women or those who have recently given birth or are breast-feeding. Safety and health are only one aspect of this directive, which includes risk assessments in areas where this type of employee is working. It also provides a ban on night work and prevention of exposure to specified chemicals, physical and biological agents, and processes. Effective October 1994. Directive 92/85/EEC.
- **CE Marking** — Indicates that all EU directives must be met before the product can bear the CE marking. The directive provides consistent act of rules for use of CE mark through all product directives. Effective January 1995. Directive 93/68/EEC.
- **Explosives** — Includes standards for civil use of explosives. Places obligation on member states to ensure that essential safety requirements are met. Effective January 1995. Directive 93/15/EEC.
- **Machinery** — Harmonies safety requirements for machinery manufactured for use in the EU. Effective January 1995. Directive 93/44/EEC.
- **Products** — States requirements for manufacturers and all in the supply chain to assure consumers of safe products and provide information on those products. Effective June 1994. Directive 92/59/EEC.

### Proposed/Recommended/Draft Directives

- **Industrial Diseases** — Provides a list of industrial diseases liable for compensation (recommended).
- **Physical Agents** — Outlines exposure requirements to physical agents, including noise, vibration, nonionizing electromagnetic radiation (proposed).
- **Chemical Agents** — Regulation on control of worker exposure to chemical risks (proposed).
- **Explosive Atmospheres** — Regulation for equipment and protective systems used in explosive atmospheres (proposed). Also regulations for worker safety in explosive atmospheres (draft).
- **Working Time** — Covers regulations for weekly, daily, and work rest breaks, annual leave, and night work (proposed).
- **Young Workers** — Provides restriction for young workers: risk assessments, medical evaluation (if applicable), and specific hazard avoidance (proposed).

The state of these new directives continues to evolve. Therefore, a section on English-speaking EU resources follows for future reference.

## EUROPEAN UNION RESOURCES

The following list provides resources available to assist you in further understanding health and safety legislation in the EU:

- **Institution of Occupational Health and Safety (IOSH)** — Professional health and safety organization — can provide a list of *qualified* safety consultants.

The Grange
Highfield Drive
Wigston
Leicester E18 0NN
England

Tel: 011–44–53–357–1399
Fax: 011–44–53–371–1451

- **Health and Safety Executive (U.K.)** — Information Service (OSHA's U.K. counterpart) will provide individual HSE contact for EU directive in question; also provides database of U.K.-HSE information.

Sheffield Information Centre
Broad Lane
Sheffield
England

Tel: 011–44–74–289–2345
Fax: 011–44–74–289–2333 (if not urgent)

- **Directives General V (DG-V)** — EU body for employee industrial relations and social affairs. DG-V will direct you to individual EU member state contacts.

Directorate V-F
Commission of European Committees
Batiment Jean Monnet
Rue Alcide de Gasperi
Luxembourg
L-2920

Fax: 011–352430–1-34511

- **European Information Office** (open 2–5 p.m. Monday–Friday)

Jean Monnet House
8 Storey's Gate
London SW1P 3AT

Tel: 011–071–973–1992
Fax: 011–071–973–1904

- **U.S. Department of Labor** (contact government office in your area)

- **European Commission Office**

  2100 M Street, N.W. (7th Floor)
  Washington, D.C. 20037

  Tel: 202–862–9500
  Fax: 202–429–1766

- **Eurosafety**

  Industrial Relations Services
  18–20 Highbury Place
  London N5 1QP
  England

  Tel: 011–44–71–354–5858
  Fax: 011–44–71–359–4000

## REFERENCES AND ADDITIONAL READING

IDS European Special, February 1992.
EC legislative state of play, *Eurosafety,* Issue 3, 4–7, 1993.
Neal, A. C. and Wright, F. B. *European Community's Health and Safety Legislation,* Vol. II. London: Chapman & Hall, 1992.
*Croner's Health and Safety Directory 1991/92.* Kingston Upon Thames, England: Croner Publications, 1991.

# INTERNATIONAL ORGANIZATION FOR STANDARDS (ISO) 9000: THE EFFECT ON THE GLOBAL SAFETY COMMUNITY

# 39B

## Mark D. Hansen

## TABLE OF CONTENTS

1-56670-054-X/96/$0.00+$.50
© 1996 by CRC Press, Inc.

## HISTORY AND INTRODUCTION

For those of you who have no idea what ISO 9000 is or its purpose, we will begin with a brief history to illustrate its roots. Today many safety engineers live (and die) through the use of standards, whether they are military standards, Environmental Protection Agency (EPA) standards, Occupational Safety and Health Administration (OSHA) standards, or something very similar. These standards are often inconsistent, and especially when it comes to international trade they are not well suited for widespread use. In many cases terminology in these standards as well as in commercial and industrial practice was inconsistent and confusing.

The publication of the ISO 9000 series and the terminology standard (ISO 8402) in 1987 has attempted to bring some harmonization on an international scale. It has supported the growing impact of quality as a factor in international trade. This is not a fad or a passing "buzzword" of the day, as many have viewed some other passing fads. The ISO 9000 series has been adopted by many nations and regional bodies and is rapidly supplanting prior national- and industry-based standards.

The ISO 9000–9004 and the American National Standards Institute (ANSI)/ American Society for Quality Control (ASQC) Q90–Q94 series documents contain information relevant to systematic management for product and service development, design, production, and installation activities, including safety, health, and environmental aspects. These standards represent an international movement to establish worldwide "quality system" standards for hardware, software, and processed products and services. Although strongly driven by the European Community originally, the total quality management (TQM) vision represented by these standards will most probably drive U.S. business to become compliant and require certified practitioners (e.g., Certified Safety Professionals [CSPs]) in the near future in order to conduct business internationally. The ISO 9000 series documents will directly challenge each of us as safety professionals.

## SUMMARY OF ISO 9000

ISO 9000 has set forth four strategic goals, which are universal acceptance, current compatibility, forward compatibility, and forward flexibility. ISO has set tests for these goals which are detailed below.

**Universal Acceptance** — The tests for universal acceptance include (1) the standards are widely adopted and used worldwide, (2) there are few customer complaints in proportion to the volume of use, and (3) few sector-specific supplementary or derivative standards are under development.

**Current Compatibility** — The tests for current compatibility include (1) "part number" supplements to existing standards do not change or conflict with requirements in the existing or parent document, (2) the numbering and the clause structure of a supplement facilitate combined use of the parent document and the supplement, and (3) supplements are not stand-alone documents but are to be used with the parent document.

**Forward Compatibility** — The tests for forward compatibility include (1) revisions affecting requirements in existing standards are few in number and minor or narrow in scope, and (2) revisions are accepted for existing as well as new contracts.

**Forward Flexibility** — The tests for forward flexibility include (1) supplements are few in number but can be combined as needed to meet the needs of virtually any industry/economic sector or generic category of products, and (2) supplement or addendum architecture allows new features or requirements to be consolidated into the parent document at a subsequent revision if the supplement's provisions are found to be used (almost) universally.

## ISO 9000 Product Categories

**Hardware** — Hardware is considered to be products consisting of manufactured pieces and parts, or assemblies of parts.

**Software** — Software is considered to be products, such as computer software, consisting of written or otherwise recordable information, concepts, transactions, or procedures.

**Processed Materials** — Processed materials are considered to be products (final to intermediate) consisting of solids, liquids, gases, or combinations thereof, including particulate materials, ingots, filaments, or sheet structures. Processed materials are typically delivered (packaged) in containers such as drums, bags, tanks, cans, pipelines, or rolls.

**Services** — Services are considered to be intangible products which may be the entire or principal offering, or incorporated features of the offering, relating to activities such as planning, selling, directing, delivering, improving, evaluating, training, operating, or servicing for a tangible product.

For each of the product categories discussed above there is a quality loop that interacts with all activities pertinent to the quality of the product or service. It involves all phases from initial identification to final satisfaction and customer expectations, as shown in Figure 39B-1. For service functions a more detailed quality loop is provided in Figure 39B-2.

## Organization of ISO 9000 Documentation

ISO 9000 is organized into four separate documents. They are 9001: Quality Systems — Model for Assurance in Design/Development, Production, Installation, and Servicing; 9002: Quality Systems — Model for Assurance in Production and Installation; 9003: Quality Systems — Model for Assurance Final Inspection and Test; and 9004: Quality Management and Quality System Elements — Guidelines. The relationship between ISO 9000 documentation and producers/suppliers and purchasers is shown in Figure 39B-3.

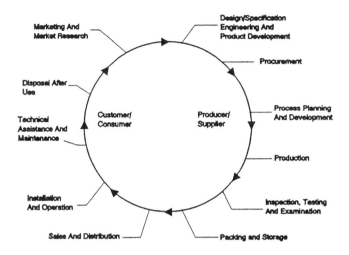

**Figure 39B-1** Quality loop.

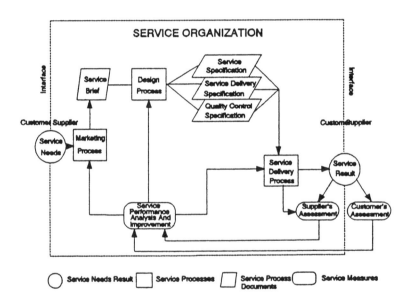

**Figure 39B-2** Service organization quality loop.

## ISO 9000 Definitions

Some key ISO 9000 concepts are defined below. The relationships between these concepts are illustrated in Figure 39B-4.

**Quality Policy** — Quality policy is the overall quality intentions and direction of an organization as it concerns quality, as formally expressed by top management. The quality policy forms one element of the corporate policy and is authorized by top management.

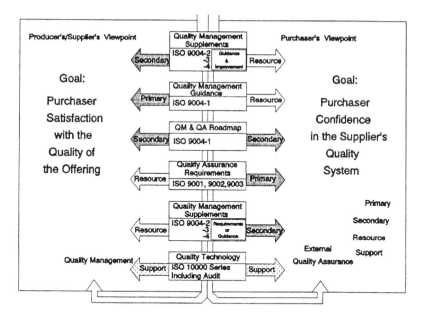

**Figure 39B-3**  Relationship between ISO 9000 documentation and producers/suppliers and purchasers.

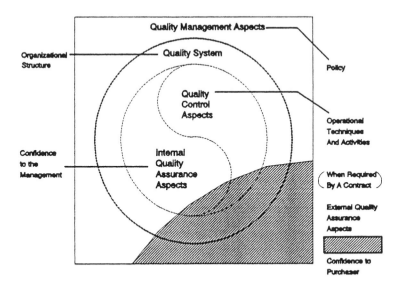

**Figure 39B-4**  Relationships of ISO 9000 key concepts.

**Quality Management** — Quality management is that aspect of the overall management function that determines and implements the quality policy. The attainment of the desired quality requires commitment and participation of all members of the organization, whereas the responsibility for quality management

**Figure 39B-5**  Key aspects of ISO 9000.

belongs to top management. Quality management includes strategic planning, allocation of resources, and other systematic activities for quality, such as quality planning, operations, and evaluations.

**Quality System** — A quality system consists of the organization structure, responsibilities, procedures, processes, and resources for implementing quality management.

**Quality Control** — Quality control consists of the operational techniques and activities that are used to fulfill requirements for quality. This includes monitoring processes to eliminate the causes of unsatisfactory performance in order to continually improve the processes.

**Quality Assurance** — Quality assurance consists of all those planned and systematic actions necessary to provide adequate confidence that a product or service will satisfy the given requirements for quality. Some of the related aspects include continuing evaluation of the design or specification and audits of production, installation, and inspection operations. These activities provide confidence to management.

## Key Aspects of ISO 9000

There are three key aspects of ISO 9000: management responsibility, the quality system structure, and personnel and material resources. Figure 39B-5 illustrates these key aspects and shows that the customer is at the center of these activities. It also illustrates that customer satisfaction can only be achieved when there is harmony between these aspects. These three aspects are discussed below.

**Management Responsibility** — Management is responsible for establishing a policy for quality and customer satisfaction. Successful implementation of this policy is dependent on management's commitment to the development and effective operation of a quality system. This includes having quality objectives, defining clear responsibility and authority guidelines, and having periodic reviews.

**Personnel and Material Resources** — Personnel is one of the most — if not the most — important resource any organization has. Personnel issues include motivation, training and development, and communication. Material resources are also important for any organization and may include provisioning of equipment, transportation, instrumentation, and documentation.

**Quality System Structure** — The quality system structure should focus on the output of the organization, which may be different for hardware, software, processed materials, and services. The structure should assure adequate control over all operational processes affecting quality. The quality system structure should also focus on preventive actions that avoid occurrence of problems while not sacrificing the ability to respond to and correct failures, should they occur.

## ISO 9000 CONTENT ANALYSIS RELATED TO ENVIRONMENTAL, SAFETY, AND HEALTH PROFESSIONALS

Now that ISO 9000 has been properly introduced, it is time to discuss the safety-related aspects of ISO 9000. ISO 9000–9004 and ANSI/ASQC Q90–Q94 series documents contain information relevant to systematic management for product and service development, design, production, and installation activities, including safety, health, and environmental aspects. The following descriptions provide a summary of key sections of ISO 9000–9004 and ANSI/ASQC Q90–Q94 which contain the words "safety", "health", "liability", or "environment". The attached excerpts include references to the established or draft standards. Additionally, many sections not cited contain elements directly applicable to managing safety into products and services, by design.

As you know, these standards represent an international movement to establish worldwide "quality system" standards for hardware, software, and processed materials products, as well as for services. Although strongly driven by the European Community originally, the total quality management vision represented by these standards will most probably drive U.S. businesses to become compliant and certified practitioners in the near future in order to conduct business internationally.

This summary provides the "flavor" of the direct challenge to our business that can be best related to traditional aerospace industry product safety, system safety, and product liability efforts.

- **Quality:** The totality of features and characteristics of a product or service that bear on its ability to satisfy stated or implied needs.... Needs are usually translated into features and characteristics with specific criteria (which) may include aspects of usability, **safety,** availability, maintainability, economics, and **environment** (ISO 8402–1986, Section 3.1, draft ISO/DIS 8402, Section 2.1 and ANSI/ASQC A3–1987, Section 2.1).
- **Design Review:** A formal, documented, comprehensive, and systematic examination of a design to evaluate the design requirements and the capability of the design to meet these requirements and to identify problems and propose solutions.... The capability of the design encompasses such things as fitness for purpose, feasibility, manufacturability, measurability, performance, reliability, maintainability, **safety, environmental** aspects, time scale, and life cycle cost (ISO 8402–1986, Section 3.13 and ANSI/ASQC A3–1987, Section 3.8).
- **Requirements of Society:** Obligations resulting from laws, regulations, rules, codes, statutes, and other considerations (which) include protection of the **environment, health, safety,** security, conservation of energy and natural resources (draft ISO/DIS 8402, Section 2.4 and ISO 9004:1987, Section 3.3).
- **Safety:** The state in which the risk of harm (to persons) or damage is limited to an acceptable level. NOTE: **Safety** is one of the aspects of quality (draft ISO/DIS 8402, Section 2.8).
- **Terms and Definitions ... Product Liability; Service Liability:** A generic term used to describe the onus on a producer or others to make restitution for loss related to personal injury, property damage, or other harm caused by a product or service (ISO 8402–1986, Section 3.19, ISO/DIS 8402, Section 2.12 and ANSI/ASQC A3–1987, Section 3.12).
- **Selection of Model for Quality Assurance ... Selection Factors ...** In addition to functional criteria detailed in 8.2.1.a) to 8.3.1.c), the following six factors are considered fundamental for selecting the appropriate model for a product or service ... **e) Product or service safety.** This factor deals with the risk of occurrence of failure and the consequence of such failure (ISO 9000:1987, Section 8.2.3.e and ANSI/ASQC Q90–1987, Section 8.2.3.e).
- **Selection of Model for Quality Assurance ... Demonstration and Documentation ...** The nature and degree of demonstration may vary from one situation to another in accordance with such factors as ... e) the safety requirements of the product or service (ISO 9000: 1987, Section 8.3.e and ANSI/ASQC Q90–1987, Section 8.3.e).
- **Annex ... Cross Reference List of Quality System Elements ... 19 ...** Product Safety and Liability (ISO 9000:1987 and ANSI/ASQC Q90–1987).
- **Design Control ... General ...** The essential quality aspects, such as safety, performance, and dependability of a product (whether hardware, software, services, or processed materials), are established during design and development phase (draft ISO/DIS 9000–2, Section 4.4.1).
- **Design Control ... Design and Development Planning ...** The supplier should establish procedures for design and development planning that include plans for evaluating **safety,** performance, and dependability incorporated in the product design (draft ISO/DIS 9000–2, Section 4.4.2).

- **Design Control** ... **Design Verification** ... Design reviews for the purpose of design verification can consider questions such as ... **are safety considerations covered?** (draft ISO/DIS 9000–2, Section 4.4.5).
- **Design Control** ... **Design Changes** ... Design of a product may be changed or modified for a number of reasons. For example: **safety,** regulatory, or other requirements have been changed (draft ISO/DIS 9000–2, Section 4.4.6).
- **Purchaser's Requirements Specification** ... **General** ... In order to proceed with **software development,** the supplier should have a complete, unambiguous set of functional requirements (which) should include all aspects necessary to satisfy the purchaser's need. These may include, but are not limited to, the following: performance, **safety,** reliability, security, and privacy (ISO 9000–3:1991, Section 5.3.1).
- **Development Planning** ... **Output from Development Phases.** The required output from each (software) development phase should be defined and documented. The output from each development phase should be verified and should ... d) identify those **characteristics of the product that are crucial to its safe and proper functioning** (ISO 9000–3:1991, Section 5.4.5.d).
- **Process Control** ... **General** ... The supplier shall identify and plan the production and, where applicable, installation processes which directly affect quality and shall ensure that these procedures are carried out under controlled conditions. Controlled conditions shall include the following: a) ... use of suitable production and installation equipment, suitable working environment, compliance with reference standards/codes, and quality plans (ISO 9001:1987, Section 4.9.1.a) and ANSI/ASQC Q91–1987, Section 4.9.1.a)).
- **Process Control** ... **General** ... The supplier shall identify and plan the production and, where applicable, installation processes which directly affect quality and shall ensure that these procedures are carried out under controlled conditions. Controlled conditions shall include the following: a) ... use of suitable production and installation equipment, suitable working environment, compliance with reference standards/codes, and quality plans (ISO 9002: 1987, Section 4.8.1.a) and ANSI/ASQC Q92–1987, Section 4.8.1.a)).
- **Risk Considerations** ... **For the Customer** ... Consideration has to be given to risks such as those pertaining to the health and safety of people, dissatisfaction with goods and services, ... (ISO 9004:1987, Section 0.4.2.2 and ANSI/ASQC Q94–1987, Section 0.4.2.2).
- **Management Responsibility** ... **Quality Objectives** ... For the corporate quality policy, management should define objectives pertaining to key elements of quality, such as fitness for use, performance, safety, and reliability (ISO 9004:1987, Section 4.3.1 and ANSI/ASQC Q94, Section 4.3.1).
- **Economics** ... **Types of Quality-Related Costs** ... **Operating Quality Costs** ... Operating quality costs are those incurred by a business in order to attain and ensure specified quality levels. These include the following: b) Failure costs (or losses) ... external failure: costs resulting from a product or service failing to meet quality requirements after delivery (e.g., ... **liabil-**

**ity costs**) (ISO 9000.4, Section 6.3.2.b) and ANSI/ASQC Q94–1987, Section 6.3.2.b)).

• **Quality in Specification and Design ... Design Planning and Objectives (defining the project)** ... In addition to customer needs, the designer should give due consideration to the requirements relating to **safety, environmental and other regulations,** including items in the company's quality policy which go beyond existing statutory requirements (ISO 9004:1987, Section 8.2.4 and ANSI/ASQC Q94–1987, Section 8.2.4).

• **Quality in Specification and Design ... Design Planning and Objectives (defining the project)** ... The quality aspects of the design should be unambiguous and adequately define characteristics important to quality, such as the acceptance and rejection criteria. Both fitness for purpose and safeguards against misuse should be considered. Product definition may also include reliability,... **including benign failure and safe disposability,** ... (ISO 9004:1987, Section 8.2.5 and ANSI/ASQC Q94–1987, Section 8.2.5).

• **Quality in Specification and Design ... Design Qualification and Validation** ... The design process should provide periodic evaluation of the design at significant stages. Such evaluation can take the form of analytical methods, such as **Failure Mode and Effects Analysis (FMEA), fault tree analysis or risk assessment,** as well as inspection or test of prototype models and/or actual production samples. ... The tests should include the following activities: a) evaluation of performance, durability, **safety,** reliability, and maintainability under expected storage and operational conditions; ... (ISO 9004:1987, Section 8.4 and ANSI/ASQC Q94–1987, Section 8.4).

• **Design Review ... Elements of Design Reviews** ... As appropriate to the design phase and product, the following elements outlined below should be considered: **a) Items pertaining to customer needs and satisfaction** ... 5) safety and environmental compatibility; ... **b) Items pertaining to product specification and service requirements** ... 4) ... disposability; 5) benign failure and fail-safe characteristics; 7) failure modes and effects analyses, and fault tree analysis; ... 9) Labeling, warnings, ... and user instructions; **c) Items pertaining to process specifications and service requirements.** 3) Specification of materials, components, ... including approved supplies. ... 4) Packaging, handling, and shelf-life requirements, especially safety factors related to incoming and outgoing items (ISO 9004:1987, Section 8.5.2 and ANSI/ASQC Q94–1987, Section 8.5.2).

• **Corrective Action ... Disposition of Nonconforming Items** ... Recall decisions are affected by considerations of safety, product liability, and customer satisfaction (ISO 9004:1987, Section 15.8 and ANSI/ASQC Q94–1987, Section 15.8).

• **Handling and Post-Production Functions ... Installation** ... Instructional documents should contribute to proper installations and should include provisions which preclude improper installation or factors degrading the quality, reliability, **safety,** and performance of any product or material (ISO 9004:1987, Section 16.1.5 and ANSI/ASQC Q94–1987, Section 16.1.5).

- **Product Safety and Liability** … The safety aspects of product or service quality should be identified with the aim of enhancing product safety and minimizing product liability. Steps should be taken to both limit the risk of product liability and to minimize the number of cases by a) **identifying relevant safety standards** in order to make the formulation of product or service specifications more effective; b) **carrying out design evaluation tests and prototype (or model) testing** for safety and documenting the test results; c) **analyzing instructions and warnings** to the user, **maintenance manuals** and **labeling and promotional material** in order to minimize misinterpretation; d) **developing a means of traceability to facilitate product recall** if features are discovered which compromise safety and to allow a planned investigation of products or services suspected of having unsafe features (see 15.4, Investigation of Possible Causes, and 16.1.3, Identification). … (ISO 9004:1987, Section 19 and ANSI/ASQC Q94–1987, Section 19).
- **Use of Statistical Methods** … **Statistical Techniques** … Specific statistical methods and applications available include, but are not limited to, the following: … c) **safety evaluation**/risk analysis (ISO 9004:1987, Section 20.2 and ANSI/ASQC Q94–1987, Section 20.2).
- **Appendix B — Product Liability and User Safety** … This Appendix is not part of ANSI/ASQC Q94–1987 … and is not included in ISO 9004, but is included for information purposes only … (ANSI/ASQC Q94–1987, Appendix B).
- **Characteristics of Service** … **Service and Service Delivery Characteristics** … Examples of characteristics that might be specified in requirement documents include: … **hygiene, safety,** reliability, and security (ISO 9004–2:1991, Section 4.1).
- **Quality System Operational Elements** … **Marketing (Services) Process** … **Quality in Market Research and Analysis** … Management should establish procedures for planning and implementing market activities. Elements associated with quality in marketing should include: … **review of legislation** (e.g. **health, safety,** and **environmental**) and **relevant national and international standards and codes;** … (ISO 9004–2:1991, Section 6.1.1).
- **Quality System Operational Elements** … **Design Process** … **Service Delivery Specification** … **General** … The service delivery specification should take account of the aims, policies, and capabilities of the service organization, as well as any **health, safety, environmental, or other legal requirements** (ISO 9004–2:1991, Section 6.2.4.1).

## ENSURING THAT ISO 9000 IS INTEGRATED INTO THE ENGINEERING PROCESS AND MANAGING THE PAPER CHASE

Compliance with ISO 9000 is difficult for many reasons. Four of the most compelling reasons are (1) ensuring that ISO 9000 activities are truly integrated into the engineering process, (2) managing continual changes, (3)

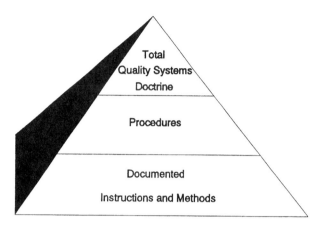

**Figure 39B-6**  Building blocks of a total quality system.

ensuring repeatability, and (4) ensuring rapid access to required information. Most companies that have to deal with ISO are forced to grapple with numerous processes, products, and organizational changes and often volumes of documentation. This includes understanding the intent of ISO 9000 as well as having to refer to various parts, sometimes on a daily basis. Attempting to manage this mass of critical data in paper form — where it may reside in diverse locations on shelves, in drawers, or at job sites — is a frustrating and error-prone endeavor. Further, it is almost impossible to have rapid and accurate access to current information when it is in paper format. For this reason it is important for many companies to build a quality system based on ISO 9000 terminology. The basic building blocks of the total quality system doctrine are illustrated in Figure 39B-6.

The problem is that environmental, safety, and health (ES&H) professionals and managers already spend an inordinate amount of time digging through regulations and filing reports rather than focusing on the ES&H job. There is literally no time to worry about ISO 9000 processes and procedures unless you have excess time on your hands.

However, by automating this process, the ES&H professional can spend more time out in the field implementing ES&H programs and less time worrying about ISO requirements; they are interwoven into the process. Examples include having material safety data sheets (MSDSs) available on computer at sites requiring employee review, managing the document flow through the review process using computers, integrating engineering drawings and other documents using computers, and using computers to manage and track all of the training requirements required for OSHA, EPA, and ISO compliance.

With the prices of powerful IBM-compatible personal computers (PCs) and Macintosh computers falling below $2000, most ES&H professionals have access to more computing power than have ever had before. Ironically, with all this computing power available, rarely are these machines fully utilized. Most

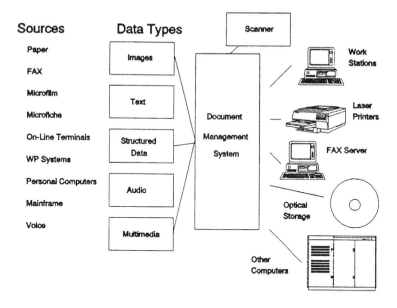

**Figure 39B-7** Typical components of a document imaging system.

of the time we use computers for routine tasks such as word processing, spreadsheets, or databases.

There are new state-of-the-art technologies now available that can be implemented on single-user systems all the way up to enterprise-wide applications. These technologies typically use stand-alone PCs either as a single user or as the front end on a client-server network. These new technologies are "document imaging" of forms (e.g., MSDSs) and business process automation using "workflow" software.

## Document Imaging

Document imaging is the core of automating the ES&H process for OSHA, EPA, and ISO compliance. Document imaging is central to the progression toward the "paperless" office. Typical components of a document imaging system are shown in Figure 39B-7. The "paper" from other organizations, such as documents and engineering drawings, can be scanned in and the paper literally thrown away. The reasons the paper is thrown away are fourfold. First, the electronic copy is now legally considered to be the original. Second, the electronic copy is often better than the original (especially when dealing with MSDSs). Third, one of the main reasons for using document imaging is to get rid of rooms full of filing cabinets containing engineering drawings and documents. Finally, document imaging can be used to capture signatures of all kinds of documents (e.g., engineering drawings, OSHA-related testing, documents, etc.)

There are two distinct ways to scan in information. One way is to capture the physical image (or picture) of the document using a document scanner. This image is usually stored in a .TIFF format. The information on this image is usually information that cannot be changed, information from outside the system, that is to be displayed or printed when necessary.

The other way is to scan in documents where the text is converted to ASCII data that can be shared in a database. The conversion process is accomplished using optical character recognition (OCR) software. This is important when dealing with data (e.g., MSDSs) that has to be stored, retrieved and manipulated from a database. For example, searching for all chemicals at a particular site, a particular building, room, etc. with specific characteristics (e.g., boiling point, etc.).

If the original is of maximum quality there are two ways to significantly improve the quality and readability. One way is to use an intelligent character recognizer (ICR). This software recognizes and matches each character based on the image scanned in.

Another software program accomplishes this and, additionally, fills in the image character with a gray color so it is more readable. This is especially important for managing documents from multiple vendors, such as MSDSs.

## Business Process Automation Using Workflow

Business process automation is the mechanism for integrating document imaging and workflow technology to make computers even more useful in the 1990s.

### *Business Process Automation*

Business process automation refers to automating manual processes using computer technology. For example, physical filing of documents in a filing cabinet is automated in a database of some sort, or the manual review of a document is automated so that the document is routed electronically and sequentially to each person in the process. Business process automation includes integrating all electronic data in a compatible form. This involves converting different types of electronic images (e.g., .TIF, .GIF, .WPG, etc.,) and word processing data (WordPerfect®, Multimate®, Word®, ASCII, etc.) to a common form. However, merely automating a manual process is usually not sufficient. Using workflow the manual process can be routed electronically and sequentially rather than manually.

### *Workflow*

Workflow refers to the electronic routing of documents and engineering drawings through a computer network. Workflow allows users to comment, approve, disapprove, etc., and send the information to the next person in the

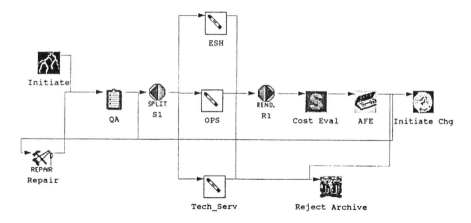

**Figure 39B-8**  An example work-flow map for automating the OSHA, EPA, and ISO process.

review process or back for editing. An example workflow is shown in Figure 39B-8. Workflow can be used to route information, documents, engineering drawings, etc. from local to enterprise-wide distributions.

To walk you through this example, let us say a new MSDS for a product you use in your manufacturing process arrives from a vendor (the Initiate box in the workflow map in Figure 39B-8). The first thing that you might do is assess the impact in the QA (Quality Assurance box in the workflow map shown in Figure 39B-8). If information is missing they may call the vendor for more information; if not it is electronically forwarded to ESH (the Environmental Safety and Health Department), OPS (Operations), and Tech_Serv (Technical Services). The ES&H Department may evaluate the MSDS for impact on how the product is to be used in the manufacturing process and the environmental ramifications of use, waste, and disposal. Operations may or may not use the product but needs it for reference purposes. Technical Services would use it to update all documentation for all operations and maintenance procedures. Next, it would be forwarded to Cost Eval (Cost Evaluation) to determine the cost impact of the new MSDS on the company. If the new MSDS has a cost impact, it would be forwarded to AFE (Authority for Expenditure) to make disbursements accordingly. If it requires more review from comments it would be rerouted through the repair function and iterated until it is acceptable to proceed. At this point, if no action is decided for the new MSDS, it would be forwarded to the Reject Archive function and archived. On the other hand, if action is required, and once everything is completed, it would be forwarded on to initiate the change and archived in the Reject Archive function.

## Benefits of Document Imaging and Workflow

Document imaging and workflow allow companies to work smarter rather than work harder. This includes reducing the number of steps in a process, reducing the amount of time needed to do a task, doing the task with less people, reducing the storage space to store all kinds of paper, etc. Reducing the amount of time to do a task is accomplished by having all of the appropriate information to complete a task electronically attached and immediately available. For example, engineering drawings stored electronically can be quickly accessed for review prior to the ES&H professional and/or manager signing off for changes or modifications. This is in stark contrast to the time-consuming manual process of searching for engineering drawings in filing cabinets. Using this method also precludes the need to enter the same data several times into the system. Once it is entered, it never has to be reentered. It follows that reducing the number of steps also provides an opportunity to reduce the amount of time needed to do a particular task. If this is done properly, manpower can often be reduced. In this day and age of "right-sizing" to maintain a competitive advantage, this allows a company to reduce manpower without sacrificing the quality of work performed.

The individuals responsible for assembling and maintaining workflow maps (like the one shown in Figure 39B-8) are not involved in the process. By removing them from the process the integrity of the process is maintained. This is crucial because if the person responsible for maintaining the workflow map is also part of the process, different activities could be circumvented by modifying the workflow map. However, if the person responsible for the workflow map is not involved in the process, no activity can be circumvented by anyone in the process. This is important for OSHA, EPA, and ISO.

When a problem is being worked or an assembly is being accomplished, it is repeatable. It is repeatable because it is not a paper process that can circumvent any activity or organization (e.g., safety and quality). The electronic mechanism inherent in workflow also lends itself to ease of auditing because everything done leaves a time/date audit trail. This is important not only for auditing but also for getting the latest copy of documents, engineering drawings, etc.

These technologies and processes are most successfully implemented in a computing environment that is flexible and open to various computing platforms and which allows growth from small applications. To accomplish this many companies are moving to a client/server architecture.

## Defining Client/Server Computing

Client/server computing is a form of distributed processing where an application is split in a way that allows a front end (the client) to request services of a back end (the server). Figure 39B-9 presents the definitions. Typically, the front ends reside in the end-user desktop systems (PCs, Macs, UNIX, workstations, etc.). Back ends can reside in any network server ranging from a local database server to the largest mainframe. The front-end and back-end portions of the application reside on different processing systems, usually

**Client:** A single-user workstation that provides presentation services and the appropriate computing, connectivity and database services relevant to the business need.

**Server:** One or more multi-user processors with shared memory that provide computing, connectivity and database services and interfaces relevant to the business need.

**Client/server Computing:** An environment that appropriately allocates application processing between the client and the server. The environment typically is heterogeneous, with the client and the server communicating through a well-defined set of standard application program interfaces and remote procedure calls.

**Figure 39B-9**  Client/server definitions.

separated by a network. Neither the application's front end its nor back end are complete applications themselves. Rather, they complement each other to form a complete application.

Client/server database systems are constructed such that the database runs on a database server while the database users interact with their own desktop systems (client), which are responsible for handling the user interface, including windowing and data presentation. Figure 39B-10 illustrates the component relationships in the client/server model.

### Summary of the Impact

Staying compliant with OSHA, EPA, and in some cases ISO 9000 in the 1990s certainly poses a challenge for ES&H professionals and managers. Automating the process to reduce time, steps, and manpower — while not only maintaining compliance but achieving superior performance — will help ensure company survivability in the 1990s. Document imaging, business process automation, and reengineering implemented on client/server architectures brings state-of-the-art technology to the desktop of the ES&H professional. Instead of spending an inordinate amount of time wrestling with regulations, documents, engineering drawings, and filing reports, ES&H professionals can now focus on the job for which they were hired.

### CONCLUSIONS

For companies that are presently or are planning on conducting business abroad, ISO 9000 is rapidly becoming the worldwide quality "de facto" standard. It is flexible enough to grow with technology and the times, while ensuring quality "by design" and repeatability of processes.

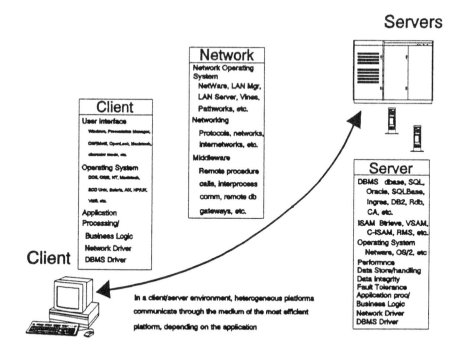

**Figure 39B-10** Client/server architectural relationships.

Managing the morass of critical data in paper form may jeopardize a company's ability to meet the intent of ISO 9000. Implementing some kind of computer technology appears to be the best bet for ensuring compliance with ISO as well as OSHA and EPA regulations.

## REFERENCES

9001: Quality Systems — Model for Assurance in Design/Development, Production, Installation and Servicing.

9002: Quality Systems — Model for Assurance in Production and Installation.

9003: Quality Systems — Model for Assurance Final Inspection and Test.

9004: Quality Management and Quality System Elements — Guidelines.

Baum, D., Computerworld, *Client/Server Development Tools for Windows,* April 26, 1993, pp. 73–75.

O'Lone, E. J., Datapro, Document Imaging Systems, Management Issues, *Client/Server Computing,* June 1993, pp. 1–8.

# Section X
# Standards of Competence

# 40

# STANDARDS OF COMPETENCE

**Roger L. Brauer**

## TABLE OF CONTENTS

1-56670-054-X/96/$0.00+$.50
© 1996 by CRC Press, Inc.

## INTRODUCTION

Competency means having the qualifications or capability to perform some task. Standards of competency and occupational credentialing assist employers and the public in determining who is competent. This chapter reviews competency standards and practices for safety and health professions.

Competence in the safety profession is based on many things. If a profession has clearly defined itself, one can begin to define competency for practice in it. The functions, tasks, knowledge, and skills can be described. Competency also depends on job experience in the profession. Accreditation contributes to competency because it evaluates whether academic programs teach subjects required for a profession. Licensing and certification measure individual competency more directly through examinations and by evaluating a candidate's education, training, and job experience.

Competency for the safety profession has its basis in several documents. The American Society of Safety Engineers (ASSE) maintains an official definition for the safety profession in *The Scope and Functions of the Professional Safety Position.* This brochure defines the primary tasks of the safety profession. This task description differentiates it from other disciplines. Competence in safety professional practice must be evaluated against tasks or functions of the safety profession.

A description of the knowledge required for the tasks and functions of safety professional practice also forms a basis for evaluating competency. Curriculum standards[1] for safety degrees define the knowledge required for safety profession tasks. Academic program accreditation for safety degrees derives its curriculum criteria[2] from this standard. The Board of Certified Safety Professionals (BCSP) uses the same curriculum standard to evaluate the academic preparation of individuals who seek the Certified Safety Professional (CSP) designation.

In England the Occupational Health and Safety Lead Body[3] is developing competency standards for several safety and health job positions. The positions include general safety, radiation protection, and occupational hygiene. The competency standards have task analyses and knowledge standards for different levels of each position.

## PROFESSIONAL COMPETENCE

Figure 40-1 illustrates the general process by which competency is assessed. The overall goal is to ensure that certain functions or tasks are performed knowledgeably and effectively. The process may address capabilities for general functions or focus on specific tasks.

Predicting actual performance is difficult. However, education and training, experience, and knowledge and skills are often measured since they

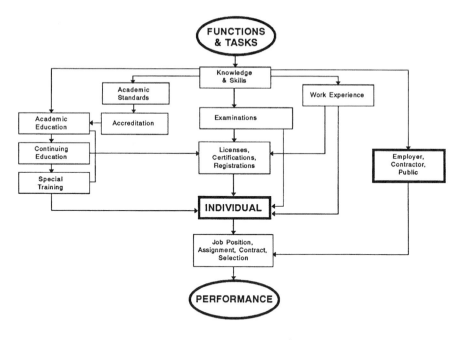

**Figure 40-1** Competency process.

contribute to competence. The evaluation process begins with a definition of the tasks and functions that are to be performed. An individual or a panel of people familiar with the tasks usually prepares the definition. For education and training programs, course descriptions describe the knowledge and skills to be developed. The specifications for an examination, which describe its length and the kinds and numbers of questions, are derived from tasks and knowledge descriptions.

Evaluations of the education, training, and experience of individuals help predict performance. An employer may prepare written standards for a position. Licensing, certification, and registration boards often have formal procedures for evaluating the education and experience of a candidate. Professional credentialing usually requires that candidates be graduates of accredited degree programs. This places a burden on program accreditation to ensure that an educational program meets standards for a profession. Both the portion of time devoted to key tasks and the breadth and level of job duties are important considerations when evaluating work experience.

Most often competency is left to employers. They define job tasks and the knowledge and skills required for them. Then they hire individuals who appear to meet those needs. The employer looks at knowledge and skills, training, experience, and academic or professional credentials. Sometimes companies, organizations, or supervisors do not go through formal procedures, but simply select a person to perform a task or assign responsibility to someone who is already available. Sometimes tasks are defined in contract specifications for contractors or their employees.

The public may have trouble determining who is qualified to perform some task. The public may not know what knowledge, skills, or abilities are required. In such cases, governments often establish licensing to protect the public and ensure that those offering to perform certain tasks are competent. The licensing process includes formal definitions of knowledge and skills, training standards, and procedures for evaluating individuals. If an individual offering to perform a task has a license for it, the public can expect competent performance. Because governments do not license all occupations and professions, the public, employers, and others may depend on peer organizations' certifications to evaluate practitioners and provide some assurance that individuals meet minimal competency standards.

## CURRICULUM STANDARDS

Part of competency assessment is evaluating whether those in a profession are properly trained to perform the profession's functions. This starts with a definition of academic requirements for the profession.

Since the late 1960s, both the American Society of Safety Engineers and the Board of Certified Safety Professionals have maintained guidelines for baccalaureate degrees in safety. In 1991 the two organizations began publishing joint standards.[4] The requirements of this standard are summarized in Table 40-1. This standard also forms the basis for safety degree accreditation criteria. They are published by the Related Accreditation Commission of the Accreditation Board for Engineering and Technology (RAC/ABET), which accredits baccalaureate and master's safety degrees at United States colleges and universities.

In 1993 the two organizations completed a similar standard for master's degrees. The number of people entering the safety profession through baccalaureate safety degrees increased significantly in the last decade. However, the majority still enter the field from other disciplines. Many use master's degrees and/or professional safety experience to complete preparation for the safety profession. Additional standards for academic safety education are under development. They include standards for associate degrees in safety, degrees in safety engineering, engineering technology, and industrial technology, and safety options within business, technology, and engineering degrees.

In the 1980s the industrial hygiene community moved to begin accreditation of industrial hygiene degrees. The American Academy of Industrial Hygiene serves as the lead organization. A standard was prepared for master's degrees in industrial hygiene and is now incorporated into RAC/ABET industrial hygiene program accreditation criteria. Table 40-2 summarizes these requirements. A curriculum standard for baccalaureate degrees is under development.

**Table 40-1 Summary of Baccalaureate Safety Degree Requirements**

| Category | Courses or Subjects |
| --- | --- |
| Preparatory | Mathematics |
| |   Through introductory calculus |
| |   Statistics |
| |   Basic computer skills |
| | Physical, chemical, and other sciences |
| |   Physics with laboratory |
| |   Chemistry (including an introduction to organic |
| |     chemistry) with laboratory |
| |   Physiology, anatomy, or biology |
| |   Psychology |
| | Other |
| |   Introductory business |
| |   Written and oral communication |
| |   Applied mechanics |
| |   Industrial or manufacturing processes |
| Professional core courses | Introduction to safety and health |
|  (full courses) | Safety and health program management |
| | Design of engineering hazard controls |
| | Industrial hygiene and toxicology with laboratories |
| | Fire protection with laboratories |
| | Ergonomics with laboratories |
| | Environmental safety and health |
| | System safety and other analytical methods for safety |
| Required professional | Measurement of safety performance |
|  subjects (may be combined | Accident/incident investigation |
|  in various courses) | Behavioral aspects of safety |
| | Product safety |
| | Construction safety |
| | Educational and training methods for safety |
| Electives | Professional electives |
| | General electives |
| Experiential education | Internship or cooperative program with course credit |

From *Curriculum Standards for Baccalaureate Degrees in Safety,* Joint Report No. 1, American Society of Safety Engineers and Board of Certified Safety Professionals, August 1991. With permission.

## COMPETENCY FOR SAFETY AND HEALTH PROFESSIONALS

The Board of Certified Safety Professionals[5] (BCSP) and the American Board of Industrial Hygiene[6] (ABIH) are the two major organizations in the United States which assess the competency of safety and health professionals. These organizations offer the Certified Safety Professional (CSP) and Certified Industrial Hygienist (CIH) designations, respectively. Together, they also operate the ABIH/BCSP Joint Committee for Occupation and Health and Safety Technologists,[7] which offers the Occupational Health and Safety Technologist (OHST) certification. By the end of 1994, BCSP had awarded the CSP to about 13,000 individuals. ABIH had awarded the CIH to over 6100 people, and about 1300 individuals received the OHST designation. Near the end of 1994, the Joint Committee added the Construction Health and Safety Technician, primarily for people handling construction job-site safety.

**Table 40-2  Requirements for Industrial Hygiene Master's Degrees**

| Factor | Criteria |
|---|---|
| Candidate requirements | 120 semester hour or greater, bachelor's degree; At least 60 hours in science, mathematics, engineering, and technology |
| Industrial hygiene master's degree | At least 30 semester hours |
| Curriculum | Industrial hygiene sciences: |
| |   Principles and practices of industrial hygiene |
| |   Principles and practice of environmental sciences |
| |   Epidemiology and biostatistics |
| | Industrial hygiene practice: |
| |   Control of physical and chemical hazards |
| |   Environmental health |
| |   Occupational safety |
| Curriculum distribution | Industrial hygiene sciences and practice |
| | 18 semester hours |
| | Unspecified hours |
| | 12 semester hours |

## Certified Safety Professional (CSP)

To be a CSP, an individual must meet an academic requirement, an experience requirement, and an examination requirement. In summary, one must have an accredited baccalaureate degree in safety and 4 years of professional safety experience. Candidates may substitute certain combinations of other degrees and additional professional safety experience for the academic requirement. Table 40-3 lists the units of credit BCSP awards for various degrees. Shortages in the academic requirement can be resolved with additional professional safety experience at 1 unit per month of employment.

At least 50% of an applicant's job duties must be at a professional level and include a breadth of acceptable safety activities. Graduate degrees can count for up to 2 years of the experience requirement. Credit for particular graduate degrees is also proportional to that found in Table 40-3 for baccalaureate degrees.

After meeting the academic requirement, an applicant is allowed to sit for the Safety Fundamentals Examination. It tests basic knowledge of safety

**Table 40-3  Board of Certified Safety Professionals Academic Credit Schedule**

| Baccalaureate Degree | Units of Credit Allowed |
|---|---|
| ABET-accredited safety or safety technology degree | 48 |
| ABET-accredited engineering degree | 42 |
| Safety or safety technology degree that is not accredited by ABET | 36 |
| Engineering technology degree accredited by ABET | 30 |
| Physical and natural science degree | 30 |
| Engineering technology and industrial technology degree not accredited by ABET | 24 |
| Business administration, industrial education or psychology degree | 18 |
| Majors not listed above | 12 |

**Table 40-4  Subjects Covered on Certified Safety Professional Examinations**

| Examination Section | Examination Subject |
| --- | --- |
| **Safety Fundamentals Examination** | |
| 1. Basic and Applied Sciences | Mathematics, physics, chemistry, biological sciences, behavioral sciences, ergonomics, engineering and technology, epidemiology |
| 2. Program Management and Evaluation | Organization, planning, and communication; legal and regulatory considerations; program evaluation; disaster and contingency planning; professional conduct and ethics |
| 3. Fire Prevention and Protection | Structural design standards; detection and control systems and procedures; fire prevention |
| 4. Equipment and Facilities | Facilities and equipment design, mechanical hazards, pressures, electrical hazards, transportation, materials handling, illumination |
| 5. Environmental Aspects | Toxic materials, environmental hazards, noise, radiation, thermal hazards, control methods |
| 6. System Safety and Product Safety | Techniques of system safety analysis, design considerations, product liability, reliability, and quality control |
| **Comprehensive Practice and Current Specialty Examinations** | |
| 1. Engineering | Safety engineering, fire protection engineering, occupational health engineering, product and system safety engineering, environmental engineering |
| 2. Management | Applied management fundamentals, business insurance and risk management, industrial and public relations, organizational theory and organizational behavior, quantitative methods for safety management |
| 3. Applied Sciences | Chemistry, physics, life sciences, behavioral sciences |
| 4. Legal/Regulatory Aspects and Professional Conduct and Affairs | Legal aspects, regulatory aspects, professional conduct and affairs |

professional practice. After a person passes this examination an interim designation, called the Associate Safety Professional (ASP), is awarded, but individuals must complete the CSP process within certain time limits. After meeting both the academic and experience requirements, an applicant may sit for the Comprehensive Practice Examination or one of several specialty examinations, which focus on professional practice. After successfully completing an examination at this level, one receives the CSP designation. In 1997 all CSP candidates must pass the Comprehensive Practice Examination to receive the designation. Specialty examinations will apply subsequent to the CSP. Table 40-4 summarizes the knowledge areas covered by CSP examinations.

## Certified Industrial Hygienist (CIH)

To be a CIH, one must meet academic, experience, and examination requirements. An applicant must have an acceptable degree in industrial hygiene, chemistry, physics, chemical, mechanical, or sanitary engineering, medicine, or biology. Other bachelor's degrees are considered on the basis of their basic science content, not including social sciences.

An applicant must have 5 years of full-time (greater than 50% of job duties) employment experience in industrial hygiene. Experience is evaluated for the broad practice of industrial hygiene or in particular aspects of industrial hygiene. The kind of experience determines eligibility for particular examinations. Acceptable graduate degrees may be substituted for up to 2 years of professional experience.

Candidates must pass two examinations. The first examination is called the Core Examination and covers basic knowledge and skills related to the functions of industrial hygiene. Those who successfully complete the Core Examination receive the interim designation of Industrial Hygienist in Training (IHIT). The second examination is called the Comprehensive Examination. It includes detailed questions on advanced knowledge and skills related to the functions of industrial hygiene. Those who successfully pass the Comprehensive Practice Examination receive the CIH designation. Candidates may seek certification in a specialty area after passing the Comprehensive Examination. These examinations are limited to knowledge and skills relating to the functions of the specialty.

Those holding the CIH credential must be recertified every 6 years through a certification maintenance program. Recertification is possible by being reexamined and passing the Comprehensive Examination or a specialty examination. Recertification can also be accomplished though a point system that emphasizes continuing education in the industrial hygiene field.

## Occupational Health and Safety Technologist (OHST)

The Occupational Health and Safety Technologist certification is operated jointly by ABIH and BCSP through the ABIH/BCSP Joint Committee. Candidates for the OHST designation must have 5 years of experience in occupational health or safety. At least 35% of job duties must be in safety or health. Up to 2 years of experience is waived for candidates with associate degrees in safety or health or other technical and scientific fields or with a baccalaureate degree. Candidates must pass a 7-hour examination. Candidates with degrees in industrial hygiene, occupational or industrial safety, occupational health, or environmental science may sit for the examination before meeting all of the experience requirement. The certification is not awarded until all requirements are met.

The OHST examination covers seven subject areas:

1. Basic and applied sciences
2. Laws, regulations, and standards
3. Control concepts
4. Investigation (post-event)
5. Survey and inspection techniques (pre-event)
6. Data computation and record keeping
7. Education, training, and instruction

## Construction Health and Safety Technician (CHST)

Beginning in 1994, the Construction Health and Safety Technician certification has also been operated by the ABIH/BCSP Joint Committee. The typical candidate is one who oversees job site safety on significant construction projects, usually involving several contractors. People in this position may have academic training for the position, but often advance into these positions from building crafts. Training/education and experience requirements reflect these two routes. Candidates must have 3 years of experience in construction and have at least 2 years of supervisory or management experience. A minimum of 40 hours of safety and health training are required, with an associate degree in safety and health being the ideal academic preparation.

The examination is built around the job tasks listed below, and knowledge areas important for these tasks are also identified for candidates:

1. Inspections
2. General safety training and safety orientation
3. Safety and health record keeping
4. Hazard communication compliance
5. Safety analysis and planning
6. Accident investigations
7. Program management and administration
8. Occupational Safety and Health Administration (OSHA) and other inspections

## Other Certifications

There are other certifications in the safety and health field or in related fields. For example, the American Board of Health Physics[9] offers the Certified Health Physicist (CHP) designation. There are also certain competency requirements established by federal and state agencies. One of many possible examples is the requirement by the U.S. Environmental Protection Agency that those involved in radon mitigation pass an examination. Many states license people involved in asbestos mitigation. Two states (California and Massachusetts) have a safety engineering specialty within their professional engineer licensing programs. Massachusetts uses certain CSP examinations for the second level examination (Engineering Principles and Practices), while

California maintains its own Principles and Practices Examination for safety engineering.

There are also some foreign competency programs for safety professionals. In Canada the Association for Canadian Registered Safety Professionals[10] offers the Canadian Registered Safety Professional (CRSP) designation. The designation is awarded to those candidates meeting education and experience requirements and passing an examination. In the United Kingdom, the National Examination Board in Safety and Health (NEBOSH)[11] offers a National Diploma in Safety and Health to applicants who meet educational and experience standards and pass an examination. Safety organizations in Mexico are working to establish a credentialing process for safety professionals.

## Occupational Health and Safety Lead Body Competency Standards

In the United Kingdom a national effort addressed the vocational qualifications for a wide range of occupations. A primary activity involved the development of standards of competency. The Occupational Health and Safety Lead Body was responsible for developing standards for six areas: general safety, occupational hygiene, radiation protection, enforcement (HSE inspectors/environmental health officers), occupational health physicians, and occupational health nurses. A task analysis identified units and subunits and their applicability to different levels. Units of Competence for general safety, occupational hygiene, and radiation protection include the following functions:

A. Provide support for the control of risk.
B. Establish and maintain occupational health and safety policies.
C. Control risks
    1. Establish, maintain, and develop occupational health and safety (OHS) systems.
    2. Assess requirements for risk control.
    3. Plan, implement, and develop risk control.
D. Establish and maintain a culture of OHS awareness.
E. Regulate OHS operations.
F. Contribute to organizational OHS competence.
G. Manage OHS operations.
H. Contribute to advances in OHS.

The competency standards may apply or apply in different ways to each of five levels of jobs:

Level 1. Competence in the performance of a range of varied work activities, most of which may be routine and predictable.
Level 2. Competence in a significant range of varied work activities, performed in a variety of contexts. Some of the activities are complex or nonroutine,

and there is some individual responsibility or autonomy. Collaboration with others, perhaps through membership in a work group or team, may often be a requirement.

Level 3.  Competence in a broad range of varied work activities performed in a wide variety of contexts, most which are complex and nonroutine. There is considerable responsibility and autonomy, and control or guidance of others is often required.

Level 4.  Competence in a broad range of complex, technical, or professional work activities performed in a wide variety of contexts and with a substantial degree of personal responsibility and autonomy. Responsibility for the work of others and the allocation of resources is often present.

Level 5.  Competence which involves the application of a significant range of fundamental principles and complex techniques across a wide and often unpredictable variety of contexts. Very substantial personal autonomy and often significant responsibility for the work of others and for the allocation of substantial resources feature strongly, as do personal accountabilities for analysis and diagnosis, design, planning, execution, and evaluation.

## CONTINUING PROFESSIONAL DEVELOPMENT

Today information, knowledge, methods, standards, and equipment affecting many professions are changing rapidly. Someone has estimated that the half–life of an engineer's knowledge is 5 years and that by the time today's kindergarteners graduate from high school the amount of knowledge in the world will have doubled four times.[12] As a result, the public and employers expect people to keep up with the changes in their profession. This rapid rate of change affects safety professionals, too. For professional licensing and certification credentials, recertification or continuing education requirements are the norm.

To retain the CSP and CIH designation, both BCSP and ABIH require recertification. Both have a point system that involves continuing education in safety and health, continued professional practice, reexamination, advanced degrees, and other means for staying current with the profession. Those who do not meet the requirements at the end of a cycle lose their CSP or CIH designation.

The OHST implemented a recertification requirement in 1994. The CHST will probably have a recertification requirement in the future. Recertification is required for the CRSP (Canada) and NEBOSH National Diploma in Safety and Health (United Kingdom).

## NATIONAL STANDARDS FOR CERTIFICATION PROGRAMS AND EXAMINATIONS

It is easy to create and award a credential for a discipline or vocation. Employers, contractors, and the public may be confused by credentialing

**Table 40-5  Summary of Major National Occupational Credentialing Standards**

| Category | Requirements |
|---|---|
| Agency and operations | Is the agency not-for-profit? |
| | Is the agency national in scope? |
| | Is the agency separate and independent from any associated educational body? |
| | Does at least one public member serve as a member of the governing board? |
| | Are financial statements published? Are they audited by a recognized auditing agency? |
| Eligibility and examinations | Is the certification open to those who are *not* members of a professional association in the field or *not* members of the certifying agency? |
| | Is eligibility logically related to relevant job requirements? |
| | Is the examination free of bias and nondiscriminatory with published demographic data? |
| | Are pass/fail cutoff scores established using recognized and psychometrically sound procedures? |
| | Can anyone be granted the credential without examination and achieving a passing score? |
| | Are reliability statistics produced after each examination administration? |
| | Has the validity of the examination been established by conducting a national job analysis survey or other psychometrically sound validation survey? |
| Recertification | Is there a recertification program to insure continued competence? |
| | What is the program? |

systems and have difficulty knowing which credentials have strict evaluation procedures and quality examinations.

Several organizations exist in the United States to help evaluate and accredit examinations, credentials, and credentialing organizations. These organizations have standards which help assure government agencies, employers, and the public that occupational and professional credentials and the examinations associated with them have quality. Typically, competency organizations apply for recognition by these organizations and are evaluated for compliance with standards before recognition is given. The standards typically cover the occupational competency organization and its activities and examinations. Table 40-5 summarizes the key elements of these national occupational credentialing standards. Most standards require occupational credentialing bodies to be reevaluated periodically. In the future meeting national standards for competency organizations and examinations will become more important.

## Council on Licensure, Enforcement, and Regulation (CLEAR)[13]

In the United States occupational licensing and regulation are the responsibility of state governments. The Council on Licensure, Enforcement, and Regulation is an association of state and provincial officials and administrators involved with occupational licensing and regulation issues. Its mission is to improve the quality and understanding of professional and occupational regulation. It maintains a directory of state licensing officials and offers information on occupational licensing. It maintains the National Disciplinary Information System that tracks disciplinary actions taken by licensing boards against practitioners in various licensed occupations and professions. Membership is open to individuals, licensing boards, associations of licensing boards and other organizations. Membership does not infer compliance with examination practice standards.

CLEAR and the National Organization for Competency Assurance (NOCA) jointly publish guidelines for testing procedures and practices.[14] These principles cover such things as content validity, format, setting passing points, reporting results, conduct of examinations, security, facilities, scoring, appeals, confidentiality of test results, reexamination procedures, and many other factors affecting the quality of credentialing and credentialing examinations.

At present safety and health functions licensed by states are not directly involved with CLEAR. However, boards and state employees who govern certain safety functions may participate in CLEAR activities.

## National Organization for Competency Assurance (NOCA) and National Commission for Certifying Agencies (NCCA)[15]

Another organization that has standards for certification is the National Organization for Competency Assurance. NOCA publishes detailed standards governing certification bodies and examinations. The standards address fair and psychometrically sound practices for certification operations, organization and structure, financial operations, candidate evaluation, examinations, recertification, and other factors in certification operations. NOCA membership is open to any organization that is interested in certification and related credentialing.

The National Commission for Certifying Agencies is a branch of NOCA that evaluates and enforces the standards of NOCA. Organizations offering certification apply to NCCA and are evaluated against the NOCA standards. Only those organizations which fully comply with the standards can become members of NCCA. Member organizations must be reevaluated periodically to retain memberships.

The Board of Certified Safety Professionals is a member of NCCA and meets NOCA standards.

## Council of Engineering Specialty Boards (CESB)[16]

The Council of Engineering Specialty Boards sets standards for those credentialing activities of engineering, technology, and related fields not governed by state engineering licensing boards. CESB was organized in the late 1980s and publishes its standards for credentialing activities. Member organizations are evaluated periodically against the CESB standards and must comply with the standards to retain membership.

The Board of Certified Safety Professionals and the CSP designation are accredited by CESB and the NCCA. The OHST credentialing activity has been evaluated by CESB. A few modifications, especially with regard to recertification, are being made to achieve compliance.

### National Public Service Accreditation Board[17]

Another organization involved in evaluating competency examinations is the National Public Service Accreditation Board. It was formed in the early 1990s by the International City/County Management Association, an association of managers of local governments. Because many local governments use examinations of other organizations in determining qualifications of applicants for local government positions, there was a need to establish some way to ensure that the examinations met minimal standards. While examinations for public safety (police and fire departments) positions may be accredited by this organization, to date safety professional and industrial hygiene examinations have not been submitted for evaluation.

### SUMMARY

The competency of safety and health professionals is improving. The improvement stems from a clear definition of what the profession is. It is improving because there are academic standards for degree programs and programs are being accredited. It is improving because more individuals are being evaluated against academic and experience standards and examined on knowledge required for professional practice in safety and health. Competency is improving because safety and health professionals are required to keep up with changes. Improvement of competency in the profession will continue and the profession will be better for it.

### ENDNOTES

1. *Curriculum Standards for Baccalaureate Degrees in Safety,* Joint Report No. 1, American Society of Safety Engineers and Board of Certified Safety Professionals, August 1991; *Curriculum Standards for Master's Degrees in Safety,* Joint Report No. 2, American Society of Safety Engineers and Board of Certified Safety Professionals, March 1994.
2. The Related Accreditation Commission of the Accreditation Board for Engineering and Technology (RAC/ABET, 111 Market Place, Suite 1050, Baltimore, MD 21202) publishes curriculum criteria for safety degrees used in accreditation of baccalaureate and master's degrees in safety. The curricular criteria are derived from Joint Reports 1 and 2 referenced above.

3. Occupational Health and Safety Lead Body, Seventh Floor North, Rose Court, 2 Southwork Bridge, London, SE1 9HS, England.

4. *Curriculum Standards for Baccalaureate Degrees in Safety,* Joint Report No. 1, American Society of Safety Engineers and Board of Certified Safety Professionals, August 1991; *Curriculum Standards for Master's Degrees in Safety,* Joint Report No. 2, American Society of Safety Engineers and Board of Certified Safety Professionals, March 1994.

5. Board of Certified Safety Professionals, 208 Burwash Avenue, Savoy, IL 61874.

6. American Board of Industrial Hygiene, 4600 W. Saginaw, Suite 101, Lansing, MI 48917.

7. ABIH/BCSP Joint Committee for Certification of Occupational Health and Safety Technologists, 208 Burwash Avenue, Savoy, IL 61874.

8. Under current procedures, the Comprehensive Practice and Specialty Examinations all lead to the CSP designation and cover the same subjects, but differ somewhat in the number of questions devoted to each subject. Beginning in 1997, specialty examinations will follow the CSP designation and focus only on knowledge applicable to the specialty.

9. American Board of Health Physics, 8000 Westpark Drive #400, McLean, VA 22102-3101.

10. Association for Canadian Registered Safety Professionals, 6519B Mississauga Road, Mississauga, Ontario L5N 1AL, Canada.

11. National Examination Board in Safety and Health, 222 Upingham Road, Leicester LE5 0QG, United Kingdom.

12. M. Cetron and O. Davies. *American Renaissance: Our Life at the Turn of the 21st Century.* St. Martins Press, New York, 1989.

13. The Council on Licensure, Enforcement, and Regulation, The Council of State Governments, 3560 Iron Works Pike, P.O. Box 11910, Lexington, KY 40578-1910.

14. *Principles of Fairness: An Examining Guide for Credentialling Boards,* February 1993 (published jointly by the Council on Licensure, Enforcement, and Regulation and the National Commission for Certifying Agencies).

15. National Commission for Certifying Agencies, 1101 Connecticut Avenue, N.W., Suite 700, Washington, D.C. 20036.

16. Council of Engineering Specialty Boards, 130 Holiday Court, Suite 100, Annapolis, MD 21041.

17. National Public Service Accreditation Board, c/o International City/County Management Association, 777 North Capitol Street, N.E., Suite 500, Washington, D.C. 2002-2401.

# Section XI
## Afterword – The Future

# 41       WHERE ARE WE GOING? — AN EDUCATOR'S PERSPECTIVE

**Janice L. Thomas**

## TABLE OF CONTENTS

## INTRODUCTION

> It must be considered that there is nothing more difficult to carry out, nor more doubtful of success, nor more dangerous to handle than to initiate a new order of things.
>
> *Machiavelli (1469–1527)*

This warning, from the Italian writer and statesman, rings true today as we stand at the threshold of a new millennium and ask the question, "Where is our profession going?" The answer is clear for those who practice futuristic techniques. Change — that "new order of things" — is afoot. We will explore where these changes are leading us.

The practice of safety and health management has advanced greatly in the past 20 years; it will change even more in the next 10. The education of safety and health professionals, within the United States' college and university system, is one way to measure this change. Another way is to study current trends in professional continuing education for purposes of career advancement.

A third means is to examine the catalytic events and prime agents which *appear* to influence the practice of safety and health professionals. This third measure may hold the key to the past as well as to the future. We will discuss it first.

## A BETTER EXPLANATION OF THE PROFESSION'S EVOLUTION

The common belief is that the safety and health profession grew up out of a demand for "specialists" who could help organizations respond to increasing regulatory scrutiny of hazardous conditions. In fact, the notion of "specialists" was validated during the 1970s when regulation of hazards became prolific and sophisticated to the point of dividing risks into sectors, such as occupational, environmental, consumer, or transportation. The belief that the growth in the profession in the 1970s was due *solely* to the growth in laws and regulations is very misleading.

Instead, we should look at a model of professional "state-of-the-practice" that demands a higher level of organizational response to hazards and threats. This model shows that, prior to state or federal regulation, there are always "critical masses" of safety and health professionals operating at a high level of practice. This evolving state-of-the-practice becomes the benchmark for continued standardization and regulation — the models of proaction we prefer in this nation. The OSHAs and EPAs are firmly treading in the footsteps of the professionals — not the other way around!

Figure 41-1 illustrates this model by looking at several key historic events in occupational safety state-of-the-practice. If we can accept this picture, then we see that every time we individually choose to stretch professionally — to strive for "continuous quality improvement" in safety and health practice — we are leading ourselves into the future. The key here is to strive for improvement — to find the better and more effective ways to prevent or control risks.

With this model in mind, let us turn to a discussion of how the education of safety and health professionals has been conducted in the past and how this process is beginning to change.

## THE PAST AND THE FUTURE OF PROFESSIONAL EDUCATION

Unfortunately, it must be admitted that most college and university programs for entering practitioners have been unduly influenced by the presence of legislation and regulation. For example, only one or two "safety management" degrees were available prior to OSHA's inception. Most professionals, if they had the benefit of a 4-year degree, were graduating with degrees in allied professions such as engineering or science or a traditional specialization such as fire protection. Then, when OSHA and EPA came upon the scene, the educators perked up their ears, hired the few credentialed safety and health

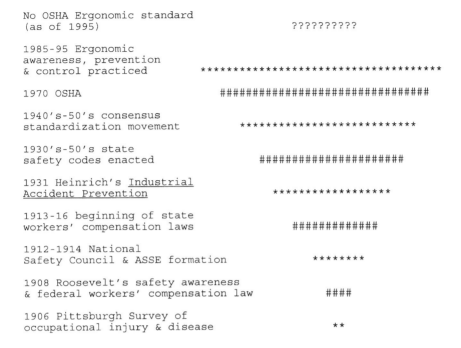

```
No OSHA Ergonomic standard
(as of 1995)                          ??????????

1985-95 Ergonomic
awareness, prevention
& control practiced      *************************************

1970 OSHA                ###############################

1940's-50's consensus
standardization movement    ***************************

1930's-50's state
safety codes enacted        #####################

1931 Heinrich's Industrial
Accident Prevention           *****************

1913-16 beginning of state
workers' compensation laws       ############

1912-1914 National
Safety Council & ASSE formation     ********

1908 Roosevelt's safety awareness
& federal workers' compensation law     ####

1906 Pittsburgh Survey of
occupational injury & disease            **
```

**Reading from the bottom up...**
**** = professional activity & practice
#### = regulatory codification
       of state-of-the-practice

**Figure 41-1**   State-of-the-practice as the catalyst for change (historic examples from occupational safety and health practice).

professionals they could find, and began to offer safety and health degrees of various sorts.

The American Society of Safety Engineers (ASSE) has identified 123 colleges or universities that collectively offer 237 degrees in safety and health. Of these institutions, 70 offer undergraduate degrees, 61 offer master's-level work, and 26 universities offer doctoral-level work. These numbers do not represent the many certification programs or 2-year associate degrees offered in the community college system. The actual names and emphasis vary greatly between these institutions.

Despite a recent movement among colleges and universities to consolidate and eliminate small programs, the number of safety and health degrees is remaining constant. Some growth has actually been observed in the number of master's and doctoral degrees since the early 1990s. This trend is important for several reasons. First, safety and health curriculums offer additional employment opportunities. Women and men may now choose to become safety and health educators, as well as practitioners within the public and private sectors. Second, educators who are truly aligned with the profession will be able to advocate for their peers. They will be able to provide that "state-of-the-practice" leadership in education, research, and writing that is so necessary to a dynamic profession.

Now that we are beginning to see a dedicated cadre of safety and health educators, we are also beginning to see a debate about the best "disciplinary" foundations from which to train practitioners. In the past, either the foundation was technical (e.g., engineering) or programs were forced to fit into an existing scheme (e.g., safety education in the physical education department).

The new debate has its origins in the National Institute of Occupational Safety and Health (NIOSH) Minerva Project, which was begun in the 1970s and which continues to promote the teaching of safety and health modules inside of business schools. Yet, business schools have traditionally been resistant to expanding their already full curriculums to include significant amounts of safety and health education.

The newest school of thought is that management sciences must be combined with technical knowledge. This combination can be achieved through several means. One way is to redesign current undergraduate degrees with a balanced emphasis. Another way is to provide technical training at the undergraduate level with management emphasis at the graduate level. Alternatively, undergraduate and postgraduate safety and health programs could be housed in management schools. Here technical knowledge could be taught with a management emphasis, running parallel with other management specializations.

The management school model for education of safety and health professionals has not yet been attempted, yet it is one whose time has arrived. After all, it is other managers who the safety and health practitioner must work with, gain respect from, and convince in order to be effective. This process is best begun while everyone is in school.

As this debate begins to take shape, another, very different discussion concerning professional education is already underway. How much education is enough? This question is not unique to the safety and health profession, but it is one which is quickly confronting many colleagues who thought an undergraduate education in safety or health or in an allied field was quite enough.

Now we find that professional certification by examination as well as postgraduate degrees may be needed for advancement. Continuing education in the form of seminars, certificate programs, and constant self-study is necessary to keep up to date, as well as to prepare for the future.

## FUTURISTIC TECHNIQUES

What does the future hold for safety and health professionals? We could gaze into a crystal ball, or we could consciously become "futurists."

Futurists are interdisciplinary thinkers. They see relationships and change where others do not. They describe the future looking 5 to 10 years ahead. They try to prescribe the appropriate action, becoming proactive in their risk control and management efforts. They also integrate futuristic techniques into every-day safety and health practice. These techniques include scanning, content analysis, data research, trend analysis, forecasting, cross impact matrix analysis, Delphi surveys, scenario building, decision tree analysis, simulation, and games. These and others tools can be found fully described in "future science" texts.

The example used in Figure 41-1 will illustrate the technique and process. In the early 1980s the issue of ergonomics was creeping into our awareness. This began when some practitioners began to take an in-depth look at a new injury phenomenon of repetitive motion illness and the old problem of back injury. Data were analyzed. Trend analysis was performed and economic and social implications were forecasted. The need for change became obvious to some. At the same time, these professionals became aware of the growing discussion amongst their colleagues concerning ergonomics. This awareness took the form of scanning and informal content analysis. Early articles began to appear; services and products became available.

Those who decided to do something "now" about ergonomic problems — not waiting for OSHA to require it — were acting as "change agents." Their individual actions developed into state-of-the-practice which will ultimately guide OSHA activity. Similar technical issues and opportunities appear on our professional horizons every year.

## SEVERAL PREDICTIONS ABOUT THE PROFESSION'S FUTURE

We began by asking, "Where are we going?" We should end by offering several predictions in response. There is no magic to this forecast. Futurists understand that the future is firmly founded in the present. Many will already be aware of their own professional growth in the direction described below.

The safety and health professional of the new millennium will:

- Be part of a multicultural and diverse population
- Be well versed in international issues and work across national boundaries
- Use sophisticated analytical skills
- Perform survey and data research as a step in risk analysis and decision making
- Serve as "generalists" within their organizations
- Work, in larger numbers, as consulting "specialists" to organizations on an "as needed" basis
- Become proficient in communication skills of writing, presentations, and training
- Use computer-based technology as a daily tool
- Be better educated and credentialed than their predecessors

Minerva, Roman goddess of wisdom and education, can lead us past Machiavelli's warning and into a predictable future. Have a safe journey!

## RESOURCES AND FURTHER READING

American Society of Safety Engineers (ASSE). *Safety and Related Degree Programs.* Des Plaines, IL: ASSE, 1994–1995 (or most recent edition).

Dickson, P. *The Future File: A Guide for People with One Foot in the 21st Century.* New York: Rawson Associates, 1977.

Fowles, J. *Handbook of Futures Research.* Westport, CT: Greenwood Press, 1978.

Holmer, O. *Looking Forward: A Guide to Futures Research.* Beverly Hills, CA: Sage Publications, 1983.

LaConte, R. T. *Teaching Tomorrow Today: A Guide to Futuristics.* New York: Bantam Books, 1975.

Martino, J. P. *Technological Forecasting for Decision Making.* New York: McGraw-Hill, 1993.

World Future Society. *The Futurist: A Journal of Forecasts, Trends, and Ideas about the Future.* Bethesda, MD: World Future Society.

# 42

# THE SAFETY AND HEALTH PROFESSIONAL OF THE FUTURE

Richard W. Lack

## TABLE OF CONTENTS

In March 1991, I presented a paper at the Annual Conference of the Safety Executives of New York entitled "The Role of the Safety Professional as We Approach the Year 2000". The research conducted for this paper, plus my ongoing interest and research on this subject, will form the foundation of this chapter.

I am grateful to the Natural Safety Council and Tillinghast Publications for publishing this material in part. The details on those publications may be found in the Further Reading section at the end of this chapter under LaDou et al. and Kloman.

As we all recognize, nothing in this great world of ours stays the same. We change as we get older and so does our environment. The process is constant, but the cycles are usually gradual so that it really becomes a progression or evolution. Every creature on Earth, including the human race, has to continuously adjust to change.

In spite of this almost relentless Universal Law, many people resist change, and instead of recognizing it as an opportunity for improvement they cling to their memories of the "good old days". Because of this reluctance to embrace change, history has shown many times that few people seem to be able to identify trends or shifts in the making. For an inquiring open mind the signs are there, but they are either ignored or misunderstood. Eventually, some dramatic

1-56670-054-X/96/$0.00+$.50
© 1996 by CRC Press, Inc.

series of events may occur to introduce sweeping change. For example, many of us brought up in the "cold war" years could not envisage the collapse of communism in Russia and its satellite states.

While attending the American Society of Safety Engineers Annual Professional Development Conference in Las Vegas, Nevada in June 1994, I listened to a fascinating keynote speech by Dan Burrus, a leading futurist. Dan pointed out that technically the next 10 years has already been invented and that what we need to do is change the way we think. He said that we must be opportunity managers and use the new tools. This takes new thinking.

Other trends Dan mentioned were that *time* will be the currency of the late 1990s. We cannot continue to do things the old way. We need to "leverage our time with technology", as he put it. "Change is opportunity," said Dan. We need to open our minds to the creative application of technology. This is the Communication Age, and one trend Dan foresees is that we are going to automate and humanize training so that it is more interactive, self-directive, and self-diagnostic. His final words were that we must think in terms of idealism, think beyond to the future problems that are about to happen, and what can we do about them. We must take the new tools and become the craftspeople of the 21st century — in other words, be preactive!

These are inspiring concepts which hopefully will provide food for thought while you are reading this chapter.

The following is a checklist of trends based on my research:

- Population explosion — China and India are the most populous nations.
- Megacities — By the year 2000 Mexico City, São Paulo, Shanghai, and Calcutta are all forecast to have populations exceeding 20 million.
- Worldwide free trade is coming — NAFTA and GATT are just the beginning.
- The Pacific Rim, South America, India, and Africa all have huge growth potential, not to mention the former Communist-controlled countries.
- Travel and tourism is forecast to be a huge growth area.
- In the world of business, entrepreneurs are creating the global economy. Intense competition will force the big corporations to break up in order to survive.
- Environmental contamination will become a worldwide problem.
- Terrorism and civil unrest will probably increase.
- Trend forecasts for the U.S. and other developed countries include
  - Population moves from the urban areas to smaller rural industrial/hi-tech centers — the so-called "Fifth Migration"
  - Renaissance of the arts
  - A more "caring" age, concerns for the elderly and less fortunate
  - Health care becoming a huge growth industry due to the aging population
  - Increasing community activity to help combat crime and other uncivil behavior

In the world of business, most of us have already felt the effects of downsizing, reengineering, and outsourcing to reduce costs and become more competitive.

The burden of costs to business arising from employee injuries and illnesses is already providing creative opportunities for safety and health professionals. Government resources are already totally unable to solve this problem, and increased legislation can only deal with the symptoms — not the root causes. In fact, government officials are already seeking ways to promote a more cooperative approach, and more and more that key word "PREVENTION" is coming up in these debates.

With all these anticipated issues and trends, my prediction is that the safety and health professional of the future will, in general, require a much broader technical knowledge and must gain increased management and communications skills in order to help the organizations they serve to set up and continuously improve effective prevention systems.

Expanding on these attributes, here is a checklist of knowledge and skills that I believe will form the job requirements for the safety and health professional of the future:

**Personal Abilities and Characteristics**
- Excellent education in language and communication skills, written and oral
- A wide area of interests
- A high level of professional and ethical conduct

**Professional Technical Knowledge**
- State-of-the-art safety management techniques
- Process safety management
- Industrial hygiene and toxicology
- Medical/health (at least the preventive aspects)
- Ergonomics
- Security
- Fire protection
- All related laws, codes, and standards

**Professional Management and Related Knowledge and Skills**
- Advanced professional management
- Computer skills
- Communicating
- Training
- Psychology/human behavior
- Civil and criminal law (at least the fundamentals)
- Risk management and insurance
- Human resource management

*Certifications* — This aspect will most likely continue expanding in the years ahead, so the *minimum* recommended certifications are
- Certified Safety Professional (CSP)
- Certified Industrial Hygienist (CIH)
- Additional certifications will be needed as position responsibilities require. Examples are
  - Certified Hazard Control Manager (CHCM)
  - Certified Hazardous Materials Manager (CHMM)

- Master of Public Health (MPH)
- Registered Nurse (RN)
- Certified Protection Professional (CPP)
- Fire Protection Engineer (FPE)
- System Safety Engineer (SSE)
- Professional Engineer (P.E.) in Safety Engineering or other disciplines
- Global responsibilities will include international certifications and memberships.

Needless to state, there are *many* examples of other technical-type certifications. Some of those are related to certain industries such as construction (Crane Instructor) or mining (MHSA Instructor) or to specific materials and chemicals such as asbestos or lead. This list is continuously growing, and needs and will depend on your specific job responsibilities.

For more information on how to obtain these certifications, please refer to Dr. Roger Brauer's chapter (Chapter 40) entitled "Standards of Competence".

## SUMMARY AND CONCLUSIONS

Continuous professional development will be absolutely vital for the safety and health professional of the future. Here is my three-step prescription for success:

- **Step I. Get out on the professional edge.** We will need to constantly update, improve, and broaden the scope of our professional knowledge and skills. This will involve taking courses, attending seminars, joining professional organizations, obtaining reference materials, and gaining certifications. At all costs, avoid "floating along with the crowd". This is the sure path to mediocrity and eventual disillusionment with your situation in life. Stay flexible, and be ready to accept and master challenges as they will face you throughout your professional career. At all costs, never stop learning!
- **Step II. Constantly develop your skills in the arts and science of professional management.** Prevention is the name of the game in terms of helping our organizations to control and avoid unintentional and intentional losses of their assets, both human and physical. We need to continuously improve our management skills by taking courses, reading professional books and journals, and becoming involved in management-related organizations such as the AMA (American Management Association), ASTD (American Society for Training and Development), and HRC (Human Resource Council).
- **Step III. Find Ways to Help Promote Collective Cooperation and Coordination Among All Related Professional Societies and Organizations.** The old saying "united we stand, divided we fall!" was never more true for all our professions in the years ahead. Each of us must do our part to help bring our various professional societies closer together. This will not happen until we, the membership, make our desires known to the leadership of the societies. How do we do this? To me the key is *involvement* — we

must get involved in the process. Take an active role in your local chapter, serve on a committee, volunteer for projects, write, give presentations, travel to see how others have done it. There are many ways to do this, but ACTION — DOING — is the key. By getting involved in this way you will automatically be improving your skills and knowledge in many different areas, besides gaining valuable recognition from your peers.

One thing is for sure, by going *singly* on their own way, some of our societies will not survive the years ahead. They will be merged into another organization or they will just slowly fade into oblivion. *Collectively* they can become a powerful force for good in the service of the peoples of this world. It is up to us to see that this happens.

## PROFESSIONAL ORGANIZATIONS IN THE FIELD OF SAFETY AND HEALTH MANAGEMENT AND RELATED FIELDS

### American Board of Industrial Hygiene
6015 West St. Joseph, Suite 102
Lansing, MI 48917-3980
(517) 321-2638

### American Conference of Governmental Industrial Hygienists
Kemper Woods Center
1330 Kemper Meadow Drive
Cincinnati, OH 45240
(513) 742-2020

### American Industrial Hygiene Association
P.O. Box 8390
345 White Pond Drive
Akron, OH 44320
(216) 873-2442

### American Management Association
135 West 50th Street
New York, NY 10020-1201
1-800-262-9699

### American Management Association Extension Institute
P.O. Box 1026
Saranac Lake, NY 12983-9986
1-800-225-3215
 (518) 891-0065

**American Public Health Association**
1015 15th St., N.W.
Washington, D.C. 20005
(202) 789-5600

**American Society for Industrial Security**
1655 North Fort Myer Drive, Suite 1200
Arlington, VA 22209
(703) 522-5800

**American Society for Quality Control**
310 W. Wisconsin Avenue
Milwaukee, WI 53203
1-800-248-1946
(414) 272-8575

**American Society of Safety Engineers**
1800 East Oakton St.
Des Plaines, IL 60018-2187
(847) 699-2929

*Note:* Readers interested in management aspects are strongly advised to join the Management Division and possibly other divisions depending on your specialty interests.

**American Society for Training and Development**
1640 King Street
Box 1443
Alexandria, VA 22313-2043
(703) 683-8129

**Association for Quality and Participation**
801-B West 8th Street
Cincinnati, OH 45203-1607
1-800-733-3310
(513) 381-1959

**Board of Certified Safety Professionals**
208 Burwash Ave.
Savoy, IL 61874
(217) 359-9263

**Canadian Society of Safety Engineering**
330 Bay Street, Suite 602
Toronto, Ontario M5H 2S8
Canada
(416) 368-2230

**Human Factors Society**
P.O. Box 1369
Santa Monica, CA 90406-1369
(310) 394-1811

**Institute of Personnel and Development**
IPD House
Camp Road
London SW19 4UX
United Kingdom
081-946-9100

**Institution of Occupational Safety and Health**
The Grange
Highfield Drive
Wigston
Leicester LE18 INN
United Kingdom
0533-571399

**International Commission on Occupational Health (ICOH)**
Department of Community,
 Occupational and Family Medicine
National University Hospital
Lower Kent Road S. (0511)
Republic of Singapore
(65) 772-4290

**International Institute of Risk and Safety Management**
National Safety Centre
Chancellors Road
London W6 9RS
United Kingdom
081-841-1231

**International Personnel Management Association**
1617 Duke Street
Alexandria, VA 22314
(703) 549-7100

**International Professional Security Association**
IPSA House
3 Dendy Road
Paignton
S. Devon TQ4 5DB
United Kingdom
01803 554849

**National Fire Protection Association**
Batterymarch Park
P.O. Box 9101
Quincy, MA 02269-9101
1-800-344-3555

**National Safety Council**
1121 Spring Lake Drive
Itasca, IL 60143-3201
(708) 285-1121

**National Safety Management Society**
12 Pickens Lane
Weaverville, NC 28787
1-800-321-2910

**National Society for Performance and Instruction**
1300 L Street N.W., Suite 1250
Washington, D.C. 20005
(202) 408-7969

**Risk Insurance Management Society, Inc.**
205 East 42nd Street
New York, NY 10017-5779
(212) 286-9292

**Royal Society for the Prevention of Accidents**
Cannon House
The Priory Queensway
Birmingham B4 6BS
United Kingdom
0121-200-2461

**Society for Human Resource Management**
606 N. Washington St.
Alexandria, VA 22314-1997
(703) 548-3440

**System Safety Society**
Technology Trading Park
5 Export Drive, Suite A
Sterling, VA 22170

**World Future Society**
7910 Woodmont Avenue, Suite 450
Bethesda, MD 20814
1-800-989-8274
(301) 656-8274

For information on International safety and health organizations, readers can consult the International Directory of Occupational Safety and Health Institutions, which is published by the International Labour Office (ILO).

ILO Publications can be obtained through major booksellers, from ILO local offices in many countries, or directly from ILO Publications, International Labour Office, CH-1211 Geneva 22, Switzerland.

## FURTHER READING

### Books

Boyett, J. H. and Conn, H. P. *Workplace 2000.* New York: Plume Penguin Books, 1992.

Boylston, R. P. et al. *The Safety Profession Year 2000.* Des Plaines, IL: American Society of Safety Engineers, 1991.

Burrus, D. and Gittines, R. *Technotrends.* New York: Harper Collins Publishers, 1993.

Cornish, E. et al. *The 1990's and Beyond.* Bethesda, MD: World Future Society, 1990.

Drucker, P. F. *Managing for the Future — The 1990's and Beyond.* New York: Truman Talley Books/Dutton, 1992.

Drucker, P. F. *Post Capitalist Society.* New York: Harper Collins Publishers, 1993.

Fennelly, L. et al. *Security in the Year 2000 and Beyond.* Palm Springs, CA: ETC Publications, 1987.

Holt, A. St. J. et al. *Health and Safety towards the Millennium.* Leicester, U.K.: IOSH Publishing, 1987.

Knowdell, R. L., Branstead, E., and Moravec, M. *From Downsizing to Recovery: Strategic Transition Options for Organizations and Individuals.* Palo Alto, CA: CPP Books, 1994.

LaDou, J. et al. *Occupational Health and Safety.* Itasca, IL: National Safety Council, 411–420, 1994.

Manuele, F. A. *On the Practice of Safety.* New York: Van Nostrand Reinhold, 1993.

Naisbitt, J. *Global Paradox.* New York: William Morrow, 1994.

Olesen, E. *12 Steps to Mastering the Winds of Change.* New York: Rawson Associates/Macmillan, 1993.

Peck, M. S. *A World Waiting to Be Born: Civility Rediscovered.* New York: Bantam Books, 1994.

Peters, T. *The Tom Peters Seminar.* New York: Random House, 1994.

Scott, C. D. and Jaffe, D. T. *Take this Job and Love It.* New York: Simon & Schuster, 1988.

Scott, C. D. and Jaffe, D. T. *Managing Personal Change.* Los Altos, CA: Crisp Publications, 1989.

Thomas, H. G. *Safety, Work and Life — An International View.* Des Plaines, IL: American Society of Safety Engineers, 186–212, 1991.

## Other Publications

Lack, R. W. The safety-HR connection. *Personnel J.,* 71(7), 18 July 1992.

Kloman, H. F. (Ed.) Risk Management Reports — The Safety Professional and the Year 2000 (Author Lack, R.W.) 20(2), 15–27, Stanford, CT, Tillinghast Publications, March/April 1993.

# 43

## ENVIRONMENTAL SAFETY AND HEALTH COMPLIANCE IN THE 1990s: USING COMPUTER TECHNOLOGY TO MANAGE THE PAPER CHASE

**Mark D. Hansen**

### TABLE OF CONTENTS

1-56670-054-X/96/$0.00+$.50
© 1996 by CRC Press, Inc.

## INTRODUCTION

Environmental safety and health (ES&H) compliance is strongly regulated by the government. Most companies that have to deal with these regulations are forced to grapple with volumes of Occupational Safety and Health Administration (OSHA) and Environmental Protection Agency (EPA) documentation and, in some cases, that of the International Organization for Standards (ISO) if they sell their products abroad. This includes understanding the regulations as well as having to refer to various parts, sometimes on a daily basis. Being compliant with these regulations is further complicated by numerous reports, many of which overlap and are due to the government at different times during the year.

The problem is that ES&H professionals and managers spend an inordinate amount of time digging through regulations and filing reports rather than focusing on the ES&H job. By automating this process, the ES&H professional can spend more time out in the field implementing ES&H programs and less time ensuring that compliance documentation is filed on time and manually digging through regulations. Examples include having material safety data sheets (MSDSs) available on computer at sites for employee review, managing the document flow through the review process using computers, integrating engineering drawings and other documents using computers, and managing and tracking all of the training requirements necessary for OSHA, EPA, and ISO compliance using computers.

With the prices of powerful IBM-compatible personal computers (PCs) and Macintosh computers falling below $2000 most ES&H professionals have access to more computing power than they have ever had before. Ironically, with all this computing power available, rarely are these machines fully utilized. Most of the time we use computers for routine tasks such as word processing, spreadsheets, or databases.

## COMPUTER ARCHITECTURES

First of all, what is computer architecture? Computer architecture is a combination of hardware and software that when integrated works together to perform a set of specific functions. It is usually depicted in block diagrams that look like functionality diagrams. Depending on the size and function of the application, architectures may vary from single-user systems to enterprise-wide solutions.

### Single-User Systems

Single-user systems (Figure 43-1) usually consist of a central processing unit (CPU), which is a 286 on the low end up to a 586 Pentium on the high end, with clock speeds ranging from 16 megahertz to 100 megahertz. The monitor

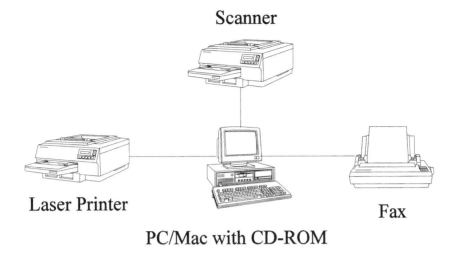

**Figure 43-1** Single-user computer architecture.

is usually a CGA at a minimum up to a SVGA, and sometimes graphics monitors are used which have 21-inch diagonal screens for computer-aided design (CAD) capabilities. Random access memory (RAM) can range anywhere from 1 megabyte up 16 megabytes depending on the computing needs. Hard drives can range from 20 megabytes to 1 gigabyte, also depending on the computing needs. Printers that can be used vary from dot matrix to near letter quality and from letter quality to laser printers. Scanners can be integrated — from hand scanners to full page scanners. Scanners usually have optical character recognition (OCR) capabilities to read in data and translate to ASCII characters. Some may even have optical storage like compact disk-read only memory (CD-ROM) to store and retrieve large amounts of information. A variation on the CD-ROM is the write once read many (WORM) drive that enables users to generate, manipulate, exchange, and store information once and retrieve it in a read-only fashion. This is helpful in total quality management, OSHA PSM, ISO 9000, and EPA compliance environments that require time-dated sequential data on information used and transferred in and through company-wide systems.

## DEFINING CLIENT/SERVER COMPUTING

Client/server computing is a form of distributed processing where an application is split in a way that allows a front end (the client) to request the services of a back end (the server). Figure 43-2 presents the definitions. Typically, the front ends reside in the end-user desktop systems (PCs, Macs, UNIX systems, workstations, etc.). Back ends can reside in any network server

**Client:** A single-user workstation that provides presentation services and the appropriate computing, connectivity and database services relevant to the business need.

**Server:** One or more multi-user processors with shared memory that provide computing, connectivity and database services and interfaces relevant to the business need.

**Client/server Computing:** An environment that appropriately allocates application processing between the client and the server. The environment typically is heterogeneous, with the client and the server communicating through a well-defined set of standard application program interfaces and remote procedure calls.

**Figure 43-2** Client/server definitions.

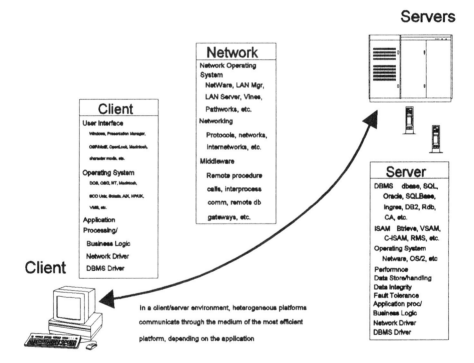

**Figure 43-3** Client/server architecture relationships.

ranging from a local database server to the largest mainframe. The front-end and back-end portions of the application reside on different processing systems, usually separated by a network. Neither the application's front ends nor its back ends are complete applications themselves. Rather, they complement each other to form a complete application.

Client/server database systems are constructed such that the database runs on a database server while the database users interact with their own desktop

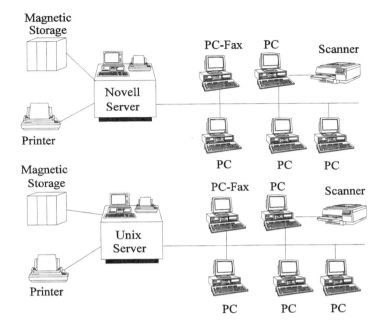

**Figure 43-4**  Simple and UNIX network architectures.

systems (client), which are responsible for handling the user interface, including windowing and data presentation. Figure 43-3 illustrates the component relationships in the client/server model.

## Simple Network/UNIX Systems

With a very basic introduction to client/server technology, Figure 43-4 illustrates a potential application for simple and UNIX networks. These architectures are usually implemented at the low end for 1 to 5 concurrent users ($25,000 to $50,000) to a high end of 10 to 50 concurrent users ($100,000 to $500,000) or department-level applications that generate, manipulate, exchange, and store common information. Some of the low end and all of the high end applications employ fourth-generation program applications software to provide ease of use. Fourth-generation languages allow users to reconfigure and manipulate the system without reprogramming the software. In the past this kind of reconfiguration relied upon programmers to recode the software.

## Enterprise-Wide Solutions

Enterprise-wide solutions are generally solutions that allow up to 5000 users across the company to communicate, generate, manipulate, exchange, and store common information that may be in several different geographic locations. These architectures are usually implemented at the low end for 50 to 100+ concurrent users ($100,000 to $500,000) to a high end of 100 to 5000 concurrent users ($500,000 to $1,000,000+). All of the low end and high end

**Figure 43-5** Environmental safety and health enterprise-wide solution.

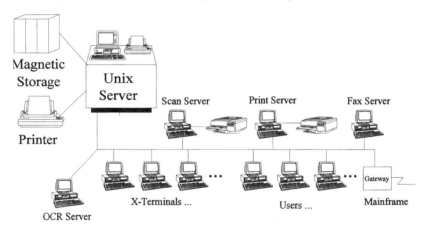

**Figure 43-6** UNIX enterprise-wide solution.

applications of the enterprise-wide systems employ fourth-generation program applications software to provide ease of use. The primary driving difference in the cost of each of these implementations is the number of concurrent users. An example of an enterprise-wide solution for ES&H compliance is shown in Figure 43-5.

Architectures for enterprise-wide solutions are shown in Figures 43-6 and 43-7. Many of the attributes of enterprise-wide solutions are similar to those of department-level solutions, except that system requirements provide larger amounts of information to be transferred and exchanged between locations and a larger number of concurrent users.

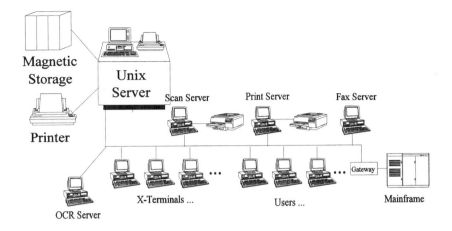

**Figure 43-7** UNIX enterprise-wide solution.

Figure 43-6 illustrates an enterprise-wide solution that uses a gateway to provide connectivity to an existing host system. The advantage of such an implementation is that if an existing system is currently being used it is not taxed or slowed by this type of implementation.

Figure 43-7 illustrates an enterprise-wide solution that also uses a gateway to provide connectivity to an existing host system. In this case, the existing system is a UNIX mainframe. Since many computing systems today are currently implemented on mainframe platforms, this example illustrates how upgrades can take place without taxing or slowing the existing system. The architectures shown in Figures 43-6 and 43-7 can also be implemented as stand-alone systems if existing systems do not exist or if companies desire to replace their outdated computing system.

## ENVIRONMENTAL SAFETY AND HEALTH SOFTWARE

There is so much software available today that it would be impossible to provide a comprehensive discussion in this chapter. Software available today ranges from word processing and spreadsheets to relational databases and applications software. In a similar manner, ES&H software is also too numerous to mention in this chapter. However, a cross-sectional summary of ES&H software will provide an idea of what is available today to help ES&H professionals do their job better. The software presented here covers general safety and health, clinic management/tracking/medical records/administration, health screening and surveillance, chemical and environmental, health promotion and fitness, on-line regulations, geographical information systems, laboratory information systems, process manufacturing, emissions control, fax-back services, and miscellaneous areas.

| Software | Uses | System Requirements |
|---|---|---|
| **General Safety and Health** | | |
| AIM-Supervisor | Historical database, procedures, instrumentation, statistical process control | VAX, Alpha |
| CHAMPS | Maintenance, management information system, equipment management, work order control, maintenance scheduling, personnel management, budget, purchasing, accounts payable | VAX, Alpha |
| ECMS | ES&H information system, record keeping, reporting for state/local regulations, relational database, compliance tracking | VAX, DECpc |
| FLOW Gemini | Medical, health physics, industrial hygiene/safety, hazard communication | Micro VAX, VAX, IBM mainframes |
| MDS/OASES | Medical surveillance, survey management and exposure, MSDS management, incident investigation, plant audits and inspections, administration and all training, reporting for Tier I, II, and R | VAX/VMS, MicroVax, PS/2, LANs, 370 Series, AS/400s |
| MediBank | Registration, medical history, employee assistance programs, physicals, insurance reports | Any CPU that runs MUMPS |
| MetroHealth™ | Core, personnel, hazardous agents, medical workplace exposure, safety and fire, and environmental reports | IBM 386 |
| MISER | Wastewater, energy management, and utilities | VAX |
| Pix/Tex | Document management for the petrochemical industry | VAX, RISC, Alpha |
| OSP | Tracking, analysis, reporting accidents and near misses | PC, IBM-PCs |
| Sentry™ | Occupational health, industrial hygiene, safety, environmental information, OSHA 200 reporting, SARA Title 3 | IBM PC/AT, PS/2, 386, HP, VAX, MicroVAX, Motorola, Unisys, NCR, UNIX, mainframes |
| SunHealth | Medical records, audiology, occupational injury/illnesses, industrial hygiene, materials agents, safety training, MSDSs | IBM 9370+, MVS, VM/CMS |
| TCF | Total compliance framework, management of compliance obligations | VAX, RISC, DECpc |
| Weston | Manage MSDSs, waste management, safety records, reports, environmental audits, document image management | VAX, DECpc |
| **Clinic Management/Tracking/Medical Records/Administration** | | |
| Medical Management System | Occupational health, employer information and reports | System/36, AS/400 |
| MediTrax | Occupational health, workers' compensation, OSHA record, hazardous chemicals exposure, safety education | IBM PS/2, 286+ |
| Micro-Hlthware™ | Occupational health, reports | PC-ATs |
| OHE System | OSHA forms, occupational health, screening appointments | 8088, 8086+ |

| Software | Uses | System Requirements |
|---|---|---|
| OMMM Sys-OTJICTS | On-the-job injury cases, reports, occupational health | 286+ |
| The Practice Manager | Occupational medical office, patient tracking, medical surveillance, accounts receivable, security | 286+ |
| ProData | Occupational health clinics | PC, AT, PS/2, LANs |
| RETURNPLUS | Disability case management, health benefits, contract provider manager, loss control | PC |
| SYSTOC | Clinics and hospitals, occupational health | 286+ |
| Trax/EH™ | Clinic administration, employee records management, OSHA, workers compensation, surveillance/tracking | 286+ |
| ULTRA | Case management, occupational medicine for clinics | 286+ |

### Health Screening and Surveillance

| | | |
|---|---|---|
| Clean Slate | Medical review, administering drug tests programs, consortiums | PC/AT, PS/2, COMPAQ, DESKPRO 286+ |
| Hearing Conservation HIV Screening | Audio maintenance, background information, HIV screening spreadsheet | PC/XT, PC/AT, PS/2, Lotus 123 spreadsheet |
| Mayo Health Station | Pulmonary, audio, hypertension databases | PS/2 |
| OMMS-Drug Screen | Administrative correspondences, record-keeping statistics | PC |
| OMI PI | Office spirometry interpretation | PC |
| OMI PS | Pulmonary test and data management | PC |
| Screening Worksheet | Test parameters and cost | Lotus 1–2–3 spreadsheet |

### Disability Management

| | | |
|---|---|---|
| CAIR | Impairment rating calculations | 8088+ |
| OSHALog.200 | Accident analysis, report form OSHA 200, summary of recordable injuries, workers compensation, menu-driven | PC |
| OSHA-200 | Reportable injuries illnesses, datafile listing | PC |
| OSHA Record Keeper | Create/maintain OSHA 101/200, reports | PC |
| Safety I™ | Accident tracking and reporting | PC |
| Safety Trax | Accident analysis and record keeping, OSHA 101/200 | 8088+ |

### Chemical and Environmental

| | | |
|---|---|---|
| CamHealth | Comply with EPA, material usage, procedures and substances, waste streams, emissions, authorizations, potential pollutants | VAX, RISC |
| CINFOdisc | OH&S CD-ROM tradenames, regulations on pesticides, management research information | PC, CD-ROM |
| CCM | Regulatory and advisory lists for EPA, OSHA, DOT, NFPA | VAX |
| CTS | Track quantities and locations of chemicals on site | VAX |
| Chemtox | Regulatory compliance, occupational medicine, | PC |

| Software | Uses | System Requirements |
|---|---|---|
| database | emergency response, industrial hygiene/ toxicologists, transportation | |
| CRISP | Procedures, statistical process control, reports, trending, water, wastewater | VAX |
| CHRIS Plus | MSDS, chemical inventory, training, reports, SARA | PC |
| Compliance Manager | OSHA hazard communication, MSDS chemical inventory, employee training | PC |
| ENFLEX | Corporate tracking, inventory reports, management reports | VAX, DECpc |
| ENFLEXINFO | Database of CFR 40, 29, 49 in 27 states | DECpc |
| Environmental Archive | Water, environmental, utilities, and oil | VAX |
| HAZMIN | Training and records retention, hazardous materials management for EPA, OSHA, DOT, IARC, NIOSH, NTP | PC, VAX |
| LogiTrac | Track MSDSs, SARA 312, right-to-know inventories | VAX, Alpha |
| The MSDS Solution | Store, display, print MSDS and tracks location of the MSDS (information must be input into the system) | VAX, Alpha |
| Medical Waste Manager | Industrial hygiene, safety management, occupational health managers, manage medical waste, tracking and manifests | PC |
| M/POWER | Predictive diagnosis and analysis system | VAX, Alpha |
| MSDS Standard | MSDS builder for 1910.1200 | PC |
| PI | Collects and stores procedural data for analysis and reporting | VAX, Alpha |
| TOMES Plus™ | Toxicology, environmental hazard data bank, oil and hazardous materials, chemical hazard response, first medical response, DOT emergency guidelines, integrated risk information | PC, CD-ROM |
| Toxic Alert | Maintain MSDSs, SARA documentation, storage tank monitoring | Call-in service |

### Decision Support/Loss Control

| | | |
|---|---|---|
| CALMS | Track observation, planned inspections, accident causes, accident investigation, PPE, personal communication, report generation | PC |
| Epi Info | Epidemiology, questionnaires, statistics, and databases | PC |
| LCE Software 38 | Low-cost epidemiology packages | PC |
| REASON | Prevention measures, cost-benefit analysis, report generation | PC |

### Health Promotion/Fitness

| | | |
|---|---|---|
| FITSCAN™/ FIREFIT | Physical fitness and health assessment | PC, Mac |
| FitTest | Fitness parameters for health surveillance screening | PC |
| HPHRA | Health risk appraisal | PC |
| PARP Spreadsheet | Health promotion and wellness, 36 questions | PC |
| Wellness Inventory | Health education and promotion, wellness | PC |

| Software | Uses | System Requirements |
|---|---|---|
| **On-Line Regulations** | | |
| Counterpoint | EPA, OSHA, DOT regulations | CD-ROM |
| FastRegs OSHA | 1900–1910.29 CFR, bookmarks, EPA, CAL-OSHA, MSHA | PC |
| HazMaster | Current and regulations and illustrations | PC |
| IHS Regulatory Products | EPA, OSHA, DOT regulations, public laws, federal regulations, final and proposed rules, industry standards | CD-ROM |
| LIS | Legislative information, preambles, and proposed regulations | PC |
| OSHA Authority | 29 CFR, OSH Act, Federal Register, compliance memos, field operations, manuals, emergency response guide, illustrations | PC |
| OSHA Trieve | OSHA, marine standard, Washington State Code, OSHA consultants manual, quarterly update | PC |
| RegMaster | Tailor regulations as needed | PC |
| RegScan | Regulations, looseleaf binder, newsletter of updates and regulatory news | PC |
| Regulation Scanning | OSHA, DOT, 1900–1910 | PC |
| SARA! | Reports for EPA, SARA 311, 312, 313 | PC |
| Virtual Media Corporation | OSH Act, Federal Register, 29 CFR 1900 series, variances, chemical sampling, field operations, manual, compliance agency memos, memoranda of understanding, emergency response guide, CFR amendments | PC |
| **Geographical Information Systems** | | |
| AIS Info Sys | Links databases to relational databases | VAX, RISC, Alpha |
| ALK-GIAP™ | Map spatial data, pipeline documents, water management | VAX, RISC, Alpha |
| DATATAB™ | Environmental, utilities, facility management; federal, state, local governments; manufacturing, industrial research and development | VAX, RISC, DECpc |
| SYSTEM 9™ | Modeling and integration of spatial data, utilities, environmental agencies | RISC |
| GDS | Administrative data, decision-making, manage resources | VAX, RISC, DECpc, Alpha |
| ARC/INFO | Automate, manipulate, and display report and map data | VAX, RISC, Alpha |
| ERDAS IMAGINE™ | Automate, manipulate, and display report and map data | RISC, Alpha |
| GENAMAP | Automate, manipulate, and display report and map data for city and county governments | VAX, RISC, Alpha |
| GOTHIC | Automate, manipulate, and display report and map data for national governments and environmental agencies, telecommunications | VAX, RISC, Alpha |
| HORIZON | Automate, manipulate, and display report and map data for national governments, environmental agencies, telecommunications | VAX, Alpha |
| INFOCAM | Automate, manipulate, and display report and map data for state and local governments, utilities | VAX |
| SPOTView™ | Automate, manipulate, and display report and map data for utilities, engineering, oil and gas | VAX, RISC, DECpc |

| Software | Uses | System Requirements |
|---|---|---|
| INFORMAP™ | Automate, manipulate, and display report and map data for utilities, gas and electric, telecommunications | VAX, RISC, DECpc |
| CARIS | Compile, store, analyze spatial data | VAX, RISC |
| Hypercabinet | Records management for MSDSs | VAX, DECpc |

### Laboratory Information Systems

| Software | Uses | System Requirements |
|---|---|---|
| CALS | Chromatography, data acquisition, analysis, instrumentation and automation | VAX, Alpha |
| VG LIMS | Sample schedule, barcode printing, lab work schedules, results entry, QA/QC, reporting | VAX, RISC, Alpha |
| VG Multichrom | Chromatography, data acquisition, analysis, instrumentation and automation | VAX, Alpha |

### Process Manufacturing

| Software | Uses | System Requirements |
|---|---|---|
| CIMPRO | Formulation, product cost, inventory, distribution, financial and regulatory issues | VAX, RISC, Alpha |
| GEMMS | Relational database, multiple language and currency requirements | VAX, RISC, Alpha |

### Emissions Monitoring

| Software | Uses | System Requirements |
|---|---|---|
| ESC/CEM | Collect pollutant information from remote sensors, convert units, store averages | VAX, DECpc |
| RAQS | Collect pollutant and meteorological information from remote sensors, convert units, store averages | VAX, DECpc, PDP-11 |
| AIR-1 | Air toxics management system, regulatory tracking and compliance management | VAX |
| Env Aide II | Continuous environmental monitoring, data acquisition, analysis, reporting, alarming, maintenance logs | DECpc, RISC |
| VECTOR | Environmental monitor and information system real-time monitoring of all types of environmental sensors | Vax, Alpha |

### Fax-Back Services

| Software | Uses | System Requirements |
|---|---|---|
| Chemtrec | Hazardous material information | Phone, fax |
| MSDS Clearinghouse | MSDS for OSHA compliance | Phone, fax |

### Miscellaneous

| Software | Uses | System Requirements |
|---|---|---|
| BIBLIO3 | Database on solvents, lung cancer, asbestos diseases | PC, Mac |
| Billboard of Imagination: Hlth Connection | Health education, information, entertainment, advertising | Atari |
| CHID: Combined Hlth Info db | AIDS, arthritis, musculoskeletal, skin disease, digestive tract, heart, diabetes, health education, kidney disease | Modem |

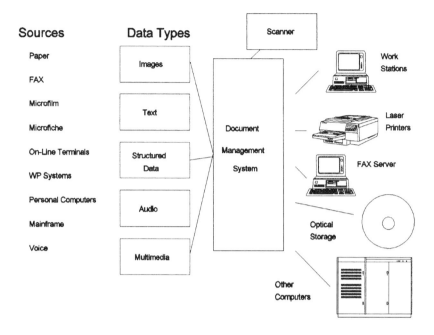

**Figure 43-8**  Typical components of a document imaging system.

## STATE-OF-THE-ART TECHNOLOGIES

There are new state-of-the-art technologies now available that can be implemented on single-user systems all the way up to enterprise-wide applications. These technologies typically use stand-alone PCs either as a single user or as the front end on a client-server network. These new technologies are "document imaging" of forms (e.g., MSDSs), business process automation using "workflow" software, and computer output to laser disk (COLD).

### Document Imaging

Document imaging is the core of automating the ES&H process for OSHA, EPA, and ISO compliance. Document imaging is central to the progression toward the "paperless" office. Typical components of a document imaging system are shown in Figure 43-8. The "paper" from other organizations, such as documents and engineering drawings, can be scanned in and the paper literally thrown away. The reasons the paper is thrown away are fourfold. First, the electronic copy is now legally considered to be the original. Second, the electronic copy is often better than the original (especially when dealing with MSDSs). Third, one of the main reasons for using document imaging is to get rid of rooms full of filing cabinets containing engineering drawings and documents.

Finally, document imaging can be used to capture signatures on all kinds of documents (e.g., engineering drawings, OSHA-related testing, documents, etc.)

There are two distinct ways to scan in information. One way is to capture the physical image (or picture) of the document using a document scanner. This image is usually stored in a .TIFF format. The information on this image is usually information that cannot be changed, information from outside the system, that is to be displayed or printed when necessary.

The other way is to scan in documents where the text is converted to ASCII data that can be shared in a database. The conversion process is accomplished using optical character recognition (OCR) software. This is important when dealing with data (e.g., MSDSs) that needs to be stored, retrieved, and manipulated from a database. For example, searching for all chemicals at a particular site, in a particular building or room, etc., with specific characteristics (e.g., boiling point, etc.).

If the original is of maximum quality there are two ways to significantly improve the quality and readability. One way is to use an intelligent character recognizer (ICR). This software recognizes and matches each character based on the image scanned in. Another software program accomplishes this and, additionally, fills in the image character with a gray color so it is more readable. This is especially important for managing documents from multiple vendors such as MSDSs.

Document imaging can also provide virtually instantaneous access to engineering drawings and documents. This is particularly important with OSHA 1919.119, Process Safety Management, which stipulates:

*"... prior to the introduction of highly hazardous chemicals to a process the PSSR must confirm that construction and equipment are in accordance with design specifications."* **Studies have shown that a document imaging system allows drawings to be updated six times faster than manual methods.**

*"... the written program provides employees and their representatives access to process hazard analysis and all other information developed as required by the Process Safety Management standards."* **Because documents are stored electronically they can be retrieved instantaneously and are less susceptible to fire and other catastrophes.**

*"... written process safety information will be compiled before conducting a process hazard analysis (PHA)."* **Document imaging enhances a company's ability to conform to this requirement by having all documents available electronically, by date of preparation, revision date, etc.**

Document imaging can be used to reduce the paper in the process both reliably and accurately, as well as to provide easy access for manipulation. Document imaging provides the mechanism for progressing to electronic data

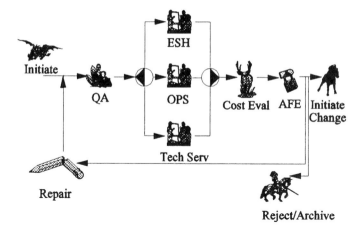

**Figure 43-9** An example of a workflow map for automating the OSHA/EPA/ISO process.

management (EDM), which is the key to survival in the 1990s both from a cost standpoint as well as for speed of access.

## Business Process Automation Using Workflow

Business process automation is the mechanism for integrating document imaging and workflow technology to make computers even more useful in the 1990s.

## Business Process Automation

Business process automation refers to automating manual processes using computer technology. For example, the physical filing of documents in a filing cabinet is automated in a database of some sort, or the manual review of a document is automated so that the document is routed electronically and sequentially to each person in the process. Business process automation includes integrating all electronic data in a compatible form. This involves converting different types of electronic images (e.g., .TIF, .GIF, .WPG, etc.) and word processing data (Word Perfect®, Multimate®, Word®, ASCII, etc.) to a common form. However, merely automating a manual process is usually not sufficient. Using workflow the manual process can be routed electronically and sequentially rather than manually.

## Workflow

Workflow refers to the electronic routing of documents and engineering drawings through a computer network. Workflow allows users to comment, approve, disapprove, etc. and send the information to the next person in the review process or back for editing. An example workflow is shown in Figure 43-9.

Workflow can be used to route information, documents, engineering drawings, etc., from local to enterprise-wide distributions.

To walk you through this example, let us say a new MSDS for a product you use in your manufacturing process arrives from a vendor (the Initiate box in the workflow map in Figure 43-9). The first thing that you might do is assess the impact in the QA (Quality Assurance box in the workflow map shown in Figure 43-9). If information is missing they may call the vendor; if not, the MSDS electronically forwarded to ESH (the Environmental Safety and Health Department), OPS (Operations), and Tech_Serv (Technical Services). ESH may evaluate the MSDS for impact on how the product is to be used in the manufacturing process and the environmental ramifications of use, waste, and disposal. Operations may or may not use the product but needs it for reference purposes. Technical Services would use it to update all documentation for all operations and maintenance procedures. Next, it would be forwarded to Cost Eval (Cost Evaluation) to determine the cost impact of the new MSDS on the company. If the new MSDS has a cost impact, it would be forwarded to AFE (Authority for Expenditure) to make disbursements accordingly. If it requires more review from comments it would be rerouted through the repair function and iterated until it is acceptable to proceed. At this point, if no action is decided for the new MSDS, it would be forwarded to the Reject Archive function and archived. On the other hand, if action is required, and once everything is completed, it would be forwarded on to initiate the change and archived in the Reject Archive function.

### *Benefits of Document Imaging and Workflow*

Document imaging and workflow allow companies to work smarter rather than work harder. This includes reducing the number of steps in a process, reducing the amount of time needed to do a task, doing the task with less people, reducing the space needed to store all kinds of paper, etc. Reducing the amount of time to do a task is accomplished by having all of the appropriate information needed to complete a task electronically attached and immediately available. For example, engineering drawings stored electronically can be quickly accessed for review prior to the ES&H professional and/or manager signing off on changes or modifications. This is in stark contrast to the time-consuming manual process of searching for engineering drawings in filing cabinets. Using this method also precludes the need to enter the same data into the system several times. Once it is entered, it never has to be reentered. It follows that reducing the number of steps also provides an opportunity to reduce the amount of time required to do a particular task. If this is done properly, manpower can often be reduced. In this day and age of "right-sizing" to maintain a competitive advantage, this allows a company to reduce manpower without sacrificing the quality of the work performed.

The individuals responsible for assembling and maintaining workflow maps (like the one shown in Figure 43-9) are not involved in the process. By removing them from the process the integrity of the process is maintained. This is crucial because if the person responsible for maintaining the workflow map is also involved in the process, different activities could be circumvented by modifying the workflow map. However, if the person responsible for the workflow map is not part of the process, no activity can be circumvented by anyone in the process. This is important for OSHA, EPA, and ISO. When a problem is being worked or an assembly is being accomplished, it is repeatable. It is repeatable because it is not a paper process that can circumvent any activity or organization (e.g., safety and quality). The electronic mechanism inherent in workflow also lends itself to ease of auditing since everything done leaves a time/date audit trail. This is important not only for auditing but also for getting the latest copy of documents, engineering drawings, etc.

These technologies and processes are most successfully implemented in a computing environment that is flexible, is open to various computing platforms, and allows growth from small applications. To accomplish this many companies are moving to a client/server architecture.

## Computer Output to Laser Disk (COLD)

It is estimated that American companies spend nearly $20 billion a year spooling from 2 million to 50 million pages of computer-generated reports per month to a printer or computer output microfilm (COM) equipment for distribution or archiving. Sorting through these vast amounts of growing paper files and fiche for OSHA, EPA, and ISO 9000 compliance is becoming more tedious, expensive, and time-consuming every day. Meanwhile, companies are looking to improve their bottom line and conserve valuable office space by harnessing this vast information store. To do this, users, and ES&H professionals in particular, need a quicker, more efficient method to access their mainframe data. Computer output to laser disk (COLD) technology represents a promising solution.

### *Introduction*

Ideally, users generally prefer on-line viewing capabilities to the mainframe. Information managers, however, contend that they cannot afford the consumption of disk resources and the impact on mainframe processing as the volume of inquiries increases. The result, management information systems (MISs), restrict the volume of available report data, as well as access to those reports, on the mainframe to the most current data only from selected reports. The reports are put out on a daily, weekly, or monthly basis. The rest of the reports have traditionally been output to printers and stored in paper files and on magnetic tape for subsequent access. As the paper files grow, access to them becomes laborious and expensive. Access from magnetic tape archives requires remounting the tape and taking up valuable mainframe resources that are better spent on current operations. To remedy the disadvantages of these traditional methods, MIS managers naturally looked to COM.

The technique of photographing micro images of computer-generated reports onto easily handled cards has become a multi-billion-dollar-a-year business. There are few MIS groups today that do not use microfiche to some extent. In comparison to paper output, microfiche output lowers the cost of storage by a factor of 10 or 20. This is in addition to the cost savings involved in reduced floor space and cabinet requirements of paper storage. Microfiche also provides users with automatic, in-line viewing via viewers and printers using automatic bar coding and readable indexes — a much more efficient method than searching through reams of paper.

COM technology, however, has its drawbacks. Once a datum is dumped from computer storage onto microfiche, it is no longer in data format. It is a photographic picture that must be redigitized before it can be brought back into the computer system so that it can be distributed throughout the network and displayed. As an image, it cannot be displayed on mainframe terminals; image-capable terminals are required.

Security is another problem. A report or other confidential information is only as secure on microfiche as it is in the envelope in which it is delivered or shipped through the mail. In addition, the use of chemical developing agents poses ongoing cost and disposal problems. Most of all, while more efficient than paper searches, retrieval time can be tedious and slow.

### Technology Perspective

The feasibility of optical disk technology as a storage and retrieval medium for document images has become widely accepted. The advantages of this technology — namely, its faster retrieval rates over microfilm, its ability to compress images, and its "write once read many" (WORM) security feature — all provide solutions to the disadvantages encountered with COM storage. Why not use optical disks for storing data and images?

A special software program that will compress and store computer-generated reports as data to an optical disk is required. This software, combined with a special processor and stand-alone optical media, makes up what is commonly called a computer-output-to-laser-disk subsystem.

As compared to optical media, the cost of storing data reports on microfiche is reported to be much less, by a factor of about ten to one. User productivity gains have been attributed to the ability to retrieve pages in 10 to 15 seconds, as compared to the much longer time it takes to manually sort through fiche pages on a shared viewer. Windows-based retrieval software has enabled users to gain the benefits from COLD with very little up-front training. For some, it is just another Windows application. Also, because the reports are stored as data, not images, there is a great deal of flexibility in developing indexing criteria, including the ability to cross-reference reports based on data in the report as well as automated indexing by downloading data from a mainframe. Many COLD vendors estimate that companies switching from paper or microfilm to COLD can reduce their current distribution and archival costs by as much as

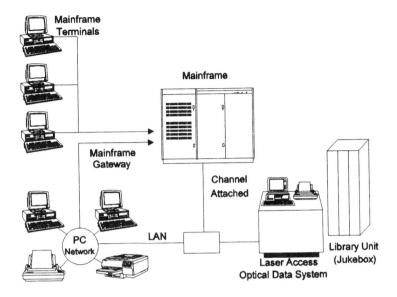

**Figure 43-10** An example of COLD system architecture.

45 to 90%. In addition, vendors report that users can typically achieve a return on their investment within 12 to 18 months after implementation.

## COLD Subsystem Characteristics

One of the reasons COLD has become such an economical replacement for COM is that COLD subsystems are available for use on stand-alone PCs, on servers on local area networks (LANs), on microcomputers which interface with mainframes, or directly channel-attached to a mainframe. Data stored through these systems can also be accessed through remote workstations via communications or a modem. PC-based systems typically cost anywhere from $30,000 to $100,000, while mainframe-based systems cost upward of $150,000.

A typical COLD system consists of a jukebox or optical disk interface program; optical media plus an onboard processor for off-loading compression, indexing storage, and retrieval functions from the mainframe; a magnetic disk for storing the index and directories for retrieving reports; and the user interface for report retrieval. Many COLD systems include global text retrieval and keyword searching capabilities, forms overlay for merging an image of a form with the data (thus recreating an accurate representation of the original form), facsimile transmission, and the ability to export the data into other applications such as databases or spreadsheets. An example of COLD system architecture is shown in Figure 43-10.

Since most of the COLD software runs on the subsystem processor, very little must be done to existing mainframe software and hardware. From the mainframe standpoint, the administrator must configure two additional devices (to the mainframe) that the optical system emulates — a printer and a tape controller. The mainframe sees the optical disk as a printer; by printing one copy to an optical disk, any number of people can access it.

The administrator must also install the optical disk interface subsystem program in the subsystem library. Finally, the indexes for retrieving the reports must be identified so that when the data are read onto an optical disk the index file can be built.

### Emerging Technologies

The advent of the writable CD-ROM technology may have a dramatic impact on the use of COM. Writable compact disks represent an ideal and low-cost alternative to microfiche for report distribution since they can store thousands of pages of ASCII text and typically conform to industry standards, such as the ISO 9660 standard for file and volume structure, allowing them to be read on any CD-ROM drive attached to a PC. In addition, writable CD-ROM technology eliminates the need to have CDs pressed at an outside service bureau, thus saving companies thousands of dollars. Large organizations that need to distribute computer-generated report data to remote locations (field offices) will benefit from this technology. The typical cost for writable CDs is less than $25 for each CD that can store 550 megabytes of data, or approximately 240,000 pages of ASCII text. One CD can also hold the equivalent of 444 fiche or 120,000 pages.

### Relationship to Imaging

Many document imaging system vendors allow data processing reports to be entered directly onto their imaging systems via COLD software. As imaging applications extend into the enterprise, the ability to handle multiple data types becomes imperative. The question is not whether to use COM, COLD, or imaging; rather, it is which method (or combination of methods) will meet an organization's requirements.

One concern is retrieval time. It takes much longer to retrieve an image of a report than it does to receive an encoded page. However, not all documents can be stored as data. Some must be preserved as images because they contain signatures, handwriting, or other noncodable information. Others, even some data reports, will be accessed so infrequently that it is not cost-effective to keep them on optical disk. These documents will be better handled by an imaging system for distribution and archiving.

Implementing a COLD system can help reduce operating, paper distribution, storage, retrieval, and archiving costs (including office space). On the other hand, implementing a COLD system can help increase user productivity, report access, distribution, storage, and retrieval time. COLD systems can be easily integrated into current computing architectures. The low cost of CDs which store large amounts of information helps make COLD technology affordable in the 1990s. The relatively short period of time for a return on investment (12 to 18 months) makes COLD a desirable mechanism for achieving office automation.

## CONCLUSIONS

Staying compliant with OSHA, EPA, and in some cases ISO 9000 in the 1990s certainly poses a challenge for ES&H professionals and managers. Automating the process to reduce time, steps, and manpower — while not only maintaining compliance but achieving superior performance — will help ensure company survivability in the 1990s. Document imaging, business process automation, reengineering, and COLD implemented on client/server architectures brings state-of-the-art technology to the desktop of the ES&H professional. Instead of spending an inordinate amount of time wrestling with regulations, documents, engineering drawings, and filing reports, ES&H professionals can now focus on the job for which they were hired.

## FURTHER READING

Addamo, F., Document imaging and windows imaging, *Imaging: A Critical Component of Process Safety Management (PSM),* May/June 1993, pp. 21–28.

9001: Quality Systems — Model for Assurance in Design/Development, Production, Installation and Servicing.

9002: Quality Systems — Model for Assurance in Production and Installation.

9003: Quality Systems — Model for Assurance Final Inspection and Test.

9004: Quality Management and Quality System Elements — Guidelines.

Baum, D. Computerworld, *Client/Server Development Tools for Windows,* April 26, 1993, pp. 73–75.

Brauer, R. L., *Directory of Computer Resources,* American Society of Safety Engineers, Des Plaines, IL, 1994.

O'Lone, E. J., Datapro, document imaging systems, management issues, *Client/Server Computing,* June 1993, pp. 1–8.

Shegala, K. and Richardson, M., Computer output to laser disk (COLD), *Datapro, Document Imaging Systems,* October 1993, pp. 5060–5080.

Stephans, R. A. and Warner, T. W., *System Safety Analysis Handbook, A Source Book for Safety Practitioners,* System Safety Society, Albuquerque, NM, July 1993.

29 Code of Federal Regulation, 1910, General Industry, U.S. Department of Labor, Occupational Safety and Health Administration, 1993.

# INDEX

# INDEX

## A

Abatement
  asbestos, 227
  hazard, 291, 296, 298
  regulatory violations, 327
    appeal, 314
ABIH, see American Board of Industrial
    Hygiene
ABIH/BCSP Joint Committee certification,
    632, see also Competency
Absenteeism, 549–550
AC, see Actively caring behavior
Acceptability level, 193, see also Process
    safety management
Acceptance
  changing environments, 414
  finding and Creative Problem-Solving
    Institute technique, 451
Accidents
  causation, 130, see also Job hazard
    analysis
  definition, 146, 147
  incident reporting, 163–167, see also Near
    miss incidents
  intervention with employees, 518–519
  investigation/reporting
    checklist, 154–156
    definitions, 146–147
    evaluation of reports, 153–154
    global health/safety management
      system, 594
    in-house and California employers,
      297–298
    noting conditions, 150
    overview, 145–146
    reporting procedure and guidelines,
      147–153
    worksite analysis, 291
  prevention programs, 283, 286–289
  problem identification, 378
  pyramid, 163, 164
  reduction and cultural interventions, 522

supermarkets and model for safety/health
    practices, 549–550
work-related and psychological risk
    factors, 513–515
Accountability
  definition of management, 373
  documentation of effectiveness of training
    courses, 569
  managers, 291, 562
  regulatory compliance, 314–315
  safety performance and safety management
    programs, 180–181
Accreditation, 626, 627, see also Competency
Achievement potential, 482, see also
    Behavior
Action
  blueprint and process safety management,
    188
  effective influence, 437
  plan
    changing safety cultures, 526, 532–533
    Creative Problem-Solving Institute
      technique, 451
Action level (AL), 226, see also Asbestos
Activators, 487, see also Activator-Behavior-
    Consequence model; Behavior
Activator-Behavior-Consequence (ABC)
    model
  designing interventions, 487
  reward/incentive programs, 490–492
  safety coaches, 497
Actively caring (AC) behavior
  cultivation of, 486
  model, 500–506
  safety coaches, 496–497
  thank you cards, 489–490
Adaption index tool, 249, see also Stress
Administration
  classification in safety and health manual,
    62
  controls and global health/safety
    management system, 593
  job hazard analyses, 131–132

683